Handbook of Geothermal Energy

Handbook of Geothermal Energy

Editors:
L. M. Edwards
G. V. Chilingar
H. H. Rieke III
W. H. Fertl

Gulf Publishing Company
Book Division
Houston, London, Paris, Tokyo

The editors dedicate this book to Ronald Reagan, President of the United States, in appreciation of his interest and support in developing America's energy resources.

Handbook of Geothermal Energy

Library of Congress Cataloging in Publication Data

Handbook of geothermal energy.
  Includes bibliographies and index.
    1. Geothermal engineering—Handbooks, manuals, etc.
2. Geothermal resources—Handbooks, manuals, etc.
I. Edwards, L.M.
TJ280.7.H36   621.44   81-20246
ISBN 0-87201-322-7   AACR2

Contents

Silicic Magma Bodies. Large Silicic Magma Bodies of Uncertain Tectonic Association. Relatively Small Magma Bodies of Intermediate to Silicic Composition. The Distribution of Andesitic Composite Volcanoes. Geothermal Resources in Regions Dominated by Basaltic Volcanism. Geothermal Areas Associated with Continental Rifts. Geothermal Resources in Tectonically Active Regions. Geopressured Geothermal Resources. Low-Temperature, Radiogenically-Derived Geothermal Resources. Appendix E—Classification of Common Igneous Rocks.

Seeps and Fossil Seeps. Geological Techniques for Exploration. Geochemical Techniques for Exploration. Geophysical Techniques for Exploration. Geoelectromagnetic Techniques. Seismic Methods in Geothermal Exploration.

Drilling Rigs and Rig Equipment. Casing Design. Cementing and Casing. Drilling Bits. Muffler Systems. Drilling Fluids. Case Histories.

Geothermal Well Completion Technology. Casing Design Procedure. Design Procedure for Casing Strings. Design for Thermal Stress. Tubing. Seals. Blowout Preventers. Cementing.

Well Design. Importance of Obtaining Complete Casing Cementing. Cementing Compositions. Cement Mixing Methods. Material Handling Aspects. Cement Placement Techniques.

Objectives of Geothermal Well Logging. Geophysical Well Logs. Subsurface Pressure Concepts. Formation Pressure Determination from Well Logs. Hydrocarbon Distribution in Clastic Overpressured Environments. Formation Temperature. Formation Water Salinities. Geothermal Well Log Analysis. Interpretive Concepts in Clastic and Carbonate Reservoir Rocks. Interpretative Concepts in Igneous and Metamorphic Rocks. Elastic Rock Properties.

Foreword

It has long been recognized that countries heavily dependent upon fossil energy resources have become less self-sufficient when their own domestic supplies became inadequate to supply their own demand. In the United States, this concern was growing more pronounced in the late 1960s as evidenced by increased industrial activity in solar and geothermal projects and the enactment by Congress of the Geothermal Steam Act of 1970.

The Geothermal Steam Act established provisions for leasing public lands for geothermal exploration and development. Concurrently, a variety of independent R&D projects, some funded by the government, were initiated; however, there was little coordination of the many, sometimes parallel programs. The Geothermal Research, Development, and Demonstration (RD&D) Act of 1974 established a federal program to coordinate the assessment and development of geothermal resources using a mission-oriented approach. The principal thrust was to accelerate commercial utilization of geothermal energy by reducing developmental risks through federal support of R&D, demonstration projects, and loan guarantees.

Added incentives for industrial development of geothermal energy were provided by the National Energy Act of 1978, which allowed certain tax advantages, and thus removed several impediments for development. With these incentives, federal support for much of the RD&D was not as essential, and accordingly, funding has declined as anticipated since 1979, thus effectively transferring the development programs to industry.

Great strides were made in our understanding of geothermal resources during the 1970s, and now, if the industry is to advance optimally, it is essential that this accumulated knowledge be made available. That is the main reason why this *Handbook* and others that have been published recently are so important. No single book can provide an exhaustive treatment of the entire subject matter, but they do provide, in the opinion of the authors/editors, those topics that are considered the most important lessons learned through experience.

It is the intent of the *Handbook of Geothermal Energy* to present a comprehensive review of significant developments pertaining to the location and production of geothermal energy. The ten chapters cover worldwide geothermal resources, the geology of geothermal systems, exploration for geothermal energy, drilling and completion concepts, casing and tubular design concepts, well cementing, formation evaluation, reservoir engineering concepts, and energy conversion and economic issues.

I commend the editors, George V. Chilingar, Lyman Edwards, Walter Fertl, and Herman H. Rieke III, for their authorship of several chapters, but more so for their individual and collective efforts to assemble and edit the contributions of ten scientists/practitioners, each notable in his own right. This combined knowledge and expertise makes the *Handbook of Geothermal Energy* an essential reference for anyone seriously concerned with geothermal developments.

Lawrence Ball
U. S. Department of Energy
Grand Junction, Colorado
(formerly with the Division
of Geothermal Energy)
January 1982

George V. Chilingar, University of Southern California
Lyman Edwards, Dresser Industries
Walter Fertl, Dresser Industries
Herman H. Rieke III, TRW, Inc.

1

Introduction

Introduction

This book covers all the geologic and engineering phases of the geothermal energy resource spectrum. Its scope consists of:

1. Presenting state-of-the-art knowledge about the nature and geology of geothermal systems.
2. Presenting exploration techniques employed by industry to discover these systems.
3. Discussing the design and implementation of drilling, completing, and stimulating the reservoir.
4. Assessing and evaluating the reservoir performance.
5. Assigning a monetary value to the recoverable geothermal energy in the reservoir.

Types of Geothermal Energy Systems

Although not widely understood, geothermal energy is an old concept that has assumed new importance. In the broadest sense it is the natural heat of the earth. Leibowitz (1978) defined geothermal energy as the natural heat in the earth that is trapped close enough to the surface to be extracted economically. Unfortunately, in most instances, geothermal energy is too dispersed in the outer crust to be recovered economically. Only where heat is concentrated into restricted volumes, analogous to the concentration of hydrocarbon or ore deposits, does geothermal energy have economic potential. Italy has converted it to electrical power for 75 years, New Zealand for 30 years. Sixty miles north of San Francisco, California, at the Geysers, commercial production began in 1960.

1

Observed geothermal gradients and measurements of thermal conductivities of rocks in the earth's outer crust suggest that the geothermal heat flow is relatively constant worldwide (Jones, 1970). In tectonically stable regions the heat flow or geothermal gradient values do not vary appreciably. Tectonically active regions are characterized by wide variations in heat flow values. These values are either abnormally high or low. One type of tectonically active region that shows consistently high heat flow values lies along the border of tectonic plates. Zones of young volcanism and mountain building are localized at the margins of these major plates (Figure 1-1). Fracture systems in the crust may allow the heat generated by the interaction of the plates to be transported to or near the earth's surface.

There are three major types of geothermal energy systems: *hot-igneous, conduction-dominated,* and *hydrothermal.* Whereas the first two systems may contain the largest amount of useful heat energy, advancements in extraction technology are required in order to use the stored heat commercially. Consequently, hydrothermal energy systems are of prime concern at the present time.

Hydrothermal Systems

These systems consist of high-temperature water and/or steam, which are stored in porous and permeable reservoir rocks. As a result of the convective circulation of water and/or steam through faults and fractures, the heat is

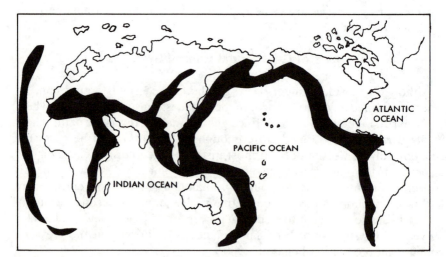

Figure 1-1. Geothermal areas of the world, based on recent volcanism and crustal-plate boundaries. (After Leibowitz, 1978; courtesy of Energy Sources.)

transported to near the earth's surface. The driving force is gravity, owing to the density difference between cold, downward-moving water and hot, upward-moving fluids. Heat stored in the geothermal reservoir rock is produced by bringing to the surface hot water and/or steam (Figure 1-2).

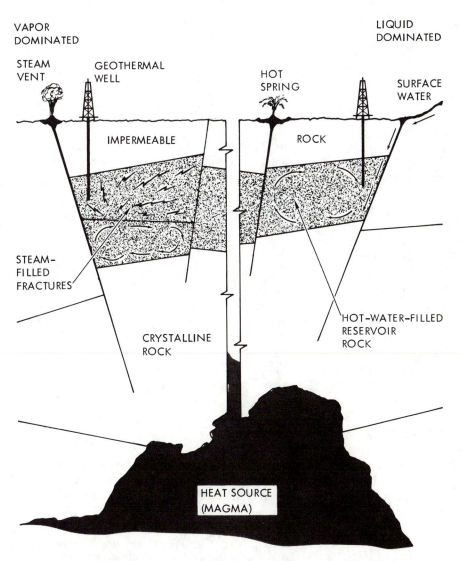

Figure 1-2. Generalized (simplified) schematic diagram of a hydrothermal reservoir. (After Leibowitz, 1978; courtesy of Energy Sources.)

Hot-water reservoirs are called *liquid-dominated,* whereas reservoirs containing steam are referred to as *vapor-dominated.* Kruger and Otte (1973) pointed out that the former reservoir types are much more numerous than the latter. Hot-water systems are usually found in permeable sedimentary or volcanic rocks and in competent rocks, such as granite, that can maintain open channels along faults or fractures (White et al., 1971). Promising areas for exploration are usually associated with one or more of the following geothermal indicators:

1. Recent volcanic activity
2. Frequent seismic activity
3. High level of conductive heat flow
4. Hot springs
5. Geysers or fumaroles

The exploration for hydrothermal resource entails considerable risk. Leibowitz (1978, p. 299) discusses the following sequence of tasks involved in exploring for geothermal energy:

1. Obtaining government permits and approval
2. Examination of hot springs and other surface indications
3. Geochemical survey
4. Geophysical survey
5. Heat flow measurements in shallow boreholes
6. Drilling deep wells and then flow testing them

The last task appears to supply the only conclusive evidence for the presence of a geothermal resource. The success rate for wildcat wells is 10% or less, and the exploration wells cost more than $1 million. After making a discovery, step-out wells are drilled in order to determine the extent of the reservoir. Flow testing, core analysis, and formation evaluation techniques are used to determine its capacity.

Hot-Igneous Systems

These systems consist of magma chambers near the earth's surface, which are created by the buoyant rise of molten rock generated deep in the earth's crust. A hot-igneous system can be divided into two major resource groupings: *hot, dry rock,* where the magma is no longer molten, but is still very hot (<650°C); and *volcanic,* where magma is still partly molten (>650°C). In the latter case, because of great depth (>3 km) and very high temperatures (650-1200°C), the heat is not recoverable with current technology. The hot, dry rocks, however, are located on the margins of the molten magma chamber. These rocks are favorable candidates for recovering heat energy. A system of hydraulic fractures can be created between special, directionally-drilled wells in order to provide circulation loops in rocks having low to very low permeability values. In general, the economic extraction of energy from this resource lies in the future, even though the test results at Los Alamos, New Mexico to date are quite encouraging (Smith et al., 1975).

Conduction-Dominated Systems

Most of the heat is transferred from great depth within the earth to its surface through thermal conduction. In order to reach temperatures of 100°C in the subsurface, one would have to drill about 5-10 km deep. The development of conduction-dominated systems is currently not economical. Presently, low-grade hydrothermal resources lying close to the surface (<3 km) are much more attractive.

Geopressured geothermal resources have been included in the conduction-dominated systems as a separate category. These systems are commonly found in Tertiary age geosynclines, such as the U.S. Gulf Coast depositional basin. In these geologically young basins large amounts of fine-grained, water-filled sediments have been deposited, rapidly buried, and compacted (Rieke and Chilingarian, 1974). The weight of the resulting overburden normally would reduce the thickness of these sedimentary layers. However, if the explusion of the connate water from the shale pores during compaction is retarded, then over-pressured fluid zones are created. The trapped pore water and the sedimentary particles exhibit low compressibility (Chilingarian et al., 1973). The total bulk volume of the sediments can only be reduced significantly by the expulsion of this pore water. Lewis and Rose (1969) showed that these overpressured zones constitute a thermal barrier. The greater the amount of water trapped, the greater is the thermal insulation of the high-fluid pressure zone. Hot water is generated, and in addition to the hydrodynamic and thermal energies of these fluids, there is combustion energy of dissolved natural gases. Matthews (1980) discusses in detail the nature and evolution of gas in geopressured reservoirs.

Chapter 2 summarizes the uses of geothermal energy worldwide, defines some terms and concepts for the resource base, gives resource and reserve estimations, and selects an interim methodology for such evaluations. Table 1-1 depicts the various types of geothermal resources, their typical lithologies, and associated drilling problems.

Although care was used in estimating the potential of the geothermal resource base, inasmuch as it is a relatively new energy base, only limited data was available from actual geothermal drilling operations. Hence, accurate quantitative evaluations are difficult to verify. Estimates of energy

Table 1-1

Geothermal Resources, Typical Lithologies, and Associated Drilling Problems

Type of Resource	Major Considerations
Hydrothermal	Involves higher temperature than in drilling for oil and gas (up to 350°C). Usually found in hard abrasive volcanic rocks, although there are some resources in sedimentary rocks.
Geopressured	Regularly detected in coastal areas of Gulf of Mexico when drilling for oil or gas. Usually found in sedimentary basins at considerable depth (down to a depth of 25,000 ft.).
Hot dry rock	Advances made in the field of hydrothermal energy can be directly applied for this resource. Less dangerous to drill than drilling into steam or superheated water owing to less corrosion, erosion, and hazard from high-velocity flow.
Magma	Top of the molten rock heat resource usually at depths greater than 3 km. Drilling to 3-6 km and at 650°-1,200°C as would be required is not now possible. Collapse of the hole will be a major problem along with introduction of heat exchange equipment. Sticking of the drillpipe is also to be expected.
Normal gradient	Must be drilled to considerable depths. Can utilize advances made in the field of hydrothermal energy.

resources from other sources such as petroleum, natural gas, uranium, coal, water supplies, and minerals are easier to quantify more accurately because of the vast amounts of data that have been accumulated over the years. Only a very small area of the earth's surface has been explored to detect and locate geothermal anomalies. Most deep-seated geothermal sources require surface-operated geophysical methods to locate and define the areal extent of the potential heat sources. Even then, some exploratory drilling is required to define and measure the potential output of a reservoir.

In addition to sources of steam and hot water, normally used for the production of electrical energy, one should not ignore medium to low resources of hot water (low process heat) that is very important for space heating, industrial processing, crop drying, and paper manufacturing. In the case of a new energy source such as geothermal, the changing economies, improvements in technology, and increases in the number of wells drilled will open new vistas for its use.

Calculation of Geothermal Resource Base

Examples for calculating the geothermal resource base are presented in Chapter 2. Determination of the resource base requires the assignment of quantitative values to the geothermal gradient, heat flow in the subsurface, and heat capacity of the rocks. It should be realized, however, that even in the most intensely investigated geothermal areas only a few of these variables are known and, then, often only within an order of magnitude.

Temperature Gradients and Geothermal Resources

Temperature gradient measurements are useful in exploration for geothermal resources, since they allow ready detection of thermal anomalies and estimation of their areal extent. Vaught (1980) pointed out that caution must be exercised in using gradients to project temperatures below the depth of measurement, for three reasons:

1. Temperature gradients vary with rock type. Shales and unconsolidated sediments have considerably lower conductivity than dolomites and well-cemented sandstones. Inasmuch as conductivity affects temperature gradients, projection of temperatures to depth must rely on a knowledge of geology.
2. Thermal conductivity values increase with depth because of increased compaction and cementation so that gradients decrease with depth. Thus, linear projection of gradients below observation points may predict temperatures much higher than those which actually exist.
3. Gradient measurements made in shallow holes are strongly influenced by near-surface effects such as precipitation and movement of groundwater. It has long been recognized that anomalously high bottomhole temperatures (and thus, elevated gradients) often occur in shallow wells (Vaught, 1980). Even in wells up to thousands of feet deep, movement of groundwater can alter the geothermal gradient.

Examples for using geothermal temperature gradients are provided for calculating the potential heat in a cooling granitic body and in the resource base determination of the northern Appalachian basin of the U.S.

The potential energy that could be extracted from 40 cubic miles of igneous rock (granite) in cooling from 200°C (360°F) to 0°C (32°F) is 1.71×10^{19} cal (see Figure 1-3).

The sensible heat stored in the granitic mass was calculated using the equation:

$$Q = C\gamma V \Delta T \tag{1-1}$$

where Q is the stored heat, cal (or Btu); C is the thermal capacity of the granite, cal/g $-$ °C (or Btu/lb $-$ °F); γ is the specific weight of the granite, g/cm^3 (or lb/ft^3); V is the volume of the mass under consideration, cm^3 (or

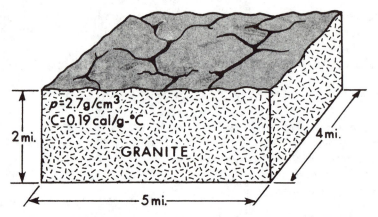

Figure 1-3. Simplistic model of a 40-cubic-mile granitic mass.

ft^3); and ΔT is the difference between an *in-situ* temperature at some depth and the mean surface temperature or some arbitrary base temperature.

The approach can be made more sophisticated as in the example of determining the geothermal resource base for the northern Appalachian region given by Rieke and Skidmore (1974) and Chuop (1980). These studies covered portions of the states of New York, Pennsylvania, North Carolina, Maryland, Virginia, Ohio, and West Virginia. Total development of geothermal resources in this region could provide additional amounts of energy for generating electrical power and supplying commercial and process heat, valuable minerals, and desalinated water for the eastern U.S.A.

Detailed geothermal gradient data that cover portions of northern Appalachia have been published only recently. Grisafi et al. (1973) utilized non-equilibrium bottomhole temperature data from formation evaluation studies in order to calculate geothermal gradients for northern West Virginia. Temperature calculations for generating isothermal mapping data can be made by employing the following linear equation:

$$T_f = (T_{bh} - T_{ms}) (D_f/D) + T_{ms} \qquad (1\text{-}2)$$

where T_f is the temperature of the formation in question, °F; T_{bh} is the temperature at the bottom of the well, °F; T_{ms} is the mean surface temperature, °F; D is the total depth of the well, ft; and D_f is the depth of the well to the formation in question, ft. Rainis et al. (1974) demonstrated the feasibility of a computational method for determining segmental and overall gradients and heat flow values from well log and geologic data.

Geothermal temperature calculations in this example were based on the following equation:

$$G = \frac{T_{bh} - T_{ms}}{D} \qquad (1\text{-}3)$$

where G is the geothermal gradient, F/ft; T_{bh} is the bottomhole temperature of the well, °F; T_{ms} is the mean surface temperature, °F; and D is the total depth of the well, ft.

The American Association of Petroleum Geologists (AAPG) and the U.S. Geological Survey jointly published several maps showing regional variations in subsurface temperatures. These maps may be of only limited use in geothermal exploration because of arbitrary corrections applied to calculated temperature gradients. Any gradient values more than two standard deviations from the mean were excluded because of suspected error. Because of these deletions, some true geothermal anomalies may not show up on the published maps. Although the AAPG-USGS maps are widely used by geothermal workers in the east, their local accuracy is questionable (Grisafi et al., 1973; Vaught, 1980).

Vaught (1980) pointed out that the AAPG-USGS data set is valuable as a first approximation of geothermal potential because of its comprehensive nature. However, it must be used with caution and with an appreciation of the quality of the data, especially in interpreting data from the shallow wells. Differentiation between shallow-hole data and deep-hole data is of utmost importance in interpreting the geothermal potential of an area. Even if the shallow-hole gradients are real, projected temperatures in deep wells are not likely to be high enough to provide economically usable thermal fluids (Vaught, 1980).

Finally, the AAPG-USGS temperature-gradient maps can be useful guides for preliminary geothermal exploration. But various studies have shown that they cannot be used as conclusive indicators of geothermal potential unless substantiated by other geologic data.

Figure 1-4 illustrates a reevaluated distribution of geothermal gradient values for northern Appalachia as derived from shallow and deep wells. Unfortunately, these data show only a broad range in the variation of these values (0.7°F/100 ft to 2.1°F/100 ft). Localized high and low gradient values could readily occur within the contours of the map.

The global average heat flow is approximately 1.5 HFU (1 HFU (heat flow unit) = 1 μcal/cm^2-sec). It has been estimated previously from published measurements that the heat flux in the Appalachian system is about 1 HFU (Smith, 1972). Reiter and Costain (1973) found that a linear relationship between surface heat flow and subsurface heat production exists for the folded Appalachian mountains in southwestern Virginia, northeastern Tennessee, and northwestern North Carolina. This relation-

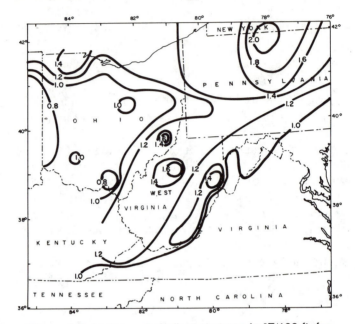

Figure 1-4. Isothermal gradient contours in °F/100 ft for northern Appalachia. (After Rieke and Skidmore, 1973.)

ship suggests a correlation between heat flow and the geological, as well as chemical, attributes of the Appalachian region. Smith et al. (1981) provided 23 new values of heat flow in the southern Applachian Mountains and southeastern and coastal plains. These ranged from 0.3 HFU in the Piedmont and the valley and ridge provinces to 1.6 HFU's in the Atlantic Coastal Plains.

Natural heat transfer measurements provide good information about the amount of geothermal energy available in a region: heat discharged from the earth can be defined as the heat flowing from some depth to the surface in unit time interval through a defined area. Horizontal heat transfer between adjoining rock bodies is ignored. In evaluating the potential resource base of a region, some of the more important interrelations used are those between geothermal gradient, heat flow, and thermal conductivity of the rocks. This relationship is conveniently expressed by the following equation: Conduction of heat from the earth is calculated by the following equation:

$$Q = -KA \, (dT/dD) \tag{1-4}$$

where Q is the heat flow through the ground, HFU; K is the thermal conductivity of the rock formations, cal/sec-cm °C; A is the cross-sectional area of interest, cm² (unit area is assumed in present calculations); and dT/dD is the geothermal gradient in the direction of the heat flow, °C/cm. The minus sign is inserted to satisfy the second law of thermodynamics.

The geothermal gradient is directly proportional to heat flow, but inversely proportional to conductivity in Equation 1-4. Heat flux cannot be accurately measured at low levels; therefore, it is common to calculate it from the geothermal gradient and conductivity values measured in the borehole. Heat flow values were obtained from the literature, or calculated by using Equation 1-4. Thermal resistivities (reciprocal of conductivity) were estimated for various lithological units in Appalachia. Horizontal heat transfer between adjoining rock bodies was ignored. Figure 1-4 illustrates the variation of the heat flux for this region, which ranges from 0.6 to 1.5 HFU (Combs and Simon, 1973; Diment and Robertson, 1963; Diment and Werre, 1960, 1964; Joyner, 1960; Reiter and Costain, 1973; Roy et al., 1968). One area bordering southern West Virginia and western Virginia contains anomalous high and low heat flux values.

The heat capacity determinations are based on three different rock-type models (limestone, sandstone, and shale). Values of the thermal capacities and densities for the models were obtained from Somerton's (1958) data (see Table 1-2). A water saturation of 100% was assumed to exist in the pore space.

Table 1-2
Porosity Values, Heat Capacities and Bulk Specific Weights for Water-Saturated Rock Types*

	φ	Specific Weight (lb/ft³)	C (Btu/lb-°F)	T (°F)	p (psia)	Q (cal)
Sandstone	0.196	142	0.252	90	14.7	5.44×10^{22}
Shale	0.071	149	0.213	90	14.7	4.83×10^{22}
Limestone	0.186	149	0.266	90	14.7	6.04×10^{22}

*Mean surface temperature = 55°F.

The resource base of sensible heat stored in a pure limestone model was calculated to be 6.04×10^{22} cal. The sandstone model gave 5.44×10^{22} cal and the shale model gave the least amount of 4.83×10^{22} cal. A composite model (45% sandstone, 35% limestone, and 20% shale) that is more representative of the different rock types in the northern Appalachian region gives 5.53×10^{22} cal of stored heat.

Equation 1-1 has to be modified by correcting the heat capacity and density terms to include heat stored in the pore water mass:

$$Q = [\phi \gamma_w C_w + (1 - \phi) \gamma_r C_r] \, V \Delta T \qquad (1\text{-}5)$$

where ϕ = fractional porosity; γ_w = specific weight of water; and γ_r = specific weight of rock (see Equation 1-1).

This analysis of the geothermal resource base of the northern Appalachia region has shown that the region has some areas with high geothermal gradients and high heat flow rates. Further geothermal exploration could clearly define the potential and delineate any local intense heat cells. Appalachia apparently lacks the areas of extremely high temperature in association with dry steam at shallow depths, which is now considered necessary for supplying a significant amount of energy for production of electricity. The resource base of sensible heat energy stored in the sedimentary rocks of the region was estimated to range from 4.83×10^{22} cal to 6.04×10^{22} cal. This thermal energy range is equivalent, on a heat-of-combustion basis, to approximately 30 to 35 trillion barrels of crude oil.

What proportion of any geothermal resource base for a given region can be considered as a resource depends upon a number of factors:

1. Depth of extraction
2. Effective porosity and permeability of the geothermal reservoirs
3. Physical state of the fluid
4. Chemical nature of the fluid
5. Availability of technology
6. Economics

This book provides the reader the methodologies and information to help make such assessments.

Geology of Geothermal Sources

Chapter 3 examines the nature of the geologic and tectonic settings of geothermal energy sources. This is followed by the geographic distribution and description of individual cases. Principal topics covered in Chapter 3 are:

1. Large silicic magma bodies (plutons and batholiths)
2. Intermediate composition plutonic bodies
3. Regions of predominantly basaltic volcanism
4. "Near-normal gradient" sources
5. Closed sedimentary basins

One of the basic problems in understanding the geology of geothermal systems is coping with unfamiliar lithologies. Geothermal reservoirs are often composed of crystalline igneous and metamorphic rocks, vesicular volcanic rocks, glassy or crystalline volcanic rocks, volcanic ashes, and welded volcanic rock materials. In order to clarify what types of lithologies are involved, rock classification schemes for igneous, volcanic (pyroclastic), and sedimentary rocks are presented in Tables 1-3 and 1-4, and for hydrothermal and metamorphic rocks in Table 1-5.

The igneous rocks are those that have formed by the cooling and hardening of molten rock (magma) on or at various depths below the earth's surface. Depending upon whether the magma cooled and hardened below the surface or was poured out on top before hardening, the resulting rocks are classified as either intrusive or extrusive, respectively. This influences the rock's texture and structure. The most common intrusive igneous rock is granite, whereas the most common extrusive one is basalt. Originally, all dark gray and black lavas were called basalts and were divided into olivine basalts, feldspathic basalts, etc. on the basis of the visible mineral crystals present.

Granite is a light-colored igneous rock mainly composed of a mixture of mineral grains or crystals of orthoclase feldspar and quartz. Usually, mica is also present. Sometimes one or more other minerals are also present, but they are not essential. The orthoclase feldspar is usually pink, white, or gray in color, and the quartz is usually of the smoky gray variety. Depending upon the predominance of a certain mineral having distinct color, the granite may be called pink, white, or gray. Basalt is a very fine-grained dark gray to black igneous rock. The individual grains consist of a mixture of several dark-colored minerals, which are too small to be identified with the naked eye.

Pyroclastic rocks are those materials that have been explosively or aerially ejected from a volcanic vent (Table 1-3). Lava is a molten magma that is brought up from the earth's crust and is poured out on the surface. It may produce great flows or may pile-up in thick masses to build volcanic mountains.

When products of the disintegration and decomposition of any rock type are transported, redeposited, and partly or fully consolidated or cemented into a new rock type, the resulting material is classified as a sedimentary rock (Table 1-3). Thus sedimentary rocks have been formed in beds or layers by the hardening (lithification) of sediments that dropped out of a fluid.

The low-grade metamorphic rocks are represented by serpentine and talc. These rocks and minerals originate from hydrothermal alteration of ferromagnesian minerals by late magmatic fluids or by fluids of extraneous origin. Alteration by late magmatic fluids is termed "autometamorphism," which is a form of the igneous process. These rocks are formed in the zone of remelting and are mixtures of igneous and metamorphic rocks. Serpentine,

Table 1-3

A Generalized Classification of Igneous and Volcanic Rocks

	Potash Feldspar >2/3 Total Feldspar		Potash Feldspar 1/3-2/3 Total Feldspar		Plagioclase Feldspar >2/3 Total Feldspar			Pyroxene, Uralite, Olivine	Little or No Feldspar
					Potash Feldspar >10% Total Feldspar	Sodic Plagioclase	Calcic Plagioclase		Chiefly Pyroxene and/or Olivine
Essential Minerals	Quartz >10%	Quartz <10% Feldspathoid <10%	Quartz >10%	Quartz <10% Feldspathoid <10%	Potash Feldspar >10% Total Feldspar Quartz >10%	Quartz >10%	Quartz <10% Feldspathoid <10%	Pyroxene, Uralite, Olivine	Chiefly Pyroxene and/or Olivine
Accessory Minerals	Hornblende, Biotite, Pyroxene, Muscovite		Hornblende, Biotite, Pyroxene		Hornblende, Biotite, Pyroxene				Serpentine, Iron Ore
Igneous rocks									
phaneritic	granite	syenite	quartz monzonite	monzonite	grano-diorite	quartz diorite (tonalite)	diorite	gabbro	peridotite
porphyritic (phaneritic groundmass)	granite porphyry	syenite porphyry	quartz monzonite porphyry	monzonite porphyry	grano-diorite porphyry	quartz diorite porphyry	diorite porphyry	gabbro porphyry	peridotite porphyry
porphyritic (aphanitic groundmass)	rhyolite porphyry	trachyte porphyry	quartz latite porphyry	latite porphyry	dacite porphyry		andesite porphyry	basalt porphyry	
aphanitic (micro-crystalline)	rhyolite	trachyte	quartz latite	latite	dacite		andesite	basalt	
Volcanic rocks									
aphanitic (flows)	rhyolite lavas	trachyte lavas					andesite lavas	basalt lavas	
(ejecta)					obsidian pumice scoria				

diabase (dolerite)

Homogeniety

After Travis, 1955

Table 1-4
Sedimentary Rocks Classified by Grain Size and Origin

Grain Size	Origin			
	Chemical	Biochemical	Clastic	Evaporitic
Coarse-grained >½ mm in diameter	limestone ($CaCO_3$) dolomite [$CaMg(CO_3)_2$]	limestone	conglomerate sandstone orthoquartzite arkose graywacke	
Medium-grained ½-¼ mm	limestone dolomite	limestone	oolitic limestones sandstone arkose graywacke	
Fine-grained ¼-⅛ mm	limestone dolomite	limestone diatomite chalk	sandstone calcareous arkose graywacke	
Very fine-grained (Clay-sized) <⅛ mm	limestone dolomite	limestone coal chalk	shale mudstone siltstone	halite anhydrite gypsum
Amorphous	chert			

which is a product of hydrothermal alteration, has a nondirectional structure and is fine-grained. Although normally found as localized contact metamorphic products, serpentinized bodies of wide extent also exist as evidenced by the Franciscan Formation in California. Migmatites (rock formations that originally were sedimentary rock, such as sandstone, which after intense heating sometimes resemble granite) probably originate by partial granitization or metamorphic differentiation. These rocks have a lineate or foliate structure and are products of plutonic processes (deep seated magma that solidifies slowly in the earth's crust).

Metamorphic rocks are products of alteration by heat, pressure, or both (Table 1-5). Originally they may have been igneous, sedimentary, pyroclastic, or another kind of metamorphic rock. Upon being subjected to great pressures arising within the earth along with frictional heat and chemical action, these rocks may experience numerous alterations while still remaining as solid materials. These changes may occur as a result of folding and buckling of the rocks, contact with hot magma, or deep burial beneath overlying rocks. The effect may be that of flattening the grains of the original rock, shearing of the grains so that they slide past each other, compacting and cementing of the grains more firmly together, or formation of new minerals that are squeezed in. If the latter formations require more space than is available for them, minerals are broken down atomically, and the atoms rearrange themselves into different minerals that are denser and do

Table 1-5

A Generalized Classification of Metamorphic/Hydrothermal Rocks

Color	Nondirectional Structure Massive or Granulose		Mechanical Metamorphism	Directional Structure (lineate or foliate)				Plutonic Metamorphism
	Contact Metamorphism			Regional Metamorphism				
	Aphanitic	Phaneritic	Cataclastic	Slaty	Phyllitic	Schistose	Gneissose	Migmatitic
Light	metaquartzite marble soapstone	metaquartzite marble	mylonite	slate	phyllite	schist	gneiss granulite	migmatite
Intermediate (includes red or brown)	metaquartzite marble serpentine	metaquartzite marble serpentine amphibolite						
Dark (includes green)						schist amphibolite	gneiss	

After Travis, 1955

not require as much space. Nothing is added or subtracted from the rock — the simple elements are merely rearranged into denser kinds of minerals. For example, if a granite is metamorphosed, some of the atoms of quartz and feldspar may be squeezed out and be recombined into garnet crystals. The garnet contains the same simple elements that are present in the other two minerals, but it is quite different because the proportion and arrangement of its atoms are different.

The most widespread metamorphism takes place during movements of the earth's crust, particularly at times when vast mountain ranges are built by buckling and folding of the rocks. Metamorphism is also caused by hot magma coming into contact with rocks and baking them. Sometimes hot water and steam are forced through the pores of rocks, causing the grains to become fused together and making the rock very dense and solid. All original kinds of igneous and sedimentary rocks have their metamorphic counterparts, and these are subdivided into two large groups, called the *foliated* and *nonfoliated*. (Table 1-4). The foliated ones have their mineral grains flattened or arranged in streaks, making the rock appear to be composed of thin layers, often as thin as the book pages. These are not true layers like the beds of sedimentary rocks, although the rocks often split easily along these foliations. In the nonfoliated group, the rocks are dense and massive and do not have such foliations.

Exploration for Geothermal Energy

Because of the relative scarcity of commercial geothermal systems, the volume of results and comparison of exploration systems needed to fully evaluate the desired exploration methods are insufficient. Generally speaking, when a large area of interest is to be evaluated, rapid low-cost reconnaissance techniques are employed. When the search has been narrowed, then more precise and somewhat more expensive techniques are used to define the anomaly. The final test is to drill a well to test the prospect.

In addition to the more sophisticated methods of exploration such as infrared aerial surveys, magneto-tellurics, and TDEM (Time Domain Electro Magnetic), one should not overlook the less sophisticated methods of "seepology," e.g., fumaroles, hot or warm springs, geysers, fossil seeps, and mud volcanoes. Chapter 4 examines the current state of geothermal energy exploration.

Drilling and Completing Geothermal Wells

The methods used in drilling and completing geothermal wells are based on the knowledge gained in performing the same operations when searching for and producing oil and gas. There are some special procedures, however, applicable to drilling and completing geothermal wells, as Chapter 5 points out. The equipment and methods have been adapted from an industry that regularly drills into sedimentary formations. As noted before, however, geothermal resources more commonly exist in areas having igneous and metamorphic rocks. Hydrothermal resources, for example, are likely to be found in resource areas dominated by young volcanic formations in tectonically active areas. Geopressured resources are associated with sedimentary formations having alternating sand and shale strata (Table 1-6).

Most geothermal drilling (other than geopressured wells) is performed at low pressures. Also, most geothermal wells, which are relatively shallow in depth and have high formation temperatures (except in some Gulf Coast areas), are drilled through or into igneous and metamorphic formations. Occasionally, geothermal wells are drilled in sedimentary basins. Features of a reservoir that control the drilling cost include the depth of the well, the designed diameter of the well (which is determined by the temperature and probable flow rate of the well), and the approximate hardness of the overlying rocks. The end result of all of these factors is that drilling and completion of the well may progress very slowly. The penetration rate achieved in a particular formation is, of course, a function of a number of factors, but is dominated by the way the rock hardness and abrasiveness affect the bit performance. The compressive strength of the rock and its elastic properties are also influential. Chapter 5 also covers drilling rigs, rig equipment, bit and hydraulics programs, muffler systems, use of drilling fluids, and drilling case histories.

Casing and Tubular Design Concepts

Chapter 6 presents the state-of-the-art in design of casing and tubular goods for geothermal and geopressured-geothermal wells. Once a potential

Table 1-6
Geothermal Drilling Characteristics of Selected International Sites

Location	Resource	Rock Type
Larderello, Italy	Steam	Fractured limestone
The Geysers, California	Steam	Fractured graywacke
New Zealand	Hot water	Acid volcanics
Japan		
Pathe, Mexico	Hot water	Fractured middle Tertiary volcanics
Cerro Prieto, Mexico	Hot water	River delta sediments
Niland, California		
Iceland	Hot water	Fractured cavernous basaltic lavas
Northern Taiwan	Hot water	Acid volcanics and sedimentary rocks

geothermal system is located, verification of its existence must be made by drilling, completing, and testing wells. This chapter covers basic well configurations, casing design procedures utilizing the maximum load design concept, corrosion, scaling, wellbore stability, and temperature effects.

Geothermal Well Cementing

The procedures for cementing a geothermal well are similar to those for cementing an oil and/or gas well; however, many of the conditions encountered in a geothermal well are quite different. Temperature-stabilized drilling fluids are required, as are high-grade steel casing strings with special threaded couplings. The cement used is generally temperature-stabilized. Chapter 7 covers these as well as the following topics: cement mixing methods, compositions, and placement techniques.

Formation Evaluation

Chapter 8 describes the basic principles of geophysical well logs and the parameters they measure. Properties to be determined for geothermal evaluation are lithology variations, thickness, porosity, nature and dimensions of fracture systems, formation pressures, temperature, salinity of aqueous phase, and entry points and quality of steam or hot water. Although many of the logs run in wells drilled for oil and gas are also used in geothermal wells, many require different interpretations to assist in the understanding of formations and fluids in geothermal wells, and exploration for geothermal resources.

The Geothermal Log Interpretation Steering Committee of DOE developed the following list of priorities for needed development of log measurements and interpretation techniques to be used to characterize the entire reservoir or the resource:

1. True formation temperature profiles by time-lapse temperature measurements.
2. Lithology, depth, and thickness of formations.
3. Intergranular and fracture permeabilities.
4. Intergranular and fracture porosities.
5. Further characterization of fracture systems.
6. Borehole geometry.
7. Fluid composition.
8. Thermal conductivity and heat capacity.
9. Elastic moduli of rock.

This chapter also explains parameters measured and interpretations best adapted for determining the geothermal properties of the well.

All available data has to be collected in order to answer questions such as:

1. What is the origin and nature of the system?
2. What is the system's heat content?
3. How much heat is recoverable?
4. At what rate can heat be extracted?
5. What type of energy usage is envisioned?
6. How long will the system last?
7. How will the production decline?
8. What type of surface facilities are required?
9. Is the project economically viable?

Most of these questions can be answered only as accurately as the facts that are known about the system under consideration. Even the requirements for accuracy may vary from reservoir to reservoir. For instance, if power generation is desired, the question of economic viability is more critical for a given reservoir with 175°C water, than if the same reservoir contained 250°C water (Sanyal et al., 1980).

An excellent appraisal of the measurement requirements and methods for geothermal reservoir system parameters has been performed (Lamers, 1979). It is very difficult to establish both the need and measurement performance requirements for geothermal energy applications. Owing to the infancy of geothermal energy development, except vapor-dominated reservoirs such as the Geysers in California, U.S.A., most requirements and efforts to measure process parameters have been associated with reservoir testing (i.e., well flow and interference tests). As such, the need and requirements for process plant start-up, operation, and maintenance must be estimated and will vary depending on the specific reservoir, its fluid properties, and the type of energy conversion process employed. This is further amplified by the large number of different fluid flow conditions within a given process plant.

In order to appraise fluid properties and composition, pressure, and formation and producing zone parameters and measurement requirements, Lamers (1979) prepared an overall perspective of the range of borehole and pipe sizes and associated access restrictions, fluid flow rates, viscosities, tolerable pressure losses, corrosion, abrasion and scaling constraints, etc. that must be considered. Tables 1-7, 1-8, and 1-9 delineate some of these key constraints. For those interested in estimating the power production of a geothermal well, Karamarakar and Cheng (1980) discuss a theoretical assessment of James' empirical method for determining the wellhead

Table 1-7
Geothermal Fluid Measurement Conditions

Downhole Conditions

Well diameters	— 5¾ in. to 14 in.
	— Production wells typically will be greater than 9 in. in diameter.
	— Open hole washouts can range up to 30 in.
Obstruction	— High pressure wellhead valves for tool access
	— Will range down to 3 in., typically <6 in.
Borehole deviation	— Up to 45° (typically less than 20°)

Process Pipelines

Orientation	— 0 to 90°
Diameter	— 6 to 48 in. (less than 24 in. for liquid)

Fluid Flow Rate

Wellhead (flowing well)	— $<2(10)^6$ lb/hr
Process plant	— $<10^7$ lb/hr
Shut-in well	— Can have downhole flow (up or down) beween two zones or fractures (measure down to 10 gpm)

Maximum Pressure Drops to be Imparted by a Sensor

Liquid lines <5 psi
Steam lines <2 psi

Fluid Viscosity (temperature dependent)

Liquid: $0.05 \leq \mu \leq 1.4$ centipoises
Steam: $0.01 \leq \mu \leq 0.025$ centipoise

After Lamers, 1979.

discharge characteristics, such as flow rate, stagnation, enthalpy, and steam quality.

Fluid temperature is always a high priority measurement parameter in geothermal energy systems from early exploration through process plant operations. Downhole temperature measurements are required primarily during exploration, well formation test evaluation, and start-up monitor phases of geothermal operations. For plant operations, there is a demand for reliable electronic readout sensors with remote monitor capability (Lamers, 1979).

Temperature is an important factor to know when evaluating a direct process parameter, but it is also required when calibrating measurements from pressure and flow rate sensors. Whereas accurate and straightforward measurement techniques exist for obtaining process fluid temperatures, one limitation has been stressed. The limitation is that of scaling and associated thermal insulation on pipelines and thermowells inserted in pipelines, resulting in loss of calibration and/or lack of confidence in measurements.

Table 1-8

Measurement Performance Requirements for Fluid Pressure

Performance	Interference testing*	Downhole Other applications	Wellhead and power conversion plant
Range	20 psia < p <5000 psia**	20 psia < p <5000 psia	20 psia < p <500 psia natural flow 20 psia < p <3000 psia pumped flow
Accuracy	≤0.1% of full scale (FS)	≤0.25% FS	1% of working pressure
Resolution	≤0.005% FS (≤0.1 psi desired)	<0.05% FS	½% of working pressure
Response time	10 sec	1 sec	10 sec
Exposure time (measurement period)	10 days to 9 months	4 hrs to 4 days	Continuous
Drift (long term) (over measurement)	<0.01% FS (≤0.5 psia desired)	<0.1% FS	<1% FS

After Lamers, 1979.

* Downhole interference pressure measurements will be in a constant high temperature field (±3°C).

	Current	Future
** Liquid-dominated hydrothermal	p <5,000 psia	p > 5,000 psia
Vapor-dominated hydrothermal	p <700	p >700 (can go to 1500 psia w/liquid build-up at bottom)
Hot dry rock	p <7,000	p >10,000
Geopressured	p <15,000	p >15,000

Table 1-9
Formation and Producing Zone Parameters and Measurement Requirements

	Range	Accuracy	Resolution	Comments
Formation permeability:k (Darcys)	$10^{-3} < k < 3$	5% of reading	—	Typically obtained from pressure transient tests
Location and identification of permeable producing and thief zones	Small changes in vertical flow when shut in (up and down)	—	—	High priority measurement
Producing formation, grain size (microns)	1 to 100	—	—	Can be obtained from cutting samples
Fracture size: σ (millimeters)	$0.5 < \sigma < 10$ typical (can be larger)	±0.5	±0.5	Some fractures in geysers reported to be ~ ¾ in.
Fracture spacing:Δ (number per meter)	$1 < \Delta < 100$	5%	5%	Can be only one producing fracture in producing zone
Fracture orientation with borehole: θ_B	0 to 90°	2°	1°	Very high priority measurement
Fracture orientation in-situ formation: θ_F	0 to 45°	2°	1°	Very high priority measurement
Formation porosity	1 to 30%	±1%	±1%	High priority measurement
Formation temperature: T_F (°C)	$100 < T_F < 400$	±3°C	±2°C	Very high priority measurement during drilling operations
Vertical heat flow: Q_h ($Q_h = q/A = kb$) (μcal/sec − cm²) HFU's	$0.5 < Q_h < 20$ nominal ~ 6	±10% of reading	—	—
Thermal conductivity: K	$3(10)^{-3} < K < 1(10)^{-2}$	±10% of reading	—	Usually obtained from core samples
Vertical temperature gradient: b (dt/dL, °C/Km)	$40 < b < 1000$ nominal ~ 100	—	—	—

After Lamers, 1979.

Reservoir Engineering Concepts

Chapter 9 covers the basic principles of a geothermal reservoir. Reservoir engineering, well test analysis, and mathematical reservoir simulation are covered in detail. A brief history of experience in classic studies of geothermal reservoir engineering is given. This information covers well location, well logging, drilling and flow measurements, identification of the production mechanism and, finally, performance prediction to find the optimum production conditions leading to maximum economic heat recovery. An attempt should be made to answer the following questions:

1. What is the optimum development plan for the reservoir?
2. How many wells and what kind of pattern will be required?
3. What will the rate of production be?
4. How much heat will be recovered?
5. What will be the variation of temperature versus time?
6. Would it be feasible to implement an enhanced recovery process to recover additional heat?

The desirability of developing a conceptual model of the reservoir and assessing its physical and thermal properties are emphasized in this chapter.

A review of the state-of-the-art in geothermal reservoir modeling is appropriate. Pinder (1979) pointed out that there are several distinct but interrelated elements of geothermal reservoir modeling. The most fundamental element is the conceptual model of the reservoir. Whereas field data is relatively scarce and, at least in part, not freely available, there is nevertheless a general consensus of opinion on the fundamental aspects of the geothermal reservoir. It has been demonstrated in some reservoirs that the primary conduits of geothermal energy transport are fractures. The porous medium blocks, delineated by these fractures, act as the long-term energy suppliers feeding the fracture system.

Geothermal reservoirs can be classified on the basis of their fluid composition (Figure 1-5). The most common type of geothermal field is characterized by reservoir fluid which is predominantly water in the liquid phase. This type of field, often referred to as a hot-water system, is found at Wairakei, New Zealand, Cerro Prieto, Mexico, and many other locations around the world. Reservoirs which produce primarily steam are called "vapor-dominated". The major reservoirs of this class are found at the Geysers in California, at Larderello in Italy, and at the Matsukawa field in Japan. Hot-water systems characteristically produce from 70 to 90% of their total mass as water at the surface, whereas vapor-dominated systems produce dry to superheated steam (Toronyi and Farouq Ali, 1977). The pressures of vapor-

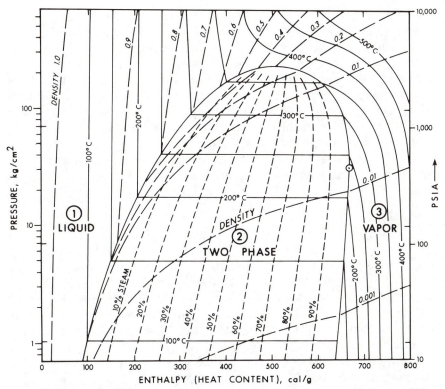

Figure 1-5. Pressure versus enthalpy diagram for pure water and steam with thermodynamic fluid reservoir conditions: (1) compressed water, (2) two-phase steam and water, and (3) superheated steam. Data computed from Keenan and Keyes (1936) and White et al. (1971). Open circle indicates maximum enthalpy of saturated steam, i.e., 670 cal/g at 236°C and 31.8 kg/cm². Critical point—3,206.2 psia at 705.4°F.

dominated systems are below hydrostatic. The initial temperatures and pressures of this system are very near those corresponding to the maximum enthalpy of saturated steam, i.e., 236°C and 31.8 kg/cm² (see Figure 1-5). The properties of water and dry saturated steam are presented in Table 1-10 in English units. The enthalpy is arbitrarily set at zero for saturated water at 32°F. The distribution of fluids within a reservoir is essentially unknown.

Thermal properties of water — steam are the basis for the energy contained in producing geothermal reservoirs. The ability of a substance to absorb heat is indicated by its *specific heat,* which in the English unit system is defined as the number of Btu's necessary to increase the temperature of one pound of mass by one degree Fahrenheit. The boiling point temperature of pure water is also known as the *saturation temperature*. After water starts to boil, any further addition of heat does not cause an increase in its temperature (water is heat saturated). This additional heat is used in vaporizing the water and is known as the *latent heat of vaporization* (two-phase region of Figure 1-5). At the critical pressure (3206.2 psia), the heat of vaporization is equal to zero. The enthalpy (sensible heat) plus the latent heat of vaporization equals the total heat (Table 1-10).

Steam quality is of prime importance in determining the amount of heat available per pound of steam, and is defined as the percent by weight of dry steam (steam mass) contained in one pound of wet steam. Quality of the steam (x) can be calculated by using the following equation:

$$x = \left(\frac{h_{fg} - h_f}{h_g - h_f} \right) \tag{1-6}$$

where h_{fg} is the enthalpy of the liquid-vapor mixture, Btu/lb; h_f is the enthalpy of the saturated liquid in equilibrium with the vapor, Btu/lb; h_g is the enthalpy of the saturated vapor in equilibrium with the liquid, Btu/lb; x is the fractional value of the steam quality. With a reduction in the quality of steam (x), the contribution of the latent heat decreases and the enthalpy of water is increased.

The specific volume of wet steam is the sum of the volumes of dry steam and suspended liquid:

$$v_{fg} = \frac{x}{100} v_g + \left(1 - \frac{x}{100} \right) v_f \tag{1-7}$$

where v_{fg} is the specific volume of wet steam, ft³/lb; v_g is the specific volume of the saturated vapor, ft³/lb; and v_f is the specific volume of the water, ft³/lb.

Ramey (1968) pointed out that the mass fraction of steam must lie between 0 and one, and is an insensitive measure of the volume fraction which is steam. Ramey (1968) related the steam quality to the vapor saturation, S_g (volume fraction of steam), as follows:

$$S_g = \frac{v_g}{v_g + \left(\dfrac{1 - x}{x} \right) v_f} \tag{1-8}$$

Inasmuch as the specific volumes depend upon temperature or pressure, S_g is related to the steam quality as a function of temperature.

The heat capacity of the reservoir rock is large compared to that of water or steam. In any reservoir, one has to account for the heat contributed by the

(text continued on page 30)

Table 1-10

Properties of Water and Dry Saturated Steam

Temp. °F T	Pressure p psia	Specific volume (ft³/lb)			Enthalpy (Btu/lb)			Entropy (Btu/lb)		
		Sat. Liquid v_l	Evap. v_{lg}	Sat. Vapor v_g	Sat. Liquid h_l	Evap.* h_{lg}	Sat.** Vapor h_g	Sat. Liquid s_l	Evap. s_{lg}	Sat. Vapor s_g
32	0.08854	0.01602	3306	3306	0.00	1075.8	1075.8	0.0000	2.1877	2.1877
35	0.09995	0.01602	2947	2947	3.02	1074.1	1077.1	0.0061	2.1709	2.1770
40	0.12170	0.01602	2444	2444	8.05	1071.4	1079.3	0.0162	2.1435	2.1597
45	0.14752	0.01602	2036.4	2036.4	13.06	1068.4	1081.5	0.0362	2.1167	2.1429
50	0.17811	0.01603	1703.2	1703.2	18.07	1066.6	1063.7	0.0361	2.0903	2.1264
60	0.2563	0.01604	1206.6	1206.7	28.06	1059.9	1088.0	0.0555	2.0393	2.0948
70	0.3631	0.01606	867.8	867.9	38.04	1054.3	1092.3	0.0745	1.9902	2.0647
80	0.5069	0.01608	633.1	633.1	48.02	1048.6	1096.6	0.0932	1.9428	2.0380
90	0.6982	0.01610	468.0	468.0	57.99	1042.9	1100.9	0.1115	1.8972	2.0087
100	0.9492	0.01613	350.3	350.4	67.97	1037.2	1105.2	0.1295	1.8531	1.9826
110	1.2743	0.01617	265.3	265.4	77.94	1031.6	1109.5	0.1471	1.8106	1.9577
120	1.6924	0.01620	203.25	203.27	87.92	1025.8	1113.7	0.1645	1.7694	1.9339
130	2.2225	0.01625	157.32	157.34	97.00	1020.0	1117.9	0.1816	1.7296	1.9112
140	2.3886	0.01639	122.99	123.01	107.89	1014.1	1122.0	0.1984	1.6910	1.8894
150	3.718	0.01634	97.06	97.07	117.89	1008.2	1126.1	0.2149	1.6837	1.8685
160	4.741	0.01639	77.27	77.29	127.89	1002.3	1130.2	0.2311	1.6174	1.8485
170	5.992	0.01645	62.04	62.06	139.90	996.3	1134.2	0.2472	1.5822	1.8293
180	7.510	0.01651	50.21	50.23	147.92	990.2	1136.1	0.2630	1.5480	1.8109
190	9.339	0.01657	40.94	40.96	157.95	984.1	1142.0	0.2785	1.5147	1.7932
200	11.526	0.01663	33.62	33.64	167.99	977.9	1145.9	0.2938	1.4824	1.7762

(table continued on next page)

Table 1-10 continued

Temp. °F T	Pressure p psia	Specific volume (ft³/lb)			Enthalpy (Btu/lb)			Entropy (Btu/lb)		
		Sat. Liquid v_l	Evap. v_{fg}	Sat. Vapor v_g	Sat. Liquid h_t	Evap.* h_{fg}	Sat.** Vapor h_g	Sat. Liquid s_t	Evap. s_{fg}	Sat. Vapor s_g
210	14.123	0.01670	27.80	27.82	178.05	971.6	1149.7	0.3090	1.4508	1.7598
212	14.696	0.01672	26.73	26.80	180.07	970.3	1150.4	0.3120	1.4446	1.7566
220	17.186	0.01677	23.13	23.15	188.13	965.2	1153.4	0.3239	1.4201	1.7440
230	20.760	0.01684	19.365	19.382	198.23	958.8	1157.0	0.3387	1.3901	1.7288
240	24.969	0.01692	16.306	16.328	208.34	952.2	1160.5	0.3531	1.3609	1.7140
250	29.825	0.01700	13.804	13.821	216.48	945.5	1164.0	0.3675	1.3323	1.6998
260	35.429	0.01703	11.746	11.763	228.64	938.7	1167.3	0.3817	1.3043	1.6880
270	41.858	0.01717	10.044	10.061	238.84	931.8	1170.6	0.3958	1.2769	1.6727
280	49.203	0.01726	8.628	8.645	248.06	924.7	1173.8	0.4096	1.2501	1.6592
290	57.556	0.01735	7.444	7.461	259.31	917.5	1176.8	0.4234	1.2238	1.6472
300	67.013	0.01745	6.449	6.466	269.59	910.1	1179.7	0.4369	1.1980	1.6350
310	77.68	0.01755	5.609	5.626	279.92	902.6	1182.5	0.4504	1.1727	1.6231
320	89.66	0.01765	4.896	4.914	290.28	894.9	1185.2	0.4637	1.1478	1.6115
330	103.06	0.01776	4.269	4.307	300.68	887.0	1187.7	0.4769	1.1233	1.6002
340	118.01	0.01767	3.770	3.788	311.13	879.0	1190.1	0.4900	1.0992	1.5891
350	134.63	0.01799	3.324	3.342	321.63	870.7	1192.3	0.5029	1.0754	1.5783
360	153.04	0.01811	2.939	2.957	332.18	862.2	1194.4	0.5158	1.0519	1.5677
370	173.37	0.01823	2.606	2.626	342.79	853.5	1196.3	0.5286	1.0287	1.5573
380	195.77	0.01836	2.317	2.335	353.45	844.6	1198.1	0.5413	1.0069	1.5471
390	220.37	0.01850	2.0651	2.0836	364.17	835.4	1199.6	0.5539	0.9832	1.5371
400	247.31	0.01864	1.8447	1.8633	374.97	826.0	1201.0	0.5664	0.9608	1.5272
410	276.75	0.01878	1.6512	1.6700	395.83	816.3	1202.1	0.5788	0.9386	1.5174

(Table 1-10 continued)

Temp. °F T	Pressure p psia	Specific volume (ft³/lb)			Enthalpy (Btu/lb)			Entropy (Btu/lb)		
		Sat. Liquid v_l	Evap. v_{fg}	Sat. Vapor v_g	Sat. Liquid h_l	Evap.* h_{fg}	Sat.** Vapor h_g	Sat. Liquid s_l	Evap. s_{fg}	Sat. Vapor s_g
420	308.83	0.01894	1.4811	1.5000	396.77	806.3	1203.1	0.5912	0.9166	1.5078
430	343.72	0.01910	1.3308	1.3499	407.79	796.0	1203.8	0.6035	0.8947	1.4982
440	381.59	0.01926	1.1979	1.2171	418.90	785.4	1204.3	0.6158	0.8730	1.4887
450	422.6	0.0194	1.0799	1.0993	430.1	774.5	1204.6	0.6280	0.8513	1.4793
460	466.9	0.0196	0.9748	0.9944	441.4	763.2	1204.6	0.6402	0.8293	1.4700
470	514.7	0.0198	0.8811	0.9009	452.8	751.5	1204.3	0.6523	0.8083	1.4606
480	566.1	0.0200	0.7972	0.8172	464.4	739.4	1203.7	0.6615	0.7868	1.4513
490	621.4	0.0202	0.7221	0.7423	476.0	726.8	1202.8	0.6766	0.7653	1.4419
500	680.8	0.0204	0.6545	0.6749	487.9	713.9	1201.7	0.6887	0.7438	1.4325
520	812.4	0.0209	0.5385	0.5594	511.9	686.4	1193.2	0.7130	0.7006	1.4136
540	962.5	0.0215	0.4434	0.4649	536.6	656.6	1193.2	0.7374	0.6668	1.3942
560	1133.1	0.0321	0.3647	0.3868	562.2	624.2	1166.4	0.7621	0.6121	1.3742
580	1326.8	0.0226	0.2989	0.3217	586.9	588.4	1177.3	0.7872	0.5659	1.3532
600	1542.9	0.0236	0.2432	0.2668	617.0	548.5	1165.3	0.8131	0.5176	1.3307
620	1736.6	0.0247	0.1955	0.2201	646.7	503.6	1150.3	0.8398	0.4664	1.3062
640	2059.7	0.0260	0.1538	0.1798	678.6	452.0	1130.5	0.8679	0.4110	1.2789
660	2365.4	0.0275	0.1165	0.1442	714.2	390.2	1104.4	0.8987	0.3485	1.2472
680	2708.1	0.0305	0.0810	0.1115	757.3	309.9	1067.2	0.9351	0.2719	1.2071
700	3093.7	0.0369	0.0392	0.0761	823.3	172.1	995.4	0.9905	0.1484	1.1389
705.4	3206.2	0.0503	0	0.0503	902.7	0	902.7	1.0580	0	1.0580

After Keenan and Keyes, 1936; courtesy of John Wiley and Sons.
*Latent heat of evaporation.
**Total heat of steam.

(text continued from page 26)

rock and the fluids in the pore space. The following example is presented to illustrate this point.

Given: a geothermal reservoir contains mostly pure hot water. The porosity is 0.15 and the pore space contains approximately 20% by volume of steam. Initial temperature is 450°F.

1. Calculate the total heat, Q_T, available from the geothermal system above a datum of 250°F (inlet temperature), in terms of Btu/ac-ft of the reservoir. Assume the C_r is 0.25 Btu/lb-°F, ρ_r is 2.6 g/cc, and $\rho_w = 1.0$ g/cc. (See Equations 1-1 and 1-5, and Table 1-10.)

$$Q_T = Q_r + Q_w + Q_s \tag{1-9}$$

$$\frac{Btu}{ac\text{-}ft}(total) = \frac{Btu}{ac\text{-}ft}(rock) + \frac{Btu}{ac\text{-}ft}(water) + \frac{Btu}{ac\text{-}ft}(steam)$$

$$Q_T = [(1-\phi)V_bC_r\Delta T\rho_r] + [\phi V_b\left(\frac{1-V_g}{v_{f450}}\right)(h_{f450} - h_{f250})]$$

$$+ [\phi V_bV_g\left(\frac{1}{v_{g450}}\right)(h_{g450} - h_{g250})]$$

where V_b = bulk volume (43,560 ft³/ac-ft), V_g = volume of steam in pore space, C_r = thermal capacity of rock v_f = specific volume of fluid, v_g = specific volume of vapor, h_f = enthalpy of fluid, h_g = enthalpy of vapor, ρ_r = density of rock, and ρ_w = density of fresh water. Thus:

$Q_T = 300 \times 10^6$ Btu/ac-ft + 57 $\times$ 10⁶ Btu/ac-ft + 0.048 $\times$ 10⁶ Btu/ac-ft
    = 357.048 $\times$ 10⁶ Btu/ac-ft

2. Calculate the mass of the fluids in place in the reservoir in lb/ac-ft of bulk volume:

$$W_T = W_w + W_g \tag{1-10}$$

$$\frac{lb}{ac\text{-}ft}(total) = \frac{lb}{ac\text{-}ft}(water) + \frac{lb}{ac\text{-}ft}(steam)$$

$$W_T = \phi V_b\left(\frac{1-V_g}{v_{f450}}\right) + \phi V_b\left(\frac{V_g}{v_{g450}}\right)$$
$$= 269,443 + 1189 = 270,632 \text{ lbs/ac-ft}$$

3. Calculate the mass of the rock in lb/ac-ft of bulk volume:

$$W_r = (1 - \phi) \, V_b \, \rho_r = 6.00 \times 10^6 \text{ lb/ac-ft}$$

The assumptions inherent in the conceptual model of the reservoir should dictate the framework of its mathematical description. In the case of geothermal reservoirs, however, the physical and mathematical foundations for multiphase mass and energy transport through fractured porous media do not exist. Consequently, all of the existing multiphase models assume the reservoir to be a porous medium. When fractures are included, they are highly idealized geometrically and, although the parameter values may differ, the same governing equations are employed as for the porous medium (Coats, 1977). Although fractured reservoir mass and energy transport has been considered in a formal way for hot-water systems, this has not yet been extended to a steam-water system (O'Neill, 1977). The governing equation for heat transfer within convective systems has been thoroughly reviewed by Cheng (1978). An up-to-date annotated bibliography has been prepared for geothermal reservoir engineering, which contains summaries and additional references not included in this book (Sudol et al., 1979).

Given the theoretical constraint previously cited, the governing flow and transport equations for geothermal reservoir simulation are obtained through one of three ways. The simplest approach is essentially a macroscopic mass balance. In other words, one assumes that the balance laws observed at the microscopic level are, with minor modification, valid for the porous medium as well. This approach does not provide insight into the micro-physics of energy transfer at the pore level, but does provide a set of governing equations not unlike those obtained using more sophisticated techniques. A second approach involves the use of mixture theory as developed in continuum mechanics. This approach is more rigorous but, while recognizing the existence of pore level interaction, it does not provide adequate insight into the nature of this interaction. The most promising approach to obtaining a rigorous formulation of the governing equations is through formal integration of the microscopic balance equations over the porous medium, possibly augmented through constitutive theory (Pinder, 1979).

Having generated an appropriate set of governing equations, one is faced with the task of solving a set of highly nonlinear partial-differential equations. In nearly all cases, this is approached numerically. There are several difficulties encountered in the numerical solution of equations for the geothermal reservoirs. The first task is to select a set of dependent variables because of existence of several possibilities. One must then decide upon a method of approximation. Currently, finite difference and finite element schemes are employed. One is confronted with the problems associated with the simulation of convection-dominated transport, namely, numerical dispersion (oscillations) and diffusion (smearing of a sharp front). However, possibly the most difficult task remains, i.e., the efficient and accurate treatment of the highly nonlinear coefficients. Every geothermal model handles this problem in a different manner.

From the reservoir engineering point of view, there are two additional factors to consider. The field application of a geothermal code requires a proper representation of the wellbore dynamics and thermodynamics. This is particularly important in the case of simulations in the immediate vicinity of the well. A second practical problem involves the reduction of the general three-dimensional system to an areal two-dimensional representation. This requires formal integration over the vertical, which should be carried out carefully so that essential elements of the reservoir physics are salvaged. A recent study of Pritchett et al. (1980) applied this approach to the study of the Wairakei geothermal field in New Zealand.

The numerical schemes employed in existing geothermal models have been thoroughly described (Pinder, 1979), and the salient features of each one are summarized in Table 1-11. The important elements of Pinder's discussion can be briefly stated as follows:

1. The sets of dependent variables employed in solving the flow and energy transport equations are (ρ_f, U_f), (p_f, h_f), (p_f, S_w), $(p_w, T; \rho_g, S_g)$ and $(\rho_w, T; S_w, T; \rho_g, T)$. The choice between (ρ_f, U_f) and (ρ_g, h_f) seems rather arbitrary because one is readily derived from the other for presentation.

2. The majority of models will accommodate one, two, and three space dimensions: the notable exceptions are Toronyi and Farouq-Ali (1977) and Huyakorn and Pinder (1977).

3. With the exception of the Toronyi and Farouq-Ali model, all simulators can handle either one- or two-phase flow.

4. Finite difference, finite element, and integrated finite difference methods have been used in spatial approximations: the majority of models employ finite difference methods.

5. All models approximate the time dimension using finite difference methods.

6. Explicit, implicit, and mixed explicit-implicit schemes are employed in the representation of the nonlinear coefficients. Most algorithms employ an implicit formulation.

7. Where an implicit formulation is used, either Newton-Raphson or the total increment method is employed to linearize the approximating equations.

8. The only vertically integrated areal model is the one developed by Faust and Mercer (1977).

9. All methods employ some form of upstream weighting for the convective term.
10. The transition across the phase boundary is accomplished in a number of ways. Most schemes involve some method of numerical damping, which stops the oscillation across this boundary. Only the model of Voss and Pinder completely resolves the phase change problem. The approach of Thomas and Pierson deserves additional study; it was difficult to evaluate based on the available literature.
11. A wellbore model is included in the models of Toronyi and Farouq-Ali (1977), Coats (1977), Thomas and Pierson (1976), and Brownell et al. (1977).

The formulation of the approximating equations is relatively straightforward. Linearization of the resulting nonlinear equations is challenging. The methodology is the treatment of the phase change and this can be important. For those problems that are dominated by the phase change phenomenon, an accurate formulation is essential. Because there is no test to sufficiently demonstrate the accuracy of geothermal reservoir simulators, one can only speculate on the adequacy (Pinder, 1979).

Flow and energy equations can be solved either sequentially or simultaneously. The sequential solution employs estimates of the energy variable when solving the flow equation and estimates of the flow variable when solving the energy equation. This uncoupling is desirable because it is more efficient to solve N equations twice than $2N$ equations once. The disadvantage is that it is generally necessary to iterate between the equations and convergence is not, in general, guaranteed. The majority of existing models solve the two equations simultaneously and employ Newton-Raphson type schemes to accommodate the nonlinearity that arises. The two-dimensional model of Faust and Mercer and the formulation of Lasseter et al.(1975) are exceptions to this general rule.

The matrix equations that arise in either approach may be solved either directly or iteratively. Direct methods are based on Gaussian elimination and are reliable when applied to a well-behaved system of equations. Iterative methods tend to be more efficient for large problems (e.g., more than 500 equations), but generally require a higher level of numerical ingenuity to program and apply effectively. The majority of iterative schemes are block iterative, and thus incorporate a direct solution module in the iterative algorithm. This is true for the models considered with the exception of that of Lasseter et al. (1975).

The primary factors to consider in the selection of a solution scheme are accuracy and efficiency. Ease of programming will probably play a second-

(text continued on page 36)

Table 1-11

Comparison of Multiphase Distributed Parameter Geothermal Reservoir Models ("ratings are subjective")

Models	Dependent Variables	Well Approximation	Equation Approximation	Dimensions	Phases	Spatial Approximation	Temporal Approximation	Vertical Approximation	Convective Term Approximation	Convective Term	Time Integration Term Approximation Unknowns	Time Integration Approximation Coefficients	Nonlinear Approximation	Phase Change Method	Solution Scheme	Matrix Solution	Availability
Toronyi and Farouq Ali	p_f, S_w	Yes	Macro	2	2	FD	FD	No	$\nabla\cdot v_f h_f$	UFD	IMP	IMP	NRA	—	SIM	D	PRIV
Lasseter, Witherspoon and Lippman	ρ_f, U_f	No	Macro	1, 2, 3	1, 2	IFD	FD	No	$\nabla\cdot v_f U_f$	UFD	0	EXP	None	None	SEQ	ITR	PUB
Brownell Garg and Pritchett	ρ_f, U_f	Yes	Mix	1, 2, 3	1, 2	FD	FD	No	$\nabla\cdot v_f U_f$	UFD	CENT	IMP	NRA	LEX	SIM	ADI	PRIV
Faust and Mercer	p_f, h_f	No	VINT	1, 2, 3	1, 2	FD (FE)	FD	Yes	$\nabla\cdot v_f h_f$	UFD	0	IMP	NRA	2D TAN —; 3D	2D SEQ; 3D SIM	2D D; 3D SSOR	PUB
Coats	p_w, T; p_g, S_g	Yes	Macro	2, 3	1, 2	FD	FD	No	$\nabla\cdot v_o h_o$	UFD / EXP	IMP	IMP	NRA	?	SIM	D	PRIV
Huyakorn and Pinder	p_f, h_f	No	Macro	1	1, 2	FE	FD	No	$\nabla\cdot v_f h_f$	UFE	IMP	IMP	NRA	TAN	SIM	D	PUB
Thomas and Pierson	p_w, T; S_w, T; p_g, T	Yes	Macro	1, 2, 3	1, 2	FD	FD	No	$\nabla\cdot v_f h_f$	?	IMP	IMP EXP	IMPES	IMP	SIM	IMPES	PRIV
Voss and Pinder	p_f, h_f	No	VINT	2, 3	1, 2	FE	FD	No	$v\nabla h_f$	UFE	0	0	TIM	SLA	SIM	BIFEPS	PUB

After Pinder, 1979.

p_f = fluid pressure
S_w = volume saturation of liquid (water)
ρ_f = density of fluid
U_f = internal energy/unit mass of fluid
H_f = enthalpy/unit mass of fluid
ρ_w = density of liquid (water)
ρ_g = density of steam (gas)
S_g = volume saturation of steam (gas)
T = temperature
p_g = pressure of steam (gas)

Dependent Variables: Variables solved for explicitly in the governing equations.
- Variables are defined in list of variables.

Well Approximation: The utilization of a model of the wellbore.

Equation Approximation: The mathematical formalism employed in obtaining the governing porous medium equations.
- Macro designates a macroscopic balance.
- Mix designates mixture theory methodology of continuum mechanics.
- VINT denotes volume integration from the microscopic level to the macroscopic level.

Dimensions: Number of space dimensions employed in example problems.

Phases: The number of phases that can coexist at any given point in space and time.

Spatial Approximation: The numerical scheme used to approximate space derivatives.
- IFD denotes integrated finite difference.
- FD denotes finite difference.
- FE denotes finite element.

Temporal Approximation: The numerical scheme used to approximate the time derivative.
- FD denotes finite difference.

Vertical Integration: The formal procedure of integrating the three-dimensional equations vertically when generating a two-dimensional areal model.

Convective Term: Form in which the convective term appears in the model.

Convective Term Approximation: Numerical scheme employed in approximating the convective term.
- UFD denotes upstream weighted finite difference.
- UFE denotes upstream weighted finite element.

Time Integration of Unknowns: Type of time derivative approximation employed.
- 0 denotes a general formulation, $(0.5 \leq 0 \leq 1.)$
- CENT denotes a Crank-Nicholson scheme (i.e., $0 = 0.5$).
- IMP denotes a backward difference approximation.

Time Integration Coefficients: The location in the time domain where the nonlinear coefficients are evaluated. Nomenclature the same as previous case.

Nonlinear Approximation: Method used to linearize nonlinear equations.
- NRA denotes Newton-Raphson iteration.
- IMPES denotes implicit pressure, explicit saturation.
- TIM denotes the total increment method.

Phase Change Method: Technique used to move numerically across the phase change boundary.
- LEX denotes limited excursion technique.
- Δt ADJ denotes a modification of Δt as the phase boundary is approached.
- TAN denotes a modification of Newton-Raphson to allow the tangent to be taken in a direction away from the phase boundary.
- IMP denotes a formulation accounting for the phase change with the equations.
- SLA denotes saturation line adjustment.

Solution Scheme: The method used to solve the two coupled governing equations.
- SEQ denotes the sequential solution of each, i.e., N equations are solved twice per iteration.
- SIM denotes the simultaneous solution of $2N$ equations at each iteration.

Matrix Solution: Technique used to solve linear algebraic equations.
- ITR denotes an iterative method.
- ADI denotes alternating direction implicit procedure.
- D denotes a direct solution scheme.
- SSOR denotes slice successive over relaxation.
- IMPES denotes implicit pressure explicit saturation method.
- BIFEPS denotes block iterative finite element preprocessed scheme.

Availability: Designation of public availability to the model.
- PUB designates models funded through public monies and therefore available to the public.
- PRIV designates model developed with private funds and thus probably proprietary.

(text continued from page 33)

ary role, because of the considerable computer costs involved in geothermal reservoir simulation. Because a comparison of the accuracy and efficiency of the models outlined in Table 1-11 has never been undertaken, one cannot select an optimal approach directly.

The complexity of geothermal reservoir physics essentially precludes the verification of existing codes using analytical solutions. One can, however, compare solutions generated by a model against other numerical solutions or experimental data. This has been done to varying degrees by the majority of modellers.

The Economics of Geothermal Energy

Chapter 10 stresses the need to match the properties of the geothermal resource to appropriate energy conversion alternatives. Difficulties involved in the estimation of such factors as drilling costs, reservoir characteristics, and long-term performance of power conversion equipment are emphasized. Some overly optimistic promoters incorrectly assess costs because they are not fully familiar with the technology. Some problems in assessment occur because this is a new industry and the necessary data to compute costs is not available. The key issues related to commercial feasibility that are discussed in this chapter are:

Table 1-12
Conversion Table

Parameter	Dimensions	English Units to Metric Units	English Units to SI Units	Metric Units to SI Units
Length	L	in. $\times$ 2.54 = cm	in. $\times$ 0.0254 = m	
		ft $\times$ 30.48 = cm	ft $\times$ 0.3048 = m	
		mi $\times$ 1.609 = km	yd $\times$ 0.9144 = m	
		ft $\times$ 3.048 $\times$ 10^{-4} = km	mi $\times$ 1.609 $\times$ 10^3 = m	
Area	L^2	in.^{2} $\times$ 6.4516 = cm^2	ft^2 $\times$ 9.290 $\times$ 10^{-2} = m^2	
		acre $\times$ 4.0468x10^{-1} = ha	yd^2 $\times$ 8.3612 $\times$ 10^{-1} = m^2	
		section $\times$ 2.5889x10^2 = ha	mi^2 $\times$ 2.5889 = km^2	
		mi^2 $\times$ 2.5889 = km^2	acre $\times$ 4.0468 $\times$ 10^3 = m^2	
Volume	L^3	in^3 $\times$ 1.639 $\times$ 10^1 = cm^3	in^3 $\times$ 1.639 $\times$ 10^{-5} = m^3	
		ft^3 $\times$ 2.832 $\times$ 10^1 = l	ft^3 $\times$ 2.832 $\times$ 10^{-2} = m^3	
		mi^3 $\times$ 4.1655 = km^3	yd^3 $\times$ 7.646 $\times$ 10^{-1} = m^3	1×10^{-3} = m^3
		U.S. gal $\times$ 3.7854 = l	U.S. gal $\times$ 3.785 $\times$ 10^{-3} = m^3	
		U.K. gal $\times$ 4.5460 = l	U.S. bbl $\times$ 1.5898 $\times$ 10^{-1} = m^3	
		acre-ft $\times$ 1.2335 $\times$ 10^{-1} = ha-m	acre-ft $\times$ 1.2335 $\times$ 10^3 = m^3	
Mass	M	oz (av) $\times$ 2.8349 $\times$ 10^1 = g		
		poundals $\times$ 1.41 $\times$ 10^1 = g	lb$_m$ $\times$ 4.536 $\times$ 10^{-1} = kg	g $\times$ 10^{-3} = kg
		U.S. ton $\times$ 9.078 $\times$ 10^{-1} = tons (metric)	U.S. ton $\times$ 9.0719 $\times$ 10^2 = kg	tons (metric) $\times$ 10^3 = kg
		U.K. ton $\times$ 1.0160 = tons (metric)	U.K. ton $\times$ 1.016 $\times$ 10^3 = kg	dynes $\times$ 1.020 $\times$ 10^{-6}= kg
		grain $\times$ 6.4798 $\times$ 10^{-2} = g		
Density	ML^{-3}	lb$_m$/ft^3 $\times$ 1.602 $\times$ 10^{-2} = g/cm^3	lb$_m$/ft^3 $\times$ 16.018 = kg/m^3	
		lb$_m$/U.S. gal $\times$ 1.1983 $\times$ 10^{-1} = g/cm^3	lb$_m$/U.S. gal $\times$ 1.1983 $\times$ 10^2 = kg/m^3	g/cm^3 $\times$ 10^{-3} = kg/cm^3
Specific Volume	L^3M^{-1}	ft^3/lb$_m$ $\times$ 6.2427 $\times$ 10^{-5} = m^3/g	ft^3/lb$_m$ $\times$ 6.2427 $\times$ 10^{-2} = m^3/kg	
		U.S. gal/lb$_m$ $\times$ 8.3454 = cm^3/g		

1. Information about the resource, such as rock type and its properties, depths to the reservoir, fluid temperatures, and chemical composition of fluids and rocks.
2. Knowledge of production capacity, lifetime of the reservoir, physical structure of the reservoir system, and the design and performance of the plant.
3. Management skills including exploration, drilling, surface plant costs, selling price of the produced energy, assessment of risk, anticipated rate of return, operating and maintenance costs, the type of ownership, and a number of regulatory issues.

Conversion Table

Table 1-12 shows most of the parameters used in heat energy calculations. It lists the parameter name, the parameter dimensions, conversion of English units to metric units, English units to Standard International units, metric units to Standard International units, English units to English units, metric units to metric units, and Hybrid mixed units. All of these can be derived and calculated from appropriate basic relationships; they are further expanded and listed here as a convenience to those making heat energy calculations.

Table 1-12, continued

English to English	Metric Units to Metric Units	Hybrid Conversions
$in \times 1/12 = ft$	$cm \times 10^{-2} = m$	
$ft^2 \times 144 = in^2$ $ft^2 \times 1.111 \times 10^{-1} = yd^2$ $mi^2 \times 2.788 \times 10^{-7} = ft^2$	$cm^2 \times 1 \times 10^{-4} = m$	
$in^3 \times 5.787 \times 10^{-4} = ft^3$ $ft^3 \times 7.4805 = U.S.\ gal$ $ft^3 \times 3.704 \times 10^{-2} = yd^3$ U.S. gal $\times 8.3267 \times 10^{-1} = U.K.\ gal$ quart $\times 3.342 \times 10^{-2} = ft^3$		$lb_mH_2O\ (15°C) \times 1.602 \times 10^{-2} = ft^3$ $lb_mH_2O\ (15°C) \times 1.198 \times 10^{-1} = gal$
oz (av) $\times 4.375 \times 10^2 = $ grains oz (av) $\times 9.115 \times 10^{-1} = $ oz (troy) poundals $\times 3.108 \times 10^{-2} = lb_m$	dynes $\times 1.020 \times 10^{-2} = mN$	
$lb_m/in^3 \times 1.728 \times 10^3 = lb_m/ft^3$		

Table 1-12 continued

Table 1-12, continued

Parameter	Dimensions	English Units to Metric Units	English Units to SI Units	Metric Units to SI Units
Concentration	L^3L^{-3}	bbl/acre-ft $\times$ 1.2889 = m³/ha-m	bbl/acre-ft $\times$ 1.2889 $\times$ 10⁻⁴ = m³/m³	
Concentration	ML^{-3}	lb$_m$/ft³ $\times$ 1.6018 $\times$ 10⁻² = g/cm³ lb$_m$/in.³ $\times$ 27.6801 = g/cm³	lb$_m$/ft³ $\times$ 1.6018 $\times$ 10¹ = kg/m³ lb$_m$/bbl $\times$ 2.8530 = kg/m³ grains/U.S. gal $\times$ 17.12 = kg/m³ grains/ft³ $\times$ 2.2883 $\times$ 10³ = kg/m³	
Pressure Stress	ML^{-2}	atm $\times$ 1.0132 = bars = N/m² atm $\times$ 1.0333 = kg/cm² lb$_f$/in² $\times$ 7.031 $\times$ 10² = kg/m² lb$_f$/ft² $\times$ 4.882 = kg/m² in Hg (32°F) $\times$ 3.453 $\times$ 10⁻² = kg/cm²	atms $\times$ 1.01324 $\times$ 10² = KPa lb/in² $\times$ 6.8947 = KPa lb$_f$/ft² $\times$ 4.7880 $\times$ 10⁻² = KPa in Hg (60°F) $\times$ 3.3768 = KPa in H₂O (60°F) $\times$ 2.4884 $\times$ 10⁻¹ = KPa	dynes/cm² $\times$ 10⁻¹ = Pa bars $\times$ 10⁵ = Pa cm H₂O (4°C) $\times$ 9.8064 $\times$ 10⁻² = KPa N/M² = Pa
Force	MLt^{-2}		lb$_f$ $\times$ 4.4482 = N	dynes $\times$ 10⁻⁵ = N
Pressure Drop/Length	$ML^{-2}L^{-1}$		psi/ft $\times$ 2.2620 $\times$ 10¹ = KPa/m	
Mass/Length	ML^{-1}	lb$_m$/in $\times$ 1.786 $\times$ 10² = g/cm	lb$_m$/ft $\times$ 1.488 = kg/m	
Dynamic Viscosity	$ML^{-1}t^{-1}$	lb$_m$/ft-s $\times$ 1.4881 = N-s/g lb$_m$/ft-s $\times$ 1.4881 $\times$ 10³ = cp	lb$_m$/ft-s $\times$ 1.4881 = Pa-s	dynes-s/cm² $\times$ 10⁻¹ = Pa-s cp $\times$ 10⁻³ = Pa-s
Kinematic Viscosity	L^2t^{-1}	ft²/s $\times$ 9.2903 $\times$ 10⁴ = mm²/s		1 cm²/s $\times$ 10² = mm²/s
Mass Flow Rate	Mt^{-1}		lb$_m$/h $\times$ 1.2599 $\times$ 10⁻⁴ = kg/s lb$_m$/yr $\times$ 1.45 $\times$ 10⁻⁸ = kg/s	
Volumetric Flow Rate	L^3t^{-1}	bbl/D $\times$ 1.8401 $\times$ 10⁻³ = l/s ft³/D $\times$ 3.2774 $\times$ 10⁻⁴ = l/s U.S. gal/min $\times$ 6.3090 $\times$ 10⁻² = l/s ft³/s $\times$ 28.317 = l/s ft³/min $\times$ 4.72 $\times$ 10² = cm³/s	bbl/D $\times$ 1.84 $\times$ 10⁻⁶ = m³/s ft³/D $\times$ 3.27 $\times$ 10⁻⁸ = m³/s U.S. gal/min $\times$ 2.271 $\times$ 10⁻¹ = m³/h	
Permeability	L^2		darcy $\times$ 9.8692 $\times$ 10⁻¹ = μm² md $\times$ 0.9869 = 10⁴μm²	
Productivity Index	$L^3M^{-1}t^{-1}$		bbl/D-psi $\times$ 3 $\times$ 10⁻⁷ = m³/KPa-s	
Moment of Inertia	ML^2		lb$_m$-ft² $\times$ 4.214 $\times$ 10⁻² = kg-m²	
Recovery/Unit Volume	L^3L^{-3}		bbl/ac-ft $\times$ 1.289 $\times$ 10⁻⁴ = m³/m³	
Energy Work	ML^2t^{-2}	Btu $\times$ 1.0550 $\times$ 10¹⁰ = ergs Btu $\times$ 2.931 $\times$ 10⁻⁴ = kW-h Btu $\times$ 252 = gcal hp-h $\times$ 7.457 $\times$ 10⁻¹ = kW-h ft-lb $\times$ 3.766 $\times$ 10⁻⁷ = kW-h Btu $\times$ 1.055 $\times$ 10³ = W/s	Btu $\times$ 1.0550 $\times$ 10³ = J hp-h $\times$ 2.6845 $\times$ 10⁶ = J ft-lb$_f$ $\times$ 1.3558 = J	ergs $\times$ 10⁻⁷ = J kg-cal $\times$ 4.187 $\times$ 10³ = J kW-h $\times$ 3.6 $\times$ 10⁶ = J W/s = J Nm = J
Power	ML^2t^{-3}	Btu/h $\times$ 7 $\times$ 10⁻² = g-cal/s hp $\times$ 1.068 $\times$ 10¹ = kg-cal/min hp $\times$ 7.4571 $\times$ 10⁹ = ergs/sec	Btu/h $\times$ 2.931 $\times$ 10⁻¹ = W hp $\times$ 7.452 $\times$ 10² = W ft-lb$_f$/min $\times$ 2.2596 $\times$ 10⁻² = W	ergs/s $\times$ 10⁻¹⁰ = kW kg-cal/min $\times$ 6.972 $\times$ 10⁻² = kW
Temperature	T	5/9 (°F − 32) = °C	5/9 (°F − 32) = K.°C 5/9 °R = K	°C = °K
Geothermal Gradient	TL^{-1}	°F/ft $\times$ 1.8228 $\times$ 10³ = °C/km	°F/100 ft $\times$ 18.227 = m°C/m	
Energy Gradient	$ML^2t^{-2}L^{-1}$		Btu/ft $\times$ 3461.3 = J/m	
Heat Content Enthalpy	L^2t^{-2}	Btu/lb$_m$ $\times$ 0.556 = cal/g	Btu/lb$_m$ $\times$ 2.326 $\times$ 10³ = J/kg	cal/g $\times$ 4.187 = kJ/kg
Heat Capacity Entropy Specific Heat	$L^2t^{-2}T^{-1}$	Btu/lb-°F = cal/g-°C	Btu/lb$_m$-°F $\times$ 4.1868 = J/g-K	cal/g-°C $\times$ 4.184 = kJ/kg-K cal/g-mol-°C $\times$ 4.184 = kJ/kmoleK

English to English	Metric Units to Metric Units	Hybrid Conversions
grains/U.S. gal $\times$ 17.12 = ppm grains/U.S. gal $\times$ 142.86 $\times$ 10^{-3} = lb$_m$/10^6 gal lb$_m$/ft^3 $\times$ 5.787 $\times$ 10^{-4} = lb/m/in^3 grains/U.K. gal $\times$ 14.25 = ppm		g/U.S. gal $\times$ 2.6417 $\times$ 10^{-1} = kg/m^3
atm $\times$ 14.6960 = lb$_f$/in^2 atm^2 $\times$ 33.90 = ft H$_2$O (15°C) lb$_f$/in^2 $\times$ 2.042 = in Hg	bars $\times$ 10^6 = dynes/cm^2 bars $\times$ 1.020 $\times$ 10^4 = kg/m^2 kg/m^2 $\times$ 9.8066 $\times$ 10^1 = dynes/cm^2 bar = N/m^2 dynes/cm^2 = ergs/mm^2	ft H$_2$O $\times$ .433 = psi
	cp $\times$ 10^{-2} = g/cm-s = poise	lb$_m$-ft $\times$ 3.108 $\times$ 10^{-2} = lb$_f$-s/ft^2
		lb$_m$H$_2$O/min $\times$ 2.674 $\times$ 10^{-4} = ft^3/s
ft^3/min $\times$ 1.247 $\times$ 10^{-1} = U.S. gal/s U.S. gal/min $\times$ 8.021 = ft^3/h		
	μm^2 $\times$ 10^{12} = Pa-s (m^3/s-m^2)(m/Pa) m^2 $\times$ 10^{-12} = μm^2	
		1,000 Btu = 0.00017 bbl oil 1,000 Btu = 0.00004 U.S. ton coal lb$_m$ coal = 12,500 Btu mcf gas = 1,000 Btu
hp-h $\times$ 2.547 $\times$ 10^3 = Btu ft-lb$_f$ $\times$ 5.050 $\times$ 10^{-7} = hp-h ft-lb$_f$ $\times$ 1.286 $\times$ 10^{-3} = Btu	g-cal $\times$ 4.184 $\times$ 10^7 = ergs g-cal $\times$ 1.162 $\times$ 10^{-6} = kW-h ergs = dyne-cm ergs $\times$ 2.773 $\times$ 10^{-11} = W-h W-h $\times$ 860.42 = cal	
ft-lb$_f$/min $\times$ 3.030 $\times$ 10^{-5} = hp ft-lb$_f$/s $\times$ 4.6274 = Btu/h hp $\times$ 70.69 $\times$ 10^{-2} = Btu/s		
		cal/lb$_m$ $\times$ 9.2241 = J/kg

Table 1-12 continued

Table 1-12, continued

Parameter	Dimensions	English Units to Metric Units	English Units to SI Units	Metric Units to SI Units
Heat Flow	Mt^{-3}	Btu/h-t^2 × 7.534 × 10^{-5} = cal/s-cm^2 Btu/h-ft^2 × 9.79 × 10^{-3} = W/cm^2	Btu/h-ft^2 × 3.1546 × 10^{-3} = kW/m^2	hfu × 4.184 × 10^1 = mW/m^2
Heat Generation	$ML^{-1}t^{-3}$	Btu/h-ft^2 × 2.472 × 10^{-6} = cal/s-cm^3	Btu/h-ft^3 × 10.35 = W/m^3	
Thermal Diffusivity	L^2t^{-1}	ft^2/h × 9.2989 × 10^{-2} = m^2/h	ft^2/h × 2.5806 × 10^1 = mm^2/s	
Thermal Conductivity	$MLt^{-3}T^{-1}$	Btu/h-ft-°F × 4.134 × 10^{-3} = cal/s-°C-cm Btu/h-ft-°F × 6.230 × 10^{-3} = kJ/h-m^2-°C	Btu/h-ft-°F × 1.7307 = W/m-K	cal/s-°C-cm × 4.187 × 10^2 = W/m-°C kcal/h-m^2-°C × 1.1622 = W/m^2-°C
Heat Transfer Coefficient	$Mt^{-3}T^{-1}$	Btu/h-ft^2-°F × 1.36 × 10^{-4} = cal/s-cm^2-°C Btu/h-ft^2-°F × 5.678 × 10^{-3} = kW/m^2-°C Btu/h-ft^2-°F × 0.4883 = cal/h-cm^2-°C	Btu/h-ft^2-°F × 5.678 × 10^{-3} = kW/m^2-K Btu/h-ft^2-°R × 5.678 × 10^{-3} = kW/m^2-K	cal/s-cm^2-°C × 4.184 × 10^1 = kW/m^2-°C kcal/h-m^2-°C × 1.162 × 10^{-3} = kW/m^2-°C
Thermal Resistivity	$M^{-1}L^{-2}t^3T$		°F-ft^2-h/Btu × 1.7611 × 10^2 = °C-m^2/kW	°C-m^2-h/kcal × 860.42 = °C-m^2/kW
Volumetric Heat Content	$ML^{-1}t^{-2}$		Btu/U.S. gal × 278.72 = kJ/m^3 Btu/ft^3 × 37.259 = kJ/m^3	k cal/m^3 × 4.184 = kJ/m^3
Irradiation	Mt^{-3}		Btu/h-ft^2 × 3.1546 × 10^{-3} = kW/m^2	
Revolutions per Minute	t^{-1}		rpm × 1.0472 × 10^{-1} = rad/s	
Area Heat Content	Mt^{-2}	Btu/ft^2 × 2.712 = kg-cal/m^2	Btu/ft^2 × 1.1356 × 10^4 = J/m^2	
Surface Tension	Mt^{-2}			dyne/cm = mN/m
Volumetric Heat Transfer Coefficient	$ML^{-1}t^{-3}T^{-1}$		Btu/h-ft^3-°F × 1.863 × 10^{-2} = kW/m^{3}K	

English to English	Metric Units to Metric Units	Hybrid Conversions
Btu/min-ft^2 × 1.22 × 10^{-1} = W/in^2		hfu = u-cal/cm^2-s
		hfu
		hgu = cal/cm^3·s
		hgu × 4.184 × 10^{12} = μW/m^3

References

Brownell, D.H., Jr., Garg, S.K. and Pritchett, J.W., 1975. Computer Simulation of Geothermal Reservoirs. 45th Annual California Regional Meeting of the Society of Petrol. Engrs., SPE 5381, Ventura, CA.

Brownell, D.H., Jr., Garg, S.K. and Pritchett, J.W., 1977. Governing Equations for Geothermal Reservoirs. *Water Resour. Res.*, 13(6):929.

Cheng, P., 1978. *Heat Transfer in Geothermal Systems. Advances in Heat Transfer.* Vol. 14, Academic Press, New York, pp. 1-105.

Chilingarian, G.V., Sawabini, C.T. and Rieke, H.H., III, 1973. Comparison Between Compressibilities of Sands and Clays. *Jour. Sed. Petrology,* 43(2):529-536.

Chuop, V., 1980. Geothermal Gradients, Heat Flow, and Internal Heat Generation in West Virginia. MS thesis, West Virginia University, Morgantown, WV, 112pp.

Coats, K.H., 1977. Geothermal Reservoir Modeling. 52nd Ann. Fall Tech. Conf. of the Soc. of Petrol. Engrs., SPE 6892, Denver, CO.

Combs, J. and Simons, G., 1973. Terrestrial Heat Flow Determinations in the North Central United States. *J. Geophys. Res.,* 78(2):441-461.

Diment, W.H. and Robertson, E.C., 1963. Temperature, Thermal Conductivity, and Heat Flow in a Drilled Hole near Oak Ridge, Tennessee. *J. Geophys. Res.,* 68:5035-5048.

Diment, W.H. and Werre, R.W., 1960. Geothermal Experiments in a Borehole at Morgantown, West Virginia. *Trans. Am. Geophys., Union,* 47:182.

Diment, W.H. and Werre, R.W., 1964. Terrestrial Heat Flow near Washington, D.C., *J. Geophys. Res.,* 69(10):2143-2149.

Faust, C.R. and Mercer, J.W., 1977a. Finite-difference Model of Two-dimensional Single- and Two-phase Heat Transport in a Porous Medium - Version I. U.S. Geol. Surv. Open File Rep., 77-234.

Faust, C.R. and Mercer, J.W., 1977b. Theoretical Analysis of Fluid Flow and Energy Transport in Hydrothermal Systems. U.S. Geol. Surv. Open File Rep., 77-60.

Garg, S.K. and Pritchett, J.W., 1977. On Pressure-work, Viscous Dissipation and the Energy Balance Relation for Geothermal Reservoirs. *Advan. Water Resc.,* 1(1):41.

Grisafi, T.W., Rieke III, H.H. and Skidmore, D.R., 1973. Geothermal Gradients in Northern West Virginia. *Bull. Am. Assoc. Petrol. Geol.,* 58(2):321-323.

Huyakorn, P.S. and Pinder, G.F., 1977. A Pressure-Enthalpy Finite Element Model for Simulating Hydrothermal Reservoirs. In: R. Vichnevetsky (ed.), *Advances in Computer Methods for Partial Differential Equations II,* IMACS, p. 284.

Jones, P.H., 1970. Geothermal Resources of the Northern Gulf of Mexico Basin. U.N. Symposium, Dev. Utilization of Geothermal Resource, Pisa, Italy, Vol. 2, pt.1:14-26.

Joyner, W.B., 1960. Heat Flow in Pennsylvania and West Virginia. *Geophys.,* 25(6):1229-1241.

Karamarakar, M., and Cheng, P., 1980. A Theoretical Assessment of James' Method for the Determination of Geothermal Wellbore Discharge Characteristics. Lawrence Berkeley Laboratory, University of California, Berkeley. Report LVL-11498, 21 pp.

Keenan, J.H. and Keyes, F.G., 1936. Thermodynamic Properties of Steam. John Wiley & Sons, New York, 89 pp.

Kruger, P. and Otte, C., 1973. *Geothermal Energy* Stanford University Press, Stanford, Calif., 360 pp.

Lamers, M.D., 1979. Measurement Requirements and Methods for Geothermal Reservoir System Parameters (an appraisal). Earth Sci. Div., *Lawrence Berkeley Laboratory, Univ. Calif.,* Berkeley, CA, Rept. LBL-9090, 33 pp.

Lasseter, T.J., Witherspoon, P.A. and Lippmann, M.J., 1975. The Numerical Simulation of Heat and Mass Transfer in Multidimensional Two-phase Geothermal Reservoirs. *Proc. 2nd U.N. Symposium on Development and Use of Geothermal Resources,* Vol. 3, p. 1715.

Leibowitz, L.P., 1978 California's Geothermal Resource Potential. *Energy Sources,* 3(3/4): 293-311.

Lewis, C.R. and Rose, S.C., 1969. A Theory Relating High Temperatures and Overpressures. SPE 2564, 44th Ann. Fall Meeting of SPE, Denver, CO, 5 pp.

Matthews, C.S., 1980. Gas Evolution from Geopressured Brines. DOE/TIC-11227 Rept., 50 pp.

O'Neill, K., 1977. The Transient Three-dimensional Transport of Liquid and Heat in Fractured Porous Media. Ph.D. dissertation, Dept. of Civil Engineering, Princeton University.

Pinder, G.F., 1979. State-of-the-Art Review of Geothermal Reservoir Modeling. Earth Sci. Div., *Lawrence Berkeley Laboratory, Univ. Calif.*, Berkeley, CA, Rept. LBL-9093, 144 pp.

Pritchett, J.W., Rice, L.F. and Garg, S.K., 1980. Reservoir Simulation Studies: Wairakei Geothermal Field, New Zealand. *Lawrence Berkeley Laboratory. Univ. Calif.*, Berkeley, CA, Report. LBL-11497, 103 pp.

Rainis, A.E., Skidmore, D.R. and Rieke, H.H., III, 1974. A Computational Method for Determining Segmental and Overall Geothermal Gradients and Geothermal Heat Flow Values. *Geotherm.*, 3(3):113-117.

Ramey, H.J., Jr., 1970. A Reservoir Engineering Study of the Geysers Geothermal Field. Testimony for the trial of Reich and Reich vs Commissioner of the Internal Revenue, Tax Court of the U.S., 52 T.C. No. 74.

Reiter, M.A. and Costain, J.K., 1973. Heat Flow in Southwestern Virginia. *J. Geophys. Res.*, 78(8):1323-1333.

Rieke, H.H., III and Skidmore, D.R., 1973. Geothermal Resource Base in Appalachia. SPE Ann. Eastern Reg. Meet., SPE 4708, Pittsburgh, PA, 8 pp.

Rieke, H.H., III and Skidmore, D.R., 1974. Geothermal Energy Potential in Northern Appalachia. *Jour. Petrol. Tech.*, 29(9):1005-1006.

Rieke, H.H., III and Chilingarian, G.V., 1974. *Compaction of Argillaceous Sediments.* Elsevier Scientific Publishing Co., Amsterdam, The Netherlands, 424 p.

Roy, R.F., Decker, E.R., Blackwell, D.D. and Birch, F., 1968. Heat Flow in the United States. *J. Geophys. Res.*, 73(16):5207-5221.

Sanyal, S.K., Juprasert, S. and Jusbasche, M., 1980. An Evaluation of a Rhyolite-Basalt-Volcanic Ash Sequence from Well Logs. *The Log Analyst* XXI(1):3-9.

Smith, M.C., 1972. The Development of Dry Geothermal Energy Sources. AEC Report.

Smith, M.C., Aamodt, R.L. Potter, R.M. and Brown, D.W., 1975. Man-made Geothermal Reservoirs. In: *Second United Nations Symposium on the Development and Use of Geothermal Resources.* Vol. 3, pp. 1781-1787.

Smith, D.L., Gregory, R.G. and Emhof, J.W., 1981. Geothermal Measurements in the Southern Appalachian Mountains and Southeastern Coastal Plains. *American Journal Science,* 281(3):282-298.

Somerton, W.H., 1958. Some Thermal Characteristics of Porous Rocks. *Trans.* AIME, 213:375-378.

Sudol, G.A., Harrison, R.F. and Ramey, H.J., Jr., 1979. Annotated Research Bibliography for Geothermal Reservoir Engineering. *Lawrence Berkeley Laboratory, Univ. Calif.*, Berkeley, CA, Rept. LBL-8664, 150 pp.

Thomas, L.K. and Pierson, R., 1976. Three-dimensional Geothermal Reservoir Simulation. 51st Annual Fall Tech. Conf. of the Soc. of Pet. Engrs., SPE 6104, New Orleans, LA.

Toronyi, R.M. and Farouq Ali, S.M., 1977. Two-phase, Two-dimensional Simulation of a Geothermal Reservoir. *Soc. Petrol. Engr. Jour.*, 17(3):171-183.

Travis, R.B., 1955. Classification of Rocks. *Quart. Col. Sch. Mines,* 50(1):1-98.

Vaught, T.L., 1980. Temperature Gradients in a Portion of Michigan: A Review of the Usefulness of Data from the AAA Geothermal Survey of North America. DOE/NV/10072-1 Rept., 39 pp.

Voss, C.I. and Pinder, G.F., 1978. Block Iterative Finite Element Preprocessed Scheme for Simulation of Large Non-linear Problems. *Int. J. Num. Methods Engrg.,* (in press).

White, D.E., Muffler, L.J.P. and Truesdell, A.H., 1971. Vapor-dominated Hydrothermal Systems Compared with Hot-water Systems. *Econ. Geol.*, 66:75-97.

John C. Rowley, Los Alamos National Laboratory

2

Worldwide Geothermal Resources

Introduction

The present status of the worldwide use of geothermal energy is summarized in this chapter. The terms and concepts for resource base and reserve estimations are defined and an interim methodology for such evaluations is selected. It is pointed out that large elements of uncertainty exist in estimating subsurface resources. Accurate quantitative evaluations depend on the availability of large amounts of historical, exploration, recovery, production, and especially drilling data. Geothermal resource development, however, is only in its initial stages and these data are largely not available. Consequently, the present estimates for geothermal energy resources are significantly less accurate than those that are made for crude oil, natural gas, uranium, coal, and water supplies.

It is very tempting to concentrate on estimating the total thermal energy stored in the earth's crust and establishing the total geothermal resource base as all the heat stored in the fluids and rocks of the earth's crust. This apparently simple concept is complicated by the incomplete knowledge of the local temperature profiles in the earth and the lack of knowledge of subsurface permeability, water content, and water recharge to fluid-dominated reservoirs. Resource base data, however, do emphasize the possibilities realizable through technology improvements.

Other elements to be factored into geothermal resource and reserve estimates are the economic and technological limitations currently placed on the transportation of heat. Its distribution is currently restricted to local use or conversion to electricity. Future developments in distribution technology, such as input to local production of a transportable hydrogen fuel, assistance in biomass conversion to pipeline methane, or purification of brackish water for industrial and irrigation uses, would expand more distant markets for geothermal developments. It is also possible to speculate on the wider application of geothermal energy to local industrial and raw material processing, e.g., ore refining, crop drying, and paper manufacturing, and, thus, in effect giving credit to the energy resource as a value added to the product.

The major restriction in making in-depth reserve estimates for geothermal energy is the limited number of geothermal wells that have been drilled. There are approximately 3,000 wells throughout the world and perhaps 1,500 wells in production. In addition, the worldwide distribution of wells that provide information on the overall reservoir type, size, and useful life of known hydrothermal fields is extremely sparse, and only provides a tantalizing glimpse of the future potential for geothermal energy. In performing reserve estimates, it is also necessary to have an accurate knowledge of present and projected economics: the costs of exploration, development, and production, and the future prices of energy (see Chapter 10).

It is especially important to recognize that local economics and institutional factors have played a dominant role in geothermal development. To date, much of the high-risk, "front-end," portions of the existing and ongoing developments have received national or international (United Nations) funding support. It may be difficult, therefore, to evaluate accurately the total economics of such developments. However, it does appear that about 18 geothermal electrical power plants, currently operating worldwide (1978), are economic successes and have been, or are in the process of, paying back their investors, whether private or public, over the amortization periods (10 to 30 years) and have produced electricity at (locally) competitive costs. It seems very probable that where high-grade hydrothermal reservoirs are found to exist, it will be economically more favorable to install a geothermal plant in the face of the uncertain future of fuel costs and supply of a plant dependent on imported oil or coal in many parts of the world. This illustrates the influence of present and future prices of traditional and alternate energy sources on future use of geothermal resources.

The objectives of this chapter are primarily (1) to record the available quantitative information on geothermal resources; (2) to provide a set of selected literature references that will serve as a useful guide for further study; and (3) to describe new concepts that appear to be rapidly emerging in geothermics. Thus, it may be useful to stress the idea that the existing and the "soon-to-be-completed" geothermal electricity generating plants, in addition to the several thousand small direct heat use projects in the world constitute geothermal "reserves." This energy is currently largely produced from hot or warm fluids derived from thermal anomalies. At the other end of the scale, natural springs, or aquifers in areas with normal thermal gradient attest to the wide variety of geothermal enterprises worldwide. The technical potential for extraction of the heat stored in rock that has little porosity or permeability is receiving increasing attention. Research and development efforts are now focused on field experiments aimed at reservoir formation technology. With continued success, this vast thermal energy resource could then contribute, in the long term, significantly to our future energy needs.

Resource Types

Several different types of geothermal resources have been recognized, explored, and developed. The general geology and tectonic setting of some resources have been rather well characterized (Chapter 3) and these background data have provided information to initiate development of improved and advanced exploration methods (Chapter 4) suited to the specific types of resources. The accumulated information is also providing the background needed for finding geothermal reservoirs in similar regional and local geological settings. This type of resource extrapolation by geological analogy (and eventual estimation of productive success probabilities) is currently only approximate, because it is still limited by incomplete experience and insufficient numbers of thoroughly documented case histories for currently developed reservoirs. The situation is similar to the one that existed during the period of early oil and gas exploration and reserve estimation. The resource data and reserve estimation procedures can be expected to mature as the development of the world's geothermal resources expands and reaches a stage of success and sophistication that has been achieved by oil and gas, coal, and uranium energy resource evaluations and reserve projections.

Four high-grade geothermal hydrothermal steam (vapor-dominated) and sixteen high-grade hydrothermal (liquid-dominated) reservoirs producing electricity are economic successes. These twenty developed reservoirs produce about 1,500 MW (e) of power worldwide. Relative to the problem reserve estimation, however, the ultimate reservoir areas or geometric volumes of the permeable subsurface rocks, details and extent of possible fluid and heat recharge, influences of fluid reinjection, and productive economic life of these fields have not yet been determined (or at least not well documented). Evidence indicates that these types of high-grade hydrothermal resources are most likely to be found in volcanic fields and limited drilling has been completed as yet in such terrains, a factor that presently restricts the precision with which high-grade hydrothermal resource development can be projected. Figure 2-1 is a representation of a hydrothermal reservoir and depicts the general geological and thermal features thought to exist in most of these systems. Indeed, relative to the usual drilling data and reservoir test results used to define commercial ore "deposits" and petroleum "fields," there is no comparable data in geother-

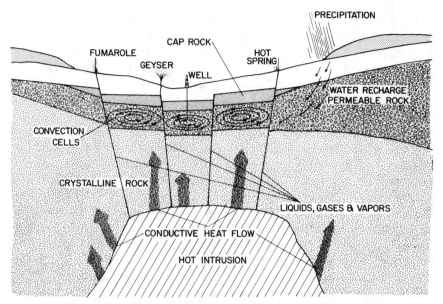

Figure 2-1. Diagram of hydrothermal geothermal reservoir.

mal energy development. It has become conventional, however, to define a hydrothermal "field" as a subsurface permeable volume, verified by the drilling of at least one productive well. It does not appear at this date that any one of about fifty high-grade hydrothermal fields* (nor the several thousand low-temperature resource developments) has received comparable attention and effort, such as drilling to define field boundaries and development of extensive reservoir test data, to support the type of methodology used to define a productive oil field and to establish crude oil reserves (Anon., 1976).

A far greater research and development effort thus far has been expended on the high-grade resources for electric power production than on the lower-temperature resources that are applicable to a broad range of energy uses. The current emphasis on electric power production is due to the transportability and flexibility of the end use of this form of energy. It is recognized, however, that energy can be conserved by using low-

---

* The Wairaki, New Zealand hydrothermal field (Haldane and Armstead, 1962) appears to be the closest analogy to an oil field by virtue of the extensive drilling, testing, and production history.

temperature geothermal waters directly for space heating — cooling and industrial processing rather than burning high-grade fuels (such as oil or coal) at 1,000°C to produce fluids ranging in temperature from 70°C to 100°C. Practical examples of the use of these low-grade resources for agriculture, spas, space heating, and various industries exist on a worldwide basis, and are estimated to exceed the current worldwide geothermal electric power production. Resource evaluations and potential reserve estimates of low-grade hydrothermal reservoirs (often termed "thermal waters"), which exist in sedimentary basins, may be more reliable than those that are possible for shallow high-temperature reservoirs, because a large number of wells have been drilled in the course of the worldwide exploration for and production of petroleum. The productive capacity testing of such stratigraphically layered hydrological systems is rather well established. A half century of experience has been largely responsible for the knowledge of the thermal structure of the world's major sedimentary basins, which has often led to the discovery and use of these aquifer thermal reservoirs.

Two types of geothermal reservoir systems that are presently under intensive investigation are the geopressured and hot dry rock systems. *Geopressured reservoirs* are located in sedimentary basins where sedimentation has proceeded rapidly over geologic time. The consequent lack of expulsion of pore fluids from formations at depths greater than 2 to 3 km results in abnormally high pore pressures, which can approach lithostatic. The geothermal gradient in the rock may also be enhanced. Increases in gradient from the normal value of 35°C to 75°C/km have been recorded in some zones at depth. In addition, geopressured zones may contain methane that may be present in economically exploitable quantities. These reservoirs hold promise of making a very large addition to the world's potential resources of low-grade geothermal energy, if the evaluation, completion, and extraction technologies can be perfected. An extensive research and demonstration effort is currently underway in the Gulf of Mexico coastal region of the United States to aid in the needed development projects (Campbell, 1977).

Hot dry rock systems are rocks of low permeability at elevated temperatures, but lacking sufficient natural fluids to convey (convect) the heat toward the surface. Figure 2-2 illustrates the general features of a hot, dry thermal system, shown adjacent to an hydrothermal reservoir. The basic technical problems to be overcome are those of detection (exploration) and extraction (reservoir rock properties). An experimental program (Los Alamos) is underway to develop drilling and fracturing technologies for such impermeable formations, circulating water in the fracture system, exchange heat, and extract the heat from the rock. Success in these efforts will make it possible to consider exploitation of hot dry rock geothermal energy over a significant portion of the surface of the earth. The demonstration of techni-

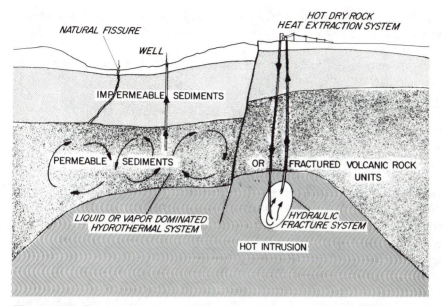

Figure 2-2. Schematic diagram of a hot, dry rock geothermal system, shown adjacent to a hydrothermal reservoir.

cal feasibility and economic viability of such reservoirs will add a very large increment to future reserves of geothermal energy. Contributions will be potentially made to both low-grade and high-temperature reserves by hot dry rock reservoirs.

It has also been speculated that it may ultimately be possible to tap the magma (Stoller and Colp, 1978) of volcanic systems. Development of hardware for insertion of heat extraction equipment into the more liquid portions of the magma is proposed as a possible mechanism for practical utilization of this super high-grade geothermal energy source (temperatures of 900°C to 1,200°C).

One recent detailed scheme for geothermal resource evaluation has been put forth by Muffler and Cataldi (1977). It may be that these definitions will prove useful for regional assessments and, with sufficient accumulation of information and experience, may ultimately be applied broadly, both by governmental agencies and commercial interests. A basic premise of Muffler and Cataldi is that resource quantities should be expressed in units of heat. It is further proposed that the geothermal *resource base* be accepted as all the stored heat beneath a given surface area, and that *accessible resource base* be that heat which is shallow enough to be "tapped by production drilling." It is in this spirit that geothermal resource discussions are formulated in this chapter.

In this sense the evaluations do not consider in detail the economic or technological factors that may exist at some future date. It seemed more appropriate at this stage of geothermal development that the status and potential be thoroughly reviewed. Indeed, there is not sufficient data worldwide to provide the assessments that have been derived for the United States (White and William, 1975) and central southern Italy (Cataldi et al., 1979).

This resource evaluation indicates that the global availability of low-temperature geothermal energy (thermal waters) is likely to be of the same order of magnitude as the quantity of energy ultimately derivable from the world's petroleum resources. The future projection of high-temperature resource magnitudes, if restricted to natural hydrothermal systems, will be limited in quantity and restricted to local geographic areas that are geologically favorable. If the hot dry rock technology develops, as is optimistically projected, then electrical power application potential will be expanded perhaps a hundred to a thousand times that of natural hydrothermal systems associated with volcanic regions. Eventually, energy price rises and steady depletion of other energy resources will cause mankind to seek the heat stored in the deep crystalline rocks of the crust.

Magnitude of Resource Base

The calculation of the thermal heat energy stored in the crust of the earth to a drillable depth is an *accessible resource base* estimate. As previously indicated, economic or useful reservoir production for electric power generation has only been achieved at about a dozen isolated hydrothermal fields. For an accessible resource base to be considered a geothermal reserve, it should be specifically identified. Then it should be determined if the heat may be extracted economically, using existing technology. This type of geothermal resource evaluation on a worldwide basis requires far more data and experience than exists at the present time. Consequently, the present discussion focuses on the accessible resource base; identified potential hydrothermal resources (drilled fields); and established high-grade producing fields (reserves).

Some significant and useful numbers can be derived, however, that place the geothermal resource in perspective. For example, the total heat stored beneath the land areas (about 1.4×10^8 km^2) of the earth is summarized in Table 2-1. What do these very large numbers portray? The value of resource base numbers is claimed to lie in: (1) comparing with resource bases derived

Table 2-1
Summary of Worldwide Geothermal Resource Base Estimates Heat Stored in Rocks and Fluids Beneath Land Areas

Crustal depth (km)	Stored thermal energy above 20°C in terajoules*			Basis
	High-grade	Low-grade	Total	
10 km	232×10^{12}	171×10^{12}	403×10^{12}	Integration of global heat flow.
3 km	20×10^{12}	15×10^{12}	35×10^{12}	High-grade areas are defined as areas with heat flow greater than 50 mW/m², which is the continental average.
3 km	5×10^{12}	36×10^{12}	41×10^{12}	Integration of thermal gradients, partitioned via the estimated areal extent of the geothermal zones (Figure 2-3). High-grade zones have thermal gradients up to 80° C/km, whereas in normal regions, it is 25°C/km.

*A terajoule (TJ) = 10^{12} joules. Inasmuch as 10^{18} joules = 1 quad, 10^{12} TJ = 10^6 quads.

for other energy resources, (2) providing a background for comparing with current energy production and use rates, and (3) assessing of technology, research, and demonstration efforts and priorities. The 40×10^{12} terajoules* (40 to 400×10^6 quads) of geothermal energy resource base can be compared with the total world energy production rate from all sources in 1975 of 258×10^6 terajoules (258 quads/year), or equivalent to the energy in about 6×10^{15} tons of oil. This large potential can, therefore, be projected to ultimately yield a large energy production rate, and it is prognosticated that in the future as much as 5×10^6 to 50×10^6 MW (e) could be produced worldwide from the high-grade geothermal resources. That is, about 10 to 100 times the current total U.S. production of electrical power could be supported by geothermal resources.

The total potential thermal energy source can also be compared to the world's "measured resources" for other energy supplies as summarized by the United Nations (Goeller et al., 1974) and presented in Table 2-2. To illustrate further the bounds of crustal stored thermal energy, one should

* One terajoule (TJ) = 10^{12} joules (J), and 10^{12} TJ = 10^{24} J = 10^6 quads.

Table 2-2
Summary of Some Measured World and
U.S. Nonrenewable Energy Resources

Energy resource	Worldwide stored energy (10^6 terajoules)†	USA stored energy (10^6 terajoules)
Fossil fuels		
Solid fuels	15,100	5,200
Crude oil	4,000	240
Natural gas	2,000	290
Tar sands and oil "shales"	11,000	6,400
Total fossil fuels	32,100	12,130
Uranium and thorium		
Non-breeding (U)	800	284
Breeders (U)	50,000	17,000
Breeders (Th)	16,600	2,700
Total fissile fuels	67,400	19,000

*Source: World Energy Conference, "Survey of World Energy Resources (1974)," See
Groeller et al., (1974), Table IX-9, pp. 253-255.
†10^6 TJ = 1 quad.

consider the fact that on the average about 2.8×10^6 terajoules (2.8 quads) are contained beneath each 1 km^2 of surface area (integrated to a depth of 10 km). This should be compared with an average surface heat flux q_s'', from the earth's interior through the same area of about $q_s'' = 50$ mW/$m^2 \times 10^6$ $m^2 = 50 \times 10^3$ J/s. This heat loss is mostly radiated into space together with most of the incoming solar heat flux. The solar flux at the earth's surface is some 4,000 times greater on the daylight areas of the earth than the geothermal flux.

The thermal energy "crustal abundance" value of 2.8×10^6 TJ/km^2 and the total stored heat of 400×10^{12} TJ (to a depth of 10 km) differ considerably from the measured resource data of Table 2-2, because the 1974 World Energy Conference attempted to assay quantities that might be close to the usual definitions of "reserves," i.e., energy resources with known locations, accessibility, technology, and waste issues that are well understood or reasonably well projected. In contrast, the stored thermal energy is known to be present at depth everywhere, but the general questions of recoverability and costs await more definitive resolution.

To provide further insight into the local nature and magnitude of the heat stored at depth, it is instructive to record the calculation of the heat energy that will be carried into the crust by the intrusion of a magma body from depth. For example, if a body of molten magma 4 km^3 in volume (approximately one cubic mile) at a temperature of 1,000°C (1,800°F) is implaced into the upper crust (Decius, 1961, and Elder, 1979), it will upon cooling to 150°C release energy in the amount of a 6.5 × 10^6 TJ (or 6.5 quads). If this heat were recovered and converted to electricity at a thermal recoverability of 100%, it would support a 200-MW (e) power plant (assuming 10% conversion efficiency) for a century.

It is also instructive to calculate the amount of heat stored in a block of rock at depth. For example, one can consider 160 km^3 (40 cubic miles) of rock, visualized as a slab 2 km thick with a surface area of 80 km^2 (an area about 5 × 8 miles in extent). The block has a mean temperature of 250°C, located at a mean depth of about 5 km in a region having a thermal gradient of 50°C/km. Cooling this block to 50°C would release about 62 × 10^6 TJ (62 quads). This magnitude of energy is comparable to the total (all sources) of 73 quads of energy used in the U.S.A. during the year 1974.

Summary

The world's geothermal energy resources, of one type or another, are available under most land areas of the globe. This includes all of the major energy consuming nations of the world. No scientific breakthroughs will be required to develop these resources. Instead, only extensions, demonstrations, and advances of already known or conceived technologies will be necessary.

Evidence from past and current worldwide geothermal exploration and development activities indicates that, in the near future, local high-grade thermal anomalies (high temperatures at shallow depths) will continue to be the primary targets for thermal energy production. These development activities have been accelerating since the 1973 OPEC (Organization of Petroleum Exporting Countries) oil embargo and can be expected to advance especially rapidly in those countries situated within the earth's geothermal (volcanic) belts (Figure 2-3). Nations and regions with few or no fossil energy resources can be expected to be especially active. It is also projected that an increasing expansion of direct use of geothermal heat for industrial and agricultural processing and for space heating will occur.

In the longer term, continued development of advanced technologies for hot dry rock reservoirs, reduced relative drilling costs, and rising world energy prices will bring forth a sustained development of the deeper crustal heat energy. It seems likely that this trend will continue as the number of successful demonstrations and use of hydrothermal resources increase and give confidence in the economic viability of a variety of surface facilities.

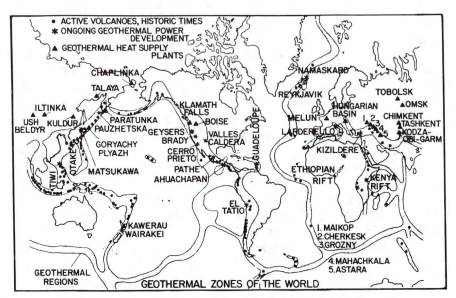

Figure 2-2. Map of "high grade" geothermal zones of the world.

With the much more widespread availability of potential sites for hot dry rock heat extraction, and with more numerous technical feasibility demonstrations of the reservoir technology, planning of new thermal energy sources for electricity production, cogeneration systems, industrial use, space conditioning, and agricultural applications can be more precisely compared with the design and costs of other projected alternative energy sources.

At present, less than 0.1% of the world's energy needs are met with geothermal energy resources. Thus, energy source and financial planners have access to very little data and expertise as compared to other alternative energy sources. As the anticipated growth of the alternate energy market will increase the percentage of world geothermal energy usage to the 5% - 10% range, many more cases will be available as firmer bases for design and financial modeling. This accelerating accumulation of data and experience can be expected to provide a positive feedback to increased selections in favor of geothermal energy.

Concentration on the more obvious high-grade resources would constrain geothermal energy development to small, but locally valuable, contributions to the energy supply. Thus, it is important to expand and extend research and development in exploration and resource evaluation techniques and reservoir formation technology of the widespread, deeper, hot basement crystalline rocks. This is most important so as to have a more reliable and complete inventory of thermal resources available and so that production projections can start for the large-scale hot dry rock resources as the world energy demands increase. The many existing hot dry wells drilled in and adjacent to developing or producing hydrothermal fields offer one initial opportunity for experimentation by the developers of such fields. These wells attest to the likely association of hot dry rock resources with nearly all hydrothermal systems.

Approach

In making resource evaluations and reserve estimates, it is especially important that the basis for the results are clearly defined and that the objectives of the exercise are understood. Inasmuch as there is only a very limited amount of organized subsurface information and little published geothermal well and field production data available at this time, the objectives of this chapter are most properly characterized as a status report, and an agenda outline for the problems that will need to be considered in the future. Also, projections of world economics, energy demand and prices, and technological evolution and innovations play an especially significant role in the future of the geothermal resource and reserve picture. Individual perspectives and the degree of personal optimism or pessimism can color both the approach and results. The presentation herein is likely to be classified as optimistic with respect to the potential for scientific and technical advances, and pessimistic with respect to rapidly rising world energy prices, the rate of growth of energy consumption, and acceleration of depletion of the reserves of nonrenewable energy resources.

A note of caution to the reader is in order. It must be realized that decisions to conduct exploration projects, initiate research and development efforts, invest in field development, expend capital for power plant construction or expansion, and other long-range decisions by international and regional institutions, nations and commercial organizations, are often based upon analyses of the type presented here. An attempt has been made, therefore, to provide a selected but definitive reference list and, most importantly, to provide a clear delineation of the assumptions and purposes of the various analyses. Considering the current, very initial stages of geothermal energy development and the high costs and risks of exploratory drilling (Chapter 5), those interested in generating geothermal resource and reserve estimates for other purposes are urged to take an independent and circumspect approach.

Previous Estimates

Descriptive Assessments

Many of the previously published surveys of worldwide geothermal resources (for example, Muffler, 1975; Proceedings 2nd U.N. Conference, 1975; Summers, 1971; Koenig, 1973; Barbier and Fannielli, 1973; Tatsch, 1976; Howard, 1975; Geertsma, 1978) have been primarily concerned with descriptions of occurrences (covered in Chapter 3) or summaries of current or planned commercial or government sponsored developments. Much of the data for these earlier reviews has apparently been developed from geological or geophysical studies of variable quality and availability; or report literature, tours, and personal communications that are difficult to trace to the original sources. This difficulty with the publication habits of writers and the local-interest nature of geothermal energy development has led to a U.S. Government Bibliographic Abstract service, which includes the efforts of the Geothermal Energy Update, the U.S. National Department of Energy and the Technical Information Center. Publications from the above sources are available from the National Technical Information Service (NTIS), Springfield, VA 22161, U.S.A. This bibliographic service has brought together a sometimes obscure and scattered world geothermal literature.

The EPRI Survey

Recently two valuable reports (Anon., 1977, and Roberts, 1978) have been compiled in the United States by the Electric Power Research Institute (EPRI). This survey provides insight into the difficulties of defining and determining a useful worldwide resource base, establishing the current status of reserves, and projecting future development activity. This study used both analytical techniques for the resource base calculation and a worldwide mail survey of the nations. To indicate the potential for the nations of the world, the study defined various categories of resource quality, ranked essentially by temperature level. It was then estimated what high-grade fraction (X) of a given area (A) of a particular region of the world or nation, lies in the so-called "geothermal belts" or zones of the earth (Figure 2-3).

Further, four temperature classes or energy "grades" are defined and applied to the heat stored in the rocks or fluids at depth. They are: Class 1, $<100°C$; Class 2, $100-150°C$; Class 3, $150-250°C$; and Class 4, $>250°C$. The partitioning of the worldwide geothermal resource base into temperature classes serves as a useful guide to the different potential uses of geothermal resources. In addition, this study assumed that 20% of the heat is stored in the subsurface fluids and 80% in the rocks. The analytical approach in these two reports restricts the drilling depth to 3 km and thus injects an idea of "accessibility" into the evaluations, and reflects a somewhat conservative view of drilling technology.

The total thermal energy stored beneath the continents is calculated to be 41×10^{12} TJ (41×10^6 quads). The low-grade regions, i.e., with areas $(1 - X)A$, contain no Class 2, 3, or 4 resources because with a thermal gradient of 25°C/km, temperature at a depth of 3 km will reach only 75°C. This general approach to resource base evaluation is similar to counting all the deuterium molecules in the streams, lakes, and oceans of the world, to determine the fusion energy resource. In the case of solar energy, the comparable quantity would be the total solar flux intercepted by the earth's diametrical plane, i.e., 177×10^9 MW or a total annual input at the earth's surface of about 4×10^{12} TJ.

The concepts of "resource base," "resource", and "reserve" obviously need careful definition in geothermal energy resource evaluations, because they are often subject to different interpretations. Indeed, the terms are often used very differently by different authors because the purpose of the evaluations differ, ranging from broad national energy resource appraisals performed by government agencies to detailed investment risk evaluations conducted by commercial interests. The EPRI work was an attempt to evaluate resources of each country and required completing data not supplied by the replies to the national assessments questionnaire. In general, it seems that detailed resource base evaluations might better be applied to local, national, or regional resource assessments, and reserved especially for those regions where adequate heat flow data are available in sufficient detail.

Regional and National Assessments

Several excellent regional and national geothermal resource assessments are available: Axtell, 1975; White and Williams, 1975; Donaldson and Grant, 1978; Cataldi et al., 1977; McNitt, 1978; Carle and Van Rooyen, 1969; and Muffler, 1979. These surveys are usually provided by national government agencies, from nations that have national energy development activities or by the United Nations. No known examples of a geothermal resource evaluation by private commercial firms have been published.

Due to the diversity of viewpoints and purposes of geothermal resource assessments, and especially those relative to reserve estimates, definitions used in this review are explained in the next section.

Definitions and Methodology

The approaches taken to evaluate the status and to predict the future development of geothermal resources are quite different depending upon whether a geological, economical, commercial, governmental, or academic viewpoint is being expressed. This has led to a wide range of terms used and approaches taken for evaluations and projections of the geothermal potential, and reflects the borrowing of terminology from those used for different types of natural resources (see for example Campbell et al., 1970; Vogely, 1976; White and Williams, 1976). This has led to attempts (Blondel and Lasky, 1956) to organize and systematize terminology in resource and reserve estimation — projection studies. A clear set of definitions for regional geothermal resource evaluations and assessments has been established (Muffler, 1978). Appendix A at the end of this chapter presents the details of these recommended definitions for regional assessments of geothermal resources.

Geothermal Reservoir Types

The five commonly defined geothermal reservoir types (denoted as resource types* in some evaluations) are distinguished in this chapter. They are usually divided into four categories: hydrothermal (liquid-dominated and vapor-dominated), hot dry rock, geopressured, and magma.

Liquid-Dominated Hydrothermal

Liquid-dominated geothermal reservoirs are hydrothermal reservoirs that contain circulating liquids (water or brine) that transport the thermal energy of the rock from deep sub-regions to near-surface regions by natural fluid circulation. The temperatures of these systems are as high as 360°C. temperatures in most liquid-dominated reservoirs increase rapidly with depth (high thermal gradients; i.e., up to 100°C/km) until the temperature reaches the boiling point of water. Only slight increases in temperature with depth are then noted until the depth is reached where liquid domination ceases. Temperatures seldom exceed the boiling point of water at in-situ hydrostatic pressures. The low-temperature fluids in deep aquifers (thermal waters) in regions of normal geothermal gradient may be one of the largest, most useful, and most widely available geothermal resources (Lienau and Lund, 1977).

---

* The resource is geothermally stored heat. It might be most useful to consider these resource divisions as "resource subtypes."

Vapor-Dominated (Steam) Hydrothermal

Vapor-dominated hydrothermal reservoirs*, usually referred to as dry steam fields, are apparently quite rare. The four most successful geothermal electric power developments in the world, however, are associated with the development of vapor-dominated reservoirs (Larderello, Italy; Monte Amiata, Italy; The Geysers, California; and Matsukawa, Japan). In such reservoirs the continuous phase within the pore space of the reservoir rock is steam, whereas deeper, liquid water is (presumed to be) present. Fluid temperatures are typically in the 250 to 320°C range. Technology for production of electric power from this type of reservoir is readily available.

Hot Dry Rock

Temperatures in the earth increase with depth whether a hydrothermal convection system exists or not. Relatively simple calculations can show that at common porosities much more heat is stored in the rock matrix than in circulating water found in natural geothermal (hydrothermal) systems. Porosity generally decreases with depth and, therefore, vast volumes of hot dry rock exist within the earth's crust. Research efforts (Smith, 1975; and Hill, 1978) are underway to develop reservoirs and methods of introducing cold water into such hot dry rock systems of fractures and extracting heated water. Thus, vast regions of the world may be developed for their stored heat. It has been suggested that the heat stored in known geothermal systems is much larger than the heat contained in the fluids only. Thus, many of the known hydrothermal systems may become depleted of fluids long before the heat stored in the associated rocks has been exhausted. It is possible that many liquid-dominated hydrothermal reservoirs may be further exploited as hot dry rock reservoirs after depletion of their natural fluids. Hot, dry holes were located by drilling in many regions that were known, or thought to contain, hydrothermal reservoirs.

Geopressured

Geopressured reservoirs are usually found in deep sedimentary basins where insufficient sediment compaction (undercompaction) has taken place over geologic time and where an effective caprock (e.g., shale) exists. Compaction leads to conditions whereby water is squeezed out of the interstitial pore space of clay sediments (and later into adjacent sand bo-

---

* See White et al., 1971 for a detailed discussion of steam and hot water (hydrothermal) geothermal reservoirs.

dies). An internal pressure greater than the normal hydrostatic pressure at that depth is imparted to the geofluids due to undercompaction.

In some cases of geopressure, fluid pressures approach those of the overall weight of the overlying rocks (termed lithostatic pressures). This overpressured water system can be associated with higher-than-normal temperature gradients because of the increased specific heat capacity of the overpressured rock—water system. Temperatures up to 237°C have been recorded in some geopressured zones in the Gulf Coast of the United States, and wellhead pressures in excess of one MPa (11,000 psi) are common. In addition, geopressured fluids often contain anomalously high concentrations of dissolved methane gas. Nearly all large synclinal basins of the world contain some geopressured zones. In the United States, geopressured geothermal resources cover an area of more than 200,000 km² in the states of Texas and Louisiana, and initial test results of these geopressured sands have been completed.

Magma

The near-surface magma bodies located in the volcanic zones of the earth obviously contain large quantities of heat. The technology to locate, drill, and produce these heat sources has not been developed; however, a basic research project has been established in the U.S. to define the resource and initiate research on the drilling and extraction technologies (Stoller and Colp, 1978).

Reserve and Resource Definitions and Discussion

The terminology and definitions, as well as the philosophy of the present analyses, are largely adopted from the volume by Schurr (1960). This book sets the stage rather well by illustrating (1) the existing confusion and diversity of terms, (2) purposes and results that are available in the reserve and resource evaluation study area, and (3) the confusion that exists between various discussions and studies of mineral and energy resource types.

The following terms and definitions are introduced and adhered to in the remainder of this chapter:

Geothermal resource base includes all the stored heat, in both rock and fluids, above 15°C, within a portion of the earth's crust. In this chapter, depth is restricted to 10 km into the crust. This resource base is better defined as an *accessible resource base,* as in the definitions adopted by

Muffler and Cataldi (1978). All geothermal resource and reserve estimates are subsets of the accessible resource base.

Geothermal resource is a term used to define that quantity of geothermal heat that is likely to become available if certain technologic and economic conditions are met at some future (specified) time. The approach and detailed definitions, limitations, and stipulations will depend on the purpose of the particular projection and assessment being made.

Geothermal reserves consist of identified economic resources that are recoverable at a cost that is competitive now with other commercially developed energy sources.

Demonstrated reserve is defined here as a geothermal area or field that has at least one borehole or well drilled into it and for which quantitative estimates of the areal size of the deposit of heat are available.

These general concepts and definitions are illustrated in Figure 2-4. Although these ideas are rather simply defined and relatively unambiguous, it is possible that the quantitative evaluation of the resource base may be difficult and the lack, or uncertainty, of the data may be frustrating. The foregoing organization and approach allows valuable insights to be drawn

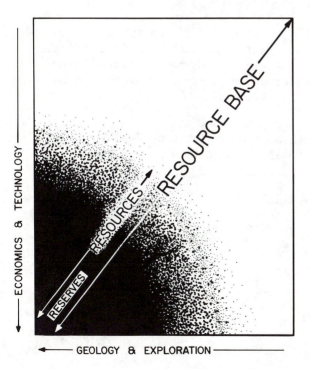

Figure 2-4. Diagram of resource base, resource, and reserve concepts.

about the research needed to be developed to project future worldwide geothermal energy reserve estimates.

Resource-Base Concept

The utility of the concept of resource base has been well expressed by Schurr (1960) as follows:

> The perspective afforded by the resource-base concept, for example, emphasizes the truly large possibilities awaiting realization through technology. With thinking based on the traditional reserve concept, even those attempts to take a longer view with latitude for technological advance are apt to remain in the general vicinity of the (present) reserve magnitude. . . .

The concept of resource base, therefore, is most valuable when targeting research and development efforts and planning long-term energy development strategies. The utility of deriving a worldwide geothermal resource base is that it is valuable as a comparison with estimates of the resource bases for other alternative energy resources.

The inclusion of technologic factors and improvements of reserve and resource projections has not been extensive, although the concept-to-terminology relationships are easily visualized as shown in Table 2-3 and Figure 2-4. Most geothermal resource assessments have thus far concentrated on geologic and economic factors. New exploration technologies, however, clearly will influence knowledge of occurrences, whereas more efficient development (drilling) and conversion technologies will influence production costs. Demonstrations of feasibility of new types of resources (e.g., hot dry rock reservoirs and geopressured formations) will affect the

Table 2-3
Relationships Between Resource–Reserve Terminology and Basic Aspects of Energy Sources

Aspects and factors	Occurrence: geologic knowledge	Economic: production costs and energy prices	Technologic: extrapolations and extensions of present and new concepts
Terms			
Reserves	Known	Present cost	Currently feasible
Resources	Known	Any cost level	Currently feasible and later feasibility indicated
Resource base	Known + unknown	Irrelevant	Feasible + infeasible

occurrence aspect and quantitative evaluations of potentially producible resources, i.e., extend portions of the resource base into the resource and reserve categories.

The U.S. Department of Interior, Geological Survey, and the Bureau of Mines (McKelvey, 1974) have attempted to establish a number of subcategories within the reserve and resource definitions, which have proven to be useful for resource evaluations for the mature and well-developed mineral resources and metals industry (Blondel and Lasky, 1956; and Vogely, 1976). The petroleum industry (Campbell, 1970) also provides a range of reserve definitions that are derived from the very nature of the commercial development of petroleum and that retain much of the flavor and thrust of that industry's current and near-term economics. These types of refinements were not appropriate in the present study.

The United States Geological Survey (White and Williams, 1976) has prepared a resource appraisal of the geothermal prospects for developments in the United States. In this publication, the resource projections were derived and resources were estimated both from a calculated resource base and the known geologic trends, i.e., hot springs and hot spots known or inferred to exist. A useful set of definitions is presented in that report. Recently, definitions and methodology have been provided that appear to have wide applicability to regional or national geothermal resource evaluations (Muffler, 1979; and Muffler and Cataldi, 1978).

Further comparisons of terminology, definitions, the implied logic, and energy resource evaluation schemes are presented in Appendix B at the end of this chapter.

Resource Diagrams

One of the more useful conceptual techniques developed is usually termed a "McKelvey diagram" (McKelvey, 1974). The essence of this approach is to plot reserves, resources, and resource base on an areal diagram with a vertical axis reflecting economics and a horizontal axis depicting geological assurance. Both axes can reflect advances in technology and thus project increases in geologic assurance through advances in exploration techniques, accumulated experience in resource finding, etc. Projections of improved. economics may also reflect estimates of future technological improvements (e.g., in drilling) that lead to lower development costs.

Figure 2-5 illustrates the use of a McKelvey diagram for U.S. crude oil resources, and especially notes resources that may partially become reserves through improved tertiary recovery methods and through increased world prices of oil. Figure 2-6 is an illustration of a conceptual McKelvey diagram for U.S. geothermal resources (Muffler, 1979) where the locations of four types of geothermal reservoirs have been sketched on the diagram. Com-

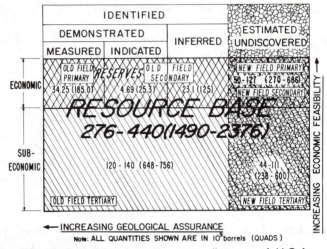

Figure 2-5. McKelvey resource diagram of U.S.A. crude oil resources, for 1975, in billions (10⁹) of barrels (and quads). Total historical U.S. cumulative production on December 31, 1974, was 106 billion barrels (or 572 quads).

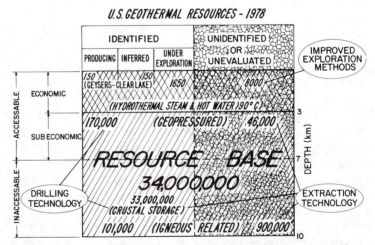

Figure 2-6. Conceptual McKelvey diagram for U.S.A. geothermal resources; values are in quads (Muffler, 1979). Compare with Figure 2-5.

parison of these two diagrams indicates that U.S. geothermal resources are larger on an energy basis than those for crude oil. A more direct comparison might have resulted if the diagram for oil resources (Figure 2-5) had been constructed with the inclusion of other known sources of oil, i.e., tar sands, oil sands, and oil shales.

Units

It has been suggested that all geothermal resource and reserve values be calculated and recorded as heat producible (or potentially producible) at the wellhead. Such figures will not take into account the losses associated with a particular application, conversion efficiency, transport loss, etc. This procedure is in accordance with the definitions and practice in the fossil fuel and mineral areas.

In this chapter the energy units and equivalency factors listed in Tables 2-4 and 2-5 are used. They are consistent with those generally used to express the larger units of energy and were adapted from various sources (Crabbe and McBride, 1978; Morgan, 1975). The presentations use the SI system (metric) of units and the basic conversion: 1 Btu = 1054.8 joules (J). This conversion is often, for the present purposes, rounded to 1 Btu = 10^3 J. Therefore, the unit of energy quad (= 10^{15} Btu) is about equal to 10^{18} joules (= 10^6 terajoules, TJ).

It is usually assumed that 1 quad (direct energy equivalent) of geothermal heat content in the ground is about equal to 190 million (1.90 $\times$ 10^8) barrels (42-gal barrel) of oil.

Worldwide Resource Base Estimates

Total Global Heat Content

By direct analogy with other mineral and energy resource base estimates, it would be necessary to calculate the total heat energy stored in the mass of the earth. The purpose here, however, is (1) to derive a resource base estimate that is useful as a perspective for comparison with the magnitude of other energy resources; (2) to provide a focus for targeting research and development efforts; and (3) to give a conceptual base for evaluations of future assessments of global geothermal resources. It will be satisfactory, therefore, to provide various estimates of the heat content of the earth's crust down to a depth that is reasonably accessible to drilling. It seems most useful to concentrate on an accessible resource base, to a depth of 10 km because that is generally considered the practical current limit of ultra-deep drilling technology.

Table 2-4
Conversion of Selected Energy Units

Multiply:	joule	kWh	MWy(t)	MWcen(t)	GWy(t)	Btu	quad	terajoule
				To obtain:				
Joule	1	2.778×10^{-7}	3.17×10^{-14}	3.17×10^{-16}	3.17×10^{-17}	9.48×10^{-4}	9.48×10^{-19}	10^{-12}
kWh	3.6×10^{6}	1	1.14×12^{-7}	1.14×10^{-9}	1.14×10^{-10}	3.415×10^{3}	3.415×10^{-12}	3.6×10^{-6}
MWy(t)	3.15×10^{13}	8.77×10^{6}	1	10^{-2}	10^{3}	2.98×10^{10}	2.98×10^{-5}	3.15×10^{3}
MW cen(t)	3.15×10^{15}	8.77^{4}	10^{2}	1	0.1	2.98×10^{12}	2.98×10^{-3}	3.15×10^{4}
GWy(t)	3.15×10^{16}	8.77×10^{5}	10^{3}	10	1	2.98×10^{13}	2.98×10^{-2}	3.15×10^{4}
Btu	1055	2.928×10^{-4}	3.36×10^{-11}	3.36×10^{-13}	3.36×10^{-14}	1	10^{-15}	1.055×10^{-9}
quad	1.055×10^{18}	2.928×10^{-19}	3.36×10^{4}	336	33.6	10^{15}	1	1.055×10^{6}
Terajoule	10^{12}	2.778×10^{5}	3.17×10^{2}	3.17×10^{-4}	3.17×10^{-5}	9.48×10^{8}	9.48×10^{-7}	1

Table 2-5
Approximate Equivalency of Some Energy Sources

Unit indicated is equal to:	Coal (tonnes)	Coal (ton)	Oil (bbls)	Natural gas (10^3ft^3)	Gasoline (U.S. gals)	Solar flux (m^2 h)	Btu	kWh
1 tonne of coal	1	1.1	47	27	220	9.1×10^3	27×10^6	79×10^6
1 ton of coal	0.9	1	43	25	200	8.3×10^3	25×10^6	73×10^6
1 bbl oil (42 U.S. gal)	0.21	0.23	1	5.8	45.5	2×10^3	5.4×10^6	17×10^6
10^3ft^3 natural gas	0.036	0.04	0.17	1	7.9	300	10^6	293×10^6
1 U.S. gal gasoline	0.0045	0.005	0.22	0.127	1	47.3	127×10^3	372×10^3
1 m^2 hour solar flux	1.1×10^{-4}		5.1×10^{-4}	3×10^{-3}	0.024	1	3×10^3	8.79×10^3

One quad (10^{15}Btu) = 10^6 terajoules = 185×10^6 bbls of oil (assumed specific gravity of 0.858 and combustion energy of 10^4 cal/g)

= 10^6 tonnes of oil = 42×10^{12} Btu

= 1 MWcent(t) = 5.5×10^5 bbls of oil

The heat content Q underlying a region of the earth, with a surface area A is calculated from the integral

$$Q(Z_f) = A \int_0^{Z_f} \rho(Z) \, C \, (Z) \, T \, (Z) \, dZ \tag{2-1}$$

where: Z = Vertical depth beneath the surface
Z_f = depth for desired computation
$\rho(Z)$ = density as a function of depth
$C(Z)$ = effective specific heat of crust as a function of depth
$T(Z)$ = temperature as a function of depth.

Such an evaluation as presented in Equation 2-1 gives only the accessible heat stored in the crust to the depth Z_f, and is not concerned with the practical aspects, such as reservoir recoverability and conversion or utilization efficiencies at the surface. To provide a detailed and accurate calculation, the spacial distribution of temperature $T(Z)$ at depth must be known at each point on the surface of the earth. Figure 2-7 shows schematically several types of temperature variations with depth depending upon the geothermal regime at depth, i.e., steam, hot water, or hot dry rock reservoir conditions. As shown in Figure 2-7, a variety of subsurface conditions can strongly influence the surface and near-surface thermal gradient $\partial T/\partial Z$ evaluated at $Z = 0$; the quantity is usually measured in heat-flow surveys.

To provide a practical calculational format for this discussion, two restrictions are made. First, the undersea or offshore resources are neglected. It should be pointed out, however, that areas of the oceans and lakes that have been sufficiently well studied (Williams, 1976; Herman et al., 1977; Morgan et al., 1974; Anderson, 1979; Martin and Welday, 1977; Girdler and Evans, 1977; Lawyer, 1975; and Stefanon, 1972) do indicate that there are potential geothermal resources beneath the lakes and oceans of the earth. Subsea resources of geothermal energy that are located great distances from land will be of little economic promise for a very long time. Those located in coastal areas, however, may indeed have a large potential use as energy supplies for such activities as electric power generation, desalination of sea water, solution mining of sulfur and potash, heavy refining of sea-delivered ores, and secondary recovery (water flooding) for offshore oil fields. This survey, however, will not concern itself further with offshore prospects and potential. Even though the resource may be significant, the developments will undoubtedly follow those on land. These resources clearly warrant more study.

The second restriction is more specific. It is necessary to consider only average values for the crustal properties in Equation 2-1. On a global basis, the temperature profile is taken to be linear with depth. Thus, the detailed local nature and extent of the high-grade hydrothermal deposits are to a

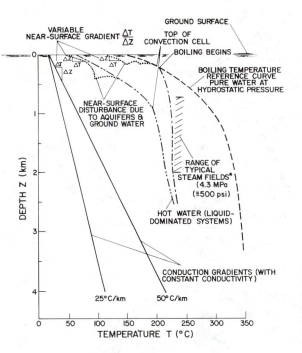

Figure 2-7. Diagram of several temperature profiles, T(Z), and corresponding influence on the geothermal gradient at the earth's surface.

large extent averaged out. It is possible, however, to account for these local heat flow variations and surface expressions in some regional settings or for individual national resource base appraisals, as shown later.

Resource Base via Global Heat Flow Data

The first geothermal resource base calculation developed here uses the global heat flow data* synthesis (Pollack and Chapman, 1977; and Chapman and Pollack, 1975), as depicted in Figure 2-8. The earth's land areas were divided into the major regions as shown in Table 2-6. The continental heat flow average is 50 mw/m² (very close to the global average). The world land masses were further subdivided into those with heat flow greater than the average (higher grade regions) and those below 50 mW/m² (lower grade regions), as shown in Figure 2-9.

The linear scale and resolution of these heat flow data are about 5° of latitude and longitude and, as shown subsequently, much of the finer scale,

---

* Approximately 5000 heat flow determination points are available; however, they are irregularly spaced on the globe with twice as many in the oceans as on land. Many of the more promising geothermal zones of the earth have not yet been surveyed (Chapman and Pollack, 1975).

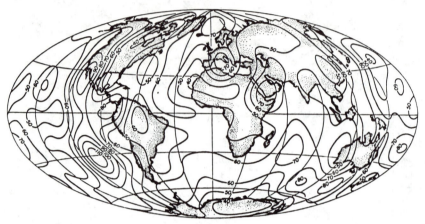

Figure 2-8. Global heat flow contours; units are in mW/m².

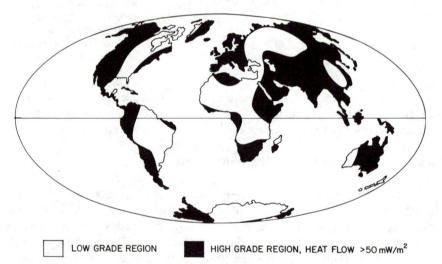

☐ LOW GRADE REGION ■ HIGH GRADE REGION, HEAT FLOW >50 mW/m²

Figure 2-9. Division of the continental heat flow regions into high-and low-grade regions.

including the majority of the so-called geothermal zones of the earth are averaged out at a scale of approximately 550 × 550 km squares. The value of this approach is that the data are consistently derived, based upon a reliable (and growing) data set, and do provide a good evaluation of the broad scale hot dry rock resource base of the earth's crust. In addition, no subjective

Table 2-6
Global Geothermal Resource Base Estimate
Based Upon Global Heat Flow Data Synthesis — 10 km depth

Region	Land area* 10^3 km² (%)	High-grade fraction, η	Resource base (10^6 quads)		
			High-grade	Low-grade	Total
Africa	30,319(22.3)	0.086	8.40	74.49	82.89
Asia	44,411(32.7)	0.729	101.76	31.71	133.47
Europe†	10,508(7.7)	0.561	21.34	12.13	33.47
North America††	24,011(17.7)	0.653	71.77	21.45	93.22
Oceania§	8,510(6.3)	0.80	8.17	1.70	9.87
South America	18,070(13.3)	0.343	20.48	29.93	50.41
Total $A_t =$	135,829** ($\sim$1.36 $\times$ 10^8 km²)		231.9	171.4	403.3

*U.N. Demographic Yearbook, 1976 (Anon., 1977).

**The Antarctica and some small island areas were not considered. See *The World in Figures* (Showers, 1973, p. 10) where total land area is given as 149,906,000 km² and the Antarctica land area is given as 18,975,000 km².

†Includes Iceland.

††Includes Greenland.

§Australia, Japan, Pacific Islands, New Zealand, and Hawaii.

judgments as to the distribution of thermal gradient values within a region or country need to be made.

Table 2-6 and Figures 2-8 and 2-9 indicate that vast regions of the continental land mass are underlain by high-grade resources, on a global scale. As shown, high- and low-grade resources are about equally prevalent. Except for Africa*, the regions have a surprisingly high proportion of high-grade areas.

To proceed with this initial calculation, the following average crustal characteristics and properties are assumed:

1. Average surface heat flow, $q_s'' = 50$ mW/m²
2. Average linear thermal gradient $\nabla T = 25°$/km
3. Average crustal density, $\bar{\rho} = 2500$ kg/m³
4. Average specific heat, $C = 770$ J/kg °C

* In this type of perspective, the major, high-grade hydrothermal resources known to exist along the East African Rift are masked out by the 5° × 5° scale of the global heat flow data.

5. Average volumetric heat capacity, $\bar{C}_v = \overline{\rho c} = 1.93 \times 10^6$ J/m³ °C
$= 1.93 \times 10^{15}$ J/km³ °C
6. Average ambient surface temperature, $\bar{T}_A = 15°C$
7. Average thermal conductivity, $\bar{\lambda} = 2.0$ W/m °C

The resource base was then calculated by using the following relations:

$$Q_{hg} = \eta A_r \int_Z^{Z(10 \text{ km})} \rho(Z) C(Z) \, T_{hg}(Z) \, dZ \tag{2-2}$$

where η is the high-grade portion of the area as tabulated in Table 2-6. The low-grade portion of the land area contains a resource base equal to:

$$Q_{lg} = (1-\eta)A_r \int_{Z_o}^{Z_f} \rho(Z) \, C(Z) \, T_{lg}(Z) \, dZ \tag{2-3}$$

where: hg = high-grade
lg = low-grade
A_r = area of region (km²)

Thus, the stored energy, Q_A, underlying an area A (km²) is equal to:

$$Q_A = 1.9 \times 10^{15} A_r \int_0^{10} (\nabla \bar{T} Z + 15) \, dZ \text{ (joules)}$$

$$Q_A = 1.9 \times 10^{-3} A_r \int_0^{10} (\nabla \bar{T} Z + 15) \, dZ \text{ (quads, see Table 2-5)} \tag{2-4}$$

where $\nabla \bar{T}$ = the average gradient associated with the area A_r

Then the global accessible stored geothermal energy, $\bar{Q}$, underlying the earth's land surface (Table 2-6) is given by:

$$\bar{Q} = 2.0 \times 10^{-3} A_t \int_0^{10} (25Z + 15) \, dZ \text{ (quads)} \tag{2-5}$$

where $A_t = 1.36 \times 10^8$ km², and therefore

$\bar{Q}_t$ = 400 × 10^6 quads.

The average "crustal abundance" of geothermal energy calculated on a continental areal basis is equal to:

Q_t/A_t = 2.8 quads/km^2.

Thus, on the average each square kilometer of land area is underlain by about three quads of heat energy stored in the rocks or fluids down to a depth of 10 km.

Although the above resource base estimates were derived and based on a consistent data set (the global heat flow data), the calculations neglect the smaller-scale geologic and tectonic phenomena that have led to the generally known localization of geothermal zones and deposits of heat at shallow depths in the earth's crust.

Geothermal Zones

The boundaries of many of the crustal plate margins have been recognized as the source of our more obvious high-grade geothermal resources. Figure 2-3 is a world map with a generalized sketch of these geothermal belts or zones illustrated. As shown, nearly all current geothermal development is associated with these zones. The difficulty for the worldwide evaluation of a resource base is that these zones are on a scale of a degree or so and not resolvable at the scale of the heat flow data used above.

The world map of geothermal zones (Figure 2-3) corresponds to a map of geologic regions (e.g., sedimentary basins) where oil, gas, or coal are most likely, or known, to be found. The difficulty in performing a quantitatively reproducible and reliable estimate is the subjective approximations of the areal extent of the zones themselves and the estimated extent and proportion of enhanced heat flow areas within the zones.

The resource base estimation has been performed (Roberts, 1978) and the heat stored beneath the land areas of one hundred and seven nations to a depth of 3 km were tabulated, as recorded in detail in Appendix C at the end of this chapter. To perform these calculations of resource base related to the geothermal zones, it was assumed that:

1. Normal areas have a linear temperature profile with a gradient of ∇T = 25 °C/km.
2. The geothermal zones have a higher than normal heat flow and a part X of a given country with land area A is *estimated* to be in a geothermal zone.

3. Of the higher than normal gradient area (XA) in a country, 90% $(0.9 \times A)$ has a gradient of $\nabla T = 40\,°C/km$ and 18% $(0.18 \times A)$ is assumed to have a gradient of $\nabla T = 80\,°C/km$.
4. The depth of consideration is restricted to 3 km and the average surface temperature everywhere is 15°C.
5. Four temperature classes of the resource were defined: Class 1, 100°C; Class 2, 100-150°C; Class 3, 150-250°C; and Class 4, 250°C.
6. In any one of the four temperature classes, 20% of the total resource base is stored in fluids (water and steam), whereas the remainder is stored in the rock matrix.
7. The average rock and fluid proportions and properties are taken to be such that $\bar{C}_v = 2.5\,J/cm^3\,°C$ (compare to $\bar{C}_v$ on page 72).

The resulting resource base calculations (tabulated in detail in Appendix C) are summarized in Table 2-7 in the same units (10^6 quads) used in Table 2-6. The high-grade (>150°C) resource base is evaluated to be equal to about 10^6 quads.

Comparing the regional summary in Table 2-7 with the previous estimates in Table 2-6, one sees that the regional distribution (relative proportions) of the energy are substantially the same as portrayed in Figure 2-10, except for the resource bases estimated for Europe. The indicated difference is due to the large area of the geothermal zone in Western Europe (mainly in Italy) and the zone projected up into Eastern Europe (the Hungarian Basin). The order of magnitude difference in worldwide totals is due to different depths of integration. This is illustrated in Table 2-8 where the results of the earlier global heat flow calculation, with a 3-km depth substituted for the previous 10-km depth, are compared.

Comparison of Resource Base with Other Energy Sources

The utility of resource base estimates lies in their comparison with other energy resource estimates to provide a backdrop for the assessment of priorities in technology, research and development. Values of geothermal resource base ranging from 40 to 400 $\times$ 10^6 quads can be compared to a worldwide total (all sources) energy production rate of $\dot{Q} = 5,730 \times 10^{16}$ tons of oil equivalent per year in 1975, as shown in Table 2-9. Converted to energy units, this amounts to $Q = 250$ quads/year (250×10^6 terajoules/year). Thus, a resource base of the order of a million quads is a very significant energy resource when compared with worldwide yearly energy consumption. Worldwide production of geothermal energy is expected to be developed only at a relatively small rate of about 2500 MW(e) of electric power in 1980 and perhaps 10^4 MW(t) of direct usage, or a total of 30 $\times$ 10^3

Table 2-7
Worldwide Geothermal Resource Base Estimate*
Based Upon Geographic Distribution of Geothermal Zones**
(3-km integration depth)

Region	(10⁶ QUADS)				
	<100°C	100-150°C	150-250°C	>250°C	Totals
North America	7.10	1.20	0.32	0.02	8.64
Central America	0.15	0.09	0.03	0.0	0.27
South America	4.6	0.83	0.23	0.01	5.67
Western Europe	1.5	0.07	0.02	0.0	1.59
Eastern Europe	6.7	0.25	0.06	0.0	7.01
Asia	7.5	0.79	0.21	0.01	8.51
Africa	5.2	0.33	0.07	0.01	5.61
Pacific Islands	3.3	0.26	0.17	0.01	3.74
Totals	36.05	3.82	1.11	0.06	41.04

*$C_v = C = 2.5 \times 10^{15}$ J/km³ °C;
**Requires a subjective evaluation of the extent of area in geothermal zone.

Table 2-8
Comparison of Worldwide Geothermal
Resource Base Calculations

Type of resource base calculation	Total geothermal energy stored (10⁶ quads)
Global heat flow (10-km depth)	403
Global heat flow (3-km depth)	35
Geothermal zones (3-km depth)	41

MW(t) of heat production per year. Reduced to a common energy unit base, this is 2.60×10^8 MWh of thermal energy per year or about one quad of worldwide geothermal heat energy produced yearly (1 quad = 33.6×10^3 MW year of thermal heat production).

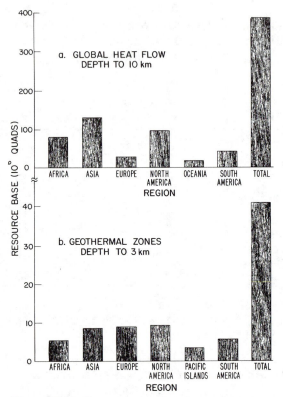

Figure 2-10. Comparison of geothermal resource base calculations based on (a) global heat flow data, and (b) geothermal zones of the world.

Attempts to derive reserves or production rate scenarios from resource base estimates have been made. These are usually accomplished by applying a "recoverability fraction" and an energy conversion factor. The results, however, are generally unsatisfactory because there is no historical basis for a worldwide recovery factor and the electric conversion and direct thermal use efficiencies will be established as a historical matter only as technology improves and much wider use occurs. It is considered, therefore, far better to concentrate attention on the basic nature of the geothermal resources, i.e., heat and not to attempt to estimate end-use conversion or direct usage efficiencies. This is a direct analogy to other resources, where the quantity of the resource is reported, i.e., oil in tons or barrels and coal in tons.

Regional and National Resource Base Estimates

Some regions or countries have been studied sufficiently to enable them to prepare detailed heat flow maps. Notable are those for the USSR (Gavtina

Table 2-9
World Energy Production by Region — 1975
(quads/year)*

Region	Solid fuel	Liquid fuel	Natural gas	Hydro/ nuclear electricity	Total
USA	16.54	20.08	20.36	1.69	58.68
Canada	0.66	3.47	2.85	0.76	7.73
Latin America	0.37	9.76	1.48	0.48	12.09
Western Europe	9.78	1.38	6.20	1.80	19.16
Middle East	0.03	41.56	1.47	0.02	43.09
Africa	2.17	10.32	0.53	0.13	13.15
Asia	4.74	3.70	0.89	0.27	9.60
Japan	0.55	0.03	0.11	0.39	1.08
Australia	2.10	0.92	0.20	0.12	3.34
USSR	14.62	20.79	11.60	0.49	47.55
Eastern Europe	2.35	0.87	1.88	0.09	13.76
China	13.54	3.39	0.13	0.13	17.20
Total	76.05	116.23	47.76	6.38	246.4

Source: McRae and Dudos (1977).
*246.4 quads/year $= 5.73 \times 10^{15}$ tons ($= 45.6 \times 10^{11}$ bbl) of oil/year energy
equivalent; and 10^6 tons oil $= 0.043$ quads.

and Makarenko, 1975), Europe (Cermak and Rybach, 1979), Canada (Glass, 1977), and the United States (White and Williams, 1976; and Muffler, 1979). Additionally, some countries or regions have invested in extensive exploration efforts and, therefore, have developed detailed evaluations of the hydrothermal production potential. Thus, they have produced scenarios for electric power production development. Such evaluations concentrate on hot fluid production and have relied strongly on known technology. These resource estimates are closely tied to well-established production experience and the established potentially producing fields. The extent of these areas are often evaluated by geophysical and geochemical surveys.

Most instructive are the geothermal resource estimates for the two nations that are currently the largest producers of geothermal electric power, i.e., Italy and the United States. The two cases are quite different, however, because the resource assessment and development in Italy is completely controlled by the federal government. In the United States the development is the province of private industry, but the resource assessment has been done by a U.S. government agency, the Geological Survey of the Department of Interior. In both nations the hydrothermal portion of resource base estimates are projected to resources. These two assessments are reviewed briefly here because it is predicted that the methods used, or refinements of these techniques, will become much more widely used in the future.

U.S. Geothermal Resource Assessments

The studies by White and Williams (1975) and Muffler (1979) contain the most exhaustive and detailed national assessment available at this time. The assessment defined four categories of geothermal resources as shown in Table 2-10. For the identified hydrothermal systems, the known steam (3) and hot-water (215) systems in the U.S. were evaluated and the total heat stored was estimated. Tabulated as to heat content and frequency distribution relative to estimated reservoir temperature range, the same data (Table 2-10) are depicted on Figure 2-11. An identified, accessible hydrothermal resource base of 1,650 quads is estimated and the resource evaluation indicates that perhaps four times this amount (8,000 quads) of hydrothermally stored heat may exist if presently unidentified hydrothermal reservoirs are included. Figure 2-12 summarizes the U.S. Hydrothermal resource assessment. The reduction of the hydrothermal resource base evaluation to resources and reserve projections is indicated in Table 2-10 and depicted in Figure 2-12.

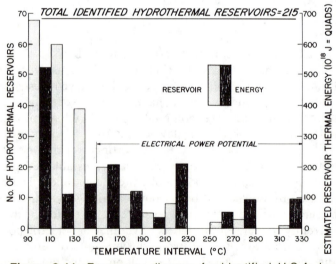

Figure 2-11. Frequency diagram for identified U.S.A. hydrothermal resources and estimated energy content vs. temperature range.

Table 2-10
Summary of U.S. Geothermal Resource Estimates
(USGS, 1978)

Type and category of resource/reservoir	Accessible resource base (quads)	Heat recovery factor	Resource (quads)	Electricity conversion factor	Potential electricity production GW y(e)	Potential direct use (quads)
1. Conduction environments (to 10-km depth)	33 × 10⁶	?	Very large	**	Very large	Large
2. Igneous related: (to 10-km depth)						
Evaluated	101,000					
Unevaluated	>900,000					
Total (59)	>1,000,000	?	Large	**	Large	Large
3. Geopressured: (on & offshore, fluid only, to 6.86-km depth)						
Thermal energy	107,000	*	270-2800	0.08	726-7,526	–
Methane energy	63,000	*	158-1640	–	–	158-1640§
Total	170,000	*	430-4400	–	–	–
4. Hydrothermal (to 3-km depth) >150°C						
Identified (52)	950	.25	237	†	690	–
Unidentified	2800-4900	.25	700-1230	†	2160-3810	–
90°C - 150°C		.25	176	–	–	–
Identified (163)	700	.25	165	–	–	42
Unidentified	5200-3100	.25	130-770	–	–	184-310
Identified	1650		400	†	690	
Unidentified	8000		2000	†	2160-3870	
Total	9650		2400		2850-4500	

  * Based upon estimate of production of fluids only and recovery scenario.
 ** Should be directly analogous to hydrothermal resource conversion efficiencies.
  † Assumed linear 0.07 to 0.13 from 150°C to 325°C.
 †† Based upon 3.73×10^7 J/m³ energy content for methane.
  § Use factor is 0.24.

The crustal storage (hot dry rock) resource base is divided into two subcategories: (1) resource associated with the identified major heat flow provinces, i.e., the broad normal conductive heat flow gradients typified as Eastern, Basin and Range, and Sierra Nevada regions; and (2) heat stored near 59 volcanic systems (termed igneous-related systems). Two maps of the heat flow data for the U.S. are shown in Figures 2-13 and 2-14. In Figure 2-13, heat flow of 19 provinces is derived from heat flow data and geological and tectonic evaluations of the 50 states of the U.S.

Figure 2-14 records geothermal gradients for the U.S. developed from temperature data from oil and gas wells. The geopressured (onshore) regions of the U.S. Gulf Coast area were similarly evaluated (see Figure 2-15) in detail (Wallace et al., 1979).

The total U.S. resource base estimates (summarized in Table 2-10) include a regional conductive resource base of 33×10^6 quads that is some three thousand times greater than the hydrothermal resource base. The methodology of the USGS approach treats the localized heat deposits of the identified (and undiscovered) hydrothermal systems, and the volcano centers, as "hotspots" superimposed on the regional conduction provinces.

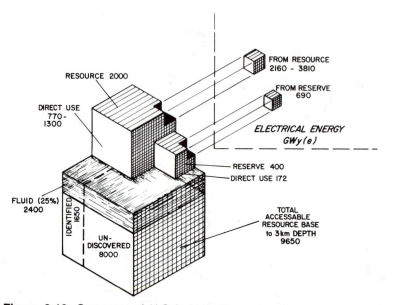

Figure 2-12. Summary of U.S.A. hydrothermal resource assessment; values are in quads. Reserve has been interpreted to be the identified reservoirs.

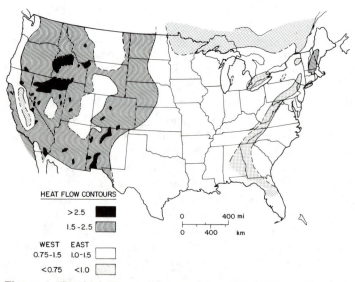

Figure 2-13. Heat flow provinces of the U.S.A. The units are in HFU (1 HFU = 41.8 mW/m²). Based upon USGS evaluations of heat flow data.

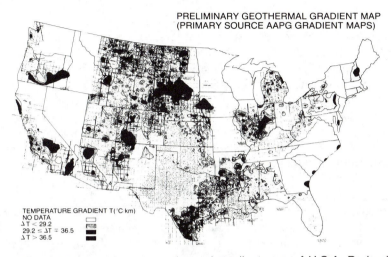

PRELIMINARY GEOTHERMAL GRADIENT MAP
(PRIMARY SOURCE AAPG GRADIENT MAPS)

TEMPERATURE GRADIENT T(°C km)
NO DATA
$\Delta T < 29.2$
$29.2 \leq \Delta T = 36.5$
$\Delta T > 36.5$

Figure 2-14. Preliminary geothermal gradient map of U.S.A. Derived primarily from well log data. (Courtesy of The American Association of Petroleum Geologists.)

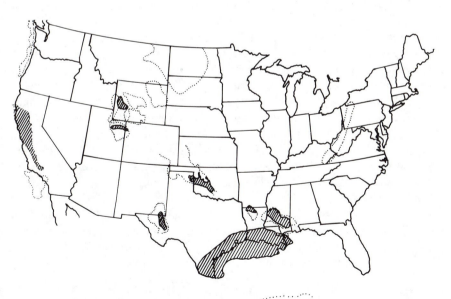

APPROXIMATE OUTLINE OF BASIN

APPROXIMATE EXTENT OF GEOPRESSURED AREA

Figure 2-15. Geopressured resource areas of the U.S.A. An intensive evaluation of this resource is underway in the Northern Gulf of Mexico region.

Many of these identified hot spots are located on boundaries of the provinces. It is further noted that the hot dry rock resource (contained in Category 4 of Table 2-10) omits consideration of the possible existence of hot dry rock types of reservoirs that may be found in association with many hydrothermal systems. Such a consideration would seem to be in analogy to resources estimated when a "dry" system is associated with one of the 59 volcanic centers. This latter circumstance is indeed the case for the U.S. research and development effort to develop the hot dry rock technology on the west flank of the Valles Caldera (New Mexico). In this instance a hydrothermal field is under active development within the caldera (Dondanville, 1978), whereas the hot dry rock resource has been found (Cummins et al., 1979) on the outer rim of this inactive volcano's caldera.

In the U.S., "dusters" (dry wells) were present in nearly all hydrothermal fields that have been drilled. These wells are sometimes found within the established hydrothermal reservoir limits. Presumably the wells are dry because they failed to intersect any significant extent of the fracture permeability. Such dry wells are excellent candidates for well stimulation projects (Nicholson, 1979). Other hot, dry wells have been drilled outside the margins of most active hydrothermal reservoirs, which indicates that hot dry rock reservoirs are associated with many hydrothermal systems. These hot, step-out hydrothermal "dusters", therefore, may in fact prove to be valuable hot, dry rock "discovery wells."

The information presented in Table 2-10 provides some insight into the relative magnitudes of the U.S. resource base estimates: the energy content of the hydrothermal convection systems is estimated to be about 0.04% and the igneous related resources to be 1.2% of the total heat calculated to be stored beneath the U.S. Figure 2-16 compares the geothermal resource base to the U.S. oil and coal resources and shows that the heat content of the U.S. to a depth of 10 km is probably equal to the heat energy expected to be produced from the U.S. oil and coal resources. The resource base calculations summarized in Table 2-10 can be compared to an average type of calculation using Equation 2-1. Inasmuch as the land area of the U.S. is equal to $9.36 \times 10^6 \, \text{km}^2$, $\overline{Q}_{US} = 26 \times 10^6$ quads. The value of 32×10^6 quads was calculated by the USGS, with the difference most likely accounted for by differences in the assumed values for rock properties. It is also instructive to compute the high-grade heat available, i.e., that above 150°C, below a depth of 4.5 km in a region with $\nabla \overline{T} = 25°\text{C/km}$. Thus,

$$\overline{Q} > 150°\text{C} = 2.0 \times 10^{-3} \, A_{US} \int_{4.5}^{10} [25Z + 15] \, dZ \qquad (2\text{-}6)$$

$$= 1.1 \times A_{US} = 10 \times 10^6 \text{ quads}$$

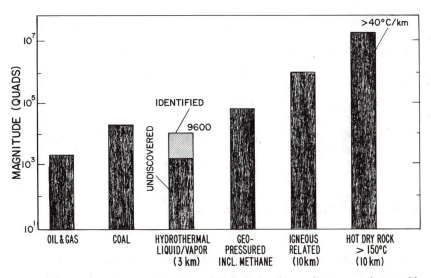

Figure 2-16. Comparison of estimated U.S.A. geothermal resource base with those of oil, gas, and coal.

which is a convenient bench mark for consideration of potential applications of the hot dry rock geothermal resource base to electric power production.

The U.S. Electric Power Research Institute (EPRI) has used the USGS 1974 evaluation (White and Williams, 1975) in an attempt to reflect the U.S. utility industry view of geothermal energy development possibilities (Anon., 1977). A summary of this interpretation is given in Figure 2-17. The EPRI analysis contains a scenario between the then current reserves (produced as electricity) of 502 MW centuries (e) and the interpretation of the USGS estimate of the resource base. The indicated total resource base stated as a potential installed capacity of 450,000 MW centuries is an output energy equivalent to about 1.4×10^3 quads, or using the USGS recoverability and conversion factors (i.e., 2 to 5%) a total heat-in-place availability ranging from 30,000 to 50,000 quads is implied.

In performing the indicated U.S. geothermal resource assessment, the USGS (White and Williams, 1975) also included arguments concerning recoverability and conversion efficiency. In these latter evaluations the USGS was constrained by a lack of historical field data. (Notable exceptions that have received intense study by the USGS are two igneous systems in the United States: Long Valley, California — Muffler and Williams, 1976; and Yellowstone, Wyoming — Smith et al., 1977.) The values derived assumed that a reservoir recovery factor and an electrical conversion efficiency could be developed from estimates of the resources in the U.S. They are derivable, however, from the experience with hydrothermal development only. The evaluation by Muffler (1978) omits this approach and concentrates on resource base and resource estimates. Energy development in the U.S. is largely a private sector enterprise and thus exploration data are usually considered to be proprietary information.

The situation is quite different in Italy where the Italian government is in control of all aspects of geothermal development. Italy has invested a large effort in geological and geophysical studies of currently producing hydrothermal fields as well as prospective areas.

Italian Geothermal Resource Evaluation

A second notable, intensive, regional geothermal resource assessment has been completed for the hydrothermal resources of central and southern Tuscany (Italy). (See Figure 2-18.) Italy had the second largest installed electric power capacity in the world, i.e., 420 MW(e) in 1978. The region of Italy studied (Cataldi et al., 1978) generally lies between Pisa in the north and about halfway to Rome in the south — a region about 150 km long (north to south) and 75 km wide. This relatively small region contains nearly all of Italy's currently producing fields and thoroughly explored hydrothermal areas. The authors did not attempt to assess all of Italy's geothermal resources. To place this area of Italy in a regional perspective, Figure 2-19 records the regional conduction heat flow (at a scale of resolution of about 1/2° longitude and latitude) and shows that regionally this part of western Europe has a maximum regional heat flow of ~110 mW/m², a value about twice the continental average. Figure 2-20 records local convection heat flow measured within the region and shows that the local "hot spots" are small scale effects that are superimposed on the regional heat flow. (This is in accord with the assessment of U.S. resources in which the igneous related and hydrothermal anomalies are superimposed on the conductive gradient of the regions or heat flow provinces identified by the USGS.)

The methodology used in this detailed regional evaluation of the steam and hot water fields (Muffler and Cataldi, 1978) involved calculation of the total volume of hot fluids and rocks expected to be in place down to a depth of 3 km at temperatures above 60°C, and for those local areas where the subsurface porosity was evaluated to be enough to contain significant fluid (>1%). The authors worked from 1:200,000 scale compilations of all available data that included drillhole, geochemical, geological, gravity, resistivity, and thermal gradient. The assessment results, as shown in Figure 2-21, apply

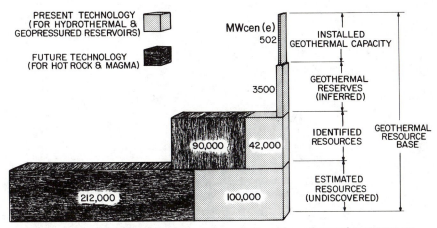

Figure 2-17. U.S.A. utility industry interpretation of geothermal resource estimates based on 1975 U.S.G.S. estimates. (After White and Muffler, 1975.)

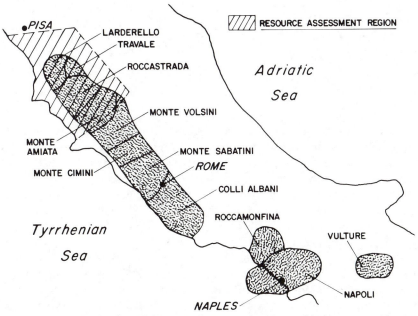

Figure 2-18. Map of geothermal regions of Italy.

Figure 2-19. Conductive heat flow contours of Italy. (After Loddo and Mongelli, 1979).

even in the restricted volume (to a depth of 3 km) of the region considered, about 557 quads are stored in the rocks and fluids of the reservoirs studied. This heat is stored in the low-porosity rocks below the two geologic zones (but above 3 km) that are considered to be the major sources of fluids. If the total heat contained beneath the 8,661-km² region under study was computed using an estimated average heat flow of 75 mW/m² and integrating to 10 km with $\nabla T = 40°C/km$, then the regional heat content is calculated to be $\overline{Q}_A \simeq 35 \times 10^3$ quads.

The authors further evaluated the total "hydrothermal resource" by considering the direct use of heat in the fluid >60°C, estimated porosities and a recovery factor. The portion of this resource having temperatures >130°C was designated as that fraction of the hydrothermal resource that is useful for electric power production. Further evaluation was done via economic arguments and a reservoir recovery factor (introduced as a function of

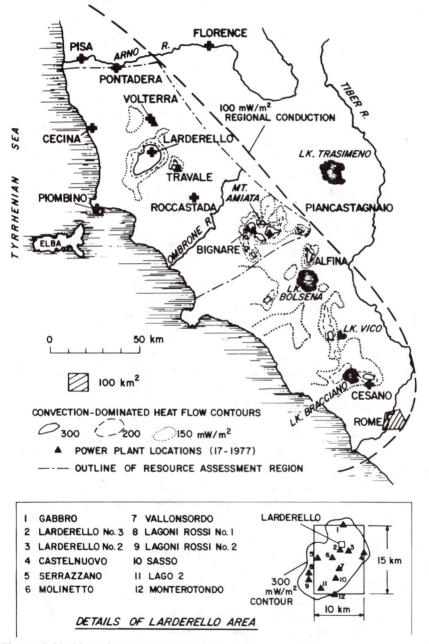

Figure 2-20. Map of local hydrothermal heat flow anomalies in the hydrothermal regions of Italy. Locations of power plants are also shown.

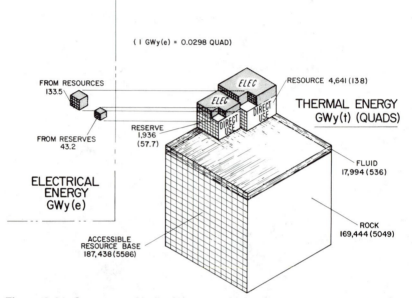

(I GWy(e) = 0.0298 QUAD)

FROM RESOURCES
133.5

FROM RESERVES
43.2

RESERVE
1,936
(57.7)

ELEC

ELEC

ELEC

DIRECT USE

DIRECT USE

RESOURCE 4,641 (138)

THERMAL ENERGY
GWy(t) (QUADS)

FLUID
17,994 (536)

ELECTRICAL
ENERGY
GWy(e)

ACCESSIBLE
RESOURCE BASE
187,438 (5586)

ROCK
169,444 (5049)

Figure 2-21. Summary of Italian resource assessment to a depth of 3 km for a region shown in Figure 2-18. Units are in GWy(t) and quads.

depth) and the fluid reserves were tabulated both for temperatures >60°C and >130°C. Finally, those resources and fluid reserves having temperatures >130°C were projected by consideration of historical data for the region and use of a dual-flash electric power plant system (Ceron, 1976). The potential for future electricity development was thus calculated, as shown in Figure 2-21, to be about 190 quads, and should be able to supply 1,800 MW(e) centuries of installed capacity.

This assessment of resource base, resource and projected reserves is clearly a valuable evaluation for the consideration and planning of near-term exploitation of Italian geothermal resources in the region studied. It might be a *modus operandi* and analysis methodology that will be followed by commercial firms interested in near-term investments in geothermal (hydrothermal) resources. From an overall resource base evaluation standpoint, however, this regional assessment essentially neglects the hot, dry rock resource. Some typical and valuable data from the studies of the Italian hydrothermal resources are summarized below:

1. Maximum localized heat flow values ≈350 mW/m²
2. Maximum surface thermal gradient 400°C/km
3. Total wells drilled = 623
4. Producing wells drilled = 398

5. Number of power plants = 17.
6. Total electrical capacity (1977) = 420.6 MW(e).
7. Drilled areas: Larderello = 200 km²; Travale = 10 km².
8. Producing areas: Larderello = 60 km²; Travale = 2 km².
9. Average production density: Larderello = 362/60 = 6 MW(e)/km²; Travale = 18/2 = 9 MW(e)/km².

It is also important to record the major subjects and problems presented by the authors as needing continuing investigation to aid in future regional assessments of hydrothermal resources:

1. Determine reservoir energy recovery factors as functions of temperature and effective porosity.
2. Evaluate fluid recharge and heat resupply.
3. Conduct analyses and experimentation to determine the extent to which a recovery factor can be enhanced by stimulation and by use of "confined circulation loops", i.e., production sweep-through-flow of a close array of reinjection wells.

These studies will require the performance of significant well and field production tests, but the results of such projects will contribute to the accuracy of geothermal resource estimates.

Finally, it is instructive to compare the resource assessments for the entire U.S. and the regional evaluation of the Italian hydrothermal resources. This comparison is summarized in Table 2-11. The most striking difference is that the small explored hydrothermal region of Italy (~12,500 km²) is estimated to have a resource of nearly 200 quads whereas the identified hydrothermal systems in the U.S. are estimated to contain only 1,650 quads, for a region some 10^3 times larger. Clearly, Italy has an extremely rich, high-grade geothermal resource, and its development for an additional 1,700 MW(e) to attain 3,500 MW(e) of installed capacity could significantly aid the Italian energy supply picture.

Rather than proceed from the above worldwide and regional resource base evaluations to future reserve estimates, the strategy followed here is to consider the current and near-term reserve picture throughout the world and then to discuss projected resources and development plans.

Estimation of Reserves and Resources

Electric Power Production

Worldwide geothermal electric power production is presently restricted to hydrothermal reservoirs. There are a total of 19 producing fields with an

Table 2-11
Comparison (1978) of U.S.A. and Italian Regional Resource Evaluation
(Energy in quads, except as indicated)

Region	Area (10^3 km²)	Identified hydrothermal* resource (3-km depth)	Crustal stored heat (1-km depth)	Installed electric power capacity 1978 (MW(e))
Total U.S.A	9.36×10^3	1650	34,000,000	502
Italy	12.5	196	35,000	421

*Temperatures >90°C were used for U.S.A. and 60°C for the Italian evaluation.

installed capacity of 1,447 MW(e) (1978), including four steam fields. Additions anticipated, as represented by orders for power plants and plants under construction, should add about 1,000 MW(e) capacity through 1981. It is difficult to determine if these projects are all completely successful economic ventures (see Chapter 10) because most of them are government sponsored. It appears, however, that nearly all of these ventures are providing economic electric power at competitive costs. It is generally reported that these 19 facilities are producing returns to the organizations that have invested in them. The total installed power production capacity in 1980 was expected to be about 2,400 MW(e). This status is summarized in Table 2-12 for 26 electric power projects that tap demonstrated high-grade hydrothermal *reserves* (see Appendix A at the end of this chapter).

Current Electric Power Production

These power plants and the projected construction plans as summarized in Table 2-12 show a distinct acceleration in growth following the 1973 oil embargo. The new power plants initiated prior to 1978 and those under construction and planned for completion in the 1979-1981 time period show that about 70% increase is expected to occur between 1978 and 1981. The historical growth trend of worldwide geothermal electric power production is depicted in Figure 2-22.

To place these geothermal electric power production capacity values in perspective, Table 2-13 compares the nine countries having geothermal power plants (projected capacity 1981) to the total electric power produced, and the thermal and nuclear plants and hydraulic (dam) contributions. As can be seen, at present geothermal capacity constitutes only a small percent-

Table 2-12
Status of Worldwide Geothermal Electric Power Plant
Capacities — 1978*

Field name and country	Present capacity MW(e)	Date of initial operation	Under construction, on-line during 1979-1980 period
1. Larderello** (Italy)	380.6	1904	
2. Wairaki (New Zealand)	192	1958	
3. Pathe (Mexico)	3	1958	
4. Geysers** (United States)	502	1960	461
5. Matsakawa** (Japan)	22	1963	
6. Monte Amiata** (Italy)	22	1967	
7. Otake (Japan)	13	1967	
8. Pauzhet (USSR)	6	1967	
9. Namafjall (Iceland)	3	1963	
10. Kawerau (New Zealand)	10	1969	
11. Tiwi (Philippines)	10	1969	110
12. Cerro Prieto (Mexico)	75	1973	
13. Onikobe (Japan)	12.5	1975	25
14. Travale (Italy)	18	1973	
15. Onuma (Japan)	10	1974	
16. Ahuachapan (El Salvador)	60	1975	30
17. Hatchobaru (Japan)	23	1977	50
18. Kakkonda (Japan)	50	1978	
19. Krafla (Iceland)	35	1978	35
20. East Mesa (United States)	0	–	58
21. Brawley (United States)	0	–	10
22. Raft River (United States)	0	–	3†
23. Heber (United States)	0	–	50
24. Roosevelt H. S. (United States)	0	–	50
25. Bulalo (Philippines)	0	–	55
26. Mori (Japan)	0	–	50
Totals	1447.1		987

1447.1 + 987 (68% of 1447.1) = 2434.1

*References for Table 2-12 are presented in Appendix D.

**Steam (vapor-dominated reservoirs; remaining are hot-water systems.)

†Several small plants which are in the planning or construction phase (Bresse, 1978), have not been included here.

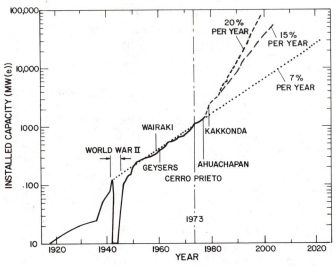

Figure 2-22. Worldwide geothermal electric power production.

Table 2-13
Comparison of Geothermal Power Plant Capacity and Electricity Produced from Other Major Fuel Sources

Country	Projected installed geothermal Capacity — 1980 MW(e)	Total electric power production — 1975* mw(e)			
		Total	Thermal	Hydro	Nuclear
USA	1,134	524,270	418,483	66,844	38,948
Italy	420	43,305	25,219	17,416	670
Japan	205.5	112,285	80,763	24,907	6,615
New Zealand	202	4,901	1,238	3,663	0
Philippines	175	3,019	1,969	1,050	0
El Salvador	90	306	198	108	0
Mexico	78	11,052	6,849	4,203	0
Iceland	73	514	120	394	0
USSR	6	217,484	171,369	40,515	5,600
Totals	2,383.5	917,136			

*Source: *World Energy Supplies,* United Nations, J-20, Dept. Econ. and Social Affairs, NY, 1977, pp. 166-177 (Anon. 1977).

age of the totals (except for El Salvador). If the higher (~20% rate of growth projected in Figure 2-22 occurs, however, then geothermal resources will contribute an increasing proportion of the world's electric power production.

The historical data indicate that prior to 1958 only Italy had any significant development. In 1963, there was about 540 MW(e) of installed capacity represented by the Larderello (Italy) and Wairaki (New Zealand) field production. The sharp decrease at 1942 shown in Figure 2-22 was caused by disruption of production in Italy due to World War II.

From about 1950 to 1979, worldwide geothermal development experienced about a 7% growth rate. During that period, The Geysers field in the United States, several fields in Japan, the Mexican development at Cerro Prieto (1973), projects in El Salvador, and the new Italian steam field at Monte Amiata were brought on line and the total capacity in 1978 was 1,450 MW(e). The planned new capacity, as represented by recorded planned projects, has put an additional 987 MW(e) on line by 1980. Thus, the pre-1973 rate of about a 7% per year growth rate, that implies a doubling every 10 years, has increased to a growth rate ranging from 15% to 20% per year leading to a possible doubling in four to five years. Projected on this basis, by 1985, worldwide capacity may reach 4,000 MW(e) and, possibly, ten times that by the end of the millenium. The reason for this acceleration is the widespread recognition that:

1. Geothermal electric plant sizes of less than 50 MW(e) are practical, and thus projects with relatively small capital outlays are possible.

2. The time to complete a power plant project can be as short as three years.

3. Ownership of a geothermal field offers an important development element in a small nation's economic growth because the geothermal resource can only be transported as electric power and, therefore, offers an opportunity for value-added activities via local industrialization.

4. A growing base of information on successful geothermal projects.

5. The rapidly increasing prices and perceived unreliability of imported fuel sources.

Such quantitative projections are hazardous at best, but it seems likely that the current steady rise of crude oil prices will maintain a pressure on the geothermal resource development rate that has apparently been in force since 1973.

Direct Uses

In addition to the use of geothermal resources for electric power production, an even larger amount is used directly for heating, cooling, and a wide variety of agricultural and industrial applications.

There is a considerable difficulty in recording the present direct use of the lower-temperature geothermal waters. The projects are often quite small and are completed without much publicity. A tabulation by Howard (1975) has summarized all the major uses as determined by a recent worldwide questionnaire and an international study (Roberts, 1978). A summary of these worldwide direct uses is presented in Table 2-14 and indicates that the current (1978) total use was estimated to range from 1,800 to 7,000 MW(t) and, therefore, may be greater than the world's geothermal electric power production.

Direct uses will probably grow much more rapidly than electric power uses of the world's high-grade geothermal resources because:

1. The more numerous lower temperature reservoirs can be used.
2. Lower capital costs are involved.
3. Efficiency of use is higher than conversion to electric power.
4. Uses of a wide range are already demonstrated to be valid commercial ventures.

The potential extent of such uses is illustrated in Figure 2-23 where a typical set of applications is matched against the approximate temperature range of use.

In most industrialized nations, a great portion of energy usage is at lower temperatures even though the original source of heat is via the burning of coal, gas, or oil at much higher temperatures. The end use of spectrum in the U.S. shown in Figure 2-24 is typical. The above circumstances would seem to argue for continued advances in direct uses around the world. Several conferences (Anon., 1978; Lienau and Lund, 1974; Lund, 1978) have stressed this potential. Current development in relation to the U.S. economic conditions (Anon., 1978) have indicated the vast potential for such energy substitution in low-grade heat applications. The most notable example of direct use is the extensive space heating facilities in Iceland (Einarsson, 1973). The lesson learned from the very successful example in Iceland for areas of the world that must depend on imported fuels for space heating is clear.

There does not appear to be any quantitative method available to predict or project the growth or worldwide availability of low-temperature (90° to 150°C) reservoirs. As shown in Figure 2-25, the "thermal waters," i.e.,

Table 2-14
Present Direct Use of Geothermal Heat*

Region	Power (J/d)	Primary uses
North America	1.5×10^{12}	District heating, industrial processes, green houses
South America	Small	Space heating, desalinization
Central America	0.9 to 1.7×10^{13}	Salt production, coffee drying
Western Europe (incl. Iceland)	4.2×10^{13}	District heating, greenhouses, industrial processing, animal husbandry, bathing
Eastern Europe	5.3 to 51.0×10^{13}	District heating, industry, greenhouses, agriculture
Asia	1.7×10^{10}	Greenhouses
Pacific Islands	4.2×10^{13}	Paper processing, district heating, fish culture, farming, greenhouses

TOTAL: 1.15-6.1 $\times 10^{14}$ J/d
or
1,800 to 7,000 MW(t)

*Summary of the worldwide low-grade geothermal reserves, 1978.

warm aquifers in the major sedimentary basins, contain energy resources that are probably far greater than the energy contained in all the petroleum resources of the earth. The thermal water resource is generally accessible through drilling deep into areas characterized by normal temperature gradients, i.e., about 25°C/km.

Evaluation of Drilled Fields and Areas

A method for characterizing the potential for high-grade hydrothermal areas has been developed by McNitt (1978). This methodology requires a literature review to determine the size of the hydrothermal reservoirs that have at least one wellbore drilled into the fluid producing formation. This size is taken to be the projected surface area (km²) enclosing the subsurface 180°C isotherm. The review starts with the 19 fields listed in Table 2-12 that are the high-grade worldwide *producing hydrothermal reserves*. To establish some idea of the future economic component of the resource, one can look at the additional identified and *drilled* high-grade reservoirs. These are areas that have been drilled but are not yet producing electric power, however, enough data or information exists in the literature to estimate the reservoir size by the above definitions. *(text continued on page 98)*

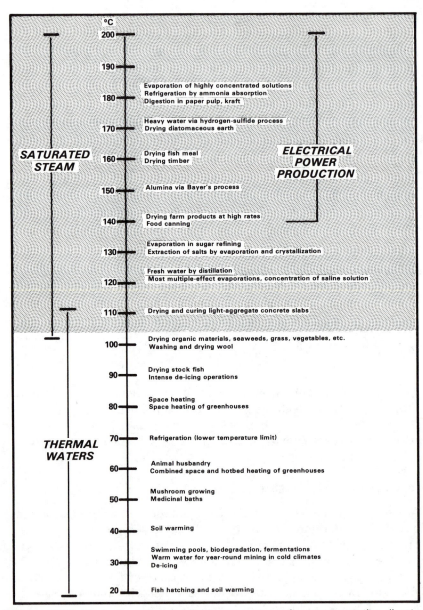

Figure 2-23. Typical geothermal fluid temperatures for representative direct-use applications.

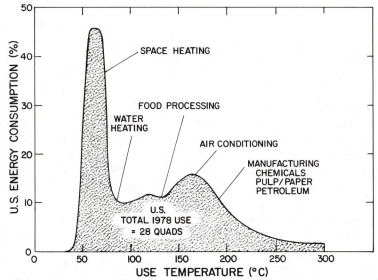

Figure 2-24. Percentage distribution spectrum of U.S.A. direct use of thermal energy in 1978.

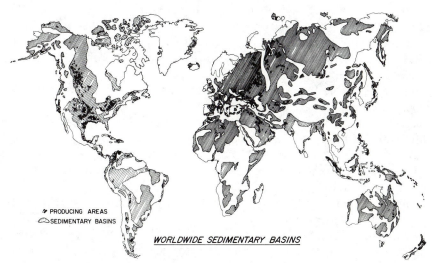

Figure 2-25. World map of major sedimentary basins and oil fields.

(text continued from page 95)

enough data or information exists in the literature to estimate the reservoir size by the above definitions.

In the terminology suggested by Muffler and Cataldi (1978) (see Appendix A) these additional drilled areas might be classified as *indicated* (drilled) *economic* (>180°C) *hydrothermal resources,* or *demonstrated* (measured and indicated) *high-grade hydrothermal reserves.* This designation is perhaps not quite as definitive as the term "proven reserves," used in petroleum development, primarily due to a lack of wider geothermal production experience and current limitations in geothermal reservoir engineering methodology. It is also a fact that no producing "field" (with the possible exception of Wairaki in New Zealand) has yet been drilled to a sufficient extent to define the boundaries of the field.

The literature search yielded 47 fields and areas that have been drilled and for which sufficient data were available to provide a reasonably accurate estimate of the area of the reservoir (Table 2-15). Five of the producing fields listed in Table 2-12 did not have sufficient data recorded in the literature for this determination, so the remaining 14 plus 33 additional fields and areas represent the current "worldwide identified potential high-grade hydrothermal reserves." For each, Table 2-15 describes reservoir size, reservoir "rock type," heat source categorization, power production potential from the wells, and the literature (Appendix D) used to evaluate the reservoir size.

The reservoir sizes have been ranked in magnitude and their cumulative percentage was determined using the usual approach in the evaluation of data in terms of a log normal frequency distribution (McCray, 1975). This total population of fields and areas is plotted on a log vs. probability graph in Figure 2-26 where sedimentary and volcanic rock type reservoir categories suggested in Table 2-15 have been distinguished. The two reservoir rock types were categorized as sedimentary (S) and volcanic (V) in Table 2-15, as judged from the reported reservoir lithology. It appears that a bimodal distribution composed of two straight line trends has resulted. As shown in Figure 2-27, this trend is confirmed when separating the data into the volcanic and sedimentary reservoir rock types. Thus, there is the possibility of the existence of two populations.

The log normal distribution has been used in geologic studies to characterize a variety of size variations in sedimentary concentrations and geochemical accumulations (Hartbough et al., 1975; and David, 1977). As shown in Figure 2-27, and further summarized as histograms in Figure 2-28, it appears that the size of high-grade hydrothermal geothermal areas may prove to follow log-normal distribution. The properties of the two distributions are summarized in Table 2-16. At this stage, the small data set available is only suggestive that a log-normal distribution may exist. If the bimodal population is real and not a feature of the small sample or biased or inaccurate data,

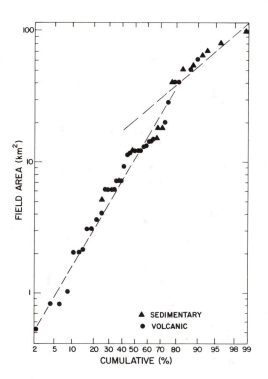

Figure 2-26. Log-normal probability plot of high-grade hydrothermal field sizes for total population of various reservoir types. Bimodal distributions are inferred from the data. The break in slope occurs for sizes between 30 and 40 km². Percentage of fields and areas "less than" a given size are plotted. (Lines are visual best fits.)

then it is difficult to explain except, possibly, by inherent differences in reservoir rocks suggested here or the regional fluid recharge characteristics. As suggested by McNitt (1978) other factors that might control reservoir area or size (such as heat source size, distance from heat source to reservoir, geometry of subsurface structure dominating fluid flow, pore geometry, and permeability characteristics of reservoir rock, age, and evolution of the system) would be anticipated to be continuously distributed variables, and thus should not give rise to the bimodal population that is observed in Figure 2-26.

The log-normal distribution has a positive skew with a tail of small counts of large values and a relative high frequency of smaller area fields. These characteristics are present in the case of these 47 fields and areas. Such log-normal plots in petroleum reservoir and ore body exploration and development are predictive. When such distributions as shown in Figure 2-28 are used to project a forecast, the probabilities of the discovery of a given size of field for each reservoir rock type can be read directly from the curves if it indeed is verified to *fall on these same two geological trends*. Thus, if it is assumed that these are found to be essentially "universal" curves, then, for example, the probability is about 70% that a sedimentary field (randomly

Table 2-15
Ranking of Reservoir Sizes for High-Grade Drilled Hydrothermal Fields and Areas*
(Demonstrated High-Grade Hydrothermal Reserves)

Item	Name of field or area	Location	Field area (km²)	Reservoir rock type V or S	Heat source type	Area of productive wells (km²)	Total Capacity of productive wells MW(e)	Power capacity MW(e)/km²	Cum.%	Partitioned Cum.% V	Partitioned Cum.% S	Ref††
1.	Tahuangtusi	Taiwan	0.50	V	B(A?)	—	—	—	2.083	2.857	—	14
2.	Bouillante	Guadeloupe	0.80	V	B	—	—	—	4.167	5.714	—	16
3.	Matsao	Taiwan	0.80	V	B	—	—	—	6.250	8.571	—	14
4.	Reykjanes	Iceland	1.0	V	C	—	—	—	8.3333	11.428	—	6
5.	Krafla**	Iceland	2.0	V	C	—	—	—	10.417	14.286	—	32
6.	Takenoya	Japan	2.0	V	B(A?)	—	—	—	12.500	17.143	—	45
7.	Matsukawa**	Japan	2.1	V	B	0.35	20	57	14.583	20.00	—	38
8.	Momotombo	Nicaragua	3.0	V	B	0.35	20	57	16.667	22.86	—	7
9.	Onikobe*	Japan	3.0	V	B	0.40	25	62	18.750	25.71	—	38,44
10.	Paujetskaya	USSR	3.5	V	B(A?)	0.30	5	17	20.833	28.57	—	17,41,42
11.	Svartsengi	Iceland	4.0	V	C	—	—	—	22.917	31.43	—	33
12.	ΔBangore	Italy	5.0	S	B(A?)	2.0	7	3.5	25.000	—	7.143	9,13
13.	Puna	U.S.	6.0	V	C	—	5	—	27.083	34.28	—	18
14.	Ahuachapan**	El Salvador	6.0	V	B	0.6	70	117	29.167	37.14	—	15,16,25
15.	Otaki**	Japan	6.0	V	B(A?)	0.03	11	367	31.250	40.00	—	31,45
16.	Hatchobura**	Japan	6.0	V	B(A?)	—	—	—	33.333	42.86	—	45
17.	Namafjall	Iceland	7.0	V	C	0.15	12.5	83	35.417	45.71	—	32,34
18.	ΔTravale**	Italy	7.0(10.0)	S	D(B?)	2.0	35.9	18	37.540	—	14.286	10,50
19.	Onuma*	Japan	7.0	V	B	—	—	—	39.583	48.56	—	35
20.	Kawerau**	New Zealand	9.0		A,B	0.3	13	43	41.667	51.43	—	37,48
21.	→Roosevelt	U.S.	11.2	V		4.4	50	12.5	43.750	54.29	—	51
22.	→Tiwi**	Philippines	11.5	V	B(A?)	5.0	120	24	45.833	57.143	—	49
23.	ΔKizildere	Iran	12.0	S	D	1.5	44	29	47.917	—	21.43	2,40

Table 2-15 continued

Item	Name of field or area	Location	Field area (km²)	Reservoir rock type V or S	Heat source type	Area of productive wells (km²)	Total Capacity of productive wells MW(e)	Power capacity MW(e)/km²	Cum.%	V	Partitioned Cum.% S	Ref††
24.	Brady	U.S.	12.0	V	D	–	–	–	50.000	60.00	–	43
25.	Olkaria	Kenya	12.0	V	D(B?)	0.13	8	62	52.083	62.857	–	30
26.	Wairakei**	New Zealand	12.0	V	A.B	1.5	148	99	54.167	65.71	–	21,22,48
27.	Broadlands	New Zealahd	13.0	V	A.B	2.6	165	63	56.25	68.57	–	8,29,48
28.	→Bulalo	Philippines	13.0	V	B(A?)	5.0	200	40	58.333	71.43	–	49
29.	El Tatio	Chile	14.0	V	B	0.25	16	64	60.417	74.28	–	28.
30.	Kawahkamajung	Indonesia	14.0	V	B(A?)	–	30	–	62.500	77.14	–	23
31.	→Tauhara	New Zealand	15.0	V		–	120	–	64.583	80.00	–	48
32.	ΔPiancajtagnoio	Italy	15.0	S	B(A?)	2.0	15	7.5	66.667	–	28.57	9,13
33.	ΔBrawley	U.S.	18.0	S	D.C.B	3.0	10	3.3	68.750	–	35.71	43
34.	ΔAlfina	Italy	18.0	S		12.	–	–	70.833	–	42.86	11,50
35.	Beowave	New Zealand	21.0	V	D	–	–	–	72.917	82.86	–	43
36.	Waiotapa	New Zealand	28.0	V	A.B	10.0	120	12	75.000	85.71	–	21
37.	ΔE. Mesa	U.S.	40.0	S	D.C.B	–	–	–	77.083	–	50.0	36,39
38.	Krisuvik	Iceland	40.0	V	C	–	–	–	79.167	88.57	–	3
39.	→Baca	U.S.	40.0	V	A	10.0	50	5	81.250	91.43	–	46,47
40.	ΔHeber	U.S.	50.0	S	D.C.B	15.0	40	2.7	83.333	–	57.143	43
41.	Hengill	Iceland	50.0	V	C	–	–	–	85.417	94.28	–	52?
42.	ΔSalton Sea	U.S.	54.0	S	D.C.B	–	–	–	87.500	–	64.28	33,43
43.	Long Valley	U.S.	60.0	V	A	–	–	–	89.583	97.14	–	24,27
44.	ΔCessna	Italy	65.0	S	B(?)	2.0	–	–	91.667	–	71.43	4,50
45.	ΔCerro Prieto*	Mexico	70.0	S	C.D	1.5	75	50	93.750	–	78.57	1,19
46.	ΔLarderello**	Italy	80.0(200)	S	D	60.0	264	3.3	95.833	–	85.71	12,13,50
47.	ΔGeysers**	U.S.	100.0( )	S	A.D	26.3	900	34	97.917	–	92.86	20,26

* Adapted from (McNitt, 1978).

** Previously listed in Table 2-12.

A = Large silicic magma body; B = relatively small body of intermediate to silicic composition; C = basaltic magma body; D = thin or disturbed crust in a tectonically active region.

†† References for Table 2-15 are recorded in Appendix D.

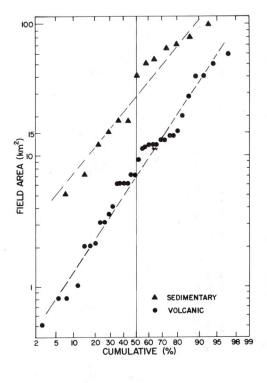

Figure 2-27. Log-normal plot of field sizes with separation of volanic and sedimentary reservoir rock types. Note: One volcanic field (*) found recently in the Azores, having an area~12 km², has been added to the distribution since the compilation of Table 2-19.

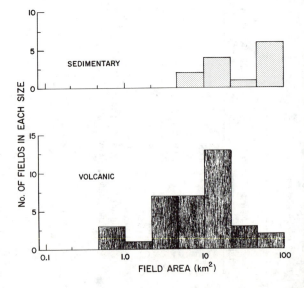

Figure 2-28. Histograms of worldwide high-grade hydrothermal field and reservoir sizes. Separate histograms are prepared for volcanic and sedimentary fields.

Table 2-16
Statistical Characteristics of High-Grade
Hydrothermal Fields and Areas

Properties of two log normal distributions in Figure 2-27.

Reservoir rock type	No. of fields or regions	Mean area, km²	68% likelihood size[a] range, km²
Volcanic	34	16	1.2 to 25.5
Sedimentary	13	50	9.6 to 89.9

Field production and 68% probability range projections

Reservoir rock type		Mean** MW(e)	68% Projected probability range or discovery size (MW(e))
Volcanic	Steam	96	7 to 113
	liquid dominated	800	52 to 1280
Sedimentary	Steam	300	60 to 720
	liquid dominated	2500	500 to 4500

*Field areas will be found that fall within this range about 70% of the time.

**Based upon data in Table 2-15, the average steam field unit areal power density is approximately 6 MW(e)/km²; and the liquid dominated fields produce about 50 MW(e)/km².

selected) will be less than 90 km² and greater than 10 km² in size* as shown in Table 2-15.

One potential use of reservoir size probabilities would be to estimate the expected electric power production capabilities of an hydrothermal area that has been surveyed by geochemical methods to establish a high-temperature subsurface temperature. As noted in Table 2-15, steam fields have a unit area productivity of about 6 MW(e)/km² (weighted average of Italian steam fields and The Geysers). The data suggest a value of about 50 MW(e)/km² for liquid-dominated fields. The use of these provisional average unit area productivities together with the log-normal derived probabilities indicates (Table 2-15) that, for example, about 70% of the discoveries in liquid-dominated fields with volcanic reservoir rocks should be found to produce from 52 to 1,280 MW(e), with a mean value of about 800 MW(e).

* As shown in Figure 2-27, the hydrothermal field recently drilled in the Azores (Meidav and Meidav, 1979) subsequent to the compilation in Table 2-15 has been plotted at the indicated reservoir size (12 km²) and falls within the 70% expectation range for volcanic fields.

Although this log-normal analysis is suggestive, it cannot be considered definitive because clearly only a limited data set is as yet available and very limited case history experience exists to test the validity of the approach. It would be possible and instructive to extend the present analysis to look at existing geophysical and geological data available for promising hydrothermal areas and, thus, to extend the projections of future capacity.

This type of study has been accomplished in New Zealand (Donaldson and Grant, 1978), is in progress in Central America (McNitt, 1978), and could be applied to the United States (see Figure 2-11). The U.S. evaluation identifies 53 hot spring areas with fluid temperatures >150°C, and if the reservoir rocks are assumed to be all volcanic, then the log-normal projection gives 42,400 MW(e). The U.S. evaluation (Muffler, 1978) used the energy-in-place method and an assumed hydrothermal field recovery factor in arriving at an estimate of 21,000 MW(e). Several large (areal size) reservoirs have been identified: e.g., Braneau-Grandview — 1,500 to 2,250 km²; Yellowstone Caldera — 375 to 1,092 km²; and the Los Banos region (where the Bulalo field is under development) south of Manila in the Philippines may have a target prospect of approximately 1,600 km². From the viewpoint of field size on a log-normal population basis, it would be expected that a few "super giant" size high-grade areas are still to be discovered and developed into producing fields.

If a reasonably uniform worldwide data set were available, then a logical step at this point would be to collect all the appropriate world geological, geochemical, and geophysical data. Then, with appropriate study, some sort of global geothermal resource projections could be attempted region by region. From the above preliminary survey of geothermal reserves, and especially in view of the range of the scale of hydrothermal reservoir sizes outlined in Table 2-15, the data should be reviewed at a linear (map) scale of 1 to 100 km. Ths situation appears to indicate that national and regional evaluations will be most valuable as guides to development and the most likely to be pursued in the near future.

Reservoir Testing, Simulation, and Size Estimate Limitations

The major analytical restriction currently imposed upon hydrothermal reservoir size evaluations is the lack of verified well and field testing and reservoir modeling methods. These reservoir engineering techniques that are applied traditionally in the development of petroleum resources are in a primitive but rapidly growing stage in geothermal energy development. There are few models available: i.e., vapor phase intergranular model, intergranular liquid flow model (Bodvarsson, 1974), and a few single plane fracture models (Bodvarsson, 1975). Most production from volcanic rock reservoirs and some sedimentary rock types (e.g., The Geysers), however, are dominated by fracture permeability and little experience and no extensively confirmed models are available that can be used with suitably developed well test procedures.*

The state-of-the-art in well testing and reservoir simulation can be judged by a review of the recent literature (Kruger and Ramey, 1975, 1976, 1977, and 1978; and Sudal, 1979), which indicates that significant progress is being made. In addition, improvements in geothermal well log interpretation techniques (Sanyal and Meidav, 1977, and Ehring et al., 1978) to provide additional insight into reservoir size and productive life is being pursued. For the purposes of the geothermal resource analyst, emphasis on establishing reservoir productive size and energy content are crucial. The majority of data developed thus far has largely relied on energy in-place (volume) evaluations (Muffler and Cataldi, 1978).

Current Status and Developments

World-Wide Developments and Plans

The years following the 1970 United Nations Geothermal Symposium (Barbier, 1970) held in Pisa, Italy, witnessed a steady expansion rate in geothermal power capacity of about 7% per year, as recorded in Figure 2-22. This growth in exploration and development was dictated by decisions and commitments made in the 1960's and reflected the power generation demands and fuel supply situation typical of the post-World War II era. This period was dominated by the increase of electric power generating capacity at The Geysers, U.S.A., from 75 to 237 MW(e) by 1972 (a 316% increment) (Koenig, 1973). The Mexican government initiated the construction of the 75 MW(e) plant at Cerro Prieto that was put in operation in 1973. There was major continued exploitation of thermal waters for space heating, agriculture, and industrial processing in Iceland, Hungary, New Zealand, and the USSR. Geothermal exploration, the necessary activity prior to development, increased in Italy, Japan, New Zealand, Iceland, the U.S.A., Indonesia, Mexico, and the Philippines. The United Nations continued support for projects in Turkey, El Salvador, Ethiopia, Nicaragua, and Kenya.

These substantial investments produced an increasing public and governmental awareness of the nature and potential of geothermal energy. They provided a basis for the surge in interest and activity that followed the events of 1973 as imported oil sharply increased in price and supplies became more

---

* One of the editors (G.V.C.) recommends the book entitled *Oil in Fractured Reservoir Rocks,* by Tkhostov, B.A., Vezirova, A.D., Vendel'shtein, B. Yu and Dobrynin, V. M. 1970, lzd. Nedra, Leningrad, 220 pp.

difficult to assure. Other energy fuel prices have subsequently advanced sharply causing industries and governments to pursue alternate or supplemental energy supplies. The effect has been especially severe on the countries or regions that have been strongly dependent on imported oil. The acceleration in worldwide geothermal development was especially evident at the 1975 United Nations Symposium where Muffler (1975) showed that the total installed electric power capacity in 1975 was 1,278.8 MW(e). (See Table 2-17.) This shows about 15% increase compared to the 1978 evaluated electric power capacity of 1,447.1 MW(e) (see Table 2-12). Fortunate indeed were those nations, regions, and institutions that have developed expertise in geothermal resource exploration and exploitation because it is evident that geothermal energy in its several forms will become a major contributor to the world's energy needs.

Brief summaries of the international status and plans for those countries, for which information and data are available (1978) in public documents and literature, are presented below. The second U.N. Symposium (Muffler, 1975) and the EPRI study (Roberts, 1978) are valuable sources; however, recent news and data come from a diverse and scattered set of government reports, newspapers, flyers, etc. Thus, the present survey may have omitted a significant development, especially in the case of exploration.

The aim of the following summaries is to provide brief information on recent national geothermal development efforts and to present the most appropriate selected references so that further details can be pursued.

North America

Canada

Canadian geothermal development is in a very preliminary state of developing an assessment of the nation's potential (Glass, 1977). It is clear, however, that western Canada is in a geothermal zone or region (Figure 2-3), which indicates that Canada has abundant resources. The Meager Mountain prospect, about 150 km north of Vancouver, B.C., has received attention as a deep-well drill site. The Prairie Provinces are considered to be a potential near-term source of hot water ($>50°C$) (Jessop, 1978). Projections for a 10 MW(t) (3.2×10^{14} J/year) supply for space heating have been made for one area. A pilot project is under way near Regina, and a double well system of the Paris (France) type is planned.

Most of the Canadian high-grade geothermal resources will be found in the western regions and will be in competition with the abundant fossil and hydraulic sources found in this area.

Table 2-17
World Geothermal Generating Capacity in MW(e)*

Country	Field	Operating	Under construction
U.S.	The Geysers	502	216
Italy	Larderello	380.6	
	Travale	15	
	Monte Amiata	22	
New Zealand	Wairakei	192	
	Kawerau	10	
Japan	Matsukawa	22	
	Otake	13	
	Onuma	10	
	Onikobe	25	
	Hatchobaru		50
Mexico	Pathe	3.5	
	Cerro Prieto	75	
El Salvador	Ahuachapan		90
Iceland	Namafjall	2.5	
	Krafla		55
Philippines	Tiwi		100
USSR	Pauzhetka	5	
	Paratunka	0.7	
Turkey	Kizildere	0.5	2.5
	Total:	1278.8	563.5

*It is instructive to compare this 1975 summary (Muffler, U.N. Symposium) with the present study (see Table 2-12).

Mexico

Mexico's principal development is at Cerro Prieto, 30 km south of the Mexico — U.S.A. border in the Lower Imperial Valley. The 75-MW(e) capacity power plant has been in operation since 1973. Construction is currently underway to build a second 75-MW(e) generating facility. The drilling of the wells for this plant began in 1972. Estimates of the ultimate capacity of the reservoir vary from 500 to 1,500 MW(e).

Geothermal drilling at Pathe (near Pachuca in the state of Hidalgo) in Mexico followed the earlier exploration there (see Figure 2-29). Wells produced only in limited quantities from what is supposed to be a fractured reservoir, and only four wells flow continuously. The installed capacity is 3.5 MW(e) from a plant established there in 1959. This reservoir is a good

I	CERRO PRIETO
2	LA PRIMAVERA
3	IXTLAN - LOS NEGRITOS
4	LOS AZUFRES
5	CUITZEO - ARARO
6	LOS HUMEROS
7	ZUNIL
8	AMATITLAN
9	CHIPILAPA
10	AHUACHAPAN
11	CHINAMECA
12	S. VICENTE
13	BERLIN
14	(HONDURAS)
15	MOMOTOMBO
16	GUANACASTE
17	CERRO PANDO
18	GUADALUPE
19	S. LUCIA
20	EL PILAR - CASANAY
21	VOLCAN RUIZ
22	(ECUADOR)
23	SALAR DE EMPEXA
24	LAGUNA COLORADA
25	PUCHULDIZA
26	EL TATIO
27	TRAPA TRAPA
28	JUJUY
29	COPAHUE

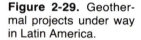

▼ EXPLOITATION
■ FEASIBILITY
● PREFEASIBILITY
▲ RECONNAISSANCE

Figure 2-29. Geothermal projects under way in Latin America.

candidate for stimulation, and a large associated hot dry rock resource may exist in the immediate locale.

Other exploration activity and some initial well drilling has proceeded in many other areas of Mexico. Alonso (1975) lists some 138 potential hydrothermal fields as shown on Table 2-18. It appears that with a large number of identified hydrothermal systems and the location of a major geothermal zone along the western regions as well as a well-established national expertise, geothermal energy has a solid growth potential in Mexico. This presumes that the new oil and gas discoveries to the south do not distract the Mexicans.

United States

A comprehensive review of U.S. geothermal development activities would require an extensive account at this time. Following the Steam Act of 1970 (implemented in 1973), many federal lands have been opened for lease to the private sector for exploration and development. Starting in 1977, the U.S. government has appropriated large amounts of funding to support research, development, and demonstration in all aspects of geothermal energy development. The formation of the U.S. Energy Research and Development Administration (ERDA), now the U.S. Department of Energy, also increased federal outlays for geothermal research and development. Further stimulation was afforded by the Energy Policy Act (October,

Table 2-18
Distribution of the Known Principal
Hydrothermal Areas of Mexico

State	Number of areas identified	Mean temperature*
Aguascalientes	5	50°C
Baja California	15	70-90°C
Chihuahua	5	60°C
Chiapas	2	70°C
Coahuila	2	40°C
Distrito Federal	1	40°C
Durango	5	60°C
Guanajuato	9	80°C
Guerrero	3	40°C
Hidalgo	6	75°C
Jalisco	16	85°C
Mexico	3	80°C
Michoacan	22	90°C
Morelos	2	60°C
Nayarit	6	70°C
Nuevo Leon	1	50°C
Oaxaca	2	50°C
Puebla	3	65°C
Queretaro	6	70°C
San Luis Potosi	3	40°C
Sinaloa	3	60°C
Tamaulipas	2	40°C
Veracruz	1	50°C
Zacatecas	2	40°C
Total	138	

*Measured surface temperatures of hot springs.

1978) that contained significant incentives for commercial development through a 22% depletion allowance, tax write-off provisions for intangible drilling costs, and an additional 10% front-end investment tax credit feature. Several states have also established laws and implemented procedures for development on state lands, and provided research and development funds.

The status of U.S. development prior to 1975 is covered by Koenig et al., (1975). At that time about 100,000 acres in California, Nevada, Oregon, and Utah were under exploration via competitive leases. In Nevada and Utah over 200,000 acres of noncompetitive leasing had been issued. Notable new hydrothermal prospects were discovered at the Valles Caldera, New Mexico; Roosevelt Hot Springs, Utah; Carson Desert, Nevada; and Heber, East Mesa, and Brawley in the Imperial Valley of California. The exploration activities in the U.S. were and are continuing to accelerate.

The three volumes of the transactions of the Geothermal Resource Council (Combs, 1977; Muffler and Yeun, 1978; and Fournier and Crosby, 1979) attest to this advance in technology, exploration, development, field production, and facilities construction. With a current capacity of 502 MW(e), the largest in the world, the Geysers has formed an example backdrop for a large number of other major U.S. energy firms to actively enter the geothermal arena. Many smaller enterprises have also been formed to exploit the promise of geothermal energy in the U.S. The extensive resource assessment efforts by the USGS (Muffler, 1978) have placed the question of potential energy developments for the U.S. in a most rational and fundamental basis. This resource assessment offers the developer a basic starting point for making judgments of the future resource potential. The results are summarized in Tables 2-10 and 2-19. This study projects the Geysers field to contain an accessible resource base of 100 quads (10^{20} J), a resource of 9.3 quads, and an expected total electrical generating capacity (for 30 years) of 1,620 MW(e). It also predicts a total U.S. hydrothermal potential of 23,000 MW(e) of electrical generating capacity and 87 quads of direct use heat energy in *identified* hydrothermal reservoirs. The undiscovered hydrothermal potential is estimated to be 95,000 to 150,000 MW(e) and 230 to 350 quads of direct use potential.

The U.S. Department of Energy (DOE) has set goals of 6,000 MW(e) capacity on-line by 1985 and 20,000 MW(e) by the year 2000. To accomplish these goals, several major projects have been established. Notable are the cost sharing demonstration power plant efforts among the DOE, the Union Oil Company and the Public Service Company of New Mexico to develop the hydrothermal reservoir at the Valles Caldera, New Mexico (Maddox and Wilbur, 1979). A major U.S. loan guarantee program (Silverman, 1978) supports all aspects of geothermal development. Major U.S. funded research and development programs to improve reservoir engineering techniques (Goransoh and Shroader, 1978), drilling and completion technology (Varnado and Stoller, 1978), well logging instrumentation (Veneruso and Stoller, 1979), low-temperature applications (<150°) and direct use demonstrations (Hornburg and Lindel, 1978), exploration technology (Salisbury et al., 1977 and Schwartz, 1978), and well log interpretation, (Sanyal et al., 1979) are providing needed cost reductions in the most critical phases of geothermal development.

Table 2-19
Resource Summary of Energies of U.S.A. Hydrothermal Convection Systems — 90°C

	Number of systems	Accessible resource base (quads)		Resource (quads)		Electrical energy (MW(e) for 30 yr)		Beneficial heat (quads)	
Identified (excluding National Parks)									
Vapor-dominated									
The Geysers	1	100	24	9.3	4.5	1630	770	–	
Hot-water (150°C)	51	850	80	210	30	21,000	3300	–	
Hot-water (90°-150°C)	163	700	110	176	55	–		42	13
Total identified	215	1650	140	400	60	23,000	3400	42	13
Undiscovered									
Vapor-dominated and hot-water (150°C)	–	2800**-4899†		700**-1230†		72,000**-127,000†		–	
Hot-water (90°-150°C)	–	5200**-3100†		1300**- 770†		–		184†-310**	
Total undiscovered		8000		2000		72,000**-127,000†		184†-310**	
Total identified (Excluding national parks) and undiscovered		9600		2400		95,000**-150,000†		230†-350**	

Source: Muffler, 1978.

*The values of means and standard deviations for accessible resource base, resource, and beneficial heat are calculated analytically. Means and standard deviations for electrical energy are obtained from Monte Carlo calculations rather than analytically because of the nonlinear dependence of available work (and thus electrical energy) on reservoir temperature. The accessible resource base in national parks, estimated to be 1,290 quads is not included in this table.

**Constant-volume assumption for undiscovered component.

†Increasing-volume assumption for undiscovered component.

Two DOE funded programs are directed at advanced reservoir extraction technological and economic evaluations. The geopressured resources (Figure 2-30) are being very actively studied in the Texas and Louisiana Gulf Coast regions (Beboot, 1978). This resource base is estimated to be 170,000 quads of thermal waters and methane. Successful attempts to assess the nature of the reservoirs and determine the recoverability and economics could provide a significant contribution to future U.S. energy needs. The efforts to define the problems and solutions include significant drilling and well testing components.

The heat extraction technology development efforts for hot, dry rock (HDR) resources are currently centered on the flanks of the Valles Caldera, New Mexico (Fenton Hill experimental site). Recent drilling, fracturing, and heat extraction experiments have successfully created (Murphy, 1979) a reservoir and demonstrated the basic technical feasibility of the HDR concepts for formation of artificial reservoirs. This initial success has encouraged the DOE to expand the research and development effort, and several other areas are under initial prospect evaluation (West and Laughlin, 1979) as potential future experimental sites. Hot, dry rock will be a valuable resource, even if only those sites associated with igneous centers are developed. The USGS estimates that the resource base is 101,000 quads of identified and 900,000 quads of undiscovered thermal energy (Tables 2-10 and 2-19).

In addition to technology-oriented projects, the U.S. government is supporting numerous efforts in the institutional, legal, environmental, market needs, and information dissemination projects. These projects may be one of the key elements in the acceleration of the growth of the U.S. geothermal industry and the near-term attainments of the goals for hydrothermal field development.

Recently, U.S. commercial activity has increased significantly. During the three-year period from 1975 to 1978, some 175 wells (Figure 2-31) were drilled in the western states, as summarized in Table 2-20 (Smith, 1977-79). The most intensive recent activity (1979) has been in northern Nevada, and consists of a considerable amount of exploratory drilling that should result in significant new field development and production drilling in the next few years.

The eastern U.S. has recently been the scene of significant exploration efforts (Costain et al., 1978; and Glover, 1979). Interest in direct use of geothermal energy in these states is increasing as the awareness of such development potential increases (Anon., 1979).

Central and South America

Central and South America are known to have extensive geothermal resources. Development would seem to be very important for the nations in these regions because it provides a native source of energy independent of imported fuel uncertainties and price escalation. Typical projects can be started with initial small capital outlay as compared to other electrical power-producing facilities. A major competitor in some of the Latin nations is hydro-generation; therefore, balanced planning must be used to assure the most effective development. The balance of water usage for irrigation and land use for agriculture are often issues to be considered in these competing developments in this region.

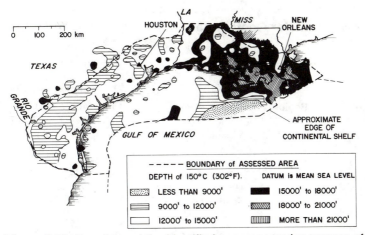

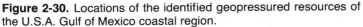

Figure 2-30. Locations of the identified geopressured resources of the U.S.A. Gulf of Mexico coastal region.

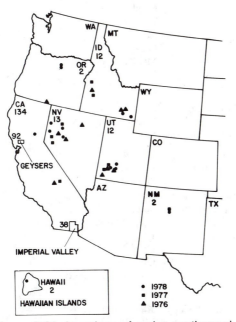

Figure 2-31. Locations of major geothermal drilling, western U.S.A., 1976-1978.

Table 2-20
Summary of Major Geothermal Drilling Activity in Western U.S.A.
(1976-1978)

State	Region	Area	Number of wells			
			1976	**1977**	**1978**	**Total**
California	Imperial Valley	Westmorland	6	1	–	7
		Brawley	2	2	4	8
		East Mesa	2	5	7	14
		Salton Sea	–	–	1	1
		Heber	6	–	–	6
		South Brawley	–	–	1	1
						37
	The Geysers	Main Geysers	18	14	15	47
		Southeast Geysers	–	13	6	19
		Northwest Geysers	–	2	1	3
		North Geysers	–	1	2	3
		Howard Hot Springs	–	1	1	2
		Borax Lake	–	–	1	1
		Middletown	3	–	1	4
		Thurston Lake	–	1	–	1
		Castle Rock	4	–	–	4
		Cloverdale	1	–	–	1
		Mt. Konaocki	1	–	–	1
		Calistoga	3	–	–	3
						89
	Mono Co.	Long Valley	1	–	–	1
	Inyo Co.	Coso Hot Springs	1	1	–	2
	N.W. California	Lassen	–	–	1	1
						4
Nevada	Churchill Co.	Desert Peak	2	–	–	2
		Stillwater	3	–	–	3
	Lander Co.	Beowawe	1	–	–	1
	Carson Sink	Soda Lake	–	1	–	1
		Allen Springs	–	1	–	1
	Pershing Co.	Rye Patch	–	1	1	2
		Dixie Valley	–	–	1	1
		Gerlach	–	–	1	1
		San Emidio Desert	–	–	1	1
						13
Oregon	Klamath Co.	Klamath Hills	1	–	–	1
		Mt. Hood	–	–	1	1
						2
Hawaii	Hawaii Island	Puna	–	–	2	2
						2

Table 2-20 continued

State	Region	Area	Number of wells			
			1976	1977	1978	Total
Idaho	Cassia Co.	Raft River	2	1	4	7
	Ada Co.	Boise	2	–	–	2
	Washington Co.	Crane Creek	–	1	–	1
	Owyhec Co.	Castle Creek	–	1	–	1
		Preston	–	–	1	1
						12
Utah	Beaver Co.	Roosevelt H.S.	3	1	1	5
		Thermo H.S.	–	–	1	2
	Millard Co.	Cove Ft.	1	1	2	4
	Iron Co.	Beryl Junction	2	–	–	2
		Lund	–	1	–	1
						14
New Mexico	Sandoval Co.	Fenton Hill	–	1	–	1
		Valles Caldera	–	–	1	1
						2
		Total	65	52	58	175

El Salvador

Exploration of the Ahuachapan (Romaguoli et al., 1975; and Urbain, 1975) field was initiated with the United Nations aid in 1970. Today, two 30 MW(e) generating units, initially in service in 1976, are on line, with a third unit of 35 MW(e) capacity being constructed in 1979. It is reported that El Salvador intends to develop at least 100 MW(e) additional geothermal power capacity by 1985. Exploration of potential fields at Chinameca, Berlin, and San Vicente is underway and drilling activity at the latter region is in progress. The El Salvador Ahuachapan project has set an excellent example of high-grade geothermal development potential for the other nations of this region.

Nicaragua

The Empresa Nacional de Luz y Fuerza (ENALUF), the National Power and Light Company, initiated the development of the Mombotombo field, 60 km west of Managua (Roberts, 1979) in 1970. Of the 19 wells drilled, 6 are productive and are estimated to be able to support a 30 MW(e) capacity plant. Prior to the 1972 earthquake and current political unrest, the nation had plans to initiate plant construction in 1978. Estimates of the field capacity range up to 800 MW(e).

There is little doubt that Nicaragua has a vast resource base of geothermal energy and could become independent of imported oil for electric power when this potential is developed. A distribution grid and hydroelectric system is already in place. The nation is continuing exploration (Goldsmith, 1979 and Teilman, 1979), with indications of the existence of a number of high-grade fields. Two areas may have been located in western Nicaragua with sizes of at least 100 km² (Anon., 1978). If political and civil conditions permit drilling and economic recovery from the December 23, 1972 earth-quake is achieved, it seems likely that the vast hydrothermal resources of Nicaragua will be confirmed and ultimately developed.

Costa Rica

Geothermal exploration is currently underway (Corrules et al., 1977) in Costa Rica in the Las Pailas Hornillis area in the northern Guanacosta province where strong heat flow areas have been delineated (Blackwell et al., 1977). Gradients up to 700°C/km have been recorded. Geochemical analyses yielded base temperatures ranging from 220 to 240°C. Exploration drilling and planning for a 100 MW(e) plant are reported to be progressing in this promising area.

Guatemala

The Guatemalan government national electric company (INDE) has explored the Moyuta region, a potential hydrothermal field located just north of the Ahuachapan field across the border in El Salvador. Simul-taneously, exploration is proceeding at two other locations, Zunil and Amatitlan, which are known to be active geothermal areas. The goal of these efforts is to find resources adequate to support a 100 MW(e) plant early in the 1980's (Roberts, 1979).

Honduras

The National Electric Authority of Honduras (ENEE) has just recently embarked on an aggressive exploration effort (Roberts, 1979). Two areas are under study: San Ignacio, northwest of the capital, Tegucigalpa, and Pavana in the southern part of the country. The objective stated is for Honduras to have a minimum of 50 MW(e) installed by 1982 and 100 MW(e) in the 1984-1985 period.

Panama

Panama is astride the major geothermal zone that is common to much of Central America and can, therefore, be presumed to have major resources. Based on geological studies, the Caldera region in Panama's Chirigui province and the Cerro Pondo region are thought to hold promise. An initial hot spring survey was reported (Merida, 1975) which indicated existence of 22 hydrothermal areas. No formal reservoir exploration efforts, however, have been initiated as yet in Panama.

Central America (Summary)

The United Nations has supported detailed studies of the Central American geothermal energy situation (Einarsson, 1975). The results of the studies indicated that the region could become independent of imported petroleum for the purposes of generation of electricity by 1985. The plan called for a regional approach with a balanced growth in both geothermal and hydro-supply generating capacity. The forecast in development yields a demand of 12,305 GWh (gig watt hour — one billion watt hours) by 1984 with a developed supply of 10,960 GWh hydro and 4,860 GWh geothermal. A regional interconnecting transmission grid is also visualized.

The Caribbean

Drilling during 1969 and 1970 on Guadeloupe in the French West Indies has found high temperatures, i.e., 224°C at a depth of 338 m. A well drilled to a depth of 1,700 m, however, did not yield any production (D'Archimban and Munier-Jolan, 1975). Planned deeper drilling may find a producing hydrothermal reservoir. It appears that a fractured reservoir may be present and stimulation could be beneficial. Obviously, a hot dry prospect exists at Bouillante in Guadeloupe.

St. Lucia

St. Lucia is an island in the Lesser Antilles that lies northwest of Venezuela in the Caribbean. The Institute of Geological Sciences (UK) is directing a project to investigate the geothermal potential of the island (Muffler, 1975). The island is entirely volcanic and there are numerous surface manifestations — boiling ponds and fumaroles. The prospective region of primary interest is the Qualibou Caldera where fumarole temperatures as high as 185°C have been measured. Geological mapping and geophysical surveys (primarily resistivity) have tentatively identified an area of low subsurface resistance. Seven heat flow and exploration holes were drilled in 1977, ranging in depths from 116 to 726 m, and significant steam flows were obtained from some holes. Further study, including a microearthquake monitoring array, is under way to define the extent of the reservoir. A 1.0 MW(e) wellhead generator for two of the steam-producing wells is under consideration.

Chile

The El Tatio hydrothermal field in northern Chile (Lansen and Trujillo, 1975) has been under development since 1968. A joint United Nations and government of Chile program conducted an exploration campaign. Initial drilling of six exploratory holes in 1973-1974 led to the further drilling of production wells to a depth of 1.8 km. Well capacity is estimated to be sufficient to support a 10 MW(e) generating capacity with the producing horizon being at a temperature of 265°C. Plans for the region include additional development drilling sufficient to support a 50 MW(e) generating plant.

Two other areas in Chile, Puchuldita and Pollo Quere in the northern part of the country, are currently under study and show potential for developments similar to El Tatio field.

Argentina

Exploration of Argentina's geothermal potential has just been initiated under the auspices of the National Secretariat of Energy. The primary area of investigation is the Copahue region in Neuguen province. The exploration activity is thus concentrated along the margins of the Andes. Hot springs are known in the Bohia and Rosaria de la Frontera areas. Although it is too early to have a quantitative idea of Argentina's geothermal resources (Roberts, 1975), it appears that the resource potential is very large.

Brazil

Geothermal exploration in Brazil has only recently been initiated. Many hot springs areas are known to exist and 23 are recorded, with the majority found in the state of Goias. Two major hot springs, Serra das Caldas and Lagua dos Peixes, are under investigation (Roberts, 1975).

A preliminary reconnaissance survey of Brazil (Hamza, 1979) has indicated that substantial quantities of low-temperature (40 to 90°C) thermal waters exist in southeastern Brazil. A gradient of approximately 36°C/km is indicated in the extensive Parana Basin. An area of 100,000 km² is underlain by an aquifer (the Botucatu Sandstone) that is estimated to contain about 9×10^{20} joules (900 quads) of energy. This stored energy, at a usable temperature range of 30 to 90°C, is equal to about 280 MW(t) years/km². A continuing resource survey and assessment is in progress.

Peru

Only reconnaissance studies have been carried out in Peru (Parod, 1975). One region near the Ubinas Volcano, however, looks very promising.

Bolivia

Reconnaissance surveys of Bolivia (Carrasco, 1975) are being carried out by the Bolivian Geological Survey. Three favorable areas have been identified based on geological characteristics and hot spring indications. Plans for more detailed exploration are being formulated.

A Latin American Initiative

Before leaving the "New World" it is important to note that the Latin American Energy Development Organization (OLADE) was formed in 1973 (Elizarraras, 1978). This organization has a charter to assist the countries of Latin America in a coordinated development of their energy resources. Located in Quito, Ecuador, OLADE has placed a priority on geothermal development. The first major activity of the new organization was to hold a workshop on geothermal exploration technology (Quito, Ecuador, March 1978). Figure 2-29, adapted from the OLADE report, indicates the locations of the activities sketched in the previous sections on Central and South America. It is evident that these development activities are concentrated mostly along the geothermal belts or zones of the western continental margin as suggested in Figure 2-3.

Oceania

There are six nations with active geothermal projects in the portion of the world usually described as Oceania. This region is notable for its active volcanic zones (Anon., 1974) and is especially rich in potential geothermal resources. Two of the countries in the region (New Zealand and Japan) were early pioneers in geothermal electric power development and have active plans for expansion of the use of their resources.

New Zealand

Currently about 8% of New Zealand's electric power is produced from the Wairakei and Kawerau fields. The generating capacity at Wairakei is at 190 MW(e) as installed in 1958 (Bolten, 1975). This facility represents the world's first successful commercial production from a liquid-dominated reservoir. The Kawerau field has been used to provide both heat and power for a pulp and paper mill. Direct use is also made of geothermal fluids to heat homes (Burrows, 1970) and to air-condition a hotel (Reynolds, 1970) in Rotorua.

Drilling activities at the Broadlands field was suspended from 1971 to 1973 because it was expected that the development of a large offshore natural gas discovery would provide for future electrical generating capacity. Since 1974, however, drilling has resumed at the Broadlands and Kawerau fields. It has been estimated that the 28 new wells at Broadlands could support a 150-MW(e) plant and construction is currently underway (1981).

The resources of New Zealand for electricity generation are ultimately estimated to be 2,000 MW(e). A recent acceleration of the New Zealand geothermal resource evaluations (Donaldson and Grant, 1978) have indicated that 13 fields (Figure 2-32) can be expected to have potential as recorded

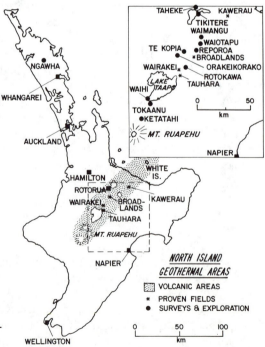

Figure 2-32. Map of hydrothermal areas in New Zealand.

Table 2-21
Assessment of New Zealand High-Grade Hydrothermal Geothermal Resources

Field**	Area (km²)	Max. temp. (°C)	Estimated electrical power potential, MW(e)*		
			Proven†	Inferred†	Speculative†
Wairakei	15	270	150		
Tauhara	14-16	280	100	80	
Broadlands	11	300	120	30	
Kawerau	6-10	290	100		30
Waiotapu-Reporoa	8-12	295		150	100
Orakei Korako	6-10	260		50	50
Rotokawa	8-12	300		50	100
Ngawha	30-50	300(?)		400	500
Tikitere-Taheke	12	270(?)		75	75
Waimangu	12	270(?)		50	100
Te Kopia	5	240		20	20
Tokaanu-Waihi	4	steam field			100
Ketetahi	?	steam field		25	
Total	140††		470	930	1075

*Installed capacity, accounting for reservoir recovery factor and conversion efficiency.
**Lack of sufficient information has precluded the inclusion of areas like Atiamuri, Mokai and Ngatamariki which are under consideration for survey.
†Measured, indicated and inferred according to the definitions proposed in Appendix A.
††Use of the log-normal analyses recorded in Table 2-16 results in a projected 110,000 MW(e) for the 136 km² of high-grade liquid-dominated reservoirs.

in Table 2-21. In 1977, "proven" resources were evaluated at 470 MW(e), "inferred" (drilled) resources at 930 MW(e), and "speculative"* resources at 1,125 MW(e) for a total of 1,525 MW(e), a value close to current total world production (Table 2-12). It seems quite clear that the increasingly improved knowledge of New Zealand's expanded geothermal resources will

* Use of the USGS proposed definitions (Appendix A) would term these categories as: measured, indicated, and inferred resources, respectively.

continuously enhance the data base for future energy planning and, thus, assure a more firm place in New Zealand's future energy supply.

New Zealand is investigating direct uses; for example, of a district heating system in the city of Auckland. It seems very probable that direct uses will expand greatly in New Zealand in the future. The developments have been a source of New Zealand's very valuable technical and scientific information for the rest of the world, and New Zealand's geothermal experts have aided developments in Mexico, Indonesia, Hawaii, and the Philippines.

Japan

Japan has launched a very vigorous and effective plan for the development of the nation's geothermal resources. This effort has been in direct response to the price escalation of fossil fuels.

The very earliest geothermal development attempts reach as far back as 1919 when a steam well was drilled at a location near Beppu, on the island of Kyushu (Axtell, 1975). Subsequent major development activities are summarized in Table 2-12. The most notable early electric power generating facility established by the Japanese is the Matsukawa plant of 22 MW(e) rated capacity. This unit, with service initiated in 1963, is located on a steam (vapor-dominated) reservoir and has served as the model and inspiration for subsequent developments at Otake (1967), Onuma (1974), Onikobe (1975), and Hatchobaru (1977). These installations are shown in Figure 2-33, together with areas under exploration. The development in Japan is rapidly expanding as a result of emphasis by the Japanese government (Anon., 1977 and 1978) and two new plant facilities at Kakkonda and Mori, both generating 50 MW(e), were operational in 1978 and 1980, respectively. All aspects of geothermal development research and development are being supported strongly in Japan and the effects of these anticipated technology advances are projected to yield 48,000 MW(e) by the year 2000 (Table 2-22). Japan has projected a significant hot, dry rock potential — 5,500 MW(e) — and it is reported that several hot, dry rock, high-temperature wells have been drilled near Mount Yakedake west of Tokyo (Figure 2-33).

In Japan, an area of very active volcanism, there are many fumaroles and hot springs; these hot springs have been used for many centuries for bathing. There are presently 1,500 hot spring resorts in Japan that serve about 150,000,000 guests yearly. Data on 2,237 hot springs have been accumulated (Sugama, 1975) in order to better define the direct use potential of the islands of Japan. Thirty sites have been identified for exploration for high-grade heat potential. The thrust toward direct use (Anon., 1978) will undoubtedly produce many future multi-use installations in Japan (Sekioka, 1979), where industrial uses such as food processing, wood drying, distilled

Table 2-22
Planned Geothermal Electric Power Generation in
Japan in Units of MW(e) — 1978

Year	Steam from shallow reservoirs	Steam from deep reservoirs	Liquid-dominated, or hot-water reservoirs	Volcanoes and hot dry rock reservoirs	Total projected capacity
1975	50	–	–	–	50
1980	220	–	2*	–	222
1985	500-1000	200	210	–	910-1410**
1990	3000	2150	910	–	6060**
2000	1950	12,000	11,000	5500	48,000
	Conventional technology	New technology (Sunshine Project) developments			

*Two 1-MW(e) binary demonstration plants (see Figure 2-33).
**Tentative prospects.

Figure 2-33. Geothermal resource areas and power plant locations in Japan—1978.

water production, space heating, greenhouse warming, eel culture, and heating of poultry and livestock barns are all under study.

The Philippines

An early (1969) power plant was established at the Tiwi hydrothermal field (near Albay) on the island of Luzon. Following this initial effort, little development activity was conducted until early in the 1970's. With power needs expanding at an accelerating rate, the Philippine government adopted a very aggressive and ambitious geothermal program (Bieser, 1978). This program has effectively spurred development at three major projects: (1) The Tiwi field, which is being rapidly drilled to accommodate 110-MW(e) capacity; (2) in the Los Banos region, the Union Oil Company is drilling to supply a 110-MW(e) facility constructed at the Bulalo field in 1979; and (3) the New Zealand firm of KTRA (Studt, 1979) is drilling a large field at Tongonan in Leyte where a 110-MW(e) facility is planned. The latter field (see Figure 2-34) has wells that produce fluids at flows that correspond to 10-16 MW(e) capacity.

★ PLANT FACILITIES UNDER CONSTRUCTION
  1. TIWI, ALBAY
  2. BULALO, LOS BAÑOS
  3. TONGONAN, LEYTE

▲ PROSPECTS UNDER ACTIVE EXPLORATION
  4, 5. M PALIMPINON-DAUIN, NEGROS
  6, 7. MANAT-MASARA, DAVAO DEL NORTE

● PROMISING PROSPECTS

LUZON

PHILIPPINE FAULT ZONE

MANILA

LEYTE

NEGROS

MINDANAO

HYDROTHERMAL DEVELOPMENT IN THE PHILIPPINES, 1979

Figure 2-34. Map of geothermal electricity generating plant sites and the development and exploration areas of the Phillippines—1979.

Table 2-23
Planned Hydrothermal Power Generation and Capacity Targets in MW(e) in the Philippines — 1978-1987

Geothermal field or area	1978	1979	1980	1981	1982	1983	1984	1985	1986	1987	Power capacity estimate*
Tiwi, Alday (drilled)	110		55	55		55	55		55	55	560
Makiling-Banahaw (Los Baños) (drilled)	55	55		55	55		55	55		55	720
Tongonan, Leyte (drilled)			55	55		55	55				Resistivity survey still in progress
Southern Negros (Palimpinon-Dauin)			55	55		55	55		55		425
Northern Negros (Mambucal)		Initial investigation of area underway									
Davao (Manat-Masara)			55	55		55	55		55		500
Total Annual	165	55	110	275	165	110	275	165	55	220	
Cumulative	165	220	330	605	770	880	1155	1320	1375	1595	
											2205*

*Estimated capability of fields or area based primarily on resistivity surveys.

The projected geothermal electric power development plan for the Philippines is summarized in Table 2-23. The total of 1,320 MW(e) for 1985 may seem to represent a very rapid expansion; however, it is not the resources that will be limiting. The Los Banos area (where the Bulalo field is being established) is about 1,600 km² in extent. From all reports, there are other very promising hydrothermal prospects within this large region. Preliminary resistivity surveys have projected other sites with large potential. There are no reports of extensive direct use in the Philippines, although some grain drying and salt production facilities do use natural geothermal sources. It seems most likely, however, that the Philippines will soon direct attention to thermal processing industrial applications.

Indonesia

Exploration in Indonesia was conducted extensively during the 1970-1975 period. This effort confirmed the very large potential of the Kavah-Kamojang region, and verified that it contains a vapor-dominated reservoir (Hochstein, 1975). The electric power potential was estimated to be about 100 to 250 MW(e). Well testing has verified the extent of the reservoir and a model was developed for this steam field.

Recent indications (Basoeki and Radja, 1978) are that a 30-MW(e) generating plant will be designed and installed at the Kamojang field; it was originally planned for completion in 1978. Other sites in Indonesia show great promise: (1) the Dieng-plateau area in central Java where drilling has verified presence of 139°C water; (2) the Salak, Pula Nukay-Rky, and Batub areas in west Java; Bali; and (3) on Sumatra where a 13-MW(e) plant is under development. It appears that the Indonesian geothermal development efforts will concentrate on the installation of small plants on different islands to supply local demands (Akil, 1975).

Taiwan

Two hydrothermal fields have been explored and drilled on the island of Taiwan (Chen, 1975). The Takun field was found to have very high-temperature fluids (293°C); however, they are extremely acid and probably will not be exploited. At Tuchong, fluid with a temperature of 173°C was discovered. Development activities are not known as of this writing, but plans are said to call for a 50-MW(e) plant (at Tuchong near Taipei) by 1985, and as much as 200 MW(e) by the year 2000. Several direct heat applications have been developed on Taiwan and there apparently are plans to expand these uses significantly.

Australia

Only very preliminary surveys of Australia have been conducted (Thomas, 1975). It appears that a region in western Victoria may hold promise due to the known presence of young volcanic fields.

Fiji

Exploration of the geothermal potential has been under way in Fiji since 1960, when a survey of hot springs was conducted (Cox, 1978). The most recent reports indicate that the Lampora area on the island of Vanua Lavce shows considerable promise as a direct use reservoir.

Europe

A valuable resource evaluation effort (Cermak, 1975) that was recently completed (Cermak and Rybach, 1979) has provided an excellent heat flow map of western Europe, with an extensive literature review. Some 812 heat flow values were used and they gave an arithmetic mean of 57.1 mW/m² for the region. The ages of major tectonic events in the region are approximately correlated to the heat flow values. The Hungarian territory with its geopressured hydrothermal deposits is characterized by a heat flow of about two times the average, i.e., 100 mW/m². Natural hot springs and warm baths have been used in the region throughout recorded history, but the extensive applications and currently active development programs occurred in Iceland and Italy.

The heat flow study and map of Cermak and Rybach (1979) indicates that much of southeastern Europe is underlain by large regions that will be found to have temperatures in excess of 40°C at a depth of 1 km. The sedimentary basins south of the Caucasus Mountains have gradients greater than 50°C/km and thus should be valuable potential regions for the development of both thermal aquifers and hot dry rock systems.

The U.S.S.R.

The Soviet Union has a long history of interest in developing the geothermal resource of that nation (Sukhorev et al., 1960). A recent review (Pryde, 1979) indicates that there has been an acceleration of development activities, especially in direct usage. An intensive study of heat flow has resulted in preparation of a thermal gradient map (Gavlina and Makarenko, 1975) (Figure 2-35) similar to the one developed for the U.S. by the USGS (Figure 2-13). There are indications of vast areas with above-normal thermal gradient (25°C/km). In addition, many local areas of intense thermal gradients (black dots on Figure 2-35) have been identified.

The development of geothermal resources for electric generation is the responsibility of the Minister of the Natural Gas Industry. The initial Soviet developments occurred early on the Kamchatka Peninsula. In 1957, drilling began near Pauzhetka on the southern tip of the Peninsula. Finally, 21 wells were completed and a 5-MW(e) plant was built and commissioned in 1967. The wells are shallow (220 to 480 m deep) and the fluid discharges at temperatures ranging from 150° to 200°C. The plant is a flashed steam turbine (15 to 20% flashover). It is reported that power is provided for lower cost than those of fossil-fuel plants in the same area.

A second plant was constructed on the Kamchatka Peninsula in 1968 at Paratunka. The resource is hot water at 80°C and, therefore, a binary nonaqueous turbine (Freon 12) plant was installed. Two 340-kW(e) turbines were used and full capacity production was attained in 1970. Extra hot water is used to heat greenhouses and for irrigation of agricultural crops.

These two power plant projects appear to have been both technical and economic successes. No further electric generating development, however,

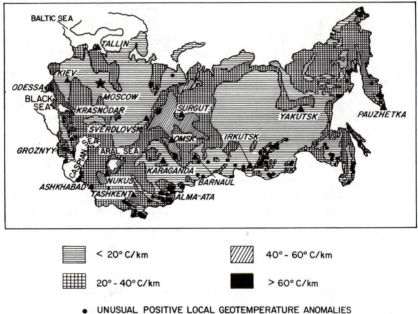

Figure 2-35. Heat flow map of the U.S.S.R.

has as yet occurred in the U.S.S.R. Possibly, emphasis and priorities in fossil fuel development and the intensive exploitation of direct use of thermal waters have taken precedence over hydrothermal resource application to electric power generation.

By far the major geothermal development in the U.S.S.R. is the direct use of the nation's low-temperature hydrothermal resource. The primary usages are: 62% for space heating, spas, and industrial processing; 25% for agriculture, mostly greenhouses; and the remaining 13% is the input to "fuel" the two geothermal power plants on the Kamchatka Peninsula. The identified hydrothermal resources (hot springs and developed wells) are summarized in Table 2-24. The major direct uses are tabulated in Table 2-25.

There is a very strong interest in the Soviet Union to expand the use of thermal waters for greenhouse warming as a means of expanding the fresh food base for the major cities. This proposal appears to be economical. The economics appear to be very favorable for many regions of the U.S.S.R. Typical costs in the more accessible regions are about one to three times less than that for greenhouses heated by fossil fuels. This type of expansion is especially important for a nation having extensive land areas in the northern latitudes. The 1976-1980 5-year plan dictated a manyfold increase in the use of thermal waters in the Soviet Union, and it was estimated that 1,500-1,600 greenhouses could be heated by geothermal energy. It was further pointed

Table 2-24
Distribution of Thermal Water Reservoirs
in the U.S.S.R.*

Region	Flow (10⁶ m³/day)	Heat Content/Year (10¹⁵ J/year)
Caucasus	2.7	145.7
Crimea and Fore-Carpathians	0.8	44.4
Other European Areas	0.7	37.7
Central Asia	2.1	102.6
Katuakstan	1.7	87.9
West Siberia	15.0	799.5
East Siberia and Far East	2.4	122.7
Kamchatka, Kurile Islands, Sakhalin	2.4	130.6
Totals	27.8	1471.1**

*These are presumed to be identified and developed hot springs, and drilled warm aquifers supplying energy for direct heat uses.

**(Equal to about 1.5 quads/year).

out that there are extensive areas of "thermal anomalies" in the southern portion of east Siberia that are in the vicinity of the route of the new Baikal-Amor Mainline (Shakad, 1977) railroad project. Use of the hydrothermal resources could measurably speed the development of this resource-rich area and aid in stretching the conventional energy resources located in, or transported to, this remote area.

Finally, there are persistent reports of hot, dry rock developments in the U.S.S.R. Two sites, the Muknovskiy and Avachinskaya volcanoes, are reported to have 100- and 300-MW(e) capabilities for such developments. It is also reported that both the Stavropol Kray and Transcarpathia regions are also under study as potential HDR sites.

Iceland

The well known developments in Iceland are the direct space heating and hot water supply systems of the capital Reykjavik and its suburbs. It is estimated that 420 MW(t) of energy is used in this manner. Iceland has large hydroelectric and geothermal resources. In the past, the majority of the electric generating capacity has been from the nation's dams. A small — 2.5-MW(e) — facility exists at the city of Namatiall.

(text continued on page 132)

Table 2-25
Direct Use of Low-Temperature Hydrothermal
Resources in the U.S.S.R.

Use	Location	Purpose	Amount	Remarks
Space heating	Dagestan Aut. Republic Capital, Marchachkda	Hot water supply	2000,000 m³/day	60% of city of 248,000 people; 60° to 70°C
	Izberbash	Space heating	–	Entire city
	Kizlyar	Space heating	35,000 m³/day With 17,000 m³/day at 100°C and 18,000 m³/day at 60°C	All new structures in city
	Kabarlian-Balka Aut. Republic, Nal'chik	Space heating	–	Use heat exchangers due to highly mineralized quality of fluids
	Chechen-Ingush Aut. Republic, Grozny	Space heating	–	
	Saburzalo Dist., Thilis	Hot-water supply	–	1/3 of city supplied
	Tyumen Oblast', Omsk and Nniezhie-vrtrosk	Space heating and greenhouse warming	–	60° to 70°C thermal waters supplied via wells drilled to 2500 m Apparently widely applied in this oil-producing region
	Kamchatka Peninsula, Paratuuka	Space heating	–	Several buildings in vicinity of power plant
Agriculture and fisheries	Tyumen Oblast'	Warming greenhouses	–	Many of the oil towns in this region use thermal waters from wells drilled in deep aquifers

Table 2-25 continued

Use	Location	Purpose	Amount	Remarks
	Kazakhstan, Chimkent	Warming green-houses and farm buildings	12,000 m²	Several state farms use hydro-thermal fluids for heating purposes
	North Caucasus, Grozny, Krasnodar	Warming greenhouses	Extensive vegetable production	
	Dagesken, Maklehkalu		–	Large collective greenhouse operations
	Georgia, Zugdidi	Warming greenhouses	–	Major vegetable production and greenhouse farms
	East Siberia, Ulan-Ude	Warming greenhouses	20,000 m²	Produce all vegetables for city
	Kamchutka Peninsula, Duratunka	Warming greenhouses	60,000 m²	1000 tons of vegetables produced per year
	Western Siberia, Omsk	Heat ponds	–	Carp breeding for regional fish supply
Chemical and industrial processing	Georgia, Zugdidi	Tea plant and pulp and paper mill	–	Major operations with national distribution of products
	Eastern Siberia	Thawing of ground for mining	–	Highly mineral-ized zones are being developed by thermal water produced by drilling to warm aquifers
Spas	Numerous in the Caucasus	Vacations and balneo-logical (bathing)	100,000 liters/ minute	Hydrothermal fluids especially mineralized ones are considered valuable for health and restorative purposes
	Throughout hydrothermal zones	Mineral water bottling plants	280 locations	

(text continued from page 129)

Due to its location astride the mid-Atlantic ridge, Iceland has a very large geothermal potential and can certainly look to energy independence via geothermal resources (and hydropower) for electricity generation, space heating, industrial processing, and food production (relief from dependence of transportation on petroleum may await economic electric vehicles). The resource evaluation and projections were made by Bodvarsson (1975), who estimates that the heat flow (conductive = 50%; convective = 20%; and magma = 30%) is equal to 250 mW/m². Iceland contains both low-tempera-ture (<150°C) and high-temperature hydrothermal reservoirs. The high-grade systems are estimated to total 400 km². This is estimated to yield 3.2 GW(e) for 50 years, or 1,600 MW(e) centuries, of potential electrical generating capacity. The results of the log-normal analysis summarized in Table 2-16 yields a mean projected capacity of 320,000 MW(e).

New district heating systems are coming on-line at Sudures (also serving the Keflavik International Airport Complex). The city of Akureyi will have a system serving its 12,000 residences. The Blondeirs District heating system was initiated in 1978. Expansion of greenhouses, swimming facilities, and industrial processing with geothermal heat will undoubtedly expand at a pace dictated only by Iceland's capital and human resources.

The first large electric power station development was initiated in 1970 (Bodvarsson, 1975) and was planned at a capacity of 70 MW(e). The volcanic activity of the region surrounding the Krafla field, however, has delayed installation of the second 35-MW(e) turbo-generation unit. Due to a nearby eruption in 1977, damage to the field has forced the plant to produce at a reduced level of about 10 MW(e).

Great Britain

The major oil and gas discoveries in the North Sea seem likely to place Great Britain on an exporting basis in the very near future. Current crude oil production (1978) is about 1.7×10^6 bbl/day (roughly equal to Kuwait's production). Rising energy prices, however, motivated the British govern-ment in 1976 to start an assessment of the hot-water and hot, dry rock potential of the British Islands. Hot aquifers have been identified in the sedimentary basins of southern and western England, and in southern Scotland. Anomalously high heat flow values were found in southwest England where gradients of 40°C/km were recorded. Initial steps in develop-ment of a hot, dry rock extraction system have been made by the Camborne School of Mines to test the feasibility of extracting (mining) the heat from the relatively old igneous intrusives of the region.

Italy

The Ente Nazionale per l' Energia Electrica (ENEL) is responsible for the development of the geothermal resources of Italy. The ENEL-operated facilities in 1975 had a total capacity of 421.5 MW(e), that was about 1.7% of the total annual electricity production of the country. A program to locate more resources, to extend producing fields and to drill exploration wells in new reservoirs is under active development in Italy. This effort is based upon the extensive experience of the nation that has led world geothermal development for many years.

The history of Italian geothermal development has been presented many times (Ceron, 1975) and extends from the initial work at the Larderello steam field, that resulted in the first installation of a 250 kW(e) generator in 1913, to the current production from three fields at Larderello, Mount Amiata, and Travale. The consensus today is that the Larderello fields have nearly reached full capacity and have been extended to an area of about 250 km². Consideration is being given to replacing the present noncondensing turbines at the field with modern units, a step that is estimated to yield about 40 MW(e) in the Larderello-Mount Amiata area (Figure 2-20).

The development and exploration situation in Italy in the period of 1969-1974 has been presented in detail (Ceron et al., 1975). Exploration in the pre-Apennine belt from Larderello to Naples, an area of about 20,000 km², was continued. A new steam field was discovered and drilled in the Travale-Radkondoli region, and 15 MW(e) of additional generating capacity was installed in 1974.

The details of the Italian developments can be followed since 1970 in the various issues of the journal *Geothermics* (Pergamon Press, New York, Oxford, and Frankfurt) founded and edited by Professor E. Tongorgia. The journal is supported by the Instituto Internazionale per le Ricerche Geothermiche, Pisa.

Since 1975, exploratory drilling has continued in that region and south of Rome to Naples. The summary of developments in 1977 are summarized in Table 2-26. The Mount Amiata field was drilled on a "blind" prospect, i.e., located by a geological, geophysical, and geochemical exploration effort. This area has especially strong and localized heat flow anomalies with geothermal gradient values as high as 400°C/km recorded over areas of 4 km in extent. As shown in Table 2-26, the subsequent exploration and drilling was of limited success, with no new steam field discoveries as yet drilled.

Exploration is continuing in the Tuscany region where the older fields are located; the Mt. Cimini and Mt. Sabatoni (North Latium) just north of Rome; the Colli Albani region just south of Rome; and the Roccamontina, Napoli, and Vulture areas surrounding Naples (see Figure 2-18).

Table 2-26
Summary of Development Results in Italy South of
Larderello (South Tuscany and North Latium) — 1977

Area*	Drilling	Result	Remarks
Roccastrada	1 well (1962)	100-120°C at 1 km depth	Geothermal gradient is ~300°C/km; under evaluation for more drilling. East of Mt. Amiata area
Alfina	8 wells	Industrial use, 150°C	Original plans were for a Freon turbine project
Cesena	4 wells	Direct use potential (?), 285°C, with heat exchangers	Highly mineralized, fluid, TDS = 350,000⁺ ppm; 20 km north of Rome
Vico	2 wells	Direct use, 85°C	May be developed for an industrial or agricultural use

*Refer to Figure 2-20.

The Italians are also seeking ways to increase production of existing fields and are studying and experimenting with reinjection to increase fluid recharge and to handle heavy brines; stimulation via hydraulic fracturing and acidizing; and establishment of a hot, dry rock heat extraction research project. New production wells are being drilled at Larderello to compensate for decreasing productivity of some of the present wells. A well has been drilled to a depth of about 7 km at Larderello to study the deep fracture systems and in preparation for possible hot, dry rock experiments. Stimulation treatments at Alfina yielded a six-fold productivity increase in one well.

Greece

The Greek government through the Public Power Corporation (PPC) and Institute of Geological and Mineral Research (IGMR) has initiated an active geothermal development program. The PPC has an agreement with the Italian government to have the experienced ENEL group support the Greek efforts to develop a promising prospect on Milos Island and to aid in continuing the survey of Greek geothermal resources (Fytikas and Kolios, 1979). Two wells are being drilled on Milos. One has been completed (1978) and is on test to evaluate the fluid and reservoir characteristics. Initial plans call for installation of two 15-MW(e) generating plants on this discovery if the test results are indicative of a sufficiently large reservoir.

The exploration activities have located six promising areas in Greece. In addition to Milos, these are: Sousaki, Methane, Visiros, the Islands of

Leskos, and the Sperchivs Riven graben (Domino and Papostanatori, 1975). It appears that Greece has an abundance of geothermal potential, a conclusion that is a reflection of its location in a geothermal zone and the presence of numerous well-known surface manifestations and historically active volcanoes.

France

France has an active geothermal development program, largely focused on the direct use of thermal waters for space heating, industrial uses, and agricultural applications. The French government is supporting a survey and assessment of the thermal waters through the Bureau de Recherches Geologiques et Minieres (BRGM). This survey (Gable, 1979) of low-enthalpy resources has supported the development of several significant space heating projects in the Paris basin. This sedimentary basin area has been thoroughly drilled and some 2,000 wells define the warm (normal gradients; see Figure 2-25) deep aquifers. There are currently four areas in the Paris region that have major geothermal space heating projects.

The largest project (Coulhois and Herault, 1975; and Gringarten and Sauky, 1975) is at Creil, 40 km north of Paris where a heating system for a 2,000-apartment complex started operating in 1976; expanded to 4000 units in 1979. These systems consist of a double pair of injection and reinjection wells, about 2 km deep, directionally drilled from a single pad. The water temperature is 57°C and the production wells have down-hole pumps. Three heat pumps and heat exchangers are used to increase the system's effectiveness. Experience to date has indicated excellent performance, with 40% of the heating requirements supplied by the thermal water. The success of this project has led to the initiation of the following three additional projects:

1. Melun: 40-km southeast of Paris, with a retrofitting of a 4,000-apartment complex, two whipstock wells drilled to a depth of 1,000 m and 900 m apart at the producing zone.
2. Villeneuve-La-Garenne: 1,700 apartments with a system similar to that at Creil.
3. Mee-Sur-Seine: 6,000-apartment complex and 50,000 m^2 of office space will be 70% supplied with a thermal water heating system.

Exploration and surveys have identified other sedimentary basins, e.g., the Aquitanian, Bresse, Limagne and Alsace basins, that have indicated significant potentially valuable sources for direct use of thermal aquifers in these areas.

The promising areas of France for high-grade resources have also been studied. Exploration efforts have concentrated on the Massif Central (Fouillac et al., 1975, and Ratigan and Linblom, 1975) where preliminary geochemical surveys indicate a possible base temperature ranging from 160 to 200°C. The Mont Dove region has been studied in detail in 1978. The exploration data, geological mapping, and gravity surveys have indicated that the caldera, 6 km in diameter, with an age of about 3.3 million years, is a potential heat source for hydrothermal and hot, dry rock reservoirs.

West Germany

At present, in the Federal Republic of Germany (FRG) geothermal resources are used directly. The major production is at Urach and Wiesbaden where a total thermal energy of about 4.5×10^{13} joules/day is exploited. A geothermal map (Bram, 1979) of West Germany was completed in 1978. It appears that the FRG has poor potential for high-grade hydrothermal resources. Only two areas have been identified: (1) Landau in the upper Rhine Valley is a known high heat flow anomaly, but low permeability has thus far prevented exploitation; (2) Urah, south of Stuttgart where a well has been drilled to a depth of about 3 km with an indicated temperature of 150°C and a 50°C/km gradient.

The FRG has shown considerable interest in hot (or perhaps warm), dry rock extraction technology. This was highlighted in September of 1979, when a bilateral agreement was signed with the U.S. to participate in the U.S. DOE-funded research and development project at the Los Alamos Scientific Laboratory's Fenton Hill experimental site. The FGR is investigating the rock mechanics, drilling technology, hydraulic fracturing, fluid flow, and heat transfer aspects of hot, dry rock heat extraction systems at a site in the Bayrishenwald Mountains.

East Germany

The German Democratic Republic (GDR) has reported results of a heat flow survey (Hurtig and Oelsner, 1979), with indications that a region of elevated heat flow exists in the southwest area of the nation. This region, the Erzebirg (Ore Mountains), may have a potential for development, with heat flux values ranging from 60 to 90 MW/m^2.

Scandinavia

In 1976, Sweden initiated a hot, dry rock geothermal project that is typical of the Scandinavian region, a region that shows little positive sign of any hydrothermal potential. Only a few warm aquifers have been identified in southern Sweden and the Baltic Islands. A research group at Chalmers University (Goteborg) has prospected the resource base in Sweden and estimated that perhaps 4.2×10^{22} joules are stored in crystalline rocks to a depth of 3 km. A detailed survey indicates that low-grade heat for space heating and other direct uses is all that can be anticipated. Prospecting (Eriksson and Malmquist, 1979) and preliminary field experiments are under way in the Bhus granites on the western coast of Sweden. Hot, dry rock study and experimentation has been initiated (Ratigan and Lindblom, 1978), and use of heat mining in deep granite masses with some sedimentary cover is thought to be an economic possibility in Sweden.

Portugal

The Portuguese Azores islands have been under study since 1974 when a Canadian drilling team from Dalhousie University and the U.S.-based Lamont-Doherty Geological Observatory drilled a well that exhibited a temperature of 200°C at a depth of 900 m. Subsequent drilling has defined a high-grade field of perhaps 12 km² in area. The field is located on the flank of a volcano, Aqua de Puri, in the center of San Miguel Island (Meidav and Meidav, 1979). It appears that Portugal will develop a co-generation system with perhaps small — about 3-MW(e) — capacity generating facilities and direct use applications. The small electric plants will phase out the existing hydrogeneration plants — 14.9-MW(e) capacity — and enable use of the much needed water for irrigation.

Spain

Spain has thus far limited geothermal exploration to the Canary Islands (Arana and Panichi, 1974). The region of intended development is at Lanzarote where very high temperatures are encountered in shallow drill holes, and there is a need to replace the diesel oil-fired plant. Spain has conducted (Albert-Beheran, 1979) a heat flow and temperature survey on the mainland. A mean value of heat flow of 82 mW/m² for the country with a value up to 120 mW/m² in the eastern (Mediterranean coastal) region indicates that a significant geothermal potential exists in Spain. At the present time, there is no known exploration activity on mainland Spain.

Switzerland

The Swiss government has conducted a thorough geologic survey of the hot, dry rock potential of the country (Rybach and Jaffe, 1975), and the potential for thermal waters. Switzerland's 6 million population depends on power and space heating from a nearly completely developed hydro-system and fossil fuels. Although there are no known surface hydrothermal manifestations in Switzerland, it is believed that thermal waters and hot, dry rock potential do exist in the Molasse basin. This valley contains the major urban and industrial centers of Switzerland. Elevated heat flows, i.e., 88 to 113 mW/m², have been measured and indications of the presence of thermal waters trapped in sediments via sampling of 1-km-deep wells have been established. It appears that the needed research and development to tap these resources will be obtained by cooperative efforts with other countries.

Austria

Apparently, Austria has just recently become aware of its geothermal potential (Zotl, 1978). Efforts to tap thermal waters for greenhouses and space heating have been initiated and several wells have been drilled. It would also appear that the western portion of the Pannonian Basin of Hungary that extends into Austria may have similar thermal water potential that has been exploited so successfully by Hungary for many years.

Hungary

The use of low-enthalpy water was initiated in 1962 (Boldizar, 1974). The Hungarian Basin produces abundant thermal waters with temperatures up to 150°C at depths up to 2 km. In 1975, it was reported that some 433 wells in Hungary produce water at temperatures greater than 35°C. These wells produce about 400 m³/min and serve a wide variety of direct uses (Balogh, 1975). Notable are 500,000 m² of greenhouses; 1.2 million m² of foil-covered vegetable growing beds; space heating for a 1,800-unit apartment complex, several municipal buildings and schools; and many agricultural applications.

In Budapest, a single well supplies heat to 5,000 apartments and the outlet serves a swimming pool. The resource estimates for Hungary's easily extracted thermal waters from a depth of 1.5-to 2.5 km are 5×10^{10} m³, with an ultimate usage of about 10^{14} joules/day. The main basin in southeastern Hungary is estimated to contain at least 50 quads of energy suitable for direct use.

Other European Countries

Exploration in the countries adjacent to Hungary indicate that Rumania (Velicium, 1977) and Poland (Radowska et al., 1977) also have significant resources of thermal waters.

Czechoslovakia (Franko and Rocicky, 1975) has conducted extensive surveys of that nation's geothermal resources. Eight drill holes 200 to 2,500-m deep yielded thermal waters with temperatures ranging from 22 to 92°C. The survey is concentrated in Suloiaka and was completed in 1980.

Geothermal manifestations have been known in Bulgaria for many centuries (McEwen, 1978). Currently, thermal water wells are used for algae production in an antibiotic plant near Sofia. Hot spring flows and wells supply spas, greenhouses, trout rearing ponds, and a heated sports stadium. Petroleum drilling near the Black Sea is encountering warm aquifers and these wells are routinely temperature logged for geothermal resource assessment purposes.

Yugoslavia is known to have geothermal resources primarily in the form of thermal waters. Geothermal gradients of at least 60°C/km have been recorded and it seems likely that these resources will eventually be tapped.

Africa

The African continent has many geothermally active zones, especially prevalent along the East African Rift. Development projects have been initiated in Kenya, Djibouti, and Ethiopia. Exploration is under way in other African countries.

South Africa

Exploration in South Africa appears to be restricted thus far to surveys of the hot springs (Kent, 1979). The hottest spring found had a water temperature of 64°C and the hottest known borehole flow is at 77°C. Use of hot springs for spas and salt production has been reported. One promising area, the Karroo rocks of southern Cape Province, is thought to be a promising region for higher-grade geothermal prospects.

Nigeria

An assessment has been made (Nwachutwu, 1975) for the potential of the Nigerian sedimentary basin that produces oil and gas. Various logs from 1,000 oil wells were the major source of data. The results indicated that overpressured zones exist in the basin and temperatures around 100°C were recorded at a depth of 3 km. The regional evaluation suggests that hot rocks of Cretaceous age very likely exist north of the oil-producing region and further study is needed to investigate this possibility.

Kenya

Three hydrothermal areas located along the African Rift Valley are known to exist in Kenya. These areas are Lake Nannington in the north, the Eburru field in the central part of the valley, and the Olkaria field south of Lake Naivasha. The United Nations has been involved in the development of the Olkaria area for electricity generation. Electric power is extremely expensive in Kenya and any contribution from geothermal resource development would be valuable (Noble and Ojiamko, 1975). Drilling at the Olkaria area has resulted in very high-temperature flows, e.g., 286°C at a depth of 1.3 km. The initial four wells drilled (1975) were considered for the use of small well-head generators due to the critical need for additional power in the area. The permeability of the reservoir appears to be quite low and well flows are able to support only 1 or 2 MW(e). Deeper drilling and stimulation are being considered.

Djibouti

The region near Lake Assal was explored (1970-1973) by the French government agency BRGM (Bosh et al., 1977). The geochemical analyses indicate the possibility of the existence of temperatures ranging from 100 to 170°C at depth. No further development of the area is recorded; however, the area is one of very active volcanism and will probably receive further attention in the future.

Uganda

Many hot springs are recorded in the western regions of Uganda. Two potentially valuable hydrothermal fields have been identified (Maasha, 1975). The Sempaya area is the most promising hydrothermal resource, and has thus far been defined by a large area of low resistivity. Fluid temperatures of at least 160°C are inferred from the hot spring water chemistry. The Kikaguta hot springs area is used as a spa and the Lake Tirtaguta geothermal area is located in a volcano crater. Temperatures in the latter reservoir are reported to be as high as 230°C. These resources are rather remote from industrial and urban development, and it seems unlikely that any significant development will occur in the near future.

Ethiopia

The Afars Rift, which occupies the northern part of Ethiopia, is a very active volcanic zone and an obvious geothermal resource-rich region. The United Nations has had an exploration program in Ethiopia for a number of years. Three regions have been found to be potentially valuable high-grade hydrothermal areas. The area of greatest potential is located in the Danakil depression at the Tendako area where several large, high-temperature, low-salinity reservoirs have been delineated. The northern Afars region, at Dallol, has very high-temperature hot springs with surface flow having temperatures as high as 115°C. Here the fluids are very highly mineralized.

The present exploration (McEven et al., 1977) is focused on the lakes north of Addis Ababa. Three very large low-resistivity anomalies are found along the main Ethiopian Rift in the vicinity of Lake Lagano. These reservoirs and the older discoveries are estimated to be very large. There is little demand for this power capacity in the region, and it seems unlikely that development will occur in the near future.

Egypt

A large number of heat flow determinations have been made in Egypt (Morgan and Swanberg, 1979) with values ranging from 42 to 175 mW/m². The geographical coverage of the heat flow survey thus far points to the Red Sea coast as the most promising for future exploration. A regional mean heat flow value of 54 mW/m² is noted. The oil well bottomhole temperatures and results of geochemical surveys indicate that the Gulf of Suez may be a good prospect for finding thermal waters. No local or specific hydrothermal systems were discovered. It is assumed that the Red Sea coastal region has generally high geothermal gradients, but that the Precambrian platform setting of this area may indicate dry, hot rock at depth. A possible application of this resource would be desalination of sea water to produce fresh water for this arid region.

Asia and the Middle East

There has been very little exploration for geothermal resources in Asia and, to date, very little electrical generation has been attempted. Turkey has, however, a long established geothermal development program. Initial exploration is also being pursued by India, China, Israel, and Jordan. Recent reports indicate that China has given a high priority to the development of geothermal energy as an alternate source to meet local needs.

Turkey

The geothermal development at the Saraykoy-Kizildere field in Turkey commenced in 1966. Currently 14 wells have been drilled with 12 producers, and a liquid-dominated reservoir of about 12 km² in area has been delineated. The flash steam production is about 10% of total well flow on the average. Plans to establish a 30-MW(e) plant have been hindered by the intense scaling problems encountered from the brines of this field. A small 0.5-MW(e) pilot plant is in operation and a pilot green-house project has been constructed. Fourteen high-grade hydrothermal areas have been identified in Turkey (Tezan, 1975; Alpan, 1975; and Kurtman and Samilgil,

1975). Exploratory drilling has been accomplished at six of these potential fields. Preliminary regional prospecting and evaluations are under way at six additional areas. The Turkish government has established a goal of producing from hydrothermal energy 10% of the nation's electrical generating needs in the year 2000. The reports of high-temperature reservoir temperatures in several of the exploration prospects indicate that Turkey may go a long way toward achieving that goal.

Israel

Prospecting for geothermal resources was started in 1974 in Israel. The Israel Geological Survey has conducted several geology mapping projects and geochemical surveys. These prospecting activities were chiefly focused on the Jordan Rift Valley. Work was also initiated to compile a heat flow map of the country (Eckstein and Simmons, 1978).

The deep fault structures related to the Jordan-Dead Sea Rift are thought to contain the major hydrothermal systems in the region. The geochemical studies of hot springs suggest a base temperature of at least 140°C, whereas the resistivity data indicate several subsurface low-resistivity areas along the Rift Valley.

The hydrology of the Negev region of Israel indicates that a warm (40°C) aquifer underlies this area and plans for greenhouse warming are under consideration. The warm spring spa area at Tiberiva in the Dead Sea region has been in use since Biblical and Roman times and is once again being developed as a tourist and vacation area. A geopressured zone has been identified near Ashkelon; however, the conditions at depth have not yet been explored in detail.

Jordan

Geothermal springs exist at Zepka Ma' on the east size of the Jordan Rift Valley in the vicinity of the Dead Sea (Levitte and Eckstein, 1979). The surveys along the rift zone indicate reservoir temperatures in the range of 90 to 115°C. The western region of the country (Eckstein, 1979) has heat flow values of about 35 to 55 mW/m², indicating that thermal waters stored in deep aquifers may be a valuable resource as well as the hydrothermal systems that supply the well-known hot springs.

India

The geothermal resources have been actively under study since 1970. Seven reports at the 2nd U.N. Symposium (Kzishnaswamy, 1975; Subramanian, 1975; Gupta and Narain, 1975; Shanker et al., 1975; Jangi et al., 1975;

and Chaturnediand, 1975) indicate the breadth and importance that India has placed on the exploration and development of the nation's natural geothermal resource potential. Efforts to 1975 had identified a number of potentially valuable regions. For example, in the Duga, Parkti and Chonathang Valleys in northern India (in the Himalayas) temperatures in boreholes 100m deep have been found to be as high as 135°C. The geochemistry of the hot springs in the area indicate a base temperature of about 250°C.

Exploration via heat flow, geotectonic analysis, and geochemistry have been under way since 1962. A hot spring survey and compilation was completed in 1973. These studies resulted in a project, partially funded by the United Nations, for an interpretive data review and study of the north India resources. The objective was to determine if a practical geothermal electric power source could be developed in this area devoid of fossil fuels and as an alternative to expensive hydro developments. Twelve other major promising geothermal provinces have been recognized, and a very wide variety of hydrothermal reservoirs has been identified. For example, in the Cabay area a geopressured geothermal reservoir is thought to be present.

With exploration activities so well advanced and the economic attractiveness of the small-size, geothermal power plants being evident, it seems very probable that India will continue geothermal activity. Soon they will commence an electric power facility installation in Kingard and expand facilities for direct use applications.

China

Recent reports from China (Finn, 1979; Anon., 1978; Fountain, 1979) indicate that the government of the Chinese Peoples Republic has a very active geothermal exploration program. In addition, there are eight small-scale geothermal power plants on line in China with a total capacity of 4.9 MW(e). They are located in seven different regions. China has about 2,500 known hot springs and eight major spa and sanatorium facilities have been developed. It is also reported that China has recently given a high priority to geothermal resource assessment and that the Geological Institute (Geothermal Division) in Peking (Beijing) is taking a lead role in these geothermal developments.

Although Chinese resources have as yet to be systematically evaluated, the eight small plants (two based on binary cycles) indicate a very strong commitment to future extensive use of the earth's heat.

Thailand

A recent report (Barr et al., 1979) indicates that Thailand has initiated a geochemical survey of hot springs in the northern part of the country. In addition, a heat flow survey in the Fang Basin, using abandoned oil wells, has revealed heat flows in the range from 90 to 220 mW/m². These results are considered to be indicative that northern Thailand may have a significant hydrothermal potential, with high base temperatures recorded via silica geothermometry. The high heat flows may also indicate the presence of hydrothermal reservoirs at relatively shallow depths.

Prospects for the Future

Short-term and long-term quantitative predictions of penetration of the world energy market by geothermal energy resources are difficult to formulate. The analytical path from resource base to reserve and the prescription for defining potential of geothermal resources have not yet been quantitatively established. The fact that this resource is practical to find and use has been amply demonstrated by the above review of current worldwide developments in 38 countries. The rate of growth and application will depend upon the complex economic factors existing in a particular region or country. It does seem that worldwide interest in local, secure energy sources is increasing rapidly.

One factor that is clear from this review is that many of the obvious, shallow, high-grade hydrothermal reservoirs — those least costly and easiest to find and develop — are located in the developing countries of the world. On the other hand, these are regions with little current local or national demand and which will have difficulty in raising the needed capital. Countries without (or with only small) supplies of indigenous fossil energy resources will undoubtedly pursue geothermal developments with increasing intensity. This will be especially true for those nations located in the so-called geothermal zones. The Philippines is an example of the response to this motivation.

Some industrial nations, with Japan being the most illustrative example, that have abundant geothermal resources and no fossil fuel reserves will, therefore, accelerate development of their geothermal resources. It is observed that a goal often set by the developed countries with high-grade geothermal resource potential is to plan to bring on-line geothermal electric generating capacity at about 10% of the presently installed fossil and nuclear generating capacity during the next 25 years or so.

Short-Term Projections

The prediction of short-term growth of installed electric generating capacity on a worldwide level is a most risky proposition. The most comprehensive recent assessment effort to make such a projection (Roberts, 1978) for

Table 2-27
Projected Hydrothermal Electricity Potential
(calculated, 3 km depth)

Country	Electric energy potential E_b (MW cen)	Electric power potential, 1000 MW(e)		
		1985	2000	2020
China	160,000	23	70	140
US	150,000	23	70	140
Indonesia	130,000	20	60	120
Peru	86,000	13	38	77
Mexico	77,000	12	36	72
USSR	72,000	11	33	65
Chile	54,000	8	25	49
Ethiopia	50,000	8	23	45
Ecuador	30,000	5	14	27
Brazil	29,000	4	13	26
Turkey	26,000	4	12	24
Japan	24,000	4	11	22
		(2)	(50)	
Kenya	24,000	4	11	22
Colombia	23,000	4	11	21
Bolivia	22,000	4	10	20
Iran	22,000	4	10	20
Philippines	20,000	3	9	18
Venezuela	12,000	2	6	11
Vietnam	11,000	2	5	10
Nicaragua	10,000	1.5	5	9
Italy	9,900	1.5	5	9
New Guinea	9,600	1.5	5	9
New Zealand	9,100	1.4	4	8
Chad	8,600	1.3	4	8
Zambia	7,500	1.1	3.4	7
Guatemala	7,200	1.1	3.2	6.5
Korea	7,200	1.1	3.2	6.5
Iceland	6,800	1	3	6
Total	1.1×10^6	169.5	502.8	998

worldwide geothermal energy electric generating capacity from high-grade hydrothermal sources is summarized in Tables 2-27 and 2-28. The 28 countries listed in Table 2-27 are those ranked by their calculated potential resources above 150°C*. They represent the nations richest in this relatively

---

* See Appendix C for a complete listing by country of the worldwide resource base; the total energy above 150°C (Classes 3 and 4) is 10.1×10^5 quads.

Table 2-28
Projected Electricity-Generating Capacity from
High-Grade Hydrothermal Resources*

Country	1976 Installed capacity MW(e)	1985 Estimated capacity MW(e)	2000 Estimated capacity MW(e)
USA	502	>3,000	>20,000
Italy	421	800	–
New Zealand	190	400	1,400
Japan	68	2,000	50,000
Mexico	75	400-1,400	1,500-10,000
USSR	5.7	–	–
Iceland	2.5	150	500
Turkey	0.5	400	1,000
Canada	–	10	–
Costa Rica	–	100	–
El Salvador	60	180	–
Guatemala	–	100	–
Honduras	–	100	–
Nicaragua	–	150-220	300-400
Panama	–	60	–
Argentina	–	20	–
Portugal	–	30	100
Spain	–	25	200
Kenya	–	30	60-90
Indonesia	–	30-100	500-6,000
Philippines	–	300	–
Taiwan	–	50	200
Total	1,325	8,335-9,335	14,775-100,000

*Based on questionnaires, from Roberts (1978).

accessible (assumed maximum depth is 3 km) resource. This calculation further presumes that only 20% of the high-grade resource base is partitioned to hydrothermal resources (the remainder is heat in potential hot, dry rock resources). This calculation also assumes an overall 2% recovery and conversion or efficiency factor. Also included in Table 2-27 is a projected growth scenario for the use of these resources in the years 1985, 2000, and 2020.

Table 2-28 is a summary of the results of the literature search (status review presented above) and the EPRI survey questionnaire responses from the 22 countries represented. Figure 2-36 illustrates these two growth predictions from these two sources. The values from national information sources indicate perhaps an order of magnitude less than the 10^6 MW(e)

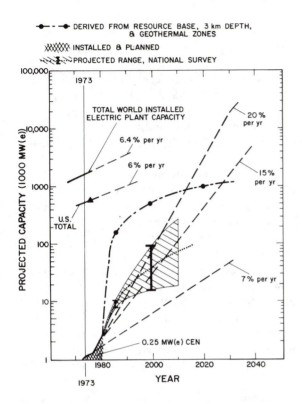

●- - -● DERIVED FROM RESOURCE BASE, 3 km DEPTH,
          & GEOTHERMAL ZONES

XXXXXX INSTALLED & PLANNED

PROJECTED RANGE, NATIONAL SURVEY

Figure 2-36. Short-term projections of worldwide hydrothermal electric power installed capacity. Note that total cumulative use in 1978 was about 0.25 MW(e) centuries per Figure 2-22.

capacity projected in the EPRI study for the 28 countries by the year 2020. A worldwide geothermal electric power generation capacity of 500,000 MW(e) can be compared to a total (1978) U.S.A. generating capacity from all sources, i.e., coal, oil, gas, hydro and nuclear, of about 560,000 MW(e), and a worldwide installed capacity of about 1.2×10^6 MW(e). The 500,000 MW(e), if fueled by oil, would represent about 40×10^6 barrels of oil per day (assuming 40% conversion).

As a further perspective, it is estimated that 100,000 MW(e) of installed geothermal capacity would represent the drilling of about 20,000 to 60,000 production wells, if it is assumed that the world average of about 5 MW(e) per well continues to hold in the future development of high-grade geothermal reservoirs.

The assumption of the world's ability to sustain a short-term geothermal development growth rate ranging from 20 to 25%, with a doubling every 4 or 5 years, is a valid one. Such growth would achieve a 10-fold increase in installed capacity over a 20- to 30-year span. It seems likely that the very sharp projected increase (10- to 100-fold) from 1978 to 1985, as shown in Figure 2-36 and derived from the resource base estimates, is perhaps too optimistic. Such optimism may result from the current pressures of the search for rapid solutions to the energy problems that many of the countries of the world experience.

As a practical matter, the development of geothermal energy will probably be impeded in the near term by a shortage of knowledgeable and experienced professionals; the lack of investment capital; limitation on available drilling equipment and plant facilities; and the competition with the accelerated development of other domestic fuel supplies. These limiting factors are of course interdependent.

There is an increasing amount of leverage in the area of availability of information, expertise, and enhancement of technology that have resulted from the recent investments in research and development around the world. In the information area, the appearance of a number of texts (Goguel, 1976; Elder, 1976; Milora and Tester, 1976; Armstead, 1978; Coolie, ed., 1978; Bierman et al., 1978; Ellis and Mahon, 1978; Rybach and Stegera, 1979) on the various aspects of geothermal development have provided access to and insights into the ideas and data of the fundamental nature of geothermal systems and the economics and technology of developing the resources. The worldwide experience base has also grown and a number of firms and institutes are available with experience and expertise in geothermal development. The United Nations has consistently supported geothermal development and provides access to funding and expertise, especially for developing nations.

It seems most probable that the most rapid advances in geothermal energy use will come from the world's low-enthalpy resources. These developments require far less commitment of capital, trained manpower, and complex hardware. It would seem very likely that a worldwide 10- or 100-fold increase in use over the next few decades of this energy resource is very probable. In 1975, Howard reported the existence of about 5×10^{14} joules/day or 7000 MW(t) of direct geothermal use. Thus, the extremely widespread geographic availability of geothermal energy in the aquifers of the world's sedimentary basins and the large resource magnitude that may total a potential energy supply of at least 4×10^{23} joules (1.5×10^8 MW centuries, th) makes very reasonable prediction of the expanded use in the year 2000 to the order of from 70 to 700 GW (i.e., a 10- to 100-fold increase) over present-day capacity. Indeed, countries or commercial interests that are not now considering or investigating applications of low-temperature geothermal resources would be very well advised to place nonelectric uses high on their agendas for future exploitation.

Long-Term Projections

How can one get a perspective on the long-term use of the world's geothermal resources? That is, what is likely to happen over the decades as the total resource base of the accessible heat energy stored in the earth's crust is used by man? The problem with writing such a scenario is that on the one hand a large resource base is derived but on the other, inasmuch as present-day production is very small, no significant amount of worldwide data on measured resources is available. Table 2-29 illustrates this point by comparing resource base, measured resource, and production rates for oil, coal, and geothermal energy. Also, inasmuch as geothermal energy resources are not transportable, in the direct sense that coal and oil are, the growth in production of high-grade hydrothermal resources for electricity generation and the local use of low-enthalpy waters will differ from oil and coal.

When hot, dry rock technology is firmly established and economic conditions dictate exploitation, some of the geographic constraints to development of geothermal energy will be relaxed. Speculation about HDR development at this time varies widely because a large number of hot, dry rock developments will be needed to provide sufficient background experience to make routine heat extraction by that technique possible. Certainly the period from 1980 to 2010 seems a likely expectation for this resolution. It also seems reasonable that high-temperature use and low-enthalpy applications will develop at different paces. This factor is somewhat analogous to the different sources of oil, i.e., crude, oil shale, coal, tars, and oil-sands; and (2) various types of solid fuels, i.e., anthracite, bituminous coal, subbituminous coal, brown coal, lignite, and peat. One other area of basic information that is still missing encompasses the *in situ* mechanisms and detailed and historical field production data for the heat and fluid recharge in geothermal reservoirs. It may be far too pessimistic and simplistic to merely assume that once the original in-place heat and fluids are removed, the resource will be depleted.

In spite of such uncertainties, a long-term projection was made and is summarized in Figure 2-37. The production rate projection methodology used is that based on the concepts proposed by R. K. Hubbert (1970). The example of Hubbert's idea is provided by the coal resource curves, where the range in maximum production rates are set at about nine times and six times the current rate (73 quads/year). They correspond to the values of the estimated ultimate resource base use, i.e., total coal resource bases of 204,000 and 115,000 quads, respectively. The areas under the production rate—time curves were adjusted to include the complete depletion of these two resource bases.

To perform quantitatively meaningful long-term projections for geothermal development, it is necessary to know the worldwide future demand for energy and to judge the development rate and balance-of-use scenarios for the alternative energy sources, i.e., coal, nuclear, hydro, oil, gas, and solar. Another constraint is the possible upper limits to practical increases in

Table 2-29
Comparison of World Crude Oil, Coal and Geothermal
Resource Base, Reserves, and Production

Resource	Resource base (quads)	Measured reserves (quads)	Production rate (quads/year)
Crude oil	11,970 to 7,700*	3,960**	131[†]
Coal	204,000 to 115,000*	15,110**	73[†]
Geothermal:			
Hydrothermal	~200,000[‖]	¶	~3[‡]
Crustal heat[§]	400×10^6 to 41×10^6	¶	

*Hubbert, 1970.

**Peck, 1974.

[†]Crabbe and McBride, 1975.

[‡]Calculated on the basis of 1,300 MW(e) installed electric generating capacity in 1976 and 7,000 MW(t) of direct use. The overall recoverability and efficiency of the in-place stored geothermal heat to power production was taken to be 2%.

[§]Total integrated heat content to a depth of 10 km, based on global heat flow data; heat stored to a depth of 3 km, based upon geothermal zones.

[‖]High-grade (temperature 150°C); to a depth of 3 km; hydrothermal only, based on world geothermal zones.

¶No detailed world evaluation of geothermal reserves has yet been attempted.

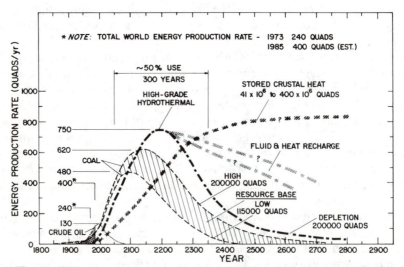

Figure 2-37. Long term projections of geothermal resource production rates in quads/year, compared with crude oil and coal yearly production.

production rates with time. In spite of these restrictions, Hubbert's methodology allows one to scope the possible use. If the analysis is restricted to high-grade, hydrothermal resources (from Table 2-29) and if it is further assumed that only future electric power production is projected, it is possible to bound the installed capacity and thus the use of the resource base. The expected duration of the use of this resource would be about 300 years if it is assumed that:

1. Rate of growth is less than 15 to 20% per year.
2. Installed worldwide capacity peaks at approximately 500,000 MW(e), which is equivalent to 750 quads/year of *in situ* heat/year production at 2% overall recovery and conversion efficiency.
3. A 1,000-fold increase in capacity over 1976 levels occurs.
4. No fluid or heat recharge occurs.

This development and use scenario is illustrated in Figure 2-37, where projected world crude oil and coal resource production rates are also shown for comparison. It is important to note that the large high-grade hydrothermal resource base is comparable to that for coal and, therefore, can be expected to be developed at a comparable rate and used for several centuries. The long-term extent of hydrothermal usage will depend upon the character of the fluid and heat recharge for such reservoirs and, as indicated in Figure 2.37, on the competition with coal resource development. The ultimate use of hot, dry rock, stored crustal heat, and thermal waters will be dependent upon the rate of depletion of transportable fossil fuels, the competing rate of increase of use of biomass-based fuels for transportation, and application of solar energy for low-grade heat use. Inasmuch as the geothermal resource base is vast, however, this energy source can be projected with confidence to make a significant contribution to the world energy needs for many hundreds of years.

References

Akil, I., 1975. Development of Geothermal Resources in Indonesia. In: *Proc. 2nd United Nations Symp. on Development and Use of Geothermal Resources,* San Francisco, CA, 11-15.

Albert-Beltran, J. F., 1979. Heat Flow and Temperature Gradient Data from Spain. In: V. Cermak and L. Ryback (eds.) *Terrestrial Heat Flow in Europe.* Springer-Verlag, NY, 226-261.

Alonzo, M., 1975. Geothermal Potential in Mexico. In: *Proc. 2nd United Nations Symp. on the Development and Use of Geothermal Resources,* San Francisco, CA, 21-24.

Alpan, S. 1975. Geothermal Exploration in Turkey. In: *Proc. 2nd United Nations Symp. on the Development and Use of Geothermal Resources,* San Francisco, CA, 25-28.

Alvarez, R., 1979. Self Potential Anomalies in Geothermal Fields: Dipolar vs. Superposed-Dipole. *GRC Trans.,* Reno, NV, 3: 11-14.

Anderson, R. N. et al., 1979. Geothermal Convection Through Oceanic Crust and Sediments in the Indian Ocean. *Science,* 204 (May 25): 828-832.

Anon., 1974. *Circum Pacific Energy and Mineral Resource Conference,* August, Honolulu, HI.

Anon., 1976. *Organization and Definitions for the Estimation of Reserves and Productive Capacity of Crude Oil.* Prepared by the Am. Pet. Inst. Statistical Dept., June, API Tech. Note No. 2 (second edition).

Anon., 1977. Geothermal Energy, Chapter in: *Unconventional Energy Resources,* prepared for the Conservation Commission, World Energy Conference, EPRI, 59-92.

Anon., 1977. *Demographic Yearbook 1976,* United Nations, New York.

Anon., 1977. Geothermal Energy—The Hot Prospect. *EPRI J.,* 3, Apr.: 6-13.

Anon., 1977. *World Energy.* U.N. Statistical Sources, Dept. Econ. and Social Affairs, NY, 166-177.

Anon., 1977. *Unconventional Energy Resources.* EPRI, prepared for the Conservation Commission, Aug., World Energy Conference.

Anon., 1977. Utilization of Geothermal Energy in Japan. *GRC Bull.,* Oct., Special Insert, Sunshine Project Promotion Headquarters.

Anon., 1977. *Reference Notes for Field Trips.* Larderello Workshop on Geothermal Resource Assessment and Reservoir Engineering, ENEL, Larderello, Italy.

Anon., 1977. *API-Reserves of Crude Oil, Natural Gas Liquids, and Natural Gas in the United States and Canada as of Dec. 31, 1976.* API Statistics Dept., 31 (May).

Anon., 1978. *Direct Utilization of Geothermal Energy. A Symposium,* Jan. 3-Feb. 2, Geothermal Resources Council, San Diego, CA CONF-780133.

Anon., 1978. *Proc. NATO-CCMS Conference on Economics of Direct Uses of Geothermal Energy,* Washington, DC, July, U.S. Dept. of Energy CONF-770681.

Anon., 1978. New Geothermal Discoveries in Nicaragua. *GRC Bull.,* p. 6.

Anon., 1978. *Geothermal Energy.* Japan Geothermal Energy Development Center (in Japanese).

Anon., 1978. Japan Makes Progress in Geothermal Energy. *Pacific Friend,* 6(11): 10-16.

Anon., 1978. *Collection of Works on Geothermal Studies.* Geothermal Research Division, Institute of Geology, Academica Sinica, Science Press, Peking, China.

Anon., 1979. Geothermal Energy to Supply Oregon-Idaho Plant. *Geothermal Energy Magazine,* 7(1): 11-16.

Anon., 1979. *Symposium of Geothermal Energy and Its Direct Uses in the Eastern United States.* Geothermal Resource Council, Spec. Report April, No. 5.

Araña, V. and Panichi, C., 1974. Isotopic Composition of Steam Samples from Lanzarote, Canary Islands. *Geothermics,* 3(4): 142-145.

Armstead, H. C. H., 1978. *Geothermal Energy.* The Holsted Press, J. Wiley and Son, New York.

Armstrong, E. L., 1977. *Geothermal Investigations of Imperial Valley, California.* Status Report, April 1977, U.S. Dept. of Interior, Bureau of Reclamation.

Axtell, L., ed., 1975. Geothermal Energy Utilization in Japan. *Geothermal Energy Magazine,* 3(2, Feb.): 5-24.

Balling, N. and Saxov, S., 1979. Low Enthalpy Geothermal Energy Resources in Denmark. In: L. Rybach and L. Stegena eds., *Geothermics and Geothermal Energy, CCRG,* No. 7: 205-212.

Balogh, J., 1975. Results achieved in Hungary in the Utilization of Geothermal Energy. In: *Proc. 2nd United Nations Symp. on the Development and Use of Geothermal Resources,* San Francisco, CA, 29-31.

Barbier, E., ed., 1970. United Nations Symposium on the Development and Utilization of Geothermal Resources. Pisa, Italy. *Geothermics,* Spec. Issue No. 2, V. 1 and 2.

Barbier, E. and Fanelli, M., 1973. Overview of Geothermal Exploration and Development in the World. *Geothermal Energy,* Pisa, Italy, National Research Council.

Barr, S. M. et al., 1979. Hot Springs and Geothermal Gradients in Northern Thailand. *Geothermics,* 8(2): 85-96.

Basoeki, I. M. and Radja, I. V. T., 1978. Recent Development of 30 mw(e) Kamojang Geothermal Power Project, West Java, Indonesia. *GRC Trans.,* 2: 35-38.

Beboot, D. G., 1978. Trend areas and Exploration Techniques. In: M. D. Campbell (ed.), *Geology of Alternate Energy Resources, Houston Geol. Soc.,* Chapter 12: 251-273.

Bierman et al., 1978. *Geothermal Energy in the Western United States-Innovation vs. Monopoly.*

Birsic, R. J., 1978. Geothermal Development in the Philippines, Iceland, and Japan. *Geothermal Energy Magazine,* 6(5): 29-34.

Blackwell, D. D. et al., 1977. Heat Flow and Geothermal Gradient Exploration of Geothermal Areas in the Cordilla de Guanacoste of Costa Rica. *GRC Trans.,* 1: 17-18.

Blondel, F. and Lasky, S. G., 1956. Mineral Reserves and Mineral Resources. *Economic Geol.,* LI(7): 686-697.

Bodvarsson, G., 1974. Geothermal Resources Energetics. *Geothermics,* 3(3): 83-92.

Bodvarsson, G., 1975. *Fracture Flow in Geothermal Reservoirs.* Workshop on Geothermal Reservoir Engineering, Stanford Univ. Press. 45-49.

Bodvarsson, 1975. Estimates of the Geothermal Resources of Iceland. In: *Proc. 2nd United Nations Symp. on Development and Use of Geothermal Resources,* San Francisco, CA., pp. 33-35.

Boldizar, T., 1974. Geothermal Energy in Hungary. In: *Multipurpose Use of Geothermal Energy,* Conf. Proc., Oregon Inst. Technology, Klamath Falls, OR, pp. 1-15.

Bolten, R. S., 1975. Recent Developments and Future Prospects for Geothermal Energy in New Zealand. In: *Proc. 2nd United Nations Symp. on Development and Use of Geothermal Resources,* San Francisco, CA, pp. 37-42.

Bosh, B. et al., 1977. The Geothermal Zone of the Lake Assal (F.T.A.I.). Geochemical and Experimental Studies. *Geothermics,* 5: 165-175.

Bowen, R. G., 1977. Net Energy Delivery from Geothermal Resources. *Geothermal Energy Magazine,* 5(2): 15-19.

Bresse, J. C. et al. (eds.), 1978. *Geothermal Pilot Study Final Report—Creating an International Geothermal Energy Community.* June 1, LBL-6860, Chapter D., Small Geothermal Power Plant Substudy.

Burrows, W., 1970. Geothermal Energy Resources for Heating and Associated Applications in Rotorum, New Zealand. UN Symp. on Development and Utilization of Geothermal Resources, Pisa 1 Proc., *Geothermics,* Spec. Issue, 2(pt. 2): 1169-1175.

Campbell, J. M., 1970. Definitions of Petroleum Reserves for Property Evaluation. *Oil and Gas Property Evaluation and Reserve Estimates,* SPE. Reprint Series, No. 3: 7.

Campbell, J. M., 1970. Reserves and Primary Performance Predictions. *Oil and Gas Property Evaluation and Reserve Estimates.* Petroleum Transaction Reprint Series, No. 3: 6-7.

Campbell, J. M. et al., 1970. *Oil and Gas Property Evaluation and Reserve Estimates.* Petroleum Transactions Reprint Series No. 3, SPE of AIME, Dallas, TX.

Campbell, M. D. (ed), 1977. *Geology of Alternate Energy Resources in the South-Central United States.* Houston Geol. Soc., Chapters 11-14: 215-297.

Carle, A. E. and Van Rooyen, A. J. M., 1969. *Further Measurements of Heat Flow in South Africa.* Geol. Soc. of Africa, Upper Mantle Project, Spec. Pub. No. 2, July: 445-449.

Carrasco, R., 1975. Preliminary Report on Bolivia's Geothermal Resource. In: *Proc. 2nd United Nations Symp. on Development and Use of Geothermal Resources,* San Francisco, CA, 45-46.

Cataldi, R. et al., 1977. Assessment of the Geothermal Potential of Central and Southern Tuscany. *Proc. of the Larderello Workshop on Geothermal Resource Assessment and Reservoir Engineering,* ENEL-ERDA, Sept., Pisa, Italy. *Geothermics,* 7(2-4): 91-132.

Cermak, V. et al., 1975. Geothermal Mapping in Central and Eastern Europe. In: *Proc. 2nd United Nations Symp. on the Development and Use of Geothermal Resources,* San Francisco, CA, 47-57.

Cermak, V. and Hurtig, E., 1977. *Preliminary Heat Flow Map of Europe.* IASPET Internal Heat Flow Commission, Geophys. Inst. Czech. Acad. Sci., Praha, and Central Earth Phys. Inst., Acad. Sci. of the GDR, Potsdam.

Cermak, V. and Rybach, L., 1979. *Terrestrial Heat Flow in Europe.* Springer Verlag, NY.

Ceron, P. et al., 1975. Progress Report on Geothermal Development in Italy from 1969 to 1976 and Future Prospects. In: *Proc. 2nd United Nations Symp. on the Development and Use of Geothermal Resources,* San Francisco, CA 59-66.

Chapman, D. S. and Pollack, H. N., 1975. Global Heat Flow: A New Look. *Earth and Planetary Sci. Letters,* 28: 23-32.

Chaturnedi, L. N. and Raymahashay, B. C., 1975. Geological Setting and Geochemical Characteristics of the Parbati Valley Geothermal Field, India. In: *Proc. 2nd United Nations Symp. of the Development and Use of Geothermal Resources,* San Francisco, CA, 1: 329-338.

Chen, C., 1975. Thermal Waters in Taiwan. A Preliminary Study. *Internat. Union Geol. and Geophys., XVI General Assembly,* Grenoble, p. 268.

Cherng, F-P., 1979. Geochemistry of the Geothermal Fields in the Slate Terrane. *Geothermal Resources Council Trans.,* 3: 107-112.

Combs, J. (ed.), 1977. Geothermal: State of the Art, *Trans. Geothermal Resources Council,* San Diego, CA. v. 1.

Coolie, M. J. (ed.), 1978. *Geothermal Energy Recent Developments.* Noyes Data Corp., ISBN: 0-8155-0727-5.

Corrules, M. F. et al., 1977. Exploration of the Guanacosta, Costa Rica Geothermal System. *GRC Trans.* 1: 57-59.

Costain, J. R. et al., 1978. *Evaluation and Targeting of Geothermal Energy Resources in the Southeastern United States.* June, VPI-SU-5648-3.

Coulhois, P. and Herault, J. P., 1975. Conditions for Competitive Use of Geothermal Energy in Home Heating. In: *Proc. 2nd United Nations Symp. on Development and Use of Geothermal Resources,* San Francisco, CA, 2104-2108.

Cox, M. A., 1978. The Lambura Area Geothermal Investigation, Figi. *GRC Trans.* 2: 121-123.

Crabbe, D. and McBride, R., 1978. *The World Energy Book.* Nichols Publ. Co., New York, Table 10, p. 245.

Cummings, R. G. et al., 1979. Mining Earth's Heat: Hot, Dry Rock Geothermal Energy. *Technology Review,* MIT Press, 81(4): 58-78.

D'Archimban, J. D. and J. P., Munier-Jolan, 1975. Geothermal Exploration Progress at Bouillante in Guadeloupe. In: *Proc. 2nd United Nations Symp. on Development and Use of Geothermal Resources,* San Francisco, CA, 105-107.

David, M., 1977. *Geostatistical Ore Reserve.* Elsevier Scientific Pub. Co., NY, 11-19.

Decius, L. C., 1961. Geological Environmental Hyperthermal Areas in the Continental U.S. and Suggested Methods for Prospecting for Thermal or Geothermal Power. Paper G148, *Proc. U.N. Conf. for New Sources of Energy,* Rome, Italy.

Domino, E. and Papostanatori, A., 1975. Characteristics of Greek Geothermal Waters. In: *Proc. 2nd United Nations Symp. on Development and Use of Geothermal Resources,* San Francisco, CA, 109-121.

Donaldson, I. G. and Grant, M. A., 1977. An Estimate of the Resource Potential of New Zealand Geothermal Fields for Power Generation. *Proc. Larderello Workshop,* 413 pp.

Donaldson, I. G. and Grant, M. H., 1978. An Estimate of the Resource Potential of New Zealand Geothermal Fields for Power Generation. Proc. Larderello Workshop on Geothermal Resource Assessment and Reservoir Engineering, ENEL Studie Ricerche, Larderello, Italy, Sept. 1977. *Geothermics,* 7(2-4): 243-252.

Dondanville, R. P., 1978. Geologic Characteristics of the Valles Caldera Geothermal System, NM. *Geothermal Resource Council, Trans.* 2(1): 157-159.

Eckstein, Y. and Simmons, G., 1978. Measurement and Interpretation of Terrestrial Heat Flow in Israel. *Geothermics,* 6: 117-142.

Eckstein, Y., 1979. Review of Heat Flow Data from the Eastern Mediterranean Region. In: L. Rybach and L. Stegena, (eds.), *Geothermics and Geothermal Energy, CCGR No. 7:* 150-159.

Eder, J. W., 1979. Magma Traps. In: L. Rybach and L. Stegena (eds.),*Geothermics and Geothermal Energy,* Basel & Stutgarte, pp. 3-33.

Ehring, T. W. et al., 1978. Formation Evaluation Concepts for Geothermal Resources. *SPWLA 19th Annu. Logging Symp.,* June, El Paso, TX.

Einarsson, S. S., 1973. Geothermal District Heating. In: H. C. H. Armstead (ed.), *Geothermal Energy,* UNESCO Press, 123-124.

Einarsson, S. S., 1975. Geothermal Energy Could Enable Central America to Eliminate Petroleum Imports for Power Generation by 1980. In: *Proc. 2nd United Nations Symp. on the Development and use of Geothermal Resources,* San Francisco, CA, 2363-2368.

Elder, J., 1976. *The Bowels of the Earth.* Oxton Univ. Press., New York.

Elizarraras, G. R., 1978. OLADE and Geothermal Energy in Latin America. *Geothermal Energy Magazine,* 6(7): 35-38.

Ellis, A. J. and Mahon, W. A. J., 1978. *Chemistry and Geothermal Systems.* ISBN 0-12-1274j0.

Eriksson, K. G. and Malmquist, D., 1979. A Review of the Past and the Present Investigations of Heat Flow in Sweden. In: V. Cermak and L. Rybach (eds.), *Terrestrial Heat Flow in Europe.* Springer-Verlag, NY, 267-277.

Eriksson, K. G. et al., 1979. Investigations for Geothermal Energy in Sweden. In: L. Rybach and L. Stegena (eds.), *Geothermics and Geothermal Energy, CCRG,* 196-204.

Finn, D. F. X., 1979. Geothermal Developments in the Republic of Indonesia-1979. *Geothermal Resources Council Trans.,* v. 3.

Finn, D. F. X., 1979. Geothermal Developments in the People's Republic of China. *GRC Trans.,* 3: 209-210.

Fouillac et al., 1975. Preliminary Geothermic Studies on Mineral Water in French Massif Central. In: *Proc. 2nd United Nations Symp. on the Development and Use of Geothermal Resources,* San Francisco, CA, 726-729.

Fournier, R. O. and G. W. Crosby, eds., 1979. Expanding the Geothermal Frontier. *Trans. Geothermal Resources Council,* Reno, NV., v. 3.

Franko, O. and Rocicky, M., 1975. Present State of Development of Geothermal Energy Resources in Czechoslovakia. In: *Proc. 2nd United Nations Symp. on Development and Use of Geothermal Resources,* San Francisco, CA, pp. 132-137.

Fytika, M. D. and Kolios, N. P., 1979. Preliminary Heat Flow Map of Greece. In: V. Cermak and L. Rybach (eds.), *Terrestrial Heat Flow in Europe,* 197-205.

Gable, R., 1979. Draft of a Geothermal Flux Map of France. In: V. Cermak and L. Rybach (eds.), *Terrestrial Heat Flow in Europe,* 172-177.

Gavtina, G. A. and Makarenko, F. A., 1975. Geothermal Map of the USSR. In: *Proc. 2nd United Nations Symp. on the Development and Use of Geothermal Resources,* May 20-29, San Francisco, CA, USA, 1013-1017.

Geertsma, J., 1978. Prospects for Geothermal Energy Recovery. *Erdoel-Erdgas Z.* (Int'l. Edition, P978) Nov.: 34-41.

Girdler, R. W. and Evans, T. R., 1977. Red Sea Heat Flow. *Geophys. J. R. Austr. Soc.,* 51: 245-251.

Glass, I. I., 1977. Prospects for Geothermal Energy Applications and Utilization in Canada. *Energy,* 2: 407-428.

Glover, L., 1979. General Geology of the East Coast with Emphasis on Potential Geothermal Energy Regions: A Detailed Summary. GRC Spec. Report No. 5, April: 9-12.

Goeller, H. E. et al. (compilers), 1974. *Survey of Energy Resources* (1974), *Proc. World Energy Conf.,* Table 1X-2, 253-255.

Goguel, J. 1976. *Geothermics.* McGraw-Hill, New York.

Goldsmith, L. M., 1979. Geothermal Exploration—A Case History. *GRC Trans.,* 3: 261-264.

Goranson, C. B. and D. C. Shroader, 1979. Site Specific Geothermal Reservoir Engineering Activities at LBL. *GRC Trans.,* 2: 265-268.

Govett, G. J. S. and Govett, M. M., (eds.), 1976. *World Mineral Supplies: Assessment and Perspectives.* Elsevier Sci. Publ. Co., NY, 268-273.

Grant, M. A., 1979. Maping Kamojang Reservoir. *GRC Trans.,* 3: 271-274.

Gringarten, A. C. and Sauky, J. P., 1975. The Effect of Reinjection on the Temperature of a Geothermal Reservoir Used for Urban Heating. In: *Proc. 2nd United Nations Symp. on the Development and Use of Geothermal Resources,* San Francisco, CA, 1370-1374.

Gupta, M., Narain, H. and Gaur, V. K. 1975. Geothermal Provinces of India as Indicated by Studies of Thermal Springs, Terrestrial Heat Flow, and Other Parameters. In: *Proc. 2nd United Nations Symp. on the Development and Use of Geothermal Resources,* San Francisco, CA, 1: 387-396.

Gupta, M. L., Saxena, V. K. and Sukhiya, B. S., 1975. An Analysis of the Hot Spring Activity of the Manikaran Area, Himachal Pradesh, India, by Geochemical Studies and Tritium Concentration of Spring Waters. In: *Proc. 2nd United Nations Symp. on the Development and Use of Geothermal Resources,* San Francisco, CA, 1: 741-755.

Haldane, T. G. N. and H. C. H. Armstead, 1962. The Geothermal Power Development at Wairake, New Zealand. *Proc. Inst. Mech. Eng.,* 176(23): 603-627.

Hamza, V. M. et al., 1975. Geothermal Energy Prospects in Brazil: A Preliminary Analysis. L. Rybach and L. Stegena (eds.) *Geothermics and Geothermal Energy,* CCRG no. 7: 180-195.

Hartbough, J. W., et al., 1975. *Probability Methods in Oil Exploration.* John Wiley, NY, 19-36.

Herman, B. M. et al., 1977. Heat Flow in the Oceanic Crust Bounding Western Africa. *Tectonophysics,* 41: 61-77.

Hill, J. H., 1978. Los Alamos Scientific Laboratory Hot Dry Rock Geothermal Energy Development Project. *Geothermal Resource Council Trans.,* 2(1, July): 275-277.

Hochstein, M. P., 1975. Geophysical Exploration of the Kowah-Kampjang Geothermal Field, West Java. In: *Proc. 2nd United Nations Symp. on the Development and Use of Geothermal Resources,* San Francisco, CA, pp. 1049-1058.

Hornburg, C. D. and B. Lindel, 1978. *Preliminary Research on Geothermal Energy Industrial Complexes.* DSS Engineers, Fort Lauderdale, FL, March IDO-1627-4.

Howard, J. M. (ed.), 1975. *Present Status and Future Prospects for Nonelectric Uses of Geothermal Resources.* U.C.R.L. 61926.

Hubbert, M. K., 1969. *Energy Resources, Resources and Man.* U. S. National Acad. Sciences, pp. 157-242.

Hurtig, E. and Oelsner, C. H., 1979. The Heat Flow Field on the Territory of the German Democratic Republic. In: V. Cermak and L. Rybach (eds.), *Terrestrial Heat Flow in Europe,* Springer-Verlag, NY, 186-190.

Jangi, B. L., Prakash, G., Dua, K. J. S., Thussu, J. L., Dimri, D. B. and Pathak, C. S., 1975. Geothermal Exploration of the Parbati Valley Geothermal Field, Kulu District, Hinwachae Pradesh, India. In: *Proc. 2nd United Nations Symp. on the Development and Use of Geothermal Resources,* San Francisco, CA, 2: 1085-1094.

Jessop, A. M., 1978. Geothermal Energy from Sedimentary Formations in Western Canada. *GRC Trans.* Hilo, HI, 2(1): 331-334.

Karakulits, O. C. and Hankins, B. E., 25-27, 1978. Chemical Analysis of Dissolved Natural Gas in Water from the World's First Geopressured Geothermal Well. *Trans. Geothermal Resources Council,* 2(1, July): 351-354.

Kent, L. E., 1969. The Thermal Waters in the Republic of South Africa. *Internat. Geol. Congr. Proc.,* 19(xxiii): 143-164.

Koenig, J., 1973. Worldwide Status of Geothermal Resources Development. In: P. Kruger and C. Otte (eds.), *Geothermal Energy,* Stanford Univ. Press, Chapter 2.

Koenig, J. et al., 1975. Exploration and Development of Geothermal Resources in the United States, 1968-1975. In: *Proc. 2nd United Nations Symp. on the Development and Use of Geothermal Resources,* San Francisco, CA, pp. 139-142.

Kruger, P. and Ramey, H. J., Jr. (eds.), 1975. *Geothermal Reservoir Engineering.* Workshop Report SGP-TR-12, Dec. 15-17, Stanford Univ., CA.

Kruger, P. and Ramey, H. J. (eds.), 1976. Geothermal Reservoir Engineering, Summaries of Second Workshop, Dec. 1-3, SGP-TR-20.

Kruger, P. and Otte, C., eds., 1976. *Geothermal Energy.* Stanford Univ. Press, CA, 23-29.

Kruger, P. and Ramey, H. J. (eds.), 1977. *Proc. Third Workshop Geothermal Reservoir Engineering,* Dec. SGP-7R-25.

Kruger, P. and Ramey, H. J. (eds.), 1978. *Geothermal Reservoir Engineering,* Summaries of Fourth Workshop, Dec. 13-15.

Kurkman, F. and Samilgil, E., 1975. Geothermal Energy Possibilities, Their Exploration and Evaluation in Turkey. In: *Proc. 2nd United Nations Symp. on the Development and Use of Geothermal Resources,* San Francisco, CA, 447-458.

Kzishnaswamy, V. S., 1975. A Review of Indian Geothermal Provinces and Their Potential for Energy Utilization. In: *Proc. 2nd United Nations Symp. on the Development and Use of Geothermal Resources,* San Francisco, CA, 143-156.

Lansen, A. and Trujillo, P., 1975. The Geothermal Field of El Tatio, Chile, In: *Proc. 2nd United Nations Symp. on the Development and Use of Geothermal Resources,* San Francisco, CA, 170-177.

Lawyer, L. A. et al., 1975. A Major Geothermal Anomaly in the Gulf of California. *Nature,* 80: 3733-3743.

Levitte, D. and Eckstein, Y., 1979. Correlation Between the Silica Concentrations and the Orifice Temperature in the Warm Springs Along the Jordan—Dead Sea Rift Valley. *Geothermics,* 7: 1-8.

Lienau, P. J. and Lund, J. W., 1974. *Proc. of the Internat. Conf.* on Oct. 7-9. *Multipurpose Use of Geothermal Energy.* Geo-Heat Utilization Center, Oct. 7-9. Oregon Inst. Technol., Klamath Falls, OR.

Lienau, P. J. and Lund, J. W. 1977. *Proc. Internat. Conf. on Multipurpose Use of Geothermal Energy,* Oct. 7-9, Oregon Inst. Technol., Klamath Falls, Oregon.

Loddo, M. and Mongelli, F., 1979. Heat Flow in Italy. In: V. Cermak and L. Rybach (eds.), *Terrestrial Heat Flow in Europe.* Springer Verlag, NY, 221-231.

Lopez, V. C., 1978. Discussion of the Hydrogeochemistry of the Eljoyo-Obraje Valley with the Momotombo Geothermal Field at Nicaragua. *Geothermal Energy Magazine,* 6(3): 7-12.

Lund, J. W., 1978. *Direct Utilization—the International Scene, Direct Utilization of Geothermal Energy. A Symposium.* San Diego, CA, April, CONF-780133: 133-140.

Maasha, N., 1975. Electrical Resistivity and Microearthquake Surveys of the Sempaya, Lake Kitagata, and the Kitagata Geothermal Anomalies, Western Uganda. In: *Proc. 2nd United Nations Symp. on the Development and Use of Geothermal Resources,* San Francisco, CA, 1183-1112.

Maddox, J. D. and A. C. Wilbur, 1979. Baca Geothermal Demonstration Plant Data Gathering, Evaluation, and Dissemination. *GRC Trans.,* Reno, NV, 3: 405-408.

Martin, R. C. and Welday, E. E., May 1977. Lake Bottom Thermal Gradient Survey at Clear Lake and Mono Lake, California. *Geothermal Resource Council,* 1: 201-203.

Mazlan, J. A. and Klei, H. E., 1975. *A Technology Assessment of Geothermal Energy Resource Development.* NSF Report (contract C-836), The Futures Group and Bechtel National.

McCray, A. W., 1975. *Petroleum Evaluation and Economic Decisions.* Prentice Hill, Inc., NJ, 114-127.

McEwen, R. B. and Abakoyas J., 1977. Geothermal Investigation of the Lake District, Ethiopia. *Geothermal Resource Council Trans.,* 1: 208-210.

McEwen, R. B., 1978. Thermal Waters of Bulgaria: Present Use and Probable Origin. *GRC Trans.,* 2: 423-426.

McKelvey, V. E., 1974. Potential Mineral Reserves. *Resource Policy,* Dec.: 75-81.

McNitt, Jr., 1978. The United Nations Approach to Geothermal Reservoir Assessment. In: *Proc. Larderello Workshop on Geothermal Resource Assessment and Reservoir Engineering,* Sept. 12-16, 1977, Pisa, Italy, *Geothermics,* 7(2-4): 231-242.

McRae, A. and Dudas, J. L., 1977. *Energy Source Book.* Aspen Systems Corp., 240-242.

Meidav, M. Z. and Meidav, T., 1979. Direct Heat Utilization of Geothermal Energy on the Island of San Miguel. *GRC Trans.,* 3: 441-444.

Merida, J., 1975. Panama's Geothermal Fields. *Geothermal Energy Magazine,* 3(3, March): 17-18.

Milora, S. L. and Tester, J. W., 1976. *Geothermal Energy as a Source of Electric Power.* The MIT Press, Cambridge, MA.

Morgan, P. et al., 1974. Heat Flow Measurements in Yellowstone Lake and the Thermal Structure of the Yellowstone Caldera. *J. Geophys. Res.,* 82(26): 3719-3732.

Morgan, M. G. (ed.), 1975. *Energy and Man.* IEEE Press, NY.

Morgan, P. and Swanberg, C. A., 1979. Heat Flow and Geothermal Potential of Egypt. In: L. Rybach and L. Stegena (eds.), *Geothermics and Geothermal Energy, CCRG* no. 7: 213-226.

Muffler, L. J. P., 1975. Present Status of Resources Development, Rapporteur Summary. In: *Proc. 2nd United Nations Symp. on the Development and Use of Geothermal Resources,* San Francisco, May 20-29, pp. xxxiii-xiiv.

Muffler, L. J. P., 1975. Present Status of Resource Development. In: *Proc. 2nd United Nations Symp. on the Development and Use of Geothermal Resources,* San Francisco, May 20-29, Sect. I: 3-279.

Muffler, L. J. P. and Williams, D. L., 1976. Geothermal Investigations of the U.S. Geological Survey in Long Valley, California, 1972-1973. *J. Geophys. Research,* 81: 721-724.

Muffler, L. J. P. and Cataldi, R., 1978. Methodology for Regional Assessment of Geothermal Resources. Proc. of the Larderello Workshop on Geothermal Resource Assessment and Reservoir Engineering, Sept. 12-16, 1977. *Geothermics,* 7(2-4): 53-90.

Muffler, L. J. P. and Yeun, D. C. (eds.), 1978. A Novelty Becomes a Resource. *Trans. Geothermal Resources Council,* Hilo, HI.

Muffler, L. J. P. (ed.), 1979. Assessment of Geothermal Resources of the United States—1978. *U.S. Geol. Surv. Circ.* 790.

Murphy, H. D., 1979. Hot, Dry Rock Geothermal Heat Extraction Experiments. *GRC Trans.,* Reno, NV, 473-476.

Nicholson, R. V. et al., 1979. Technology for Geothermal Well Stimulation. *Trans. Geothermal Resources Council,* 3(Sept.): 499-502.

Noble, J. W. and Ojiamko, S. B., 1975. Geothermal Exploration in Kenya. In: *Proc. 2nd United Nations Symp. on the Development and Use of Geothermal Resources,* San Francisco, CA, 189-204.

Nwachukwu, S. O. O., 1975. Geothermal Regime of Southern Nigeria. In: *Proc. 2nd United Nations Symp. on the Development and Use of Geothermal Resources,* San Francisco, CA, 205-212.

Palmason, G. et al., 1975. Geothermal Development in Iceland—1970-1974. In: *Proc. 2nd United Nations Symp. on the Development and Use of Geothermal Resources,* San Francisco, CA, 213-217.

Parod, I. A., 1975. Feasibility of the Development of the Geothermal Energy in Peru—1975. In: *Proc. 2nd United Nations Symp. on the Development and Use of Geothermal Resources,* San Francisco, CA, 227-231.

Peck, W. G., ed., 1974. *Survey of Energy Resources—1974.* World Energy Conf. Center, London, 253-255. (Table IX-L—Energy Content of Measured Resources of Non-Renewable Energy Resources.)

Pollack, H. N. and Chapman, D. S., 1977. The Flow of Heat from the Earth's Interior. *Sci. Am.,* Aug.: 60-76.

Pryde, P. R., 1979. Geothermal Energy Development in the Soviet Union. *Soviet Geography,* Feb.: 69-81.

Rakowska, J. M. et al., 1977. Relationships Between Terrestrial Heat Flow and Structure of the Earth's Crust in Poland. *Geothermics,* 6(1/2): 99-106.

Ratigan, J. L. and Lindblom, V. E., 1978. The Single-Hole Concept of Heat Extraction from Hot, Dry Rock Masses. *GRC Trans.,* 2: 559-565.

Reynolds, G., 1970. Cooling with Geothermal Heat. Proc. U.N. Symp. on Development and Utilization of Geothermal Resources, Pisa, Italy. *Geothermics,* Spec. Issue 2, 2(pt. 2): 1158-1161.

Roberts, V. W. (ed.), 1978. *Geothermal Energy Prospects for the Next 50 Years.* EPRI, ER-611-SR, 4-1 to 4-10.

Roberts, V. W., 1979. *Geothermal Energy Prospects for the Next 50 Years.* Feb., EPRI-611-SR, 5-5.

Romaguoli, R. et al., 1975. Aspectos Hidrogeologicos del Campo Geotermico de Ahuachapan, El Salvador. In: *Proc. 2nd United Nations Symp. on Geothermal Resources,* San Francisco, CA, 503-574.

Rybach, L. and Jaffe, F. C., 1975. Geothermal Potential in Switzerland. In: *Proc. 2nd United Nations Symp. on the Development and Use of Geothermal Resources,* San Francisco, CA, 241-243.

Rybach, L. and Stegena, L. 1979. *Geothermics and Geothermal Energy.* Birkhausen Verlag, Basel and Stuttgart.

Salisbury, J. W., et al., 1977. Geothermal Reservoir Confirmation Requirements and the Need for Federal Initiatives. *GRC Trans.,* 1: 271-274.

Sanyal, S. K. and Meidav, H. T., 1977. *Important Considerations in Geothermal Well Log Analysis.* SPE paper No. 6535, 47th Annu. Calif. Regional Meeting of the AIME, April, Bakersfield, CA.

Sanyal, S. K. et al., 1979. Classification of Geothermal Reservoirs from the Viewpoint of Log Analysis. *20th SPWLA Trans.,* Tulsa, OK, June 3-6, 1979, paper HH.

Schurr, S. H. et al., 1960. *Energy in the American Economy, 1850-1975, An Economic Study of its History and Prospects, Resources for the Future.* The John's Hopkins Press, Baltimore,Chapter V, the Conceptual Basis of Energy Supply Estimation, 295-301.

Schwartz, W., 1978. *Geothermal Exploration Technology — Annual Report 1978.* LBL-8603.

Sekioka, M. et al., 1979. Geothermal Snow Melting System in Japan. *GRC Trans.,* 3: 639-642.

Shakad, T. 1977. *Gateway to Siberian Resources: The BAM.* John Wiley and Sons.

Shanker, R., Padhi, R. N., Arora, C. L., Prakash, G., Thussu, J. L. and Dua, K. J. S., 1975. Geothermal Exploration of the Puga and Chumathang Geothermal Fields, Ladakh, India. In: *Proc. 2nd United Nations Symp. on the Development and Use of Geothermal Resources,* San Francisco, CA, 1: 245-258.

Showers, V., 1973. *The World in Figures.* Wiley-Interscience, New York.

Silverman, M., 1978. *The Geothermal Loan Guarantee Program, Direct Utilization of Geothermal Energy. A Symposium.* San Diego, CA, 105-110.

Smith, M. C. et al., 1975. Man-made Geothermal Reservoirs. In: *Proc. 2nd United Nations Conference on Development and Use of Geothermal Resources,* San Francisco, CA, 1781-1787.

Smith, R. B. et al., 1977. Yellowstone Hot Spot: Contemporary Tectonics and Crustal Properties from Earthquakes and Aeromagnetic Data. *J. Geophys. Res.,* 82(26, Sept.): 3665-3676.

Smith, J. L. et al., 1977. Summary of 1976 Geothermal Drilling—Western United States. *Geothermal Energy Magazine,* 5(5): 8-17.

Smith, J. L. et al., 1978. Summary of 1977 Geothermal Drilling—Western United States. *Geothermal Energy Magazine,* 6(5): 11-19.

Smith, J. L. et al., 1979. Summary of 1978 Geothermal Drilling—Western United States. *Geothermal Energy Magazine,* 7(5): 24-34.

Souther, J. G., 1975. *Proc. 2nd United Nations Symp. on the Development and Use of Geothermal Resources,* San Francisco, CA, pp. 259-267.

Stefanon, A., 1972. *Capture and Exploitation of Submarine Springs.* 2nd Int'l. Oceanol. Equip. and Service Exhibition and Conference, Brighton, England, March 19-24, 1972, 427-430.

Stoller, H. M. and Colp, J. L., 1978. Magma as a Geothermal Resource—A Summary. *Trans. Geothermal Resources Council,* 2(1, July): 613-615.

Studt, F. E., 1979. Summary of the Tongonan Exploration Case Study, Leyte, Philippines. *GRC Trans.,* 3: 687-688.

Subramanian, S. A., 1975. Present Status of Geothermal Resources Development in India. In: *Proc. 2nd United Nations Symp. on the Development and Use of Geothermal Resources.* San Francisco, CA, 1: 269-271.

Sudal, G. A. et al., 1979. *Annotated Research Bibliography for Geothermal Reservoir Engineering.* Terra Tek, Inc., Aug. LBL-8664.

Sugama, J. et al., 1975. Assessment of Geothermal Resources of Japan. In: *Proc. the United States—Japan Geological Surveys Panel Discussion on an Assessment of Geothermal Resources,* Tokyo, Japan, Sept.: 63-119.

Sukhoreu, et al., 1962. Geothermal Features of Caucasian Oil and Gas Deposits. *Int'l. Geol. Rev.,* 6(9): 1541-1556.

Summers, W. K., 1971. *Annotated and Indexed Bibliography of Geothermal Phenomena.* N. Mex. Bur. of Mines.

Tatsch, J. M., 1976. *Geothermal Deposits, Origin, Evolution, and Present Characteristics.* Tatsch Associates.

Teilman, N. A., 1979. A Geochemical Reconnaissance of Thermal and Non-Thermal Waters in Nicaragua. *GRC Trans.,* 3: 713-716.

Tezan, A. K., 1975. Dry Steam Possibilities in Saraykov-Kizidere Geothermal Field, Turkey. In: *Proc. 2nd United Nations Symp. on the Development and Use of Geothermal Resources,* San Francisco, CA, 1805-1813.

Thomas, L., 1975. Geothermal Resources in Australia. In: *Proc. 2nd United Nations Symp. on the Development and Use of Geothermal Resources,* San Francisco, CA, 273-774.

Tuncer, E. and Simsek, S., 1975. Geology of the Izmir-Seperihisan Geothermal Area, Western Anatolia of Turkey: Determination of Reservoirs by Means of Gradient Drilling. In: *Proc. 2nd United Nations Symp. on the Development and Use of Geothermal Resources,* San Francisco, CA, 349-362.

Urbain, C., 1975. Tapping Geothermal Formations in El Salvador Calls for Innovations. *Oil Gas J.*, Aug.: 28-32.

Varnado, S. G., and H. M. Stoller, 1978. Geothermal Drilling and Completion Technology Development. *GRC Trans.*, 2: 675-678.

Veneruso, A. F. and H. M. Stoller, 1979. High Temperature Instrumentation for Geothermal Applications. *GRC Trans.*, 2: 679-682.

Veliciu, S. et al., 1977. Preliminary Data of Heat Flow Distribution in Romania. *Geothermics*, 6(1/2): 95-98.

Vogely, W. A. (ed.), 1976. *Economics of the Mineral Industries*. 3rd edition, Am. Inst. Mining, Metallurgical, and Petroleum Engineers, Inc., New York, NY, Part 2, Central Problems Arising in the Analysis of Resources, 127-207.

Vogely, W. A., ed., 1976. *Economics of the Mineral Industry.* AIME, NM, 140-144.

Wallace, R. H. et al., 1979. Assessment of Geopressured Geothermal Resources in the Northern Gulf of Mexico Basin. In: L. J. P. Muffler (ed.), *Assessment of Geothermal Resources of the United States—1978, U.S.G.S. Circ.* 790: 132-155.

West, F. G. and A. W. Laughlin, 1979. Aquarius Mountain Area, Arizona: A Possible HDR. *Geothermal Energy Magazine,* 7(8): 13-20.

White, D. L. et al., 1971. Vapor-dominated Hydrothermal Systems Compared with Hot-water Systems. *Econ. Geol.,* 75-97.

White, D. F. and D. L. Williams, eds., 1976. Assessment of Geothermal Resources of the United States—1975. *USGS Circ.* 726, Section entitled Igneous Related Geothermal Systems, p. 58.

Williams, D. L., 1976. Submarine Geothermal Resources. *J. Volcan. and Geothermal Research*, 4: 85-100.

Williamson, K. M. and Wright, E. P., 1978. St. Lucia Geothermal Project. *GRC Trans.*, 2: 731-734.

Yanez, C. et al., 1975. Geothermal Exploration in the Numeros-Derrumbados Area. *GRC Trans.,* Reno, NV, 3: 80-1-804.

Zötl, J. G., 1978. Personal Communication.

Appendix A

Regional Assessment Definitions for Geothermal Resources*

A consistent, agreed-upon, and appropriate terminology is prerequisite for making a meaningful geothermal resource assessment. It has been proposed that a logical, sequential, and useful subdivision of the *geothermal resource base,* defined as all the heat stored in the earth's crust under a given area, measured from mean annual surface temperature, serves this need. It is suggested that the part of the resource base that is shallow enough to be tapped by production drilling be termed the *accessible resource base,* and be further divided into *useful* and *residual* components. The commercially practical or useful component, that is the heat that could reasonably be extracted or recovered at costs competitive with other forms of energy at some specified future time, is termed the *geothermal resource.* In turn, it is divided into *economic* and *subeconomic* components, using estimates based on conditions and knowledge existing at the time of the assessment. In the format of a McKelvey diagram (Figure 2-6) this partitioning of the resource base defines the vertical axis, designated as the degree of economic feasibility. These definitions are summarized in Table A-I.

The horizontal axis in Figure 2-6, shown as the degree of geologic assurance, contains the *identified* and *undiscovered* components.

Finally, the *reserve* is designated as the identified economic resource.

All categories should be expressed in units of heat, with all resource and reserve figures being calculated at wellhead conditions, prior to the losses inherent in any practical direct thermal use or in conversion to electricity.

For some resource evaluation purposes it may be possible and appropriate to further subdivide the identified category of the above definitions. For example, if one follows the general terminology of USGS (White and Williams, 1976), it may prove valuable to define the following terms:

Measured—That part of the accessible resource base, resource, or reserve whose size can be computed from drillhole data and reservoir engineering measurements.

Indicated—That part of the accessible resource base, resource, or reserve whose size can be estimated by a combination of drilling data and extrapolation using geochemical, geophysical, or geological data.

*Adapted from Muffler and Cataldi, 1978

Table A-1
Summary of Geothermal Resource Assessment Definitions

Name	Definition	Attributes and corollaries
Resource base	All of the heat in the earth's crust beneath a specified area, measured from local mean annual surface temperature.	—Stored heat at an instant in time. —Neglects transfer of heat from mantle. —Takes no regard of whether or not it would ever be technically or economically feasible to recover the heat.
Accessible resource base	All of the heat stored between the earth's surface and a specified depth in the crust, beneath a specified area and measured from local mean annual temperature.	—Stored heat at an instant in time. —Neglects transfer of heat from deeper levels. —Depth chosen for the lower limit is a matter of convenience, but must be specified in each case. —Implies that heat within the specified depth might be tapped by production drilling at some reasonable time in the future.
Inaccessible resource base	All of the heat stored between the base of the crust, and a specified depth in the crust, beneath a specified area and measured from local mean annual temperature.	—Stored heat at an instant in time. —Neglects transfer of heat from mantle. —Depth chosen for the upper limit is a matter of convenience, but must be specified in each case. —Implies that heat beneath the specified depth is unlikely to be tapped by production drilling at a reasonable time in the future.
Useful accessible resource base (= Resource)	That part of the accessible resource base that could be extracted economically and legally at some specified time in the future.	—Criterion for subdivision of accessible resource base is a subjective aggregate of predicted technology and economics at some reasonable and specified future time (≤ 100 years). —Implies knowledge of extraction efficiency, recovery factors. —Requires evaluation of heat (and fluid) recharge.
Residual accessible resource base	That part of the accessible resource base unlikely to be extracted economically and legally at some specified time in the future.	—Criterion for subdivision of accessible resource base is a subjective aggregate of predicted technology and economics at some reasonable and specified future time.

Table A-1 continued

Name	Definition	Attributes and corollaries
Economic resource	That part of the resource of a given area that can be extracted legally at a cost competitive with other commercial energy sources at the time of determination.	—Requires detailed economic analysis and commercial production history. —Projections must include evaluation of future fuel costs.
Subeconomic resource	That part of the resource of a given area that can not be extracted legally at a cost competitive with other commercial energy sources at the time of determination, but might be extracted economically and legally at some specified time in the future.	
Undiscovered economic resource	That part of the economic resource in unexplored parts of regions known to contain geothermal resources or in regions where geothermal resources are suspected but not yet discovered.	
Identified economic resource (= Reserve)	That part of the economic resource known and characterized by drilling or by geochemical, geophysical and geological evidence.	

Demonstrated = measured + indicated.

Inferred—That part of the identified accessible resource base, resource, or reserve whose size can be inferred from geochemical, geophysical, or geological evidence but for which there is little if any corroborating drillhole data.

Alternatively, it may be useful to divide the identified category into *under development* and *under exploration*. The former refers to heat in areas where production wells and utilization facilities either exist or are under construction. The latter refers to geothermal heat identified only by exploratory drilling supplemented by geophysics, chemical geothermometers, etc.

Similarly, if necessary, the undiscovered category could be divided into *in known regions* and *in new regions*. The former refers to regions where useful geothermal heat is known to exist. The latter refers to regions where useful geothermal heat is likely to exist but has not yet been positively identified. Although these categories correspond respectively to "hypothetical" and "speculative" of the USGS (White and Williams, 1976) it is suggested that these words be avoided as being insufficiently descriptive of the categories and, thus, may cause confusion and misunderstanding.

References, Appendix A

Muffler, L. J. P. and Cataldi, R., 1978. Methodology for Regional Assessment of Geothermal Resources. *Proc. Larderello Workshop on Geothermal Resource Assessment and Reservoir Engineering,* Sept. 12-16, 1977. *Geothermics,* 7(2-4): 53-90.

White, D. F. and Williams, D. L. (eds.), 1976. Igneous-related Geothermal Systems. In: *Assessment of Geothermal Resources of the United States—1975.* U.S. Geol. Surv. Circ. 726: 58

Appendix B

Definitions of Reserves and Resources

Attempts to coordinate and "regularize" the definitions and methods used in the assessments of various energy and mineral resources have been made by several workers (Blondel and Laskey, 1956; Netschert, 1976; Schurr, 1960; and McKelvey, 1972). The basic thread of the problem, which is evident in this literature, relates to the different perspectives on the resource evaluation objectives that might be briefly characterized by the "commercial" and "public policy" sectors, or as the short- or long-term views of the time dimension (Netschert and Liou, 1957).

The terminology and definitions generally used in the minerals industry are presented in Table B-1. The subdivisions listed ". . . account for only a part, generally a small part, of the resource base" (Netschert, 1976, p. 142). One can compare these designations with those used in the U.S. petroleum industry, as summarized in Table B-2 (Campbell, 1970). More details of the oil and gas commercial interests can be found in Cannon (1977).

References, Appendix B

Blondel, F. and Laskey, S. G., 1956. Mineral Reserves and Mineral Resources. *J. Econ. Geol.,* 686-697.

Netschert, B. C., ed., 1976. Central Problems Arising in the Analysis of Resources. *Economics of the Mineral Industries,* Vogely, W. A., S. W. Model Service, AIME, NY, Part 2: 125-376.

McKelvey, V. E., 1972. Mineral Resource Estimates and Public Policy. *Am. Scientist,* 60: 32-40.

Netschert, B. C. and Donor, D. M. L., 1957. Mineral Reserves and Mineral Resources. *J. Econ. Geol.,* 589-590.

Table B-1
Classes of Mineral Reserve and Resource Estimates*

Quality of the resource (cost-price dimension)	Degrees of knowledge about the resource (geologic dimension)				
	In known deposits or districts			In undiscovered districts	
	Identified Reserves			Undiscovered	Total
	Measured reserves / Proved reserves / Developed reserves	Indicated reserves / Probable reserves / Undeveloped reserves	Inferred reserves / Possible reserves		
Recoverable under market and technologic conditions	Reserves	Reserves	Reserves	Undiscovered Reserves	Potential Reserves
Recoverable at prices up to 2 times those prevailing now or with comparable advance in technology	Identified submarginal resources**			Undiscovered submarginal resources	Potential submarginal resources
Recoverable at prices 2 to 10 times those prevailing now or with comparable advance in technology	Identified latent resources**			Undiscovered latent resources	Potential latent resources
Total	Total identified resources			Undiscovered resources	Total potential resources

*The block of resources described by these estimates accounts for only a part, generally a small part, of the resource base.

**The terms measured, indicated and inferred may also be applied to known submarginal and latent resources where knowledge of them is sufficiently detailed to warrant differentiation in the degree of certainty of the estimates.

Table B-2
Definitions of Proved Reserves for Petroleum
Property Evaluation*

Proved reserves — The quantities of crude oil, natural gas and natural gas liquids which geological and engineering data demonstrate with reasonable certainty to be recoverable in the future from known oil and gas reservoirs under existing economic and operating conditions. They represent strictly technical judgments, and are not knowingly influenced by attitudes of conservatism or optimism.

   Undrilled Acreage — Both drilled and undrilled acreage of proved reservoirs are considered in the estimates of the proved reserves. The proved reserves of the undrilled acreage are limited to those drilling units immediately adjacent to the developed areas, which are virtually certain of productive development, except where the geological information on the producing formation insures continuity across other undrilled acreage.

   Fluid Injection — Additional reserves to be obtained through the application of fluid injection or other improved recovery techniques for supplementing the natural forces and mechanisms of primary recovery are included as "proved" only after testing by a pilot project or after the operation of an installed program has confirmed that increased recovery will be achieved.

When evaluating an individual property in an existing oil or gas field, the proved reserves within the framework of the above definition are those quantities indicated to be recoverable commercially from the subject property at current prices and costs, under existing regulatory practices, and with conventional methods and equipment. Depending on their development or producing status, these proved reserves are further subdivided into:

Proved developed reserves — Proved reserves to be recovered through existing wells and with existing facilities;

   Proved developed producing reserves — Proved developed reserves to be produced from completion interval(s) open to production in existing wells;

   Proved developed nonproducing reserves — Proved developed reserves behind the casing of existing wells or at minor depths below the present bottom of such wells which are expected to be produced through these wells in the predictable future. The development cost of such reserves should be relatively small compared to the cost of a new well.

Proved undeveloped reserves — Proved reserves to be recovered from new wells on undrilled acreage or from existing wells requiring a relatively major expenditure for recompletion or new facilities for fluid injection.

*Courtesy of the Society of Petroleum Engineers of AIME.

Appendix C

Calculated Worldwide Geothermal Resource Base

The Electric Power Research Institute developed a worldwide national survey and a resource assessment methodology for the Conservation Commission, World Energy Conference, held in Istanbul in 1978. A summary by Roberts (1978) includes a tabulation of geothermal resource base. The calculations used a depth of 3 km and were based upon the geothermal belts or zones of the world (Figure 2-3). The results are summarized here in Table C-I.

It must be emphasized that the quantity, X, tabulated in the third column of the table, is based upon a subjective judgment of the extent of a nation's area that is included in the world high-grade geothermal zones.

Table C-1
Calculated Worldwide Geothermal Resource Base (joules)

Country	Area (km²)	X*	Resource base in temperature classes (°C) <100 (Class 1)	100-150 (Class 2)	150-250 (Class 3)	>250 (Class 4)	Total
Algeria**	2.4 E6	0.007	6.7 E23	3.6 E21	9.6 E20	5.9 E19	6.7 E23
Angola**	1.3 E6	0	3.5 E23	0	0	0	3.5 E23
Argentina**	2.8 E6	0.03	8.0 E23	1.8 E22	5.0 E21	2.9 E20	8.4 E23
Australia	7.7 E6	0	2.2 E24	0	0	0	2.2 E24
Austria	8.4 E4	0	2.4 E22	0	0	0	2.4 E22
Bangladesh ·	1.4 E5	0	4.0 E22	0	0	0	4.0 E22
Barbados**	4.3 E2	1.00	9.2 E19	9.6 E19	2.5 E19	1.5 E18	2.1 E20
Belgium	3.1 E4	0	8.8 E21	0	0	0	8.8 E21
Bolivia**	1.1 E6	0.30	2.9 E23	7.1 E22	1.9 E22	1.2 E21	3.8 E23
Brazil**	8.5 E6	0.05	2.4 E24	9.2 E22	2.5 E22	1.5 E21	2.5 E24
Bulgaria	1.1 E5	0	3.1 E22	2.4 E20	6.3 E19	4.0 E18	3.2 E22
Burundi**	2.8 E4	1.00	5.9 E21	6.3 E21	1.6 E21	1.0 E20	1.4 E22
Cameroon**	4.8 E5	0.10	1.3 E23	1.0 E22	2.8 E21	1.7 E20	1.5 E23
Canada**	1.0 E6	0.20	2.7 E24	4.6 E23	1.2 E23	7.1 E21	3.3 E24
Chad**	1.3 E6	0.10	3.5 E23	2.8 E22	7.5 E21	4.6 E20	3.8 E23

169

Table C-1 continued

Country	Area (km²)	X*	<100 (Class 1)	100-150 (Class 2)	150-250 (Class 3)	>250 (Class 4)	Total
			\multicolumn{4}{c}{Resource base in temperature classes (°C)}		
Chile**	7.6 E5	1.00	1.6 E23	1.7 E23	4.6 E22	2.7 E21	3.8 E23
China**	9.6 E6	0.25	2.6 E24	5.4 E23	1.4 E23	8.4 E21	3.2 E24
Colombia**	1.1 E6	0.30	3.0 E23	7.5 E22	2.0 E22	1.2 E21	3.9 E23
Costa Rica	5.1 E4	1.00	1.1 E22	1.1 E22	3.0 E21	1.8 E20	2.5 E22
Cuba	1.1 E5	0	3.2 E22	0	0	0	3.2 E22
Cyprus	9.3 E3	0	2.6 E21	0	0	0	2.6 E21
Czechoslovakia**	1.3 E5	0.10	3.5 E22	2.8 E21	7.5 E20	4.6 E19	3.8 E22
Dahomey	1.2 E5	0	3.3 E22	0	0	0	3.3 E22
Denmark	4.3 E4	0	1.2 E22	0	0	0	1.2 E22
Dominican Rep.**	4.8 E4	0.25	1.3 E22	2.7 E21	7.1 E20	4.2 E19	1.6 E22
Ecuador	4.6 E5	1.00	1.0 E23	1.0 E23	2.6 E22	1.6 E21	2.3 E23
Egypt	9.5 E5	0	2.7 E23	0	0	0	2.7 E23
El Salvador	2.1 E4	1.00	4.6 E21	4.6 E21	1.2 E21	7.5 E19	1.0 E22
England	1.3 E5	0	3.7 E22	0	0	0	3.7 E22
Ethiopia	1.2 E6	0.60	3.0 E23	1.6 E23	4.2 E22	2.6 E21	5.0 E23
Fiji*	1.8 E4	0.20	5.0 E21	8.0 E20	2.1 E20	1.3 E19	5.9 E21
Finland	3.6 E5	0	1.0 E23	0	0	0	1.0 E23
France	5.5 E5	0	1.5 E23	0	0	0	1.5 E23
Germany, East	1.1 E5	0	3.1 E22	0	0	0	3.1 E22
Germany, West	2.5 E5	0	7.1 E22	0	0	0	6.7 E22
Ghana	2.4 E5	0	6.7 E22	0	0	0	6.7 E22
Greece	1.3 E5	0.30	3.5 E22	8.8 E21	2.3 E21	1.4 E20	4.6 E22
Greenland	2.2 E6	0	6.3 E23	0	0	0	6.3 E23
Guadeloupe	1.8 E3	1.00	3.8 E20	3.9 E20	1.0 E20	6.3 E18	8.8 E20
Guatemala	1.1 E5	1.00	1.4 E22	2.4 E22	6.3 E21	3.9 E20	4.6 E22
Haiti**	2.8 E4	0.025	8.0 E21	1.5 E20	4.0 E19	2.5 E18	8.0 E21
Honduras**	1.1 E5	0.50	2.5 E22	1.2 E22	3.3 E21	2.0 E20	4.2 E22
Hungary	9.3 E4	0	2.6 E22	0	0	0	2.6 E22
Iceland	1.0 E5	1.00	2.2 E22	2.3 E22	5.9 E21	3.7 E20	5.0 E22
India**	3.3 E6	0.02	9.2 E23	1.4 E22	3.8 E21	2.3 E20	9.2 E23
Indonesia	1.9 E6	1.00	4.1 E23	4.2 E22	1.1 E23	6.7 E21	9.6 E23
Iran**	1.7 E6	0.20	4.6 E23	7.1 E22	1.9 E22	1.2 E21	5.4 E23
Iraq	4.3 E5	0	1.2 E23	0	0	0	1.2 E23
Ireland	7.0 E4	0	2.0 E22	0	0	0	2.0 E22
Israel**	2.1 E4	0.025	5.9 E21	1.2 E20	3.1 E19	1.9 E18	5.9 E21
Italy	3.0 E5	0.50	7.5 E22	3.3 E22	8.8 E21	5.4 E20	1.2 E23
Ivory Coast	3.2 E5	0	9.2 E22	0	0	0	9.2 E22
Jamaica	1.1 E4	0	3.2 E21	0	0	0	3.2 E21
Japan	3.7 E5	1.00	8.0 E22	8.0 E22	2.1 E22	1.3 E21	1.8 E23
Jordan**	9.8 E4	0.025	2.8 E22	5.4 E20	1.5 E20	8.8 E18	2.8 E22

Table C-1 continued

Country	Area (km²)	X*	Resource base in temperature classes (°C)				
			<100 (Class 1)	100-150 (Class 2)	150-250 (Class 3)	>250 (Class 4)	Total
Kenya	5.8 E5	0.60	1.4 E23	7.5 E22	2.0 E22	1.3 E21	2.4 E23
Korea**							
(N & S)	2.2 E5	0.50	5.4 E22	2.4 E22	6.3 E21	3.9 E20	8.8 E22
Kuwait	2.0 E4	0	5.9 E21	0	0	0	5.9 E21
Liberia	1.1 E5	0	3.1 E22	0	0	0	3.1 E22
Libya	1.8 E6	0	5.0 E21	0	0	0	5.0 E21
Luxembourg	2.6 E3	0	7.5 E20	0	0	0	7.5 E20
Malawi**	9.5 E4	0.01	2.7 E22	2.1 E20	5.4 E19	3.4 E18	2.7 E22
Malagasy Rep.**	5.9 E5	0.005	1.6 E23	6.3 E20	1.7 E20	1.0 E19	1.6 E23
Malaysia	3.3 E5	0.02	9.2 E22	1.4 E21	3.8 E20	2.3 E19	9.2 E22
Mali	1.2 E6	0	3.4 E23	0	0	0	3.4 E23
Martinique	1.1 E3	1.00	2.4 E20	2.4 E20	6.3 E19	3.9 E18	5.4 E20
Mexico	2.0 E6	0.60	4.6 E23	2.6 E23	6.7 E22	4.2 E21	8.0 E23
Morocco	4.4 E5	0.04	1.3 E23	3.9 E21	1.0 E21	4.2 E21	8.0 E23
Nepal	1.4 E5	0.10	3.9 E22	3.1 E21	8.4 E20	5.0 E19	4.2 E22
Netherlands	3.6 E4	0	1.0 E22	0	0	0	1.0 E22
New Guinea	2.4 E5	0.60	5.9 E22	3.1 E22	8.4 E21	5.0 E20	9.6 E22
New Hebrides	1.5 E4	1.00	3.2 E21	3.3 E21	8.8 E20	5.4 E19	7.5 E21
New Zealand	2.7 E5	0.50	6.7 E22	2.9 E22	8.0 E21	4.6 E20	1.0 E23
Nicaragua	1.5 E5	1.00	3.2 E22	3.3 E22	8.8 E21	5.4 E20	7.5 E22
Nigeria	9.2 E5	0	2.6 E23	0	0	0	2.6 E23
Norway	3.2 E5	0	9.2 E22	0	0	0	9.2 E22
Pakistan	8.4 E6	0.01	2.3 E24	1.8 E22	5.0 E21	3.0 E20	2.3 E24
Panama**	7.6 E4	1.00	1.6 E22	1.7 E22	4.6 E21	2.7 E20	3.8 E22
Paraguay**	4.1 E5	0.10	1.1 E23	8.8 E21	2.4 E21	1.5 E20	1.3 E23
Peru**	1.3 E6	1.00	2.8 E23	2.8 E23	7.5 E22	4.6 E21	1.5 E23
Philippines	3.0 E5	1.00	6.3 E22	6.7 E22	1.8 E22	1.1 E21	1.5 E23
Poland	3.1 E5	0	8.8 E22	0	0	0	8.8 E22
Portugal	9.2 E4	0.05	2.6 E22	1.0 E21	2.7 E20	1.6 E19	2.7 E22
Rumania	2.4 E5	0.01	6.7 E22	5.0 E20	1.4 E20	8.4 E18	6.7 E22
Saudi Arabia**	2.3 E6	0.02	6.3 E23	1.0 E22	2.6 E21	1.6 E20	6.3 E23
Senegal	2.0 E5	0	5.4 E22	0	0	0	5.4 E22
Sierra Leone	7.2 E4	0	2.1 E22	0	0	0	2.1 E22
South Africa	1.2 E6	0	3.4 E23	0	0	0	3.4 E23
Spain	5.1 E5	0.05	1.4 E23	5.4 E21	1.5 E21	9.2 E19	1.5 E23
Sri Lanka	6.6 E4	0	1.8 E22	0	0	0	1.8 E22
Sudan	2.5 E6	0	7.1 E23	0	0	0	7.1 E23
Sweden	4.5 E5	0	1.3 E23	0	0	0	1.3 E23
Switzerland	4.1 E4	0	1.2 E22	0	0	0	1.2 E22
Taiwan	3.6 E4	1.00	7.5 E21	8.0 E21	2.1 E21	1.3 E20	1.8 E22
Tanzania	9.4 E5	0.03	2.6 E23	6.3 E21	1.6 E21	1.0 E20	2.6 E23

Table C-1 continued

Country	Area (km²)	X*	<100 (Class 1)	100-150 (Class 2)	150-250 (Class 3)	>250 (Class 4)	Total
			\multicolumn{4}{c}{**Resource base in temperature classes (°C)**}			
Thailand	5.1 E5	0	1.5 E23	0	0	0	1.5 E23
Trinidad & Tobago	5.1 E3	1.00	1.1 E21	1.1 E21	3.0 E20	1.8 E19	2.6 E21
Tunisia	1.6 E5	0.01	4.6 E22	3.6 E20	9.6 E19	5.9 E18	4.6 E22
Turkey	7.8 E5	0.50	1.9 E23	8.4 E22	2.3 E22	1.4 E21	3.1 E23
United States	9.4 E6	0.25	4.1 E24	5.0 E23	1.3 E23	8.4 E21	4.6 E24
Uruguay	1.9 E5	0.01	5.0 E22	4.1 E20	1.1 E20	6.7 E18	5.0 E22
USSR	2.2 E7	0.05	6.3 E24	2.4 E23	6.3 E22	4.0 E21	6.7 E24
Venezuela	9.1 E5	0.20	2.5 E23	0	1.0 E22	6.7 E20	3.0 E23
Vietnam (N & S)	3.3 E5	0.50	8.4 E22	3.7 E22	9.6 E21	5.9 E20	1.3 E23
Yugoslavia	2.6 E5	0	7.1 E22	0	0	0	7.1 E22
Zaire	2.3 E6	0.01	6.7 E23	5.0 E21	1.3 E21	8.4 E19	6.7 E23
Zambia	7.5 E5	0.10	2.1 E23	1.7 E22	6.7 E21	2.7 E20	2.3 E23
Totals	135,830 × 10³		3.8×10^{25}	3.8×10^{24}	9.6×10^{23}	5.0×10^{22}	4.2 E25
							$= 42 \times 10^6$ Quads

*$\times$ = Fraction of area of country in geothermal zone or belt, estimated from available knowledge of the extent of these zones.

**Country is in conjectured geothermal belt, not yet proven by exploration.

Appendix D

References for
Table 2-12 and Table 2-15

References for Table 2-12

Anon., 1975. *Union Oil Company Operations at The Geysers.* Company News Release, Apr., 1975.

Anon., 1977. *Philippine National Power Corporation 10-year Power Expansion Program,* Oct., 1977.

Anon., 1977. Utilization of geothermal energy in Japan. Sunshine Project Promotion Headquarters, *Geothermal Resources Council Bull.,* Oct., 1977.

Anon., 1978. *Geothermal Energy.* Japan Geothermal Energy Development Center.

Anon., 1978. Hawaii Plans Geothermal Power. Industrial/Research/Development, Sept., p. 70.

Axtell, L. H., 1975. Geothermal Energy Utilization in Japan. *Geothermal Energy Magazine,* 3(2, Feb.): 5-19.

Banwell, C. J., 1963. Thermal Energy from the Earth's Crust. *New Zealand J. Geol. Geophys.,* Feb.

Birsic, R. J., 1978. Geothermal Development in the Philippines, Iceland and Japan. *Geothermal Energy Magazine,* 6(5, May): 29-36.

Bodvarsson, G. and Bolton, R. S., 1971. *A Study of the Ahuachapan Field.* Internal Project Report, U.N., May.

Bodvarsson, G., 1974. Geothermal Resources Energetics. *Geothermics,* 3(3, Sept.): 83-92.

Bodvarsson, G., Dench, N. and McNitt, J., 1977. *Report of the United Nations Review Mission to Nicaragua,* 30 May to 6 June, 1977, Internal Project Report, June, U.N.

Choussy, M. E. and Penate, T. S., 1976. *Campo Geotermico de Ahuachapan Despues de un Año de Explotacion.* Paper presented at the Simposio Internacional Sobre Energia Geotermica en America Latina, Oct., Guatemala City, Guatemala.

Dominco, E., 1972. *Kizildere Geothermal Field, Estimate of Reservoir Capacity and Recharge Rate.* Internal Project Report, U.N. Aug.

Hubbert, K. M., 1969. Energy Resources. In: *Resources and Man,* Nat'l. Research Council—Nat'l. Acad. of Sci., 215-218.

James, R., 1974. *The El Tatio Geothermal Field; Test Results, Underground Theory, Power Feasibility.* Internal Project Report, U.N., Sept.

Koenig, J. B., 1973. The Potential for Geothermal Energy Development to 1980. In: P. Kruger and C. Otte (eds.), *Geothermal Energy.* Stanford Univ. Press, 15-58.

Peterson, R. F. and El-Ramly, N., 1975. The Worldwide Electric and Non-electric Geothermal Industry. *Geothermal Energy Magazine,* 3(11): 4-14.

Reed, C. S., 1978. *State of California Initiative in Geothermal Development: Its Objectives, Accomplishments and Schedules.* Proc. Geothermal Resources Council, Commercialization of Geothermal Resources. A Report, San Diego, CA, Nov.

174 Handbook of Geothermal Energy

Saint, P. K. and Jasso, A., 1976. Worldwide Geothermal Energy Resource Development. *Geothermal Energy Magazine,* 4(2, Feb.): 5-14.

Sweco and Virkir Consulting Groups, 1976. *Feasibility Report for the Olkaria Geothermal Project.* Internal Project Report, U.N., Dec.

Tatahashi, P. K. and Chen, B., 1975. Geothermal Reservoir Engineering. *Geothermal Energy Magazine,* 3,(10): 7-22.

Turner, W. J., 1969. A *Reservoir Engineering Study of Ahuachapan Geothermal Reservoir for Feasibility of Sub-Surface Disposal of Hot Saline Water.* Internal Project Report, U.N., Nov.

References for Table 2-15
(McNitt, 1978)

1. Alonso, H., 1975. Geothermal Potential of Mexico. In: *Proc. of the United Nations Symp. on the Development and Use of Geothermal Resources,* 1: 21-24.
2. Alpan, S., 1975. Geothermal Energy Exploration in Turkey. In: *Proc. of the United Nations Symp. on the Development and Use of Geothermal Resources,* 25-28.
3. Arnorsson, S. et al., 1975. Systematic Exploration of the Krisuvik High-Temperature Areas, Reykjanes Peninsula, Iceland. In: *Proc. of the United Nations Symp. on the Development and Use of Geothermal Resources,* 853-864.
4. Baldi, P. et al., 1975. Geology and Geophysics of the Cesano Geothermal Field. In: *Proc. of the United Nations Symp. on the Development and Use of Geothermal Resources,* 871-881.
5. Baldo, L. et al., 1977. Aprovechamiento del Fluido Geotermico del Campo de Ahuachapan (San Salvador). *Simposio Internacional Sobre Energia Geotermica en America Latina* (Oct. 1976), Rome, 411-431.
6. Bjornsson, S. et al., 1970. Exploration of the Reykjanes Thermal Brine Area. *Geothermics,* Spec. Issue 2: 1640-1650.
7. Bodvarsson, G. et al., 1977. *Report to the United Nations Geothermal Review Mission to Nicaragua, 30 May to 7 June, 1977.* Internal Project Report, U.N.
8. Bolton, R. S., 1975. Recent Development and Future Prospects for Geothermal Energy in New Zealand. *Proc. of the United Nations Symp. on the Development and Use of Geothermal Resources,* 37-42.
9. Burgassi, R. et al., 1965. Prospezione delle Anomalie Geotermiche e sua Applicazione alla Regione Amiatina. *l' Industria Mineraria,* Maggio.
10. Burgassi, P. et al., 1975. Recent Developments of Geothermal Exploration in the Travele-Radicondoli Area. In: *Proc. of the United Nations Symp. on the Development and Use of Geothermal Resources,* 1571-1581.
11. Cataldi, R. and Rendina, M., 1973. Recent Discovery of a new Geothermal Field in Italy, Alfina. *Geothermics,* 2(3-4): 106-116.
12. Celati, R. et al., 1973. Interactions Between Steam Reservoir and the Surrounding Aquifers in the Larderello Geothermal Field. *Geothermics,* 2(3-4): 174-185.
13. Ceron, P. et al., 1975. Progress Report on Geothermal Development in Italy from 1969 to 1974 and Future Prospects. In: *Proc. of the United Nations Symp. on the Development and Use of Geothermal Resources,* 59-66.
14. Cheng, W. T., 1970. Geophysical Exploration in the Tatun volcanic region, Taiwan. *Geothermics,* Spec. Issue 2: 262-274.
15. Choussy, M. E. and Penate, T. S., 1977. Campo Geotermico de Ahuachapan despues de un Año de Explotacion. *Simposio Internacional Sobre Energia Geotermica en America Latina,* Oct. 1976, Guatemala.

16. D'Archimbaud, J. D. and Munier-Jolain, J. P., 1975. Les Progres de l'Exploration Geothermique, à Bouillante en Guadeloupe. *Proc. of the United Nations Symp. on the Development and Use of Geothermal Resources*, 101-104.
17. Fedotov, S. A. et al., 1975. On a Possibility of Heat Utilization of the Avachinsky Volcanic Chamber. In: *Proc. of the United Nations Symp. on the Development and Use of Geothermal Resources*, 363-369.
18. Furumoto, A. S., 1977. Evaluation of Geophysical Techniques for Geothermal Exploration on an Active Basaltic Volcano (Abstract). *Trans. Am. Geophys. Union*, 58(6): 540.
19. Garcia, D. S., 1975. Estudio Geoelectrico de la Zona Geotermica de Cerro Prieto, Baja California, Mexico. In: *Proc. of the United Nations Symp. on the Development and Use of Geothermal Resources*, 1003-1009.
20. Goff, F. E. et al., 1977. Geothermal Prospecting in The Geysers—Clear Lake area, Northern California. *Geology*, 5: 509-515.
21. Healy, J., 1975. Geothermal Fields in Zones of Recent Volcanism. In: *Proc. of the United Nations Symp. on the Development and Use of Geothermal Resources*, 415-422.
22. Healy, J. and James, R., 1976. Geothermal Energy in New Zealand—Summary. *A.A.P.G., Memoir No. 25*: 130-134.
23. Hockstein, M. P., 1975. Geophysical exploration of the Kawah Kamojang Geothermal Field, West Java. In: *Proc. of the United Nations Symp. on the Development and Use of Geothermal Resources*, 1049-1058.
24. Hoover, D. B. et al., 1976. Audiomagnetotelluric Sounding as a Reconnaissance Exploration Technique in Long Valley, California. *J. Geophys. Research*, 81(5): 801-809.
25. Jimenez, A. M. and Campos, T. A. V., 1977. Perforaciones Geotermales en El Salvador. *Simposio Internacional Sobre Energia Geotermica en America Latina*, Oct. 1976, Guatemala.
26. Kilkenny, J. E., 1975. Field Guide Notes, provided by Union Oil Company.
27. Lachenbruch, A. H. et al., 1975. The Near-Surface Hydrothermal Regions of Long Valley Caldera. *J. Geophys. Research*, 81(5): 767-768.
28. Lahsen, A. and Trujillo, P., 1975. El Campo Geotermico de El Tatio, Chile. *Proc. of the United Nations Symp. on the Development and Use of Geothermal Resources*, 157-169.
29. MacDonald, W. J. P., 1975. The Useful Heat Contained in the Broadlands Geothermal Field. In: *Proc. of the United Nations Symp. on the Development and Use of Geothermal Resources*, 1113-1119.
30. Noble, J. and Ojiambo, S. B., 1975. Geothermal Exploration in Kenya. In: *Proc. of the United Nations Symp. on the Development and Use of Geothermal Resources*, 189-204.
31. Onodera, S., 1975. An Evaluation of Geothermal Potential by Resistivity Sounding Curves. In: *Proc. of the United Nations Symp. on the Development and Use of Geothermal Resources*, 1167-1173.
32. Palmason, G., 1975. Geophysical Methods in Geothermal Exploration. In: *Proc. of the United Nations Symp. on the Development and Use of Geothermal Resources*, 1175-1184.
33. Palmason, G. et al., 1975. Geothermal Energy Developments in Iceland 1970-1974. In: *Proc. of the United Nations Symp. on the Development and Use of Geothermal Resources*, 213-217.
34. Ragnars, K. et al., 1970. Development of the Namafjall Area, Northern Iceland. *Geothermics*, Spec. Issue 2: 925-935.
35. Sato, K., 1970. The Present State of Geothermal Development in Japan. *Geothermics*, Spec. Issue 2: 155-184.
36. Smith, J. L. and Matlick, J. S., 1976. Summary of 1975 Geothermal Drilling in western U.S. *Geothermal Energy Magazine*, 4(6).

37. Stilwell, W. B. et al., 1970. Ground Movement in New Zealand Geothermal Fields. In: *Proc. of the United Nations Symp. on the Development and Use of Geothermal Resources,* 1427-1434.

38. Sumi, K. and Takashima, I., 1975. Absolute Ages of the Hydrothermal Alteration Halos and Associated Volcanic Rocks in Some Japanese Geothermal Fields. In: *Proc. of the United Nations Symp. on the Development and Use of Geothermal Resources,* 625-634.

39. Swanberg, C. A., 1974. Heat Flow and Geothermal Potential of the East Mesa KGRA, Imperial Valley, California. In: *Proc. Conf. on Research for the Development of Geothermal Energy Resources.* National Science Foundation, RANN Publication No. NSF-RA-N-74-159.

40. Tezcan, A. K., 1975. Dry Steam Possibilities in Saraykoy—Kizildere Geothermal Field, Turkey. In: *Proc. of the United Nations Symp. on the Development and Use of Geothermal Resources,* 1805-1813.

41. Tikhonov, A. N. and Dvorov, I. M., 1970. Development of Research and Utilization of Geothermal Resources in the USSR. *Geothermics,* Spec. Issue 2: 1072-1078.

42. Vakin, E. A. et al., 1970. Recent Hydrothermal Systems of Kamchatka. *Geothermics,* Spec. Issue 2: 1116-1133.

43. White, D. F. and Williams, D. C. (eds.), 1975. Assessment of Geothermal Resources of the U.S.—1974. *U.S. Geol. Surv. Circ. 726.*

44. Yamada, E., 1975. Geological Development of the Onikobe Caldera and its Hydrothermal System. In: *Proc. of the United Nations Symp. on the Development and Use of Geothermal Resources,* 665-672.

45. Yamasaki, T. and Hayaski, M., 1975. Geologic Background of Otake and other Geothermal Areas in North-Central Kyushu, Southwestern Japan. In: *Proc. of the United Nations Symp. on the Development and Use of Geothermal Resources,* 673-684.

46. Townsend, D., 1977. *New Mexico Technology Review.* 4(9, Aug. 5): 1-2.

47. Dondanville, R. F., 1978. Geologic Characteristics of the Valles Caldera Geothermal System, NM. *Geothermal Resource Council, Trans.,* Hilo, HI, 2(1, Jul.): 157-159.

48. Donaldson, I. G. and Grant M. A., 1978. An Estimate of the Resource Potential of New Zealand Geothermal Field for Power. In: Proc. Larderello Workshop on Geothermal Resource Assessment and Reservoir Engineering, Sept. 12-16, 1977. *Geothermics,* 7(2-4): 243-252.

49. Birsic, R. J., 1978. Geothermal Developments in the Philippines, Iceland, and Japan. *Geothermal Energy Magazine,* 6(5, May): 29-36.

50. Annon., 1977. Reference Notes for Field Trips. *Larderello Workshop on Geothermal Resource Assessment and Reservoir Engineering,* Sept. 12-16, 1977.

51. Ross, H. P., 1980. *Case Study of the Roosevelt Hot Springs KGRA.* UURI-ESL, Report to be published.

52. Bodvarsson, G., 1975. Estimates of the Geothermal Resources of Iceland. In: *Proc. 2nd U.N. Symp. on the Development and Use of Geothermal Resources,* San Francisco, CA, May 20-29, 1975, 33-35.

53. Annon., 1973. Panel discussion on the Assessment of Geothermal Resources. In: *Proc. of the United States-Japan Geol. Surveys,* Oct. 27, Tokyo, Japan.

Grant Heiken, Los Alamos National Laboratory

3

Geology of Geothermal Systems

Introduction

This chapter identifies *sources* of heat located near the earth's surface, and then describes their genesis and tectonic setting. Some of this review examines ancient, well-exposed plutonic igneous bodies. By understanding the relationship of these bodies to remnants of volcanic fields overlying them, one may better assess the subsurface structure below modern volcanic fields that are prime targets for geothermal development.

Much of the information reviewed here concerns the hottest and most easily identified resource, i.e., geothermal systems associated with large silicic magma bodies. Until better techniques are developed to extract geothermal heat, they provide the best geothermal resource for the next few decades.

There is reason to be optimistic about geothermal energy. At the moment we really understand very little about geothermal resources, but an exciting period is beginning where anomalous sources of heat *are treated as systems*. These include all of the related processes responsible for transferring heat to the earth's surface; for example, the genesis of the magma, mechanisms of transport of melt through the crust, changes in petrologic and physical properties during its rise, interaction of melt and hot rock with meteoric water and brines, and eruption phenomena. Combined techniques from the fields of solid earth physics, petrology, hydrology, and chemistry are necessary to understand the problem. To develop geothermal energy as an important resource one must identify anomalous thermal sources and understand their genesis and geometry.

Classification of Geothermal Systems

Several classifications of geothermal systems are possible based either upon the source of the heat or the nature of the water-steam system developed over the source. In this chapter, the former is used in classifying the nature of the thermal sources. This classification shown in Table 3-1, represents a modification of the review by Muffler (1976) and is used as the outline for the chapter.

Regions with Young Volcanic Rocks

Heat from Large Silicic Magma Bodies

For tourists and scientists alike, one of the most spectacular geothermal areas in the world is the Yellowstone Plateau of Wyoming, Idaho, and Montana. Although this resource may not be developed, inasmuch as it is located within a national park, it is one of the better-known geothermal systems on earth. The hot springs and geysers of this area have intrigued man for centuries. Only recently, however, have these phenomena been shown to be the result of a very large, shallow silicic magma body located below the Yellowstone Plateau.

Much attention was paid to the chemistry of the hot spring waters early in this century (Allen and Day, 1925) and, later, to the studies of the mechanisms, chemistry, and petrology of the associated deposits, and alteration phenomena (White, 1955 and many later papers). (See Appendix E at the end of this chapter for a simplified classification of igneous rocks and their physical and chemical properties.) White's classic studies (e.g., 1957 and 1967) relate the circulation of meteoric water over a hot magma body to the existence of hot springs and geyers. Although investigators working within the Yellowstone Park area had suspected that a magma body or bodies existed below the volcanic plateau, the size and thermal state had not been defined until a multidisciplinary program of geologic mapping and geophysical measurements was initiated by the U.S. Geological Survey as part of their geothermal research program. Preliminary results of this work have been published (Eaton et al., 1975).

Volcanism within the Yellowstone plateau began with the eruption of basalt flows before the intrusion of silicic magma. The basaltic activity has continued around the plateau since that time. The earliest significant silicic activity began with the eruption of 2,500 km³ of ash 1.9 million years ago that resulted in collapse of the 30-km diameter Island Park caldera. A second cycle of activity ended with the eruption of a 250 km³ ash deposit, 1.2 million years ago. As a grand finale(?), 900 km³ of ash was erupted from the Yellowstone caldera 600,000 years ago. The composite caldera that formed during the activity is 70 x 45 km in size (Figure 3-1). Resurgent doming of the eastern segment of the caldera and small eruptions of rhyolite occurred

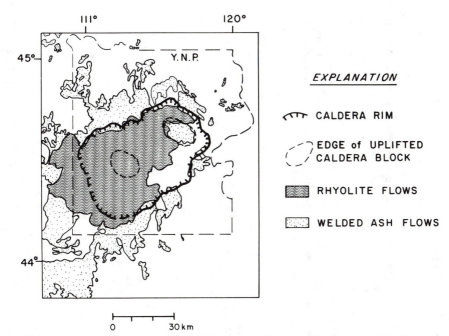

Figure 3-1. Geologic map of the Yellowstone Plateau Region, showing the upper Cenozoic rhyolite tuffs and flows, outline of the Caldera, and uplifted caldera blocks (resurgent domes). (After Eaton et al., 1975; courtesy of the AAAS.)

along ring fractures over a period of 350,000 years, following the last large eruption. A new insurgence of magma began about 150,000 years ago, with doming of a block east of Old Faithful Geyser. Much of the present hydrothermal activity at the surface is related to the recent influx of magma.

The plateau coincides with a gravity low with closure of greater than 50 mgal (Figure 3-2). This gravity anomaly is interpreted as being due to a shallow, coherent magma body. Seismic waves from local earthquakes show significant attenuation, also suggesting the presence of a magma body in the fluid state. The upper part of this body is located within a few kilometers of the surface. Delay of P-waves has been *interpreted* as a combination of a shallow silicic magma body and a deep (100 km), possibly partly molten "root" with pods of melt and thermally distributed crustal and mantle rocks (Figure 3-3).

To summarize the results presented by the U.S. Geological Survey team (Eaton et al., 1975), the recent data from the Yellowstone Plateau indicates that a large magma chamber exists at shallow (few kilometers) depth. Within the last 2 million years, breaking of the roof over this body has

Table 3-1
Summary of Geothermal Occurrences

Classification	Regions with young volcanic rocks			Tectonically active regions	Geopressurized-geothermal	Regions of normal heat flow
	Large silicic magma bodies	Magma bodies of intermediate composition	Regions with basaltic volcanism			
Description of source	Large (up to 1000 × 50 km batholiths) bodies, with smaller cupolas rising to the surface. May be thin (<10 km thick). Intrude to within 5 km of surface. Long cooling times (>10⁶ years).	Thin frameworks of dikes and sills within cones and small (<1 km³) "plugs" within central conduits. Slightly larger bodies present below volcanoes at depths of 5 - 15 km. Variable cooling times (days to 10⁶ years).	Thin networks of dikes, some sills, a few very small shallow bodies such as laccoliths. Unless rate of intrusion is high, the bodies cool very rapidly. Elongate dike systems.	High heat flow. Circulation of meteoric water down fault planes.	Hot water held in "sealed", well-insulated thick sedimentary sections.	Old silicic terrain with high radiogenic heat production, overlain by blanket of insulating sediments.
Surface manifestations	Associated with large-volume deposits of silicic ash calderas, ring-dikes. Large depressions surrounding regions of volcanic activity. Large areas of hot-spring, geyser activity.	Stratovolcanoes, domes, small calderas, some cinder cones. Short-term fumarolic activity, hot springs.	Fissures, shield volcanoes, cinder cones, associated with extension at faulting. Same fumarolic or hot spring activity.	Hot springs, normal faults. Active, recent tectonic activity.	None.	None; heat flow anomalies.

Table 3-1 continued

Classification	Regions with young volcanic rocks			Tectonically active regions	Geopressurized-geothermal	Regions of normal heat flow
	Large silicic magma bodies	Magma bodies of intermediate composition	Regions with basaltic volcanism			
Tectonic setting	– Subduction zones. – "plumes". – Offsets in major extensional faults.	– Subduction zones. – Some "plumes" and fault intersections.	– Spreading centers. – Continental rifts. – Areas with extensional faults.	Crustal uplift, faulting on continents.	Continental margins with sedimentation rates.	Stable continent.
Possible geothermal reservoirs and extraction techniques that might be used.	Heat transfer by convection and conduction. Hot-water and steam reservoirs overlie cooling bodies. All techniques may be used, depending upon age, nature of hydrology within rocks over the source.	Heat transfer from larger bodies by convection and conduction. Smaller bodies-conduction. Small hot-water or steam systems. Might use HDR or "magma-tap" techniques.	Hot-water-steam systems maintained in regions with high rate of intrusion.	Hot water for "low-grade" geothermal use.	Tap hot water steam, methane from permeable sediments.	Hot-dry rock, geothermal.

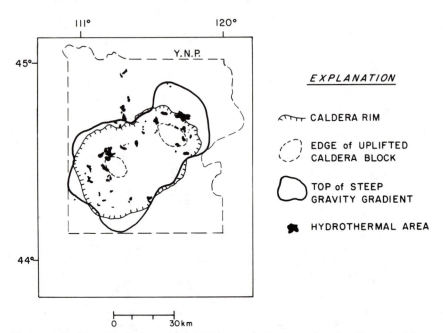

Figure 3-2. Simplified map of the Yellowstone Region, showing the coincidence of the edge of a steep Bouguer gravity gradient and the caldera rim. Also illustrated are the surface hydrothermal areas. (After Eaton et al., 1975; courtesy of the AAAS.)

resulted in the eruption of large volumes of silicic volcanic ash and catastrophic collapse of the roof rocks to form calderas.

It is apparent that large magma body or bodies are an enormous resource at Yellowstone. Smith and Shaw (1975) estimated the volume of these chambers at 45,000 km³ and that they have a ΔQ total of 26,875 x 10¹⁸ cal. Large magma chambers are the most obvious "high-grade" geothermal resources. Although little is really understood about them, by analogy with older exposed systems and study of surface phenomena, i.e., volcanic fields, some of the knowledge necessary to locate them and to understand their origins is available.

Shape and Size of Silicic Magma Bodies

On the continents, there are hundreds, perhaps thousands of older, well-exposed silicic batholiths. One of the best exposed batholiths is the Coastal Batholith of Peru, located in the western Andes. This intrusive complex rose toward the surface during upper Cretaceous to lower Tertiary time, intrud-

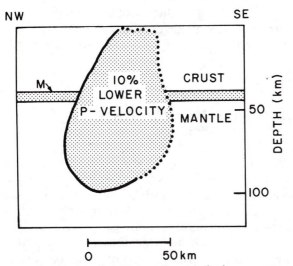

Figure 3-3. Northwest trending vertical cross-section (across the maps of Figure 3-1 and 3-2) of a hypothetical low-velocity volume creating a pattern of P-wave residuals. The heavy and dotted lines locate the boundary of the low-velocity material. (After Eaton et al., 1975; courtesy of the AAAS.)

ing a heterogeneous mass of its own volcanic rocks. It is a 1,100-km x 50-km complex of gabbro, tonalite, granodiorite, adamellite, and granite bodies.

The plutons are elongate, parallel to the tectonic fabric of the region (major faults and folds) or are equant when the surrounding country rocks are more homogeneous. Some of the larger bodies are up to 45 km x 15 km in size. Meyers (1976) and Cobbing and Pitcher (1972) observed that most of these bodies had steep walls and flat roofs. Cobbing and Pitcher (1972) felt that the plutons tapered downward and were teardrop-shaped. Francis and Rundle (1974) also believed that the ancient and recent magma bodies below the Andes had teardrop shapes and were rooted to a depth of 30 km. Based on models of the Boulder Batholith of Montana and other examples, Hamilton and Meyers (1967) proposed that batholiths were thin (≈5 km thick) bodies spread laterally at shallow depths (4-7 km), crystallizing beneath a cover of their own volcanic ejecta.

The depth to which the large silicic bodies rise and at which they crystallize is generally believed to be shallow. Based on the thermal metamorphism of country rocks, Meyers (1976) argued that the "roof" for most of the bodies within the batholith was 2 to 3 km thick. Cobbing and Pitcher noted that roof rocks were generally less than 3 km thick and as noted earlier, Hamilton and Meyers (1967) believed that most of these bodies reached depths of less than 7 km.

Relation of Large Silicic Magma Bodies to Overlying Volcanic Fields

It is generally believed that large silicic magma bodies rise buoyantly in the same manner as salt diapirs rise through a sedimentary basin. As these magma bodies approach the earth's surface, what are the effects? What are the geologic and physiographic criteria that may be used to identify such bodies? Steven and his coworkers (1974, 1975), after many years of careful mapping and laboratory work, have identified a large batholith underlying the volcanic field of the San Juan Mountains of Colorado. Although this body is too old to be a major geothermal resource area, it helps to understand younger features that are discussed later.

The dissected volcanic plateau of the San Juan Mountains is the remnant of a very large volcanic field formed largely between 35 and 26 million years ago (Steven, 1975). Early eruptions of intermediate-composition lavas formed a complex of stratovolcanoes. Activity culminated with the eruption of thousands of cubic kilometers of silicic ash-flow tuffs and collapse of vents to form a group of large calderas (Figure 3-4). The culminating eruptions occurred as the large silicic body approached the earth's surface (Figure 3-5). The area of this volcanic field is approximately coincident with a large gravity low (Figure 3-4) that has been interpreted as reflecting the presence of a shallow batholith (Plouff and Pakiser, 1972). Steven (1975) noted that the gravity low cannot be explained by the distribution of early intermediate-composition volcanoes, but may reflect the presence of a silicic batholith.

Individual calderas (vents) probably represent the surficial features associated with cupolas rising off of the main batholith. The cupolas, if they are approximately of the same width as the calderas, are nearly circular features ranging in diameter from 10 km to 40 km. The main body may be about 50 km x 100 km in size (Lipman, 1978) (Figure 3-5). This is comparable in size to the well-exposed Boulder Batholith of Montana. Reasons for the rise of individual cupolas, off of a much larger mass, to form epizonal plutons and perhaps calderas, are not understood. It may be related to regional structure of the overlying rocks, instabilities developed within the bodies, or both. As a source of near-surface heat, the epizonal plutons or cupolas are most important.

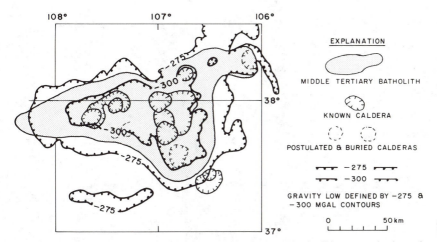

Figure 3-4. Map showing the relation of a postulated batholith to gravity lows in the San Juan Mountains, Colorado. The calderas may represent areas where cupolas rising off of the main mass of the batholith reached the surface and erupted silicic ash-flow tuffs and flows. (After Steven, 1975 and Steven et al., 1974.)

The structure and lithologies located between the ground surface and the top of a pluton are complex and the subject of voluminous books. Briefly summarized, a cupola rising to the surface, quite often through its own pile of volcanic rocks, generates concentric and radial fractures at its apex. Volatile phases, coming out of solution at shallower depths, concentrate at the top of this body. When the gas pressure exceeds that of the lithostatic pressure, the top of the body is erupted rapidly along the concentric fracture system. This is believed to be the case for the vents at Yellowstone and in the Jemez Mountains, New Mexico, another high-grade geothermal area (Smith and Bailey, 1968) (Figure 3-6). As magma is emptied from the chamber, collapse along the ring fractures begins and, after the eruption has ceased, there is circular depression partly filled with ash and coarse sediment from the caldera walls. If the area of the caldera is large enough, magma will again flow upward, to reestablish isostatic equilibrium, swelling the caldera to form a resurgent dome (Smith and Bailey, 1968). Smaller eruptions may occur during the resurgent phase along the ring fractures and could also be an important source of heat, if young enough. As shown later, the main body of the pluton is a source of heat that will last for several million years. The ring dikes along vent and caldera boundaries are thermal sources that will last for only a few thousands or tens of thousands of years. Tuffs and sediments within the caldera often serve as a good reservoir rock for hydrothermal systems. Circulation of meteoric water may well be confined within the sediment-filled cylinder, bounded by caldera walls.

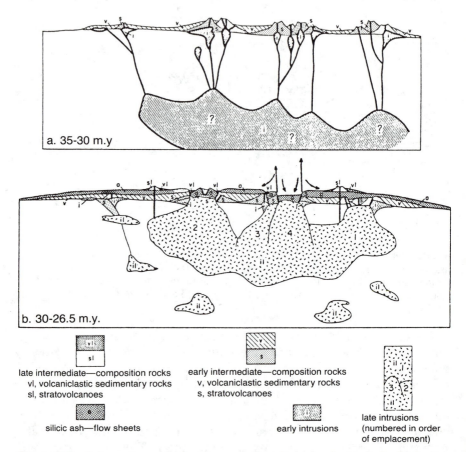

Figure 3-5. Schematic model for the evolution of a subvolcanic batholith in the San Juan Mountains, Colorado. A. Eruption of the intermediate-composition lavas and tephra to form composite cones, surrounded by aprons of volcaniclastic debris. It is uncertain that the large batholitic mass has yet developed. B. The main batholitic mass has accumulated at shallow depth. As cupolas off of the main mass reached the ground surface, there were massive ash-flow eruptions and caldera collapses. Many caldera collapses were followed by resurgent doming, indicating flow and shallow intrusion of silicic magma. (After Lipman et al., 1978; courtesy of the Geological Society of America.)

The connections of ring-dikes from intrusive bodies to the surface have long been observed at well-exposed bodies, for example in Great Britain (Anderson, 1936), Australia (Branch, 1966), and South America (Cobbing and Pitcher, 1972). Caldera walls and circular lines of vents from which silicic lavas have erupted roughly define the outline an intermediate size pluton or cupola. This is a relatively easy target to define and most of the young features of this kind, within the United States, have been identified as

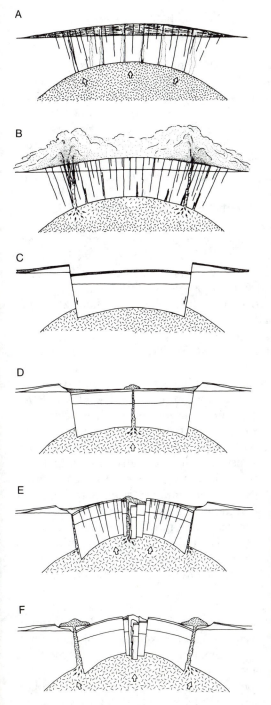

Figure 3-6. Stages in the resurgent cauldron cycle. A. Regional swelling and generation of ring fractures as the silicic magma body approaches the surface. B. Caldera-forming eruption of silicic tephra. C. Caldera-collapse. D. Preresurgence volcanism and sedimentation within the caldera. E. Movement of the new magma and resurgent doming. F. Major ring-fracture volcanism. (After Smith and Bailey, 1968; courtesy of the Geological Society of America.)

known geothermal resource areas (Smith and Shaw, 1975). Is it possible to define the outline of a shallow (less than 10 km) batholith? It is generally believed that large negative Bouguer anomalies associated with young silicic volcanic fields reflect the shape of an underlying batholith. As is the case at Yellowstone, S-wave attenuation, lack of seismic activity, and magnetic measurements may be used to roughly identify the shape of large anomalous bodies of hot rock or melt located within the crust (Eaton et al., 1975).

Many young volcanic fields are located within large depressions that are visible on small-scale maps or satellite images. There are many examples, but one of the most spectacular lies along the eastern coast of Kamchatka, U.S.S.R. The young volcanic fields of Kamchatka are located in an overlapping chain of circular to ovoid depressions located within older, dissected terrane (Figures 3-7 and 3-8). Within the depression, identified on Landsat Satellite imagery, is the "caldera line" of Kamchatka (Vlasov, 1964). If this 2.3° long depression does reflect the size of underlying batholiths, they are similar in geometry, scale, and tectonic setting to the older, well-exposed coastal batholith of Peru and its volcanic deposits (Myers, 1975) (Figure 3-9). Use of satellite imagery and small-scale geologic maps may reduce the size of geophysical surveys needed to identify the large thermal source associated with a young batholith.

Thermal History of Large Silicic Magma Bodies

A simple means of classification of the "lifespan" of thermal anomalies related to plutons has been developed by Smith and Shaw (1973, 1975) (Figure 3-10). They have developed these curves on the basis of approximate volume data, age of last eruptions, and solidification time of the magma. They conclude that large (10^2-10^5 km^3) magma chambers, located in the upper 10 km of continental crust, are the most attractive targets for geothermal exploitation. In a paper on igneous geothermal systems, Smith and Shaw (1975) noted that there are no uniform criteria for rates of hydrothermal heat transfer for all systems related to magma bodies. They have calculated heat contents for many young intrusive bodies, located in the United States, on a "dry" basis. Their thermal calculations are also based on the assumption that later gains of magma by a body are ignored. For data on individual systems within the United States, the Smith and Shaw (1975) paper is a useful reference.

For heat loss by conduction alone, the cooling of large intrusive bodies is very slow. Smith and Shaw (1973) demonstrated that a 5-km thick chamber with horizontal slab-like geometry, takes 2 million years to cool from initial magma temperature ($\approx 850°C$) to ambient temperature. A 10-km thick chamber would require 10 million years to cool. They note that most of the largest systems, such as the Yellowstone, will be preserved for 10 million years, even with an allowance for cooling by hydrothermal activity.

Figure 3-7. Landsat image (E-1117-23581) of the southern tip of the Kamchatka Peninsula, USSR, showing the major depressions associated with Holocene volcanic field. There are chains of cinder cones and composite cones *adjacent* to the large fields (to the west).

Numerous numerical thermal models of plutons have been published, dealing with cooling both by conduction and convective circulation of fluids. Fluid circulation by convection is inevitable wherever magmas are emplaced in the upper 10 km of the earth's crust (Norton and Knight, 1977). The cooling rate for a large magma body, however, is not greatly shortened unless the body has a permeability of $>10^{14}$ cm^2. Heat transfer by circulating fluids is significant over a large region above and near the intrusive body. Large volumes of rock remain at 200° to 400°C for long periods of time (Norton and Knight, 1977). Simulation of fluid circulation by convection around a batholith-size pluton 54 km in diameter and emplaced to a depth of 5 km will maintain a temperature of 400°C over a 60 km^2 region above the batholith for 10^6 years (Norton and Knight, 1977). A batholith with irregular cupolas developed on top (more realistic than a large, flat-roofed batholith) will develop secondary convective cells around each cupola.

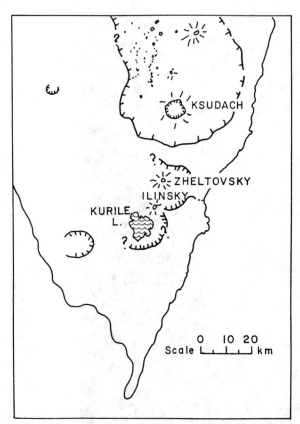

Figure 3-8. Sketch map of the Landsat photograph showing large composite cones (circles with radial lines), small cones (circles), calderas (lines with small hachures), and major depressions (solid lines with large hachures). The major depressions might reflect subsurface extent of the plutons feeding the volcanic fields.

Using some real examples, it is believed that the magma chamber located below the Valles Caldera, New Mexico, where the last major eruption occurred 1.2 million years ago, may still retain temperatures of 600°C at a depth of 7 km (Kolstad and McGetchin, 1978). Temperatures observed in deep drill holes adjacent to the caldera are consistent with the isotherms predicted for a pluton radius of 12 km and burial depth of 3 km (Figure 3-11).

At Long Valley, California, a 17 km x 32 km caldera was formed 0.7 million years ago during eruption of the silicic Bishop Tuff. Eruptions of rhyolitic magma as recently as 450 years ago suggest that magma is present now. Hydrothermal activity, which began at least 300,000 years ago, has

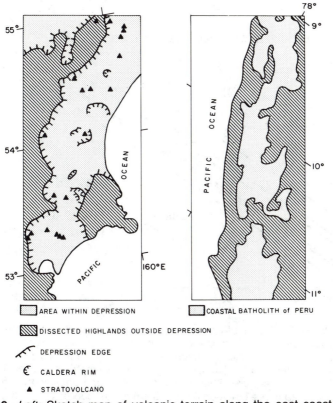

Figure 3-9. *Left*: Sketch map of volcanic terrain along the east coast of Kamchatka, USSR, with outlines of large depressions surrounding volcanic fields (based on Landsat images). *Right*: To compare the areal extent and geometry of an exposed batholith, located in a similar tectonic situation, a sketch map of the coastal batholith of Peru is presented at the same scale (adopted from Myers, 1975). (Redrawn from Heiken, 1976; courtesy of the Geological Society of America.)

declined somewhat since then due to self-sealing of surface sediments by hydrothermal alteration (Bailey et al., 1976). There is, however, a well-developed hot-water system within the caldera, with minimum temperatures of 180°C. A tremendous amount of heat remains after 0.7 million years. Lachenbruch et al., (1976) suggested that the magma body below Long Valley, California, could not have stayed molten for the 2-million year eruptive history of the volcanic field unless it were *resupplied* with magma from deep crustal or subcrustal sources. Replenishment of heat within large silicic magma bodies by hotter basaltic magma from below has also been suggested by Eichelberger and Gooley (1977), who use petrologic argu-

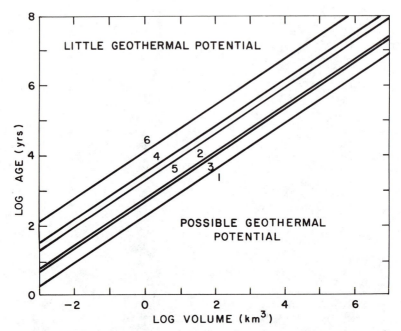

Figure 3-10. Graph of theoretical cooling time versus volume of magma bodies. Lines are drawn to represent cooling models. Those systems now approaching ambient temperatures are above lines 5 or 6; lines 5 (slab) and 6 (cube) represent an estimate of time before reaching a temperature of 300°C. Systems now approaching the post-magmatic stage are between lines 3 (slab) and 4 (cube); it is assumed that cooling occurs entirely by conduction. Lines 1 (slab) and 2 (cube) assume systems with a large molten fraction, with cooling by internal convection until solidification. (Redrawn from Smith and Shaw, 1975.)

ments for the reheating and convective mixing of the two magma types *within* the magma body. The reheating by convective mixing within a chamber will maintain high temperatures far beyond that predicted by any of the models discussed here, which is very encouraging for the geothermal industry.

The Geographic Location of Large Silicic Magma Bodies

Subduction Zones

Thousands of pages have been written on the generation of magma within subduction zones. A clear and simple statement of the concept was made by Gilluly (1971), who stated: "Magmas are developed at both ends of the moving-sidewalk-like tread of migrating plate, both where new crust is formed and along the Benioff Zone." (Figure 3-12).

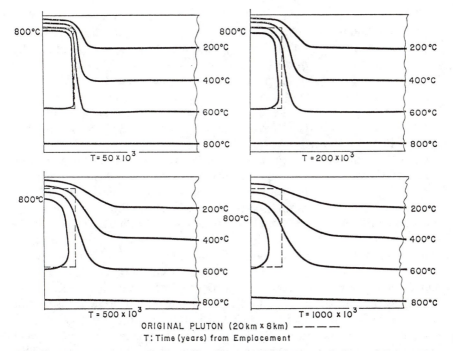

Figure 3-11. Typical evolution of isotherms with time, modeling a 20km × 8km silicic pluton. Half-width of pluton is outlined with dashed line. (After Kolstad and McGetchin, 1978; courtesy of Elsevier Scientific Publishing Company).

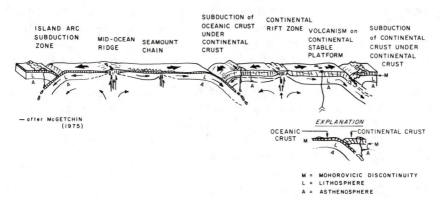

Figure 3-12. Schematic drawing of tectonic settings of most of the earth's volcanoes, plutons, and active tectonic regions. (After a sketch of McGetchin, unpublished)

Silicic magmas are generated within subduction zones or convergent plate boundaries, and there are many mechanisms proposed for this process. The hypotheses include partial fusion of mantle peridotite and subsequent differentiation of rising diapirs of basaltic or andesitic magma, partial fusion of crustal material, and melting at the base or within the continental crust by much hotter pods of basalt (summarized in Eichelberger and Gooley, 1977). Whatever the process may be, silicic magma *is* generated along subduction zones, a fact noted by Hess (1939) and later supported by petrologic data (Kuno, 1959; and James, 1971).

Once the melt body was formed, it rises buoyantly, transferring enormous amounts of heat within the crust, resulting in large-volume silicic volcanism and driving convective hydrothermal systems. Large bodies of this sort are generated only in areas of thick continental crust, e.g., the crust beneath the Andean Cordillera, where countless large magma bodies have risen to the surface since Triassic time, is at least 70 km thick (James, 1971). The rate of generation of these bodies is dependent upon the rate at which the oceanic plate moves and is thrust under the continental plate (James, 1971; Francis and Rundle, 1976). The young (Pleistocene to Recent) volcanoes of the Altiplano of the Andes include dozens of unstudied large calderas and large-volume ash-flow tuff fields (identified on satellite imagery, e.g., by Brockman and Kussmaul, 1973; and Friedman and Heiken, 1977), which are volcanic features that reflect the presence of large and probably still hot or partly molten silicic magma bodies. Because the Altiplano is sparsely populated, there has been little effort to study or develop the potential resource that exists there.

Other regions include: Kamchatka, where many calderas and large ash-flow tuff deposits of Recent to Pleistocene age are located (Vlasov, 1964; and Fedotov et al., 1976); Indonesia, the location of large volcanoes of silicic composition that have erupted in the Recent time (e.g., Toba, Krakatau); and the Taupo depression in New Zealand, the site of the Recent eruptions of large volumes of silicic tephra and of caldera collapse (e.g., Healy, 1976). Areas where large silicic bodies have developed over subduction zones, have risen to the surface during the Pleistocene to Recent time, and are potential geothermal resources as outlined in Figure 3-13. Only in New Zealand has the geothermal potential been utilized and developed over one of these young batholith-size intrusive bodies. Detailed studies of the geothermal fields within the Taupo Volcanic Zone of the North Island of New Zealand are summarized in Healy (1976).

"Hot Spots"

The concept of rigid plates moving over hot spots or mantle plumes was proposed by Morgan (1971) to explain the age relations of volcanic chains,

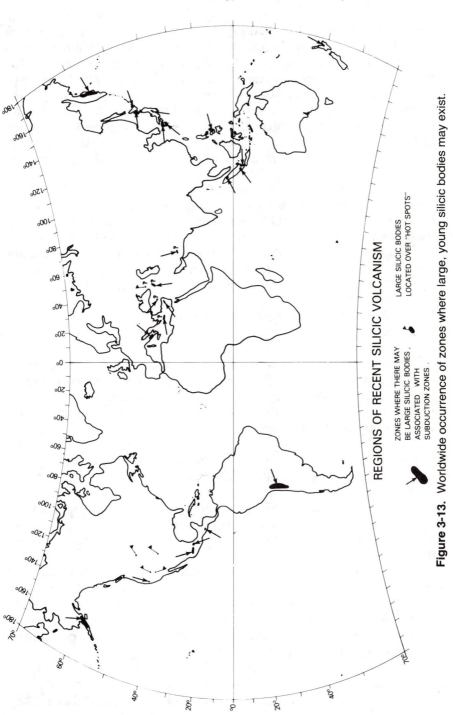

REGIONS OF RECENT SILICIC VOLCANISM

LARGE SILICIC BODIES
LOCATED OVER "HOT SPOTS"

ZONES WHERE THERE MAY
BE LARGE SILICIC BODIES,
ASSOCIATED WITH
SUBDUCTION ZONES

Figure 3-13. Worldwide occurrence of zones where large, young silicic bodies may exist.

such as Hawaii (an oceanic example) and the Snake River Plain Yellowstone volcanic fields (continental example). In these cases the plate moves over an anomalous hot plume, causing melting at the lower crustal boundary and the buoyant rise of melt bodies to the surface. This accounts for the "younging" of volcanic rocks from one end of a chain to the other. For example, the Yellowstone Plateau has been interpreted as the youngest point (now located over the plume) at the end of a northeast-trending line of volcanoes and has "navigated" along a fundamental structure in the lithosphere (Eaton et al., 1975). This trend may also be related to a much older structure within the crust (Lipman, 1978).

One of the best known geothermal areas in the world, The Geysers, California, is most likely heated by a large silicic body located below the Clear Lake Volcanic field (Goff and Donnelly, 1977). This volcanic center is located within the San Andreas transform fault system and is progressively younger from south to north (Hearn et al., 1976, 1978). It is proposed that the field is the surface manifestation of a mantle hot spot, across which the North American plate is moving southward (Hearn et al., 1978).

Large Silicic Magma Bodies of Uncertain Tectonic Association

A number of large silicic bodies are not easily explained as being formed by a plume or along a subduction zone. One of the examples cited earlier (an older volcanic field) is that of the San Juan Volcanic field, Colorado. Christiansen and Lipman (1972) have proposed that the magma body was generated by subduction of the western margin of the American plate. Gilluly (1971), commenting on an earlier manuscript on this model, disagrees with this interpretation, noting the great distance between the continental margin and the San Juan Volcanic field.

The location of Jemez volcanic field of New Mexico is also difficult to explain. It may be part of the "Jemez lineament," a line of volcanic fields stretching from eastern Arizona to western Oklahoma or may have formed at a lateral offset of the Rio Grande Rift, where basalt pooled at the base of the crust.

Other major silicic bodies and their volcanic fields are located within areas of extension, such as the Great Basin of the western United States. The Long Valley and Coso geothermal areas of California are located at the western margin of the Basin and Range Province, along major normal faults, (such as the one that forms the precipitous eastern slope of the Sierra Nevada) (Bailey et al., 1975; and Duffield, 1975), and especially where these faults are offset intersected by other major faults (Figure 3-14).

These fault intersections or offsets could be the locus for "pooling" of basalt magma at the base of the crust, melting of silicic crust, and subsequent

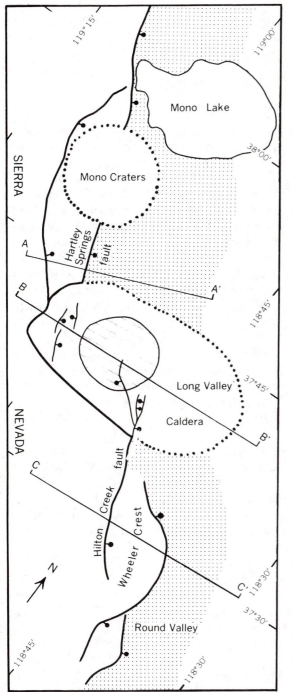

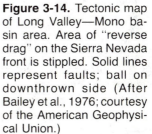

Figure 3-14. Tectonic map of Long Valley—Mono basin area. Area of "reverse drag" on the Sierra Nevada front is stippled. Solid lines represent faults; ball on downthrown side (After Bailey et al., 1976; courtesy of the American Geophysical Union.)

rise of large silicic bodies. Whatever their origins, there is yet much to be learned about large silicic bodies.

Relatively Small Magma Bodies of Intermediate to Silicic Composition

If one mentions volcanoes as sources of geothermal energy, large composite cones like Mt. Rainier in Washington or Fuji in Japan come to mind. This section deals with the smaller magma bodies associated with composite cones, their utility as a geothermal resource, and their distribution.

Composite cones consist of interbedded tephra, lava flows, dikes and sills. Ranging from a few hundreds of meters to over 4,000 m in height, the steep sided cones could not exist without the internal framework of dikes and sills that impart strength to them (Crowe and Nolf, 1977; and Crowe, et al., 1978). These cones evolve as simple cones composed of cinders and bombs, followed by the intrusion of dikes and sills, caldera collapse, and emplacement of conduit plug(s) (Crowe and Nolf, 1977) (Figure 3-15).

Some of the magma that is not erupted as lava or tephra is intruded into the cone as sills or dikes a few meters to tens of meters thick. The thin sills and dikes are cooled rapidly by conduction and interaction with water flowing through permeable tephra deposits of the cone. Some reach the flanks and are erupted at satellitic vents or fissures. The dike-sill systems are probably not an important source of geothermal heat.

A much more important thermal source is the "plug" developed within the main vent of mature composite cones. The plug is generally believed to be a cylindrical mass approximately the width of the summit crater.

At St. Augustine Volcano, Alaska, Kienle (1976) noted that the initial blast removed 0.1 km³ of dacite magma and that the cylindrical void was re-filled within a few hours by fresh magma. Surface phenomena associated with central conduits or plugs include small ash flows, air-fall tephra, viscous domes (such as those dominating Mt. Shasta and Mt. Lassen Peak, California) and numerous mud-flow deposits. Shallow, near-surface bodies such as these may also interact with meteoric water to cause intense fumarolic alteration of the summit of a composite cone.

Little is known of the rate at which a conduit or plug, located within a large composite cone, will cool. If the magma is replaced every few tens to hundreds of years, then it would be a reasonable target for geothermal development. There would be some danger from the eruptions, but that is a chance that must be taken to develop this resource. If the plug is not heated or replaced by magma from below, it will cool rapidly in a few to hundreds to a thousand of years (the cooling rate is based on a volume of 0.1 to 1.0 km, see Figure 3-10).

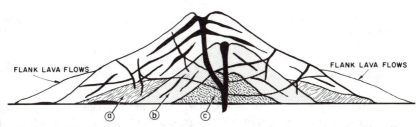

Figure 3-15. Sketch cross-section of a composite cone, showing the dike-sill systems that support the unstable structure consisting of volcaniclastic rocks and flows. (After Crowe, 1977.)

Little is known about magma chambers present below composite cones. Based upon petrologic data, mixing of silicic and basaltic magmas within crustal magma bodies is possible and may explain the genesis of many andesitic rocks typical of composite cones (summarized in Eichelberger, 1978). Based upon screening of S-waves below the Katmai volcanic range of Alaska, Kubota and Berg (1967) have identified 5 km to 50 km wide anomalies at depths of 8 km to 40 km as magma chambers. Some of these might qualify in the "large silicic magma chamber" category of the previous section, but some may feed the composite cones of the Alaskan Peninsula. Geophysical surveys of Avachinsky Volcano, an active composite cone located in southeastern Kamchatka, indicates that a 5.2 km x 2.6 km magma chamber exists below the volcano at a depth of 4 km (Fedotov et al., 1976). The last catastrophic eruption occurred 35,000 years ago, with a volume of 35 km³. Fedotov et al. calculated that the magma chamber is partly molten and will serve well as a long-term geothermal reservoir. In a model by Cathles (1977) a 1.5-km wide body, intruded into water-saturated, fractured rocks of normal geothermal gradient, will support intense geothermal activity for somewhat less than 20,000 years.

Most composite cones, due to their elevation above the surrounding terrain, have high precipitation and high infiltration of meteoric water (Healy, 1976). Fumarolic areas tend to occur near volcano summits and hot springs near the base, fitting a general model of vapor-dominated hydrothermal systems where groundwater moves downward and outward. In summary, meteoric water, in contact with the hot conduit, generates warm springs at the base of a mountain and explosively releases steam at the summit. This activity may be short-lived, started by slight movement of a magma body at the base of the volcano. Exploitation of such a source for natural steam may be difficult to justify, unless there is a steady influx of new magma at the base of the volcano and there is dependable heavy precipitation to maintain the continued influx of meteoric water. Healy (1976) cited a successful shallow well drilled in 1926 on the side of the Gandapura-Guntur volcano complex in Java, where steam is still being produced, with a wellhead temperature of 140°C.

Geographic Distribution of Andesitic Composite Volcanoes

Most andesitic volcanoes are located over subduction zones in island arcs and at active continental margins (Hatherton and Dickson, 1969). Many of these volcanic chains are associated with large silicic bodies and may precede silicic volcanism as part of long-term eruptive cycles (Lipman et al., 1978). The relation between subduction, Benioff zones, and chains of andesitic volcanoes is often cited (Fyfe and McBirney, 1975, and papers edited by Cox, 1972). The origins proposed are many, including differentiation of a basaltic melt, generation of primary andesitic melt, and mixing of rhyolitic and basaltic melts. Whatever their origin, they are associated with active plate margins such as the "Pacific Ring of Fire" (Figure 3-16). Composite cones, although often highest topographically, are generally located within depressions that run parallel to the chain (Fyfe and McBirney, 1975) (Figure 3-17). The association with regions of extension (normal faults, horsts, and grabens) is good for geothermal development. The flow of meteoric water within grabens associated with composite cones is often restricted by graben walls. Thus, heat is retained and the restricted system is easier to develop.

Much of the earth's most accessible geothermal resources are related to chains of composite cones. The great volcanic chains running through Central America and South America, the Cascade Range of the U.S., the Aleutian Islands, Japan, the Philippines, and most of Indonesia are all examples of targets for development (Figure 3-16).

One of the better-known areas of this type is El Tatio, located in northern Chile. The field is situated within a north-south graben filled with upper Cenozoic volcanic rocks, overlying sediments of Cretaceous age (Lahsen and Trujillo, 1976). A 4.24 million-year-old ash-flow tuff unit within the graben is the main producing horizon. The youngest rocks are Pleistocene to Recent rhyolite domes and andesitic composite cones that rise to an elevation of 5,000 m. Most of the heat for groundwater within the graben is believed to be supplied by the young volcanoes (Lahsen and Trujillo, 1976). A horst, located west of the graben, restricts the flow of fluids. Surface hydrothermal activity is located along NW-SE and NE-SW trending fractures within the graben (Figure 3-18).

Many of the geothermal fields of Italy are located above a subduction zone. Along the western flank of the Italian peninsula, the volcanoes of the Roman Province are mainly composite cones and small calderas (McGetchin et al., 1974). The Phlegrean Fields, Ischia, Vesuvius, etc. are all associated with NE-SW trending horsts and grabens (Cameli et al., 1976) that have also served as Pliocene-Quaternary age sedimentary basins. Older

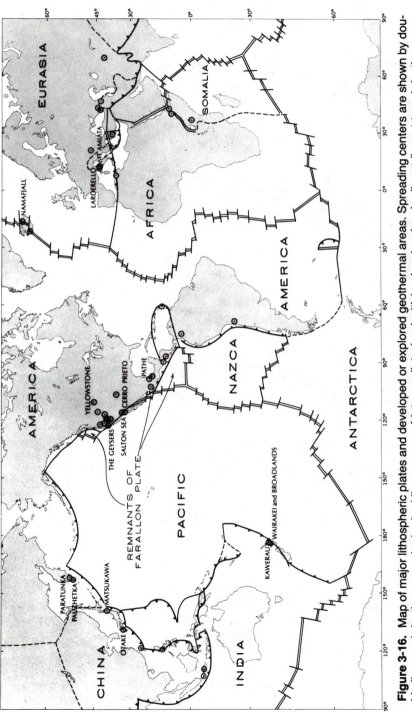

Figure 3-16. Map of major lithospheric plates and developed or explored geothermal areas. Spreading centers are shown by double lines, subduction zones by barbed lines, areas of intermediate (andesitic) volcanism by shading adjacent to subduction zone, and transform faults by single solid lines. (After Muffler, 1976.)

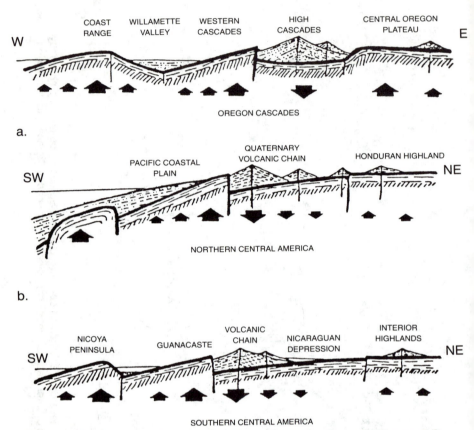

Figure 3-17. Generalized structure sections across the Cascade Range of Oregon and northern and southern Central America, showing the prominent belts of uplift and subsidence. Typical tectonic setting for andesitic volcanoes along continental margins. (After Fyfe and McBirney, 1975; courtesy of the *American Journal of Science.*)

(Mesozoic) carbonates serve as reservoir rocks, capped by less permeable late Tertiary clays, sand, and volcanic rocks of the basins. Thermal sources are cooling intrusive bodies associated with the Holocene age volcanoes. The distensive phase, active since the upper Miocene time, has always been accompanied by the eruption of intermediate and silicic volcanic rocks (Baldi et al., 1976).

The combination of the late Tertiary to Recent grabens and volcanism is also responsible for active and potential geothermal areas in Turkey (Kurt-

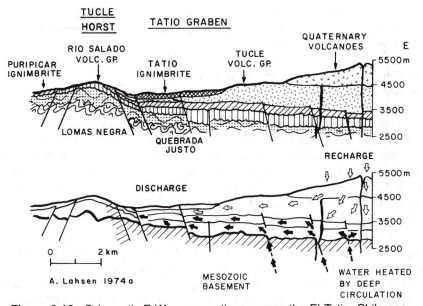

Figure 3-18. Schematic E-W cross-sections across the El Tatio, Chile geothermal field. *Top*: Geologic cross-section. *Bottom*: Movement of geothermal fluid in the system. (After Lahsen and Trujillo, 1976.)

man and Samilgil, 1976), Japan (Yamasaki and Hayashi, 1976), and the Cascade Range (Fyfe and McBirney, 1975).

There is one type of volcano that does not fit into any of the above categories (i.e., isolated shield volcanoes with a bimodal suite of rocks, basalt and rhyolite), located off to one side of well-defined volcanic chains. Examples are the shield volcanoes located east of the Cascade Range: the Medicine-Lake Highland, California, and Newberry Volcano, Oregon. These volcanoes exhibit bimodal, young (Holocene) volcanism; have shield shapes, with shallow summit calderas; are located in areas of extension (e.g., the western edge of the Basin and Range Province); and may have shallow (<5 km), small silicic bodies. Both of the cited examples are located at the junctions of fault zones (Higgins, 1973; MacLeod et al., 1976; and Heiken, 1978), where basalt melt may have pooled and melted silicic crust to form the rhyolite bodies, and are surrounded by basaltic volcanoes located along NW-SE trending normal faults. MacLeod et al., noted that Newberry Volcano is the youngest volcano in 250 km belts of Miocene to Holocene rhyolitic volcanoes that trend N75°W across eastern and central Oregon. No one yet completely understands the cause of this progressing "front" of rhyolitic volcanism. An answer would be of great value to the geothermal development of such chains.

Geothermal Resources in Regions Dominated by Basaltic Volcanism

Using an eruption on the island of Hawaii as an example, one is able to learn much about basalts on oceanic islands as geothermal resources, which in general are poor. The 1959-60 eruption of Kilauea Volcano, described by Richter et al., (1960), is an excellent example of the activity of hot (1,050°-1,100°C), low-viscosity (10^2-10^3 poises) melts (Figure 3-19). The activity began with movement of melt from a depth of 50-60 km to a reservoir 1.5-3 km below the volcano. Eruption, which began at Kilauea Iki Crater along a 460-m long fissure, soon was concentrated at a chain of vents and, then, at one main vent having dimensions of 15 m x 12 m at the surface. The summit chamber was refilled from below, but melt flowed laterally along the east rift and erupted along fissures on the flank of the volcano. During the flank eruption, there was collapse of a crater in the summit area. The surface activity and subsurface activity, monitored with geophysical measurements, indicated that the high-level reservoir was drained (it would not have been a good thermal reservoir). Most of the basalt was erupted along narrow fissures that would rapidly cool within a few years. Most of the heat was lost as erupted lava. Only the lava lake within Kilauea Iki Crater is a reasonable, easily accessible but "short-term" heat source. It is now the object of experiments that will use a heat exchanger (Colp, 1974).

Well-exposed dike complexes on Oahu are vertical, complex, intertwining sheets with an average thickness of 0.6 m, and ranging from 0.3 m to 15 m in width (MacDonald and Abbott, 1970). Although common, sills are less numerous than dikes and have an average thickness of 3 m. Dike complexes are parallel to rifts that radiate from summit calderas in a 2.4-km to 4.8-km wide zone. In the central part of the zone, there are 62 to 140 dikes per 1 km of width. This spacing density of dikes drops off to 6 to 60 per 1-km segment along the edges of the zone (MacDonald and Abbott, 1970). Individual dikes are narrow and cool rapidly. Unless the rate of dike injection were high, even the central portion of a rift would be an undependable prospect. Tremendous heat losses during short periods of time occur in the case of dikes having high surface areas (Delaney and Pollard, 1976).

If the influx of magma is continuous, then regions characterized by basaltic volcanism will have considerable potential as geothermal resources. A model of Mt. Etna, Sicily, was developed by Wadge (1977) to explain the observed eruptive activity there. He proposed that there is a main storage reservoir in the form of a 100-m wide x 2-km deep cylinder below the summit

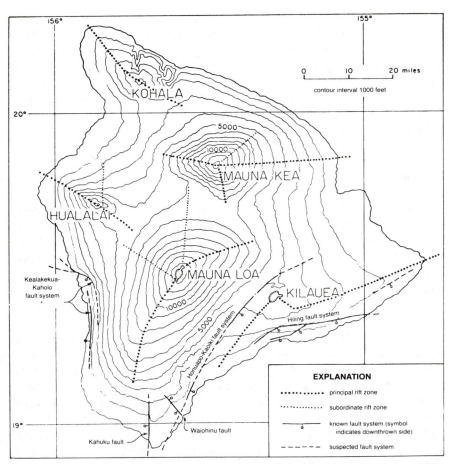

Figure 3-19. Map showing rift zones and summit calderas of the major shield volcanoes of the island of Hawaii. (After McDonald and Abbott, 1970; courtesy of the University of Hawaii Press.)

and that magma enters this reservoir at a rate of 0.35 m³/s during quiet periods. When the reservoir is full, there is some overflow at the summit and failure of fracture systems on the flanks, resulting in eruptions. This continuing process of filling, followed by hydraulic fracturing of dike-like fissures within flanks, has continued for 450 years (activity for which there are records), with eruptions occurring about every 6 years. As was described for the Hawaiian case, dike systems on the flanks, with width of 0.2m to 2m, cool rapidly and have little geothermal value. The central conduit with an approximate volume of 0.02 km³, however, may serve well as a source of heat, especially if the "magma tap" energy technology is developed (Colp, 1974).

At spreading centers, along mid-ocean ridges with a rapid spreading rate, there is a tremendous amount of heat carried to shallow levels by dikes. Due to higher wall-rock temperatures, magma reaches the surface more readily. There is less cooling, lower viscosities and, thus, higher flow rates of magma through dikes (McBirney, 1975). Within ancient spreading centers such as the one exposed in the Troodos complex of Cyprus, dikes make up more than 90% by volume of rocks and range in thickness from tens of cm to 5m (Moores and Vine, 1971). In Iceland, Bodvarsson and Walker (1964) found about 1,000 dikes, with a total thickness of 3 km, over 53 km. More than half of the Icelandic dikes failed to reach the surface at depths of 10 km. Gibson and Piper (1972) believe that the volcanic pile is composed of 100% dike rock.

Most of the earth's spreading centers are under water. There are, however, a few regions astride these ridges that are above water and accessible for extraction of geothermal heat. The best known example is Iceland, where, for many years, geothermal heat has been used for space heating, domestic hot water supplies and, most recently, for the generation of electricity.

In Iceland, which is located on the crest of the mid-Atlantic Ridge, 50% of the heat flows to the surface by conduction, 30% by magma, and 20% by thermal waters (Bodvarsson, 1976). "Low-temperature" ($<150°C$) systems of non-volcanic origin are located in the lowlands and high-temperature ($300°C$) systems are located in the neo-volcanic zone. The latter systems are located directly over or adjacent to active volcanic terrain. Hot water within aquifers, which are located along lava flow boundaries or deposits of hyaloclastic rocks, may flow laterally away from the ridge (Bodvarsson, 1976) or be ponded by dikes acting as permeability barriers. Iceland is unique in that the large central volcanoes may have up to 8% rhyolite lavas and plugs (Walker, 1959). As was discussed earlier, viscous plugs of rhyolite are excellent thermal reservoirs. There are, however, some drawbacks to geothermal development of active spreading centers. An electrical generating plant being constructed at Krafla, Iceland, has been plagued over the past year by nearby eruptions. Admittedly, the eruptions are small and damage was slight, but such a location might deter investors in such a project.

Geothermal Areas Associated with Continental Rifts

Rifting (spreading) of continental crust provides a physical and thermal setting that is considerably different than that which exists at mid-ocean ridges. Spreading rates and influxes of basaltic magmas are slower. There is, however, evidence for crustal melting, with the eruption of alkaline and silicic rocks along with basalts (Gilluly, 1971). The rifts are characterized by complex graben structures with chains of volcanic vents lined up within the grabens. They have anomalously thin crusts and high heat flow.

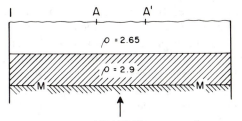

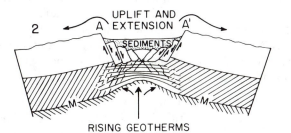

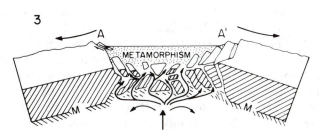

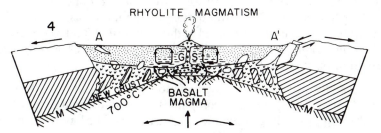

Figure 3-20. Model of rifting and magma generation during growth of a graben. (1)Two layers of crust over a mantle hot zone, M = Mohorovicic discontinuity. A-A' are reference points. (2) Upward and lateral extension. The trough is partly filled with sediments. (3) Binding trough is invaded by basaltic magma. (4) The basement rocks are melted and rhyolite magma is extruded. Ascending hot brines are generated by shallow silicic magma bodies (After Elders et al., 1972; courtesy of *Science*.)

The East African rift extends from the Red Sea to Mozambique, in the southern part of the continent. Volcanism along the rift ranges in age from Miocene time to the present (King and Chapman, 1972). The rift is filled dominantly with basaltic volcanoes, but some silicic composite cones have developed along the margins (Barberi and Varet, 1977). The Asal Rift, a branch of the complex Afar Depression of Ethiopia and Afars and Assis has been studied with geothermal development in mind (Stieltjes, 1976). The Asal opened during Plio-Pleistocene time and has since been spreading at a rate of 1.5 to 2 cm/yr. Central doming and uplift within the rift may reflect diapiric intrusion of basaltic melt into the rift. There is an anomalous gradient of 45°C/100 m, which is characteristic of the central rift. The axial valley, probably underlain by closely spaced dikes, contains an open hydrothermal system, with deep circulation of marine and meteoric waters at a temperature higher than 170°C.

Within the continental United States, two regions fit the category of rifts within continental crusts, i.e., the Rio Grande Rift, extending from Colorado into Mexico and the Salton Sea depression of California. Both are characterized by thinned crust, high heat flow, and volcanism.

The Salton Trough of southern California and Mexico has been interpreted as being part of the Gulf of California, believed to have formed by spreading of continental crust (Elders et al., 1972). The trough is part of the dynamic system that includes the San Andreas Fault. The rift valley, bordered by Mesozoic and older rocks, is filled with deltaic sediments of late Tertiary to Recent age. Geophysical models indicate that the crust is about 8 km thinner than crust adjacent to the rift (Figure 3-20). There are Holocene rhyolite domes in the rift that were intruded into sediments 16,000 to 55,000 years ago. Associated with the domes and possibly with buried plutons are thermal anomalies with temperatures >350°C at depths of 2 km (Elders et al., 1972). Interaction of the intrusions with sea water and meteoric water has set up convective flow within the sedimentary basin. Due to high concentrations of material dissolved in the geothermal fluid, there are many engineering problems with the wells of this region. See Chaper 4 for more information on fields within the Salton Trough.

Basaltic volcanism associated with mantle "plumes," discussed earlier, is important in that it precedes, accompanies (around the margins), and may follow the eruption of large rhyolitic magma bodies. The basalts themselves, erupted along thin dikes, are not an important source of heat.

Isolated basaltic vents, associated with major faults on the continents, are also anomalies that represent short-term events with little value as a heat source. The faults themselves may be more important in this sense.

Geothermal Resources in Tectonically Active Regions

Areas of young tectonic activity, either with or without volcanic activity, are often characterized by high heat flow values (Wunderlich, 1970; Lachenbruch and Sass, 1977). In regions of high heat flow, such as the Basin and Range Province of the western United States, flow of meteoric water down fault planes may result in circulatory convection systems driven by density differences. Such a system, maintained by regional heat flow, will fall to regional values after 10^3 to 10^5 years (Lachenbruch and Sass, 1977). Hot springs found along fault zones in areas of young tectonic activity may not be hot enough for generation of electricity, but may be used for space heating and creation of resorts (spas).

Cermak et al., (1976) described a zone of high heat flow associated with a belt of vertical crustal uplift and increased seismic activity crossing central and eastern Europe. Within this zone, hot water from the Tertiary Pannovian Basin of Hungary is used for heating and bathing (Boldizsar and Korim, 1976). It is an intermontane basin, formed during Alpine Mountain building, and is filled with 5,000 m of clastic sediments. The main reservoir is in the upper Pliocene sandstones, where the water temperature is 100°-150°C. Natural thermal springs occur along faults bounding the basin. Boldizsar and Korim (1976) estimated that there is 2,800 km^3 of hot water in the basin, of which 10% is economically recoverable.

Numerous hot springs, representing a "low-grade" resource, also occur in mountainous regions of Europe. Hot springs in Switzerland, with temperatures of 25°-68°C, are used for bathing (Rybach and Jaffe, 1976). Thermal waters in the western Carpathian Mountains have temperatures of up to 70° C and are also used for bathing (Franko and Racicky, 1976). Many of the hot springs listed in the remarkable document by Waring (1965) fall into the same tectonic setting as the examples listed above. Waring's report is recommended as a standard reference for anyone interested in geothermal resources.

Geopressurized Geothermal Resources

The only geopressurized geothermal resource evaluated here is located in the northern part of Gulf of Mexico basin, extending from south Texas to Louisiana (Papadopulos et al., 1975 and Jones, 1976). Fluids within the reservoirs have temperatures of 140° to 170°C and have pressures of 80 to 110 mN/m². In addition to the mechanical and thermal energy of these fluids, methane that comes out of solution may be recovered. Jones (1976) envisioned the formation of geopressurized zones as follows:

1. Rapid deposition of Tertiary to Pleistocene deltaic sediments within the Gulf of Mexico overloaded underlying fine-grained sediments.
2. Faulting sealed off many permeable units, which were thus "compartmentalized."
3. With no water loss from these zones, geothermal gradients were increased.
4. Movement of salt (with high thermal conductivity) into Mesozoic and Cenozoic sediments, like a "nest of heating rods"; the deposits were warmed by mass transfer and conduction from the substrate.

If the engineering problems are overcome, the geopressurized zones of the Gulf Coast may represent a major geothermal resource in the United States (White and Williams, 1975).

Low-temperature, Radiogenically-derived Geothermal Resources

Ancient silicic plutons and batholiths may contain enough uranium and thorium to raise the heat production within those bodies to levels adequate for geothermal utilization. Based upon this premise, a group from the Virginia Polytechnic and State University has been studying granitic plutons of late Paleozoic age in the southeastern United States. These intrusive bodies contain enough radiogenic heat-producing elements to produce usable heat at depths of 1-3 km (Costain et al., 1977). Their detailed geophysical and geochemical studies of the Liberty Hill and Winnsboro plutons, exposed in the Carolina Slate Belt, have provided information on the pluton's composition, shape, and volume. For example, the Liberty Hill pluton is characterized as a cone, tapering to a point at a depth of 5 to 6.5 km.

If silica-rich bodies, such as the Liberty Hill pluton, are buried beneath an insulating blanket of sediments having low thermal conductivities, relatively high geothermal gradients are established. High thermal gradients, associated with negative gravity anomalies in the Atlantic Coastal Plain near Georgetown, South Carolina, and Norfolk, Virginia, may be related to buried plutons of late Paleozoic age. Buried plutons in the coastal plain are prime targets for geothermal development in the eastern United States.

References

Allen, E. T. and Day, A. L., 1935. *Hot Springs of the Yellowstone Park*. Carnegie Inst. Wash. Pub. No. 466:525 pp.
Anderson, E. M., 1936. The Dynamics of the Formation of Cone Sheets, Ring-Dykes, and Cauldron Subsidences. *Royal Soc. Edinburgh Proc.*, 56: 123-157.
Bailey, R. A., Dalrymple, G. B. and Lanphere, M. A., 1976. Volcanism, Structure, and Geochronology of Long Valley Caldera, Mono County, California. *J. Geophys. Res.*, 81: 725-744.

Baldi, P., Cameli, G. M., E., Mouton, J. and Scandellari, F., 1976. Geology and Geophysics of the Cesano Geothermal Field. *Proc., Second United Nations Symposium on the Development and Use of Geothermal Resources,* San Francisco, California, 20-29 May, 1975, 1976. 2: 871-881. Three volumes, 2466 pp. (Available from the U.S. Gov't. Printing Office, Washington, D.C. 20402).

Bodvarsson, G. and Walker, G. P. L., 1964. Crustal Drift in Iceland. *Geophys. J. Roy. Astron. Soc.* 8: 285-300.

Bodvarsson, G., 1976. Estimates of the Geothermal Resources of Iceland. *Proc., Second United Nations Symposium on the Development and Use of Geothermal Resources,* San Francisco, California, 20-29 May, 1975, 1976. 1: 33-35. Three volumes, 2466 pp. (Available from the U.S. Gov't. Printing Office, Washington, D.C. 20402).

Boldizsar, T. and Korim, K., 1976, Hydrogeology of the Pannonian Geothermal Basin. *Proc., Second United Nations Symposium on the Development and Use of Geothermal Resources,* San Francisco, California, 20-29 May, 1975, 1976. 1: 297-303. Three volumes, 2466 pp. (Available from the U.S. Gov't. Printing Office, Washington, D.C. 20402).

Branch, C. D., 1966. Volcanic Cauldrons, Ring Complexes, and Associated Granites of the Georgetown Inlier, Queensland. *Bur. Min. Res., Aust. Geol. and Geophys., Bull.* 76: 159 pp.

Brockman, C. E. and Kussmaul, S., 1973. *Volcanism Subprogram: Volcanological Interpretation of the Northern Part of the Occidental Cordillera of Bolivia, Utilizing ERTS Imagery.* NASA-CR-136481: 17 pp.

Cameli, G. M., Rendina, M., Puxeddu, M., Rossi, A., Squarci, P. and Taffi, L., 1976. Geothermal research in western Campania (southern Italy): Geological and geophysical results. *Proc., Second United Nations Symposium on the Development and Use of Geothermal Resources,* San Francisco, California, 20-29 May, 1975, 1976. 1: 315-328. Three volumes, 2466 pp. (Available from the U.S. Gov't Printing Office, Washington, D.C. 20402).

Cathles, L.M., 1977. An Analysis of the Cooling of Intrusives by Ground-water Convection Which Includes Boiling. *Econ. Geol.,* 72: 304-826.

Cermak, V., Lubimova, E. A. and Stegena, Lajos, 1976. Geothermal Mapping in Central and Eastern Europe. *Proc., Second United Nations Symposium on the Development and Use of Geothermal Resources,* San Francisco, California, 20-29 May, 1975, 1976. 1: 47-57. Three volumes, 2466 pp. (Available from the U.S. Gov't. Printing Office, Washington, D.C. 20402).

Christiansen, R. L. and Lipman, P. W., 1972. Cenozoic Volcanism and Plate-Tectonic Evolution of the Western United States: II. Late Cenozoic. *Phil. Trans. R. Soc. Lond.,* A, 271: 249-284.

Cobbing, E. J. and Pitcher, S. P., 1972. The Coastal Batholith of Central Peru. *J. Geol. Soc. London,* 123: 421-460.

Colp, J. L., 1974. Magma-Tap—The Ultimate Geothermal Energy Program. In: *Circum-Pacific Energy and Min. Res. Conf.,* Honolulu, Hawaii, Aug. 28, 1974, 19 pp.

Costain, J. K., Glover, L. K. III and Sinha, A. K., 1977. *Evaluation and Targeting of Geothermal Energy Resources in the Southeastern United States.* VPI-SU-5103, Blacksburg, Virginia.

Cox, A. (ed.), 1973. *Plate Tectonics and Geomagnetic Reversals.* Freeman and Co., San Francisco, 702 pp.

Crowe, B. M., Halleck, P. M. and Nolf, B., 1978. Evolution of Composite Volcanoes: The Role of Dike and Sill Emplacement (abs.). *Geol. Soc. Am. Abs. with Programs,* 10: 101.

Crowe, B. M. and Nolf, B., 1977. Composite Cone Growth Modeled After Broken Top, A Dissected High Cascade Volcano (abs.). *Geol. Soc. Am. Abs. with Programs,* 9: 940-941.

Delaney, Paul T. and Pollard, David D., 1976. Mechanism for Development of Plug-like Intrusions from Dikes (abs.). *Geol. Soc. Am. Abs. with Programs,* 8: 833.

Duffield, W. A., 1975. Late Cenozoic Ring Faulting and Volcanism in the Coso Range of California. *Geology,* 3: 335-338. .

Eaton, G. P., Christiansen, R. L., Iyer, H. M., Pitt, A. M., Mabey, D. R., Blank, H. R., Jr., Zietz, I. and Gettings, M. E., 1975. Magma Beneath Yellowstone Park. *Science,* 188: 787-796.

Eichelberger, J. C., 1978. Andesitic Volcanism and Crustal Evolution. *Nature.*

Eichelberger, J. C. and Gooley, R., 1977. Evolution of Silicic Magma Chambers and Their Relationship to Basaltic Volcanism. In: J. Heacock (ed.), *The Earth's Crust, Am. Geophys. Union Monograph* 20: 57-77.

Elders, W. A., Rex, R. W., Meidav, T., Robinson, P. T. and Bichler, S., 1972. Crustal Spreading in Southern California. *Science,* 178: 15-24.

Fedotov, S. A., Balesta, S. T., Droznin, V. A., Masurenkov, Y. P. and Sugrobov, V. M., 1976. On a possibility of Heat Utilization of the Avachinsky Volcanic Chamber. *Proc., Second United Nations Symposium on the Development and Use of Geothermal Resources,* San Francisco, California, 20-29 May, 1975, 1976. 1: 363-376. Three volumes, 2466 pp. (Available from the U.S. Gov't. Printing Office, Washington, D.C. 20402).

Francis, P. W. and Rundle, C. C., 1976. Rates of Production of the Main Magma Types in the Central Andes. *Geol. Soc. Am. Bull.,* 87: 474-480.

Franko, O. and Racicky, M., 1976. Present State of Development of Geothermal Resources in Czechoslovakia. *Proc., Second United Nations Symposium on the Development and Use of Geothermal Resources,* San Francisco, California, 20-29 May, 1975, 1976. 1: 131-137. Three volumes, 2466 pp. (Available from the U.S. Gov't. Printing Office, Washington, D.C. 20402).

Fridleifsson, I. B., 1976. Lithology and Structure of Geothermal Reservoir Rocks in Iceland. *Proc., Second United Nations Symposium on the Development and Use of Geothermal Resources,* San Francisco, California, 20-29 May, 1975, 1976. 1: 371-376. Three volumes, 2466 pp. (Available from the U.S. Gov't. Printing Office, Washington, D.C. 20402).

Friedman, J. D. and Heiken, G., 1977. Volcanoes and Volcanic Landforms. In: *Skylab Explores the Earth,* NASA SP-380, pp. 137-171.

Fyfe, W. S. and McBirney, A. R., 1975. Subduction and the Structure of Andesitic Volcanic Belts. *Am. J. Sci.,* 275-A: 285-297.

Gibson, I. L. and Piper, J. D. A., 1972. Structure of the Icelandic Basalt Plateau and the Process of Drift. *Phil. Trans. R. Soc. London, A.* 271: 141-150.

Gilluly, J., 1971. Plate Tectonics and Magmatic Evolution. *Bull. Geol. Soc. Am.,* 82: 2382-2396.

Goff, F. E. and Donnelly, J. M., 1977. Applications of Thermal Water Chemistry in the Geysers-Clear Lake Geothermal Area, California (abs.). *Geol. Soc. Am. Abstracts with Programs,* 9: 992.

Hamilton, W. and Myers, W. B., 1967. The Nature of Batholiths. *U.S. Geol. Surv. Prof. Paper 554-C:* 30 pp.

Hatherton, T. and Dickinson, W. R., 1969, The Relationship Between Andesitic Volcanism and Seismicity in Indonesia, the Lesser Antilles and Other Island Arcs. *J. Geophys. Res.,* 74: 5301-5310.

Healy, James, 1976. Geothermal Fields in Zones of Recent Volcanism. *Proc. Second United Nations Symposium on the Development and Use of Geothermal Resources,* San Francisco, California, 20-29 May, 1975, 1976. 1: 413-422. Three volumes, 2466 pp. (Available from the U.S. Gov't. Printing Office, Washington, D.C. 20402).

Hearn, B. C., Donnelly, J. M. and Goff, F. E., 1976. Geology and Geochronology of the Clear Lake Volcanics, California. *Proc., Second United Nations Symposium on the Development and Use of Geothermal Resources,* San Francisco, California, 20-29 May, 1975, 1976, 1: 423-428. Three volumes, 2466 pp. (Available from the U.S. Gov't. Printing Office, Washington, D. C., 20402).

Hearn, B. C., Jr., Donnelly, J. M. and Goff, F. E., 1978. *Continental-Edge Volcanism at Clear Lake, California: Hot Spot, Leaky Transform, or Heated Oceanic Slab.* Unpublished abstract.

Heiken, G., 1976. Depressions Surrounding Volcanic Fields: A Reflection of Underlying Batholiths. *Geology,* 4: 568-572.

Heiken, G., 1978. Plinian-type Eruptions in the Medicine Lake Highland, California, and the Nature of the Underlying Magma. *J. Volc. Geotherm. Res.*

Hess, H. H., 1939. Iceland Arcs, Gravity Anomalies and Serpentinite Inclusions: A Contribution to the Ophiolite Problem. *Int. Geol. Congr.,* 17th, Rep. 2: 263-283.

Higgins, M. W., 1973. Petrology of Newberry Volcano, Central Oregon. *Geol. Soc. Am. Bull.,* 84: 433-488.

James, D. E., 1971. Plate Tectonic Model for the Evolution of the Central Andes. *Bull. Geol. Soc. Am.,* 82: 3325-3346.

Jones, P. H., 1976. Geothermal and Hydrodynamic Regimes in the Northern Gulf of Mexico basin, *Proc., Second United Nations Symposium on the Development and Use of Geothermal Resources,* San Francisco, California, 20-29 May, 1975, 1976. 1: 429-440. Three volumes, 2466 pp. (Available from the U.S. Gov't Printing Office, Washington, D.C. 20402).

Kienle, J., 1976. Personal communication.

King, B. C. and Chapman, G. R., 1972. Volcanism of the Kenya Rift Valley. *Phil. Trans. R. Soc. London A.,* 271: 185-208.

Kolstad, C. D. and McGetchin, T. R., 1978. Thermal Evolution Models for the Valles Caldera with Reference to a Hot-Dry-Rock Geothermal Experiment. *J. Volc. Geothermal Res.,* 3: 197-218.

Kubota, S. and Berg, E., 1967. Evidence for Magma in the Katmai Range. *Bull. Volc.,* 31: 173-214.

Kuno, H., 1959, Origin of Cenozoic Petrographic Provinces of Japan and Surrounding Areas. *Bull. Volc.,* 20: 37-76.

Kurtman, F. and Samilgil, E., 1976. Geothermal Energy Possibilities, Their Exploration and Evaluation in Turkey. *Proc., Second United Nations Symposium on the Development and Use of Geothermal Resources,* San Francisco, California, 20-29 May, 1975, 1976, 1: 447-457. Three volumes, 2466 pp. (Available from the U.S. Gov't. Printing Office, Washington, D.C. 20402).

Lachenbruch, A. H. and Sass, J. H., 1977. Heat Flow in the United States and the Thermal Regime of the Crust, (J. Heacock, ed.) In: *The Earth's Crust: Its Nature and Physical Properties Am. Geophys. Union Monograph* 20: 626-675.

Lachenbruch, A. H., Sass, J. H., Munroe, R. J. and Moses, T. H., Jr., 1976. Geothermal Setting and Simple Heat Conduction Models for the Long Valley Caldera. *J. Geophys. Res.,* 81: 769-784.

Lahsen, D. and Trujillo, P., 1976. The Geothermal Field of El Tatio, Chile. *Proc., Second United Nations Symposium on the Development and Use of Geothermal Resources,* San Francisco, California, 20-29 May, 1975, 1: 170-177. Three volumes, 2466 pp. (Available from the U.S. Gov't. Printing Office, Washington, D.C. 20402).

Lipman, P. W., 1978. Cenozoic Volcanism in Western North America; Recent Concepts and Future Problems. *Trans. Am. Geophys. Un.,* 59: 259.

Lipman, P. W., Doe, B. R., Hedge, C. E. and Steven, T. A., 1978. Petrologic Evolution of the San Juan Volcanic Field, Southwestern Colorado: Pb and Sr Isotope Evidence. *Geol. Soc. Am. Bull.,* 89: 59-82.

Lipman, P. W., 1978, Cenozoic Volcanism in Western North America; Recent Concepts and Future Problems. *Trans. Am. Geophys. Un.* 59: 259.

MacDonald, Gordon A. and Abbott, Agatin T., 1970. *Volcanoes in the Sea,* U. of Hawaii Press, Honolulu, 441 pp.

MacLeod, N. S., Walker, G. W. and McKee, E. H., 1976. Geothermal Significance of Eastward Increase in Age of Upper Cenozoic Rhyolitic Dome in Southeastern Oregon. *Proc., Second United Nations Symposium on the Development and Use of Geothermal Resources,* San Francisco, California, 20-29 May, 1975, 1: 465-474. Three volumes 2466 pp. (Available from the U.S. Gov't. Printing Office, Washington, D.C. 20402).

McBirney, A. R., 1971, Oceanic Volcanism: A Review. *Rev. Geophys. Space Phys.,* 9: 523-556.

McGetchin, T. R., Settle, M. and Chouet, B. A., 1974. Cinder Cone Growth Modeled After Northeast Crater, Mt. Etna, Sicily. *J. Geophys. Res.* 79: 3257-3272.

McLaughlin, R. J. and Stanley, W. D., 1976. Pre-Tertiary Geology and Structural Control of Geothermal Resources, The Geysers Steam Field, California. *Proc., Second United Nations Symposium on the Development and Use of Geothermal Resources,* San Francisco, California, 20-29 May, 1975, 1: 473-485. Three volumes, 2466 pp. (Available from the U.S. Gov't. Printing Office, Washington, D.C. 20402).

Meyers, John S., 1975. Cauldron Subsidence and Fluidization: Mechanisms of Intrusion of the Coastal Batholith of Peru into Its Own Volcanic Ejecta. *Geol. Soc. Am. Bull.,* 86: 1209-1220.

Moores, E. M. and Vine, F. J., 1971, The Troodas Massif, Cyprus and Other Ophiolites as Oceanic Crust, Evaluation and Implications. *Proc. Roy. Soc. London,* Ser. A., 268, 443-466.

Morgan, W. J., 1971. Convection Plumes in the Lower Mantle. *Nature,* 230: 42.

Muffler, L. J. P., 1976. Summary of Section 1—Present Status of Resources Development. *Proc., Second United Nations Symposium on the Development and Use of Geothermal Resources,* San Francisco, California, 20-29 May, 1975, 1: xxxii-xliv. Three volumes, 2466 pp. (Available from the U.S. Gov't. Printing Office, Washington, D.C. 20402).

Muffler, L. J. P., 1976. Summary of Section II—Geology Hydrology and Geothermal Systems. *Proc., United Nations Symposium on the Development and Use of Geothermal Resources,* San Francisco, California, 20-29 May, 1975, 1976. Three volumes, 2466 pp. (Available from the U.S. Gov't. Printing Office, Washington, D.C. 20402).

Muffler, L. J. P., 1976. Tectonic and Hydrologic Control of the Nature and Distribution of Geothermal Resources. *Proc., Second Nations Symposium on the Development and Use of Geothermal Resources,* San Francisco, California, 20-29 May, 1975, 1: 499-507. Three volumes, 2466 pp. (Available from the U.S. Gov't. Printing Office, Washington, D.C. 20402).

Norton, D. L. and Knight, J. E., 1977. Transport Phenomena in Hydrothermal Systems: Cooling Plutons. *Am. J. Sci.,* 277: 937-981.

Papadopulos, S. S., Wallace, R. H., Jr., Wesselman, J. B. and Taylor, R. E., 1973. Assessment of Onshore Geopressurized-Geothermal Resources in the Northern Gulf of Mexico Basin. *U.S. Geol. Surv. Circ.* 726: 125-146.

Piper and Gibson 1972. Royal Society of London.

Pitcher, W. S. and Berger, A. R., 1972, *The Geology of Donegal: A Study of Granite Emplacement and Unrolfing.* Wiley-Interscience, New York, 435 pp.

Plouff, D. and Pakiser, L. C., 1972. Gravity Study of the San Juan Mountains, Colorado. *U.S. Geol. Surv. Prof. Paper 800-B:* B183-B190.

Rast, N., 1970. The Initiation, Ascent and Emplacement of Magmas. G. Newall and N. Rast (eds.), In: *Mechanism of Igneous Intrusion.* Gallery Press, Liverpool, pp. 339-362.

Richter, D. H., Eaton, J. P., Murata, K. T., Ault, W. W. and Krivoy, H. L., 1970. Chronological Narrative of the 1959-60 Eruption of Kilauea Volcano, Hawaii. *U.S. Geol. Surv. Prof. Paper* 537-E: 73 pp.

Robertson, E. C., Fournier, R. O. and Strong, C. P., 1976. Hydrothermal Activity in Southwestern Montana. *Proc., Second United Nations Symposium on the Development and Use of Geothermal Resources,* San Francisco, California, 20-29 May, 1975, 1: 553-561. Three volumes, 2466 pp. (Available from the U.S. Gov't. Printing Office, Washington, D.C. 20402).

Rose, W. I., Jr., Anderson, A. T., Jr. and Bonis, S., 1978. The October 1974 Basaltic Tephra from Fuego Volcano, Guatemala: Description and History of the Magma Body, unpubl. ms., 103 pp.

Rybach, L. and Jaffe, F. C., 1976. Geothermal Potential in Switzerland. *Proc., Second United Nations Symposium on the Development and Use of Geothermal Resources,* San Francisco, California, 20-29 May, 1975. 1: 241-244. Three volumes, 2466 pp. (Available from the U.S. Gov't. Printing Office, Washington, D.C. 20402).

Smith, R. L. and Bailey, R. A., 1968. Resurgent Cauldrons. In: Studies in Volcanology (Editors) Coats, R. R., Hay, R. L. and Anderson, C.A., *Geol. Soc. Am. Mem.* 116: 613-662.

Smith, R. L. and Shaw, H. R., 1973. Volcanic Rocks as Geologic Guides to Geothermal Exploration and Evaluation (abs.). *Am. Geophys. Union Trans.,* 54: 1213.

Smith, R. L. and Shaw, H. R. 1975. Igneous-related Geothermal Systems. In: *Assessment of Geothermal Resources of the United States.* D. F. White and D. L. Williams (Editors). *U.S. Geol. Surv. Circ.* 726: 58-83.

Steven, T. A., 1975. Middle Tertiary Volcanic Field in the Southern Rocky Mountains. In B. F. Curtis (ed.) *Cenozoic History of the Southern Rocky Mountains, Geol. Soc. Am. Mem.* 144: 75-94.

Steven, T. A., Lipman, P. W., Hail, W. J., Jr., Barker, F. and Luedke, R. G., 1974. Geologic Map of the Durango Quadrangle, Southwestern Colorado. *U.S. Geol. Surv. Map* I-764.

Stieltjes, L., 1976. Research for a Geothermal Field in a Zone of Oceanic Spreading: Example of the Asal Rift (French Territory of the Afars and the Issas, Afar Depression, East Africa). *Proc., Second United Nations Symposium on the Development and Use of Geothermal Resources,* San Francisco, California, 20-29 May, 1975, 1: 613-623. Three volumes, 2466 pp. (Available from the U.S. Gov't. Printing Office, Washington, D.C. 20402).

Vlasov, G. M. (ed.), 1964. Geology of the USSR, V. 31, *Kamchatka, Kuril and Komandorskie Islands,* Part 1, *Geological Description.* Moscow, Nedra, 813 pp.

Wadge, G., 1977. The Storage and Release of Magma on Mt. Etna. *J. Volc. Geotherm. Res.,* 2: 361-384.

Walker, G. P. L., 1959. Geology of the Reydardfjordur Area, Eastern Iceland. *Quart. J. Geol. Soc., Lond.,* 114: 367-393.

Ward, P. L., 1971. New Interpretation of the Geology of Iceland. *Bull. Geol. Soc. Am.,* 82: 2991-3012.

Waring, G. A., 1963, Thermal Springs of the United States and Other countries of the World—A Summary. *U.S. Geol. Surv. Prof. Paper* 492: 383 pp.

White, D. E., 1955. Thermal Springs and Epithermal Ore Deposits. *Econ. Geol.,* 50th Anniv. Vol., pp. 99-154.

White, D. E., 1967. Some Principles of Geyser Activity, Mainly from Steamboat Springs, Nevada. *Am. J. Sci.,* 265: 641-684.

White, D. E. and Williams, D. L., 1975. Assessment of Geothermal Resources of the United States—1975. *U.S. Geol. Surv. Circ.* 726: 155 pp.

Wunderlich, H. G., 1970. Geothermal Resources and Present Arogenic Activity. *Geothermics,* Spec. Issue 2, 2, (Part 2): 1226-1230.

Yamasaki, T. and Hayashi, M., 1976. Geologic Background of Otake and Other Geothermal Areas in North-Central Kyushu, Southwestern Japan. *Proc., Second United Nations Symposium on the Development and Use of Geothermal Resources,* San Francisco, California, 20-29 May, 1975. Three volumes, 2466 pp. (Available from the U.S. Gov't. Printing Office, Washington, D.C. 20402).

Appendix E

Classification of Common Igneous Rocks

The creation of categories, complete with names, is artificial; in reality, igneous rocks grade into one another chemically, texturally and mineralogically.

There is a great variety of igneous rocks, but only the most common are included in this simplified classification. Rock names used for both rapidly cooled, finely crystalline or glassy volcanic rocks and slowly cooled, coarsely crystalline plutonic rocks are included in the table.

For a complete view of igneous petrology, the reader is referred to books by Carmichael, Turner, and Verhoogen, 1974; and MacDonald (1972).

References for Appendix E

Carmichael, I. S. E., Turner, F. J., and Verhoogen, J., 1974, *Igneous Petrology*. McGraw-Hill, New York.
MacDonald, G. A., 1972, *Volcanoes*. Prentice-Hall, Inc., Englewood Cliffs.
Nuckolds, S. R., 1954, Average chemical compositions of some igneous rocks. Bull. Geol. Soc. Amer., v. *65,* p. 1007-1032.
Williams, H., 1932, Geology of the Lassen Volcanic National Park. Calif. Univ. Pubs. in Geol. Sci., v. *21,* p. 195-385.

Table E-1
Simple Classification of Common Igneous Rocks and Some Physical and Chemical Properties

	Silicic	Intermediate		Mafic	
Main mineral phases	K-feldspar, quartz oligoclase	Plagioclase, biotite hornblende and pyroxenes, quartz	Plagioclase, augite, orthopyroxene, hornblende	Plagioclase, augite olivine	Feldspathoid, K-feldspar, plagioclase, hornblende, biotite
Fine-grained or glassy (volcanic)	Rhyolite	Dacite	Andesite	Basalt	Phonolite-trachyte
Coarse-grained (plutonic)	Granite	Granodiorite	Diorite	Gabbro	Syenite
SiO_2	73.66	63.58	57.7	50.83	56.90
TiO_2	0.22	0.64	0.6	2.03	0.59
Al_2O_3	13.45	16.67	17.0	14.07	20.17
Fe_2O_3	1.25	2.24	2.4	2.88	2.26
FeO	0.75	3.00	3.6	9.05	1.85
MnO	0.03	0.11	0.15	0.18	0.19
MgO	0.32	2.12	5.1	6.34	0.58
CaO	1.13	5.53	7.5	10.42	1.88
Na_2O	2.99	3.98	3.4	2.23	8.72
K_2O	5.35	1.40	1.6	0.82	5.42
Estimated extrusion**	800°–900°C	900°–1000°C	900°–1000°C	1000°–1200°	≈1000°C
Viscosities† (poises) at extrusion temperatures	≈10^{10}–10^{12}	≈10^{10}	10^5–10^7	10^2–10^4	≈10^5

*From Nuckolds, 1954 and Williams, 1932.   Average composition (oxides, wt. %)

**From Carmichael et al. 1974.

†From MacDonald, 1972.

A. William Laughlin, Los Alamos National Laboratory

4

Exploration for Geothermal Energy

Introduction

Because geothermal energy is a relatively new type of resource, appropriate exploration rationales or programs are still under development and are the source of much debate and controversy. Few geothermal systems are producing commercially, and fewer still have been adequately investigated or assessed. Unfortunately, too, when investigations have been thorough, much of the information is often considered proprietary and is unavailable to the public. As a result, it is difficult to evaluate the exploration and assessment techniques used within a field or even, in some cases, to know what techniques were used. Also, in general, redundant techniques have not been used, prohibiting comparison of exploration methods. Many more case histories must be completed to permit comparison of predictions and measured results from drilling.

The ultimate objective of any exploration program is to locate a resource that can be economically developed. Despite differences in the type of resource sought and, hence, in the geologic setting, a certain exploration philosophy has been built up over many decades (Peters, 1978). This philosophy is based on the concept that the prospector begins his search in a large area, narrowing down the area under consideration as more and more data become available until the resource is located. In the early stages of exploration, when areas to be investigated are large, rapid low-cost reconnaissance techniques are employed. As results accumulate and the search narrows, confidence increases and more expensive techniques can be utilized. In the case of geothermal exploration, this would continue until the most expensive technique, the "wildcat well," is used to test the prospect.

The objective of geothermal exploration is obviously to locate a geothermal system from which energy can be economically extracted. Because high geothermal gradients are a prerequisite for any type of geothermal system, the initial exploration effort should concentrate on defining such anomalous areas using a variety of techniques, which are discussed in this chapter. When an area of high geothermal gradient or heat flow is identified, emphasis shifts to an evaluation of the permeability and hydrology of the area. Again a variety of techniques can be used to obtain this information.

Even a casual scan of geothermal literature shows that a great number of techniques have been used in an attempt to locate geothermal systems. The use of these techniques is best summarized in the Second United Nations Symposium on the Development and Use of Geothermal Resources (1975). The techniques range in sophistication from "seepology" to modern geophysical methods such as magneto-tellurics or TDEM. Beginning with seeps, the exploration methods will be discussed individually below.

Seeps and Fossil Seeps

Leakage of fluids through the impermeable capping often occurs in natural geothermal systems. These leaks or seeps, which may produce such features as fumaroles, hot springs, warm springs, geysers, mud volcanoes, or boiling ground, are the most direct and obvious indicators of the presence of a geothermal system. In some cases, fossil seeps such as deposits of travertive or siliceous sinter may indicate that a geothermal system was present at some time in the past. These seeps commonly result from seismic activity that fractures the impermeable caprock and permits fluid escape to the surface or into near-surface groundwater.

In addition to acting as qualitative indicators of the location of geothermal systems, seeps can also be used to provide quantitative information on the nature of the reservoir and the fluids contained within it. Independent temperature estimates of the reservoir can be obtained from the concentrations of Si and the combinations Na-K and Na-K-Ca in the effluent of hot and warm springs and wells (Truesdell, 1976; Fournier and Truesdell, 1973; White, 1965; and Ellis and Mahon, 1967). Problems associated with the application of these chemical geothermometers are well treated by Ellis and Mahon (1977). To summarize briefly from Ellis and Mahon (1977), equilibrium is rapidly attained between quartz and water, providing good estimates of changing reservoir temperatures, but making the method susceptible to problems of cooling and mixing. As a result, the method is most useful when applied to hot springs and wells. The Na-K and Na-K-Ca methods, in turn, are susceptible to problems related to differences in fluid or reservoir rock composition. Acid waters or reservoir rocks with high concentrations of K, Na or Ca preclude the use of these two geothermometers (Ellis and Mahon, 1977).

Active volcanism or the presence of very young extrusive igneous rocks are also *direct* indicators of high geothermal gradient, and thus are closely related to "seeps" in their applicability to geothermal exploration. Because most of the world's commercial geothermal systems are closely associated with young volcanic rocks, especially those of silicic composition (Smith and Shaw, 1975), exploration has been concentrated in the near vicinity of these rocks. Only recently has exploration moved into other geologic terrains.

Geological Techniques for Exploration

Literature Search.

All exploration programs, reconnaissance or detailed, begin with an examination and compilation of all available data. Beginning with the geothermal data banks, the search should include the professional journals, publications of federal and state agencies, university theses and dissertations, and, in the U.S., reports from the national laboratories. Literature search is a cheap exploration technique, and the dividends from a careful search can be great.

The results of the literature search should be compiled into a preliminary conceptual model of the area under consideration. Formulation of this model should obviously emphasize the potential of the area as a geothermal prospect and should permit a decision at this time as to whether additional new work is warranted.

Mapping

Careful geologic mapping is the foundation on which any exploration program is built and all other data must be interpreted in terms of the observed geology. In geothermal exploration, mapping should emphasize young igneous rocks that could act as heat sources, potential reservoir rocks, distribution and nature of hydrothermal alteration, and the distribution, orientation, and nature of fractures and faults. In this mapping, field work should be supplemented where possible by satellite imagery and aerial photography, including low sun-angle photography. The latter technique has been proven useful for fault location in the Jemez Mountains, New Mexico (Slemmons, 1975) near Los Alamos National Laboratory's hot dry rock site at Fenton Hill.

The detail or scale of mapping should reflect the stage of the exploration program. In the reconnaissance stage, a scale of 1/125,000 to 1/62,500 is most appropriate, and much of the data required may be obtained from a literature search, compilation of existing maps, and from satellite imagery or aerial photographs, with only minimal field work. As the area is reduced in size, the map scale increases accordingly. The smallest scale suitable for detailed mapping is probably about 1/24,000 and scales as large as 1/5,000 may be necessary in complex areas. The detailed mapping should be supplemented by investigations of the orientation and frequency of fractures as discussed later in this chaper.

Petrology

Petrologic investigations of igneous rocks (associated with the thermal anomaly), the potential reservoir rocks, and impermeable capping can aid in the definition of the size of the anomaly and help predict fluid characteristics, which eventually will affect development and utilization. The investigations should include determination of modal composition, textural examinations, and fracture and alteration studies.

These petrologic investigations when combined with whole rock chemical analyses, strontium isotopic data, and geochronological data can provide significant information about the igneous sources such as: (1) differentiation of the magma, (2) depths of emplacement, (3) duration of the igneous event, (4) reaction with crustal wall rocks, and (5) the magnitude of the thermal effects on the wall rocks. Evidence of partial melting of crustal rocks can also be recognized suggesting possible increased transfer of thermal energy to near-surface wall rocks.

Investigations into the nature of fracture-filling and alteration minerals provide a better understanding of the development of self-sealed impermeable cappings and the formation of impermeable hot dry rock systems. During the reservoir assessment stage, these studies may suggest methods by which the flow of low-permeability systems may be enhanced (Laughlin and Eddy, 1977; Eddy and Laughlin, 1978). Isotopic studies of fracture-filling minerals (Brookins and Laughlin, 1977) may complement petrologic investigations and provide new data on the sources of these minerals.

Fracture Studies

Knowledge of the frequency, size, and orientation of fractures and faults is important in the exploration for and assessment of any type of geothermal reservoir. In most natural fluid-dominated systems, secondary permeability, i.e., fracture-controlled permeability, is the most important type of permeability that controls the production from the wells. Geopressured systems are typically bounded by large fractures or faults which ultimately determine the size of the reservoir. In the case of hot dry rocks, it is the presence of fractures and the nature of fracture-filling minerals that determines if this energy extraction method can be applied. For these reasons it is imperative, as part of any geothermal exploration program, to characterize the nature of fractures present within the area.

In the exploration stage, when mapping is carried out at the 1/125,000 to 1/62,500 stage, considerable information on fracture frequency and orientation can be obtained from Landsat imagery and aerial photographs (Hahman et al., 1978; Barbier and Fanelli, 1976). The data obtained can be plotted either in the form of lineament or fracture maps or as rose diagrams illustrating the frequency of observed orientations. Either presentation provides a visual representation of the preferred fracture orientations.

As the search narrows and the mapping scale increases, fracture studies should also become more detailed. It becomes important to determine the size of the fracture and the nature of fracture-filling mineralization as well as the frequency and preferred orientation of the fractures. Satellite imagery and aerial photographs can again provide information on the orientation and frequency of fractures, but careful field mapping is necessary to evaluate the size of the fractures and to determine if they are open, sealed or healed, and the nature of the fracture-filling minerals. The scale of these investigations can extend down to examinations of microfractures observed only with the scanning electron microscope (Simmons and Eddy, 1976). Results of these investigations aid in the evaluation of the secondary permeability and, in the case of dry holes, may suggest techniques for stimulation.

Geochemical Techniques for Exploration

Chemical Geothermometers

Several chemical geothermometers exist that can be used to predict reservoir temperatures prior to drilling. The most widely applicable of these are the Si and cation (Na-K, Na-K-Ca) geothermometers, which have been used for both reconnaissance and detailed exploration. Recent work by Swanberg et al. (1977), illustrated on the Preliminary Map of Geothermal Energy Resources of Arizona (Hahman et al., 1978), is a good example of the use of chemical geothermometers in reconnaissance exploration.

This map, the first of a series of state geothermal maps funded by the U.S. Department of Energy, correlates regions of high reservoir temperature as indicated by chemical geothermometers with hot spring and well locations and areas of high geothermal gradient and heat flow. Lineaments observed on Landsat photographs, the location of igneous rocks younger than 3 million years, and cinder cone locations are also shown on this map. The structural control of the location of hot springs and the location of the regions delineated by the chemical geothermometers is well illustrated on this map. Several areas that appear particularly favorable as targets for detailed geothermal exploration can be recognized by the use of this variety of geological, geochemical, and geophysical evidence.

Dellechaie (1976) reported a detailed evaluation of both the Si and Na-K-Ca geothermometers in the Santa Cruz Basin, Arizona. Results of the application of the Na-K geothermometer were presented but not discussed. Waters from 28 irrigation wells within the area were analyzed and the various geothermometers applied to the results. The Si geothermometer yielded a maximum subsurface temperature of 97°C. If mixing of hot and cold ground waters was assumed, this estimated temperature could be raised to 138°C.

The Na-K-Ca geothermometer gave results ranging from 37°C to 76°C for subsurface conditions. A 2,500-m well drilled in the area, penetrated 1,750-m of valley fill before encountering Precambrian basement rocks. A 163-hr pump test of this well produced fluids with a maximum surface temperature of 82°C and a maximum bottomhole temperature of 102°C. This measured bottomhole temperature agrees well with the maximum temperature of 97°C obtained by using the Si geothermometer without assuming mixing. The low values obtained on using the Na-K-Ca geothermometer indicate either reequilibration below 100°C or perturbation from evaporite sequences in the valley fill.

Trace-Element Investigations

Trace-element concentrations in rocks, soils, water, and plants have been widely used in the exploration for metallic deposits. Because certain trace elements, including the rare gases, are genetically related to volcanic rocks, they should serve as indicators for geothermal resources having igneous heat sources. When compared to the geophysical and geological methods, however, there are relatively few examples of the application of trace-element geochemical methods to geothermal exploration.

Because of its high mobility both in the vapor phase and in aqueous solutions, mercury is one of the most promising indicators for geothermal exploration. White (1967) recognized that mercury is commonly enriched in rocks associated with thermal springs. Matlick and Buseck (1976) and Phelps and Buseck (1978) examined mercury concentrations in rocks, soils, and spring deposits from Known Geothermal Resource Areas (KRGA's) and from areas that appeared to have high geothermal potential. Matlick and Buseck (1976) described field equipment capable of mercury analysis with a precision of 1 ppb. In this study, four areas were examined: Long Valley, California; East Mesa, California; Klamath Falls, Oregon; and Sumner Lake, Oregon. The Sumner Lake area was thought to have geothermal potential, whereas the other three areas are KGRA's. Background mercury concentrations were measured or assumed (Klamath Falls). Soil samples from within the areas were analyzed and peak-to-background ratios were calculated. These ratios were then used for the construction of contour maps. At Long Valley, Sumner Lake, and Klamath Falls there were positive correlations between mercury concentrations and presence of hot springs or geophysical anomalies indicative of geothermal activity. Particularly at Long Valley, the mercury concentrations closely defined the geothermal anomalies. Only at East Mesa did the the the method fail to yield a positive correlation.

Phelps and Buseck (1978) expanded these studies to the Yellowstone National Park, Wyoming, and Coso Hot Springs area, California. At both of these sites positive correlations were observed between mercury concentrations and thermal activity.

The gases helium and radon have also been used for geothermal exploration and assessment (Gutsalo, 1976; Dellachaie, 1977), but the results are much more tenuous and the methods are not currently in wide use.

Geochronology

Because many geothermal systems are closely associated with young igneous rocks, geochronology is important in the exploration for and evaluation of these systems (Smith and Shaw, 1975). At the present time the major interest is in geochronology as an exploration tool.

In the age range of interest in geothermal exploration the K-Ar method is by far the most useful for dating igneous rocks. In rare cases, where the rocks are extremely young (<40,000 years) and organic material can be related to the igneous activity, the C^{14} method can be used.

Naeser and Forbes (1976), Turner and Forbes (1976), and Brookins et al. (1977) have presented a unique application of the K-Ar and fission track methods in evaluating the thermal event responsible for the Jemez Mountains, New Mexico, geothermal system. As part of the Hot Dry Rock Project, two 3-km deep holes were drilled by the Los Alamos National Laboratory into Precambrian basement rocks about 2 km outside of the Valles Caldera. The caldera, which contains a fluid-dominated hydrothermal system under development by Union Oil, Public Service Co. of New Mexico, and the U.S. Dept. of Energy was formed between 1.4 and 1.1 million years ago (Doell et al., 1968) by eruption of the Bandelier Tuff. This event resulted in elevated subsurface temperatures throughout the areas; a bottomhole temperature of about 200°C was measured in the 3-km deep Los Alamos holes.

Core samples from the two deep holes were dated by the Rb-Sr., K-Ar, and fission track methods (Brookins and Laughlin, 1976; Turner and Forbes, 1976; Naeser and Forbes, 1976; and Brookins et al., 1977).

The major Precambrian unit encountered in the two wells, a metamorphic complex of banded gneiss and schist, gave a whole rock Rb-Sr isochron age of 1.66 ± 0.05 billion years (Brookins and Laughlin, 1976; Brookins et al., 1977). Intrusive into this unit is a biotite granodiorite body, which has a Rb-

Sr isochron age of 1.29 ± 0.03 billion years, and dikes of granite with an age of 1.4 ±0.2 billion years. The 1.29-billion-year age of the granodiorite may be too young because of "opening" of the Rb-Sr system (Brookins and Laughlin, 1976; Brookins et al., 1977). Mineral separates of two minerals (biotite and amphibole) from the Precambrian rocks were dated by the K-Ar method (Turner and Forbes, 1976). None of these dates approached the 1.66 billion years by using Rb-Sr method. The oldest K-Ar age is 1.42 billion years for biotite from the gneiss, and the average is 1.37 ± 0.03 (1σ). It seems clear that the K-Ar "clock" of the metamorphic complex was reset by the 1.4-billion-year event that produced the granite dikes. The biotite K-Ar age of the deepest sample (~3 km and 200°C) is 1.26 ± 0.04 billion years. (Figure 4-1). This decrease in age, although within statistical error of the granodiorite age may be the result of a second period of argon loss caused by the event that produced the Bandelier Tuff and imposed the present geothermal gradient on the area. The observed bottomhole temperature of 200° C was believed sufficient to cause this loss (Turner and Forbes, 1976). If this temperature had persisted very long, additional argon would have occurred. This implies that the present 200°C temperature is near the maximum for this depth.

The effect of the 1.4-1.1-million-year event on fission track ages is even more striking (Naeser and Forbes, 1976). Apatite separated from successively deeper samples of the Precambrian basement rocks gave progres-

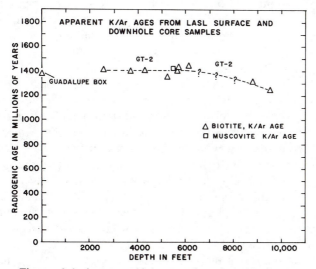

Figure 4-1. Apparent K-Ar ages from Los Alamos surface and downhole core samples, Fenton Hill, Jemez Mountains, New Mexico.

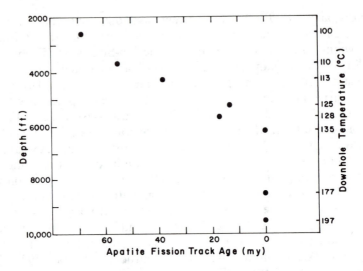

Figure 4-2. Variation of apatite fission track ages with depth in GP-2, Fenton Hill, Jemez Mountains, New Mexico.

sively younger fission track ages until zero age was reached at a depth of 1.9 km where the present temperature is 135°C (Figure 4-2). The relatively low track annealing temperature of apatite makes it extremely sensitive to thermal perturbations, suggesting that fission track dating of apatite could be developed into a geothermal exploration technique.

Hydrologic Investigations

A thorough knowledge of the hydrology is necessary for the evaluation of any type of geothermal resource. In fluid-dominated and geopressured systems, naturally-occurring aqueous fluids are the agent by which energy is transported from the reservoir to the earth's surface. This immediately implies an important role for hydrology in understanding the geothermal system. In hot dry rock systems, hydrologic methods are significant in defining the permeability of the system, predicting fluid loss, and in evaluating other exploration techniques such as heat flow measurements.

The objectives of a hydrologic investigation are to determine the source of the fluids, areas of recharge, infiltration rate, location, depth, pressure, composition, and temperature of aquifers. A number of techniques can be used to achieve these objectives. Oxygen and hydrogen isotopic and tritium analysis of the water and stream can be used to determine the source of fluids in a hydrothermal system. The local topography and

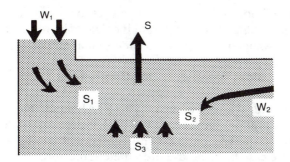

Figure 4-3. Schematic drawing of the Larderello Geothermal System with circulation patterns. W_1 = recent water from recharged areas on southern margins of the field. S_1 = steam produced from W_1. W_2 = water from the confined aquifer surrounding the field. S_2 = steam produced from W_2. S_3 = steam from deep horizons. S = total steam produced by the field. (after Petracco and Squarci, 1976.)

depths of piezometric and phreatic surfaces define the location and size of basins that serve as areas of infiltration. A quantitative estimate of the infiltration (I) can be obtained from a hydrologic balance using the relation $I = P\text{-}E\text{-}R$, where P, E, and R are precipitation, evapo-transpiration, and runoff, respectively.

An excellent review of the application of these techniques to a fluid-dominated geothermal system is provided by Petracco and Squarci (1976) for the Larderello area in Italy. They reference oxygen and hydrogen isotopic and tritium investigations, which indicate that the steam in the center of the field remains isotopically constant with time. On the margins of the field mixing with infiltrating meteoric water occurs and the steam is of variable isotopic composition. The steam from the margins is also enriched in tritium derived from the meteoric water. The tritium concentration decreases toward the center of the field indicating an age of greater than 25 years (Celati et al., 1973).

Petracco and Squarci (1976) have also performed hydrologic balance calculations based on the model shown in Figure 4-3. This model assumes that the total steam (S) from the field is produced by mixing steam derived from the infiltration of meteoric water (S_1), steam derived from confined aquifers (S_2), and steam from deep sources (S_3). Using this model and precipitation, runoff, and evapo-transpiration data, they calculated an average infiltration of 8 to 11 $\times$ 10^6 m³/yr for the Cecina River Basin in which the

Larderello field is located. This volume corresponds to about one third of the total steam produced annually from the field. The remaining two thirds must be the components S_2 and S_3 in the original model.

Geophysical Techniques for Geothermal Exploration

Although a wide variety of geophysical techniques has been used in geothermal exploration, there are few published accounts of the application of multiple techniques to an area with subsequent confirmatory drilling. Several of these case histories are presented in the Second United Nations Symposium on the Development and Use of Geothermal Resources (Baldi et al., 1976; Blackwell and Morgan, 1976; Hockstein, 1976; Swanberg, 1976; and Williams et al., 1976). Another outstanding example of a case history involving redundant geophysical methods is the series of papers on Long Valley, California, written by numerous workers in the U.S. Geological Survey and published in the *Journal of Geophysical Research,* Vol. 81, 1976.

Geophysical exploration techniques have been used successfully to locate the heat sources of geothermal systems and to characterize the permeability of the potential reservoir. The use of specific techniques are discussed below.

Gravity Measurements

Gravity mapping has proven useful in several aspects of geothermal exploration. Of primary importance is the utility of this method in locating and defining the extent of heat sources. Smith and Shaw (1975) and numerous other workers have emphasized the close association between young silicic igneous rocks and fluid-dominated geothermal systems. They stressed that formation of the relatively shallow magma chambers associated with these silicic rocks is more effective in transferring heat within the crust than is the emplacement of mafic sills and dikes. Because of the low density of the silicic plutons, they are often readily detected by gravity mapping. If magma is still present in the systems, density differences are likely to be even greater, increasing the possibility of the body being discernible on the basis of gravimetric evidence.

Negative gravity anomalies have been shown to be present beneath the Valles Caldera and the Clear Lake volcanic field associated with the Geysers geothermal area (Isherwood, 1975).

In several cases, positive gravity anomalies have been correlated with areas of high heat flow and geothermal reservoirs. Swanberg (1976) cited Biehler (1971) as correlating a positive gravity anomaly in the Imperial Valley with the Mesa geothermal anomaly. Swanberg attributed the gravity anomaly to metamorphism and cementation of the deltatic sediments when they are permeated by geothermal fluids ascending along faults. Where caprocks are formed by self-sealing in porous, permeable rocks, positive gravity anomalies should be commonly developed.

Gravity mapping is also useful in determining basement structures such as faults, which may control the location of fluid-dominated geothermal systems. Baba (1976) used gravimetric evidence to distinguish structural lines that apparently control the geothermal activity.

The Los Alamos National Laboratory is currently using residual Bouguer gravity maps by Aiken (1976) and Aiken et al. (1977) in reconnaissance exploration for hot dry rock geothermal systems. West and Laughlin (1979) have correlated a large negative anomaly in northwest Arizona with high geothermal gradients, shallow Curie point depths (Byerly and Stolt, 1977), and seismic wave attenuation.

Aeromagnetic Measurements

Aeromagnetic surveying, although not as widely used as some of the geoelectro-magnetic techniques, appears to offer considerable promise as a reconnaissance tool in geothermal exploration. The method is useful for regional structural analysis and, when combined with Curie point depth determinations, for defining large regions of the crust having elevated temperatures. This latter method is based on the fact that magnetic minerals in rocks lose their magnetism at some specific temperature referred to as the Curie temperature. For titano-magnetite, which according to Byerly and Stolt (1977) is the most common magnetic mineral in igneous rocks, the Curie temperature is less than 570°C. Conventional aeromagnetic data can be used to calculate the depth to the Curie temperature, thus providing at least a crude estimate of the average geothermal gradient.

A paper by Byerly and Stolt (1977) summarized work which, although not intended for geothermal exploration, does illustrate the utility of this method as a reconnaissance exploration tool. They analyzed existing aeromagnetic data from northern and central Arizona and calculated depths to the Curie point for this region. From these calculated depths a contour map was constructed that shows a relatively shallow zone (<10 km) of Curie point depths. Individual points within this shallow zone are as low as 2 km. If a depth of 10 km is assumed for the Curie point, the average geothermal gradient of this region must be greater than 57°C/km. This immediately indicates a large region that has high geothermal potential.

The Curie point depth determination method requires testing in a number of areas where the geothermal gradients are known from drilling and where these geophysical techniques have been employed.

Heat Flow

In theory, heat flow is the most unambiguous indicator of the presence of a geothermal system, because the property sought is being directly measured. Experience has shown, however, that there are two problems that may negate this advantage. First, the method is expensive to apply and, second, because of local hydrologic effects, it is often difficult to interpret the results.

Heat flow values are obtained by measuring the geothermal gradient (dT/dz) in a drill hole and then multiplying the result by the thermal conductivity (K). The thermal conductivity is measured on core or cuttings samples from the drillhole or more rarely *in situ*. The gradient measurements are made in holes as shallow as a few 10's of meters or as deep as several kilometers. If deep holes are used, the high cost of drilling mandates that the method must be assigned to the detailed phase of exploration unless previously existing drillholes are utilized. If deep holes are available and access can be gained to them, the method may become a reconnaissance exploration tool. Several workers including Reiter et al. (1975) and Decker and Smithson (1975) have used available drillholes to provide large amounts of data suitable for reconnaissance exploration. Reiter et al. (1975), for example, have published a heat flow map of New Mexico that indicates two areas of regionally high heat flow that may have geothermal potential (Figure 4-4). The most prominent of these is a zone with heat flow greater than 2.5 HFU roughly coinciding with the Rio Grande Rift. Within this zone, local hot spots occur in the Jemez Mountains, near Socorro and Las Cruces, New Mexico (Reiter et al., 1978). A second zone of high heat flow lies to the west, abutting the Arizona-New Mexico state line. This zone extends northeastward to near Grants, New Mexico, and encompasses much of the area of late-Cenozoic volcanism in western New Mexico. Both areas are clearly prime targets for geothermal exploration.

Turning to detailed exploration, numerous examples exist that demonstrate the utility of the heat flow method. One of the best of these is the investigation by Lachenbruch et al. (1976) of heat flow within the Long Valley Caldera.

One of the major problems with the use of the heat flow method involves distinguishing beween conductive and convective contributions to the heat flow. Local near-surface hydrologic disturbances can either mask high conductive heat flow at depth or falsely indicate high conductive heat flow. A good example of the latter effect is presented by Blackwell and Morgan (1976) for the Marysville area, Montana. They present geophysical and geological evidence that very high values of heat flow (>20 HFU) in this region result from 98°C near surface waters.

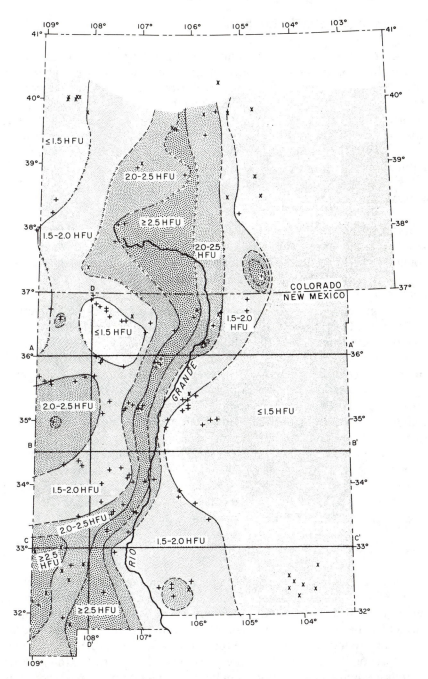

Figure 4-4. Terrestrial heat-flow contour map of New Mexico and southern Colorado. Contour interval = 0.5 HFU. (After Reiter et al., 1975; courtesy of the Geological Society of America.)

Geoelectromagnetic Techniques

At the present time, the major emphasis in geothermal exploration seems to be in the application of geoelectromagnetic methods such as magnetotellurics (MT), audiofrequency magnetotellurics (AMT), and deep electrical resistivity. A number of other geoelectromagnetic techniques have also been employed in geothermal exploration. These techniques can be broadly separated into those utilizing natural fields and those where a controlled source is used. This natural separation will be used in discussing the individual methods below.

Natural Field Methods

Self-potential (SP). This method, widely used in mineral exploration, is based on measurements of variations in the natural dc voltage which result from the interaction of ground water with conductive bodies, such as sulfide or graphite deposits; from high geothermal gradients; or from the movement of fluids. It is these latter two causes that permit the use of the self-potential method in geothermal exploration. Application of the SP method to geothermal exploration is well summarized by Corwin (1976).

Corwin attributes positive SP anomalies along survey lines in Grass Valley, Nevada, to ascending hot water and negative anomalies to descending water. Zohdy et al. (1973) detected a similar positive anomaly over ascending geothermal fluids in Yellowstone National Park, as did Combs and Wild (1976) over the Dunes anomaly in the Imperial Valley, California, and Williams et al. (1976) in the Raft River area, Idaho. Adequate explanations for these effects have not been presented.

Although the SP method has successfully located zones of ascending hot water, the method has several disadvantages: (1) the causes of the anomalies are poorly understood; (2) topographic effects may be significant (Williams et al., 1976); and (3) telluric currents may cause errors (Corwin, 1976). High noise levels are also a significant problem in the use of this method.

Telluric Methods

Two different telluric methods have been used in geothermal exploration: vector-tellurics and telluric profiling. Both methods rely on the measurement of natural electrical fields at frequencies of less than 20 Hz. Although both techniques utilize relatively simple instrumentation, the costs of data processing are high and neither method is in wide use at the present time.

Because of its sensitivity to lateral changes in resistivity, the telluric profiling method is probably most applicable to geothermal exploration. It is particularly useful when the strikes of geologic contacts are known and survey lines can be oriented perpendicular to the strike.

Magnetotelluric (MT) and Audiofrequency Magnetotelluric (AMT) Methods

These two methods cover the natural electrical and magnetic fields in the range from dc to about 20,000 Hz. The MT method, which utilizes frequencies up to about 100 Hz, is useful in defining deeper structures and conductors; whereas the AMT method, utilizing the higher frequencies, is limited to relatively shallow penetration (<2 km). Because of the low frequencies, recording time may be as long as several days at MT stations.

The great depth of penetration associated with the low frequency MT signals make the MT method useful for defining deep electrical conductors in the crust, which may ultimately control the location of geothermal fields. In the southwestern U.S., Hermance and his colleagues (Hermance and Pederson, 1976; Pederson and Hermance, 1978) and Jiracek et al. (1977) have used the MT method to locate conductive regions (partial melt zones) beneath the Rio Grande Rift. According to Jiracek et al. (1977), these conductive zones may be as shallow as 5 km in the Socorro area of New Mexico. Pederson and Hermance (1978) on the basis of MT work have suggested that a partial melt zone may underlie Santa Fe, New Mexico, at a depth of about 15 km. The Los Alamos National Laboratory is currently using commercial MT contractors to explore for deep electrical conductors in central New Mexico and Arizona, with the objective of identifying hot dry rock prospects associated with these conductive zones.

Hoover et al. (1976) have demonstrated the utility of the AMT method in delineating shallow conductive zones. Working in the Long Valley area of California, they recognized two zones of high conductivity that could be correlated with the location of thermal springs. They recorded signals in the frequency range 8-18,600 Hz and achieved depths of penetration of up to 2 km.

The MT and AMT methods offer considerable promise in both reconnaissance and detailed scale exploration. They both, however, are adversely affected by cultural features such as powerlines, pipe lines, and fences. Despite these disadvantages, the methods are being increasingly used in geothermal exploration.

Controlled Source Methods

Resistivity surveying. Several methods for evaluating vertical and lateral variations in electrical resistivity are in common use in geothermal exploration. These include the bipole-dipole, dipole-dipole, Schlumberger profiling, and Schlumberger sounding methods. All methods are based on injecting a dc current into the earth by means of a current electrode pair and measuring the resulting dc field with a second electrode pair. Electrode arrays and instrumentation for measuring the voltage between receiving electrodes vary with the method.

In the bipole-dipole method, a long current electrode pair (bipole), up to several kilometers in length, is used to inject the dc current while short receiving electrodes (dipoles) are moved about over the ground surface to produce either resistivity maps or profiles. The current injected may be as large as a few hundred amperes with potentials of several hundred volts. An excellent example of the application of this method is presented by Stanley et al. (1976) for the Long Valley area of California. In this area, four different bipole positions were used to produce a composite total field resistivity map using bipole positions 2, 3, and 4 and a total field resistivity map using only bipole 1. Differences between the two different total field maps were explained by the relative positions of the bipoles with respect to resistivity lows (Figure 4-5, Stanley et al., 1976). The composite total field map delineates the east, west, and north boundaries of the Long Valley Caldera and a prominent low resistivity area surrounding the Cashbaugh Ranch. A second major area of low resistivity is situated around the Casa Diablo Hot Springs. This resistivity low is essentially coincident with a major aeromagnetic low, which may result from a near-surface roof pendant of metasedimentary rocks or from hydrothermal destruction of magnetite. An aeromagnetic high coincident with the Cashbaugh Ranch resistivity low is unexplained.

discussed by Dey and Morrison (1977). They concluded that the results of bipole-dipole surveying are often ambiguous and that the method lacks resolution. For these reasons, the bipole-dipole method is not widely used today in geothermal exploration (Goldstein, et al., 1978).

The dipole-dipole method employs transmitter and receiver electrode pairs of equal length (a). Separations between the transmitter and receiver dipoles are integer multiples of this dipole length. In their comparison of the bipole-dipole and dipole-dipole methods, Dey and Morrison (1977) found that when there are overlying low resistivity formations, the dipole-dipole method is more effective in locating the body of interest. They concluded that in areas of complex geology the dipole-dipole method offers distinct advantages over the bipole-dipole method.

In the Schlumberger array, the four electrodes are placed in a line with the receiver pair located at the center of the transmitter pair. During the measurements, the separation of the transmitter dipole is increased to penetrate to successively greater depths. The Schlumberger array can be used for either vertical soundings at a given point or for construction of electrical

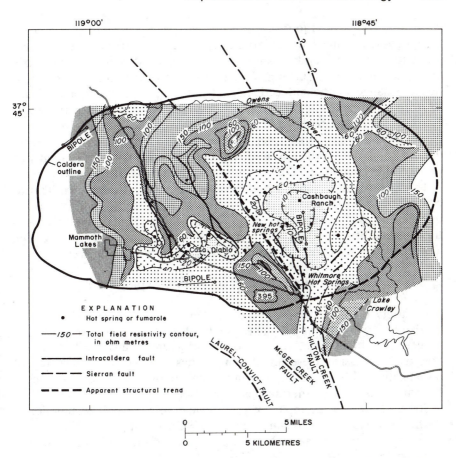

Figure 4.5. Composite total field resistivity map of Long Valley Caldera. (After Stanley et al., 1976; courtesy of the *Journal of Geophysical Research.*)

profiles by correlating between individual soundings. Stanley et al. (1976) give a good example of the use of vertical Schlumberger soundings in their paper on Long Valley.

Time-Domain Electromagnetic (TDEM) Surveying

This is a relatively new technique (Jackson and Keller, 1972; Keller and Rapolla, 1974), which appears to offer great potential in geothermal exploration.

For the TDEM method, Harthill (1976) stated that optimally a short grounded-wire source is used to transmit a current step of many frequencies. Currents of several hundred amperes are commonly used. A vertical-axis loop receiver or cryogenic magnetometer is then used to measure the time rate of change of the induced magnetic flux. The signals obtained are interpreted to yield the variation of the apparent resistivity with depth. Harthill (1976) discussed a test of the method over a geothermal prospect in Arizona. Thirty-three soundings were made with a major objective being the determination of the depth of the resistive basement. Agreement was very good between the TDEM results and the results of a gravity survey.

The Los Alamos National Laboratory has used a commercial TDEM contractor to perform soundings over an area enclosing the Laboratory technical areas on the east side of the Valles Caldera (Figure 4-6). Depths to the resistive basement again agreed with results of a gravity survey. The TDEM results were also useful in locating faults within the area.

Seismic Methods in Geothermal Exploration

A variety of seismic methods, active and passive, have been used in geothermal exploration. The active methods include both refraction and reflection profiling to determine geologic structure around geothermal reservoirs. The passive teleseismic technique has also been used in structural investigations. Micro-earthquake and seismic noise studies have been used to characterize the geothermal reservoir itself.

Some of the best examples of the application of seismic techniques to geothermal exploration result from the USGS investigation of the Long Valley, California, geothermal area (Hill, 1976; Steeples and Iyer, 1976; Iyer and Hitchcock, 1976).

Two approximately orthogonal seismic refraction profiles were run across the Long Valley caldera (Hill, 1976). Recording units were kept in fixed positions while the different shot points were used. Because of topographic and logistic problems, the recording units could not be placed in an exactly linear array, thus hindering later interpretation. Subsurface reversal of propagation was not achieved in many cases.

Velocity models were constructed along the two profiles. These models were then compared with the E-W geologic cross-section of Bailey et al. (1976). The seismic data indicates that the basement is as shallow as 1.5-2 km in the southwestern and west central portions of the caldera, deepening to 3-4 km to the north and east. A 1-2-km thick layer with velocities of 4.0-4.4 km/s overlies the basement and forms a large dome in the western part of the caldera. This agrees well with geologic evidence for post collapse resurgence in this area. As would be expected near the caldera ring faults, the refraction data indicate abrupt changes in the depth to basement. In general, agreement is good between seismic results and available gravity data.

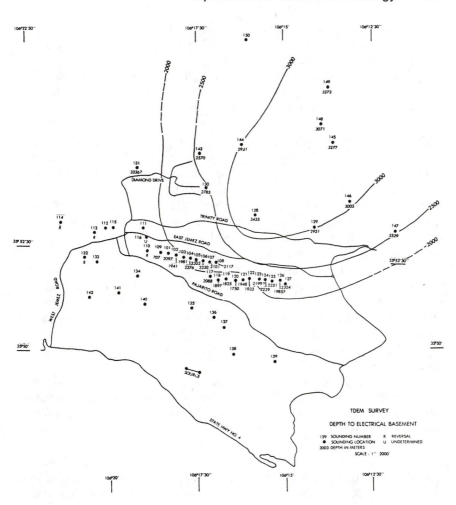

Figure 4-6. TDEM survey of the east side of the Valles Caldera, New Mexico. (Courtesy of Williston, McNeil and Associates, 1979.)

Hill (1976) interprets a set of late arrivals as resulting from reflection from a body at a depth of 7-8 km, which is characterized by decreasing velocity with depth. The calculated depth correlates well with the suggested depth to the roof of the magma chamber that produced the resurgent dome (Bailey et al., 1976).

Steeples and Iyer (1976) described the results of a seventy-eight station teleseismic survey of the Long Valley area. Recording times for this survey range from a few hours to approximately two months and a total of thirty-nine teleseisms were recorded. Reduction of the data indicates a *P* velocity contrast of at least 5% and probably 10-15% between the crust beneath and outside the caldera. The *P* velocities are low beneath the caldera. Steeples and Iyer interpreted these results as indicating a zone of anomalously high temperature beneath the west part of the caldera at a depth of at least 7 km and probably less than 25 km. A reinterpretation of gravity data, indicating a possible deep gravity low at the same location, supports their teleseismic interpretation.

The passive seismic method of microearthquake monitoring was also employed in the Long Valley area (Steeples and Pitt, 1976). Application of this seismic method is usually based on the assumption that active faults, which serve as permeable zones for fluid movement, can be delineated by microearthquake activity occurring along them. Hamilton and Muffler (1972) also suggested that some idea of the geothermal gradient could be obtained from the maximum focal depth of microearthquakes.

At Long Valley, microearthquake monitoring continued for about five weeks using 16 portable seismographs. Epicenter locations and focal depths were determined for 76 microearthquakes. Essentially all of these fell outside of the caldera, except for a cluster of seven events on a fault near the south central boundary of the caldera. The scarcity of microearthquake activity within the caldera, while activity is high outside, is in direct contrast to the observation of Ward (1972) that activity is higher within geothermal areas.

Iyer and Hitchcock (1976) reported the results of a seismic noise survey of the Long Valley area. A total of 77 stations were maintained using EV-17 seismometers. The data obtained was filtered into 0-1, 1-2, 2-4, and 4-8 Hz frequency bands and was corrected for relative noise levels by the use of reference stations.

The results of this survey were inconclusive for identifying a geothermal reservoir. A noise anomaly higher than could be explained by amplification of regional background noise is present in the eastern part of the caldera. Repeated measurements show changes in the noise level with time, which casts doubts on the validity of the method. Iyer and Hitchcock (1976) suggested that a geothermal reservoir may be responsible for the noise of frequency less than 2 Hz.

References

Aiken, C. L. V., 1976. *Analysis of Gravity Anomalies of Arizona.* PhD Dissertation.Univ. of Arizona, Tucson, AZ.

Aiken, C. L. V., Laughlin, A. W. and West, F. G., 1977. *Residual Bouguer Gravity Map of Northern New Mexico.* LA-6737-MAP, Los Alamos Scientific Laboratory, Los Alamos, NM.

Baba, K., 1976. Gravimetric Survey of Geothermal Areas in Kurikoma and Elsewhere in Japan. *Proc., Second U.N. Symposium on the Development and Use of Geothermal Resources,* San Francisco, CA, 2: 865-870.

Bailey, R. A., Dalrymple, G. B., and Lanphere, M. A., 1976. Volcanism, Structure and Geochronology of Long Valley Caldera, Mono County, California. *J. of Geophys. Res.*, 81: 725-744.

Barbier, E. and Fanelli, M., 1976. Relationships as Shown in ERTS Satellite Images Between Main Fractures and Geothermal Manifestations in Italy. *Second U.N. Symposium on the Development and Use of Geothermal Resources*, San Francisco, CA, 2: 882-888.

Bateman, P. C. and Eaton, J. P., 1967. Sierra Nevada Batholith. *Science*, 158: 1407-1417.

Biehler, S., 1971. Gravity Studies in the Imperial Valley, In: Rex, R. W. et al., *Cooperative Geological-Geophysical-Geochemical Investigations of Geothermal Resources in the Imperial Valley Area of California*, Riverside, University of California, pp. 29-42.

Blackwell, D. and Morgan P., 1976. Geological and Geophysical Exploration of the Marysville Geothermal Area, Montana, U.S.A. *Second U.N. Symposium on the Development and Use of Geothermal Resources*, San Francisco, CA, 2: 895-902.

Brookins, D. G. and Laughlin, A. W., 1976. Isotopic Evidence for Local Derivation of Strontium in Deep-Seated, Fracture-Filling Calcite from Granite Rocks in Drill Hole GT-2. Los Alamos Scientific Laboratory Dry Hot Rocks Program, *Journal of Volcanic and Geothermal Research*, 1, 193-196.

Brookins, D. G. and Laughlin, A. W., 1977. Rubidium-Strontium Geochronological Study of GT-1 and GT-2 Whole Rocks (Abs.). *Trans. Am. Geophys. Union*, 57: 352.

Brookins, D. G., Forbes, R. B., Turner, D. L., Laughlin, A. W. and Naeser, C. W., 1977. *Rb-Sr, K-Ar and Fission-Track Geochronological Studies of Samples From LASL Drill Holes GT-1, GT-2, and EE-1*. LA-6829-MS, Los Alamos Scientific Laboratory, Los Alamos, NM, 27 pp.

Byerly, P. E. and Stolt, R. H., 1977. An Attempt to Define the Curie Point Isotherm in Northern and Central Arizona. *Geophysics*, 42: 1394-1400.

Celati, R., Noto, P., Panichi, C., Squarci, P. and Taffi, L., 1973. Interactions Between the Steam Reservoir and Surrounding Aquifers in the Larderello Geothermal Field. *Geothermics*, 2:

Combs, J. and Wilt, M., 1976. Telluric Mapping, Telluric Profiling and Self-Potential Surveys of the Dunes Geothermal Anomaly, Imperial Valley, California. *Proc., Second U.N. Symposium on the Development and Use of Geothermal Resources*, San Francisco, CA, 2: 917-928.

Corwin, R. F., 1976. Self-Potential Exploration for Geothermal Reservoirs. *Proc., Second U.N. Symposium on the Development and Use of Geothermal Resources*, San Francisco, CA, 2: 937-946.

Decker, E. R. and Smithson, S. B., 1975. Heat Flow and Gravity Interpretation Across the Rio Grande Rift in Southern New Mexico and Western Texas. *J. Geophys. Res.*, 80: 2542-2552.

Dellechaie, F., 1976. A Hydrochemical Study of the South Santa Cruz Basin Near Coolidge, Arizona. *Second U.N. Symposium on the Development and Use of Geothermal Resources*, San Francisco, CA, 1: 339-348.

Dellechaie, F. and Hansen, D., 1977. Geothermal Prospecting with Helium Soil Gas Analysis. *Geol. Soc. Am. Program* with abstracts, v. 9, p. 947.

Dey, A. and Morrison, H. F., 1977. An Analysis of the Bipole-Dipole Method of Resistivity Surveying. *Geothermics*, 6: 47-82.

Doell, R. R., Dalyrimple, G. B., Smith, R. L. and Bailey, R. A., 1968. Paleomagnetism, Potassium-Argon Ages, and Geology of Rhyolites and Associated Rocks of the Valles Caldera, New Mexico. In: Studies in Volcanology. R. R. Coats, R. L. Hay and C. A. Anderson (eds.) *Geol. Soc. Am. Mem.*, 116: 211-248.

Eddy, A. and Laughlin, A. W., 1978. Core Studies Applied to Hot Dry Rock Geothermal Energy (abstr). Abstracts of papers of 144th National Meeting, 12-17 Feb., 1978, A. Herschman (ed.), p. 178, *Am Assoc. Adv. Sci.*, Washington, D.C.

Ellis, A. J. and Mahon, W. A. J., 1967. Natural Hydrothermal Systems and Experimental Hot-Water/Rock Interactions (Pt. II). *Geochim. Cosmochim.* Acta, 31: 519-538.

Ellis, A. J. and Mahon, W. A. J., 1977. *Chemistry and Geothermal Systems.* Academic Press, New York, 392 pp.

Fournier, R. O. and Truesdell, A. H., 1973. An Empirical Na-K-Ca Geothermometer for Natural Waters. *Geochim. Cosmochim.* Acta, 37: 1255-1275.

Goldstein, N. E., Norris, R. A., and Wilt, M. J., 1978. Assessment of Surface Geophysical Methods — Geothermal Exploration and Recommendations for Further Research, LBL/UC Report 6815, DOE Contract W-7405-Eng-48.

Gutsalo, L. K., 1976. Helium Isotopic Geochemistry in Thermal Water of the Kuril Island and Kamchatka. *Proc., Second U.N. Symposium in the Development and Use of Geothermal Resources,* v. 1, pp. 745-750, San Francisco, California, 20-29 May, 1975-1976. Three volumes 2466 pp. (Available from the U.S. Government Printing Office, Washington, D.C. 20402.)

Hahman, W. R., Sr., Stone, C. and Wicher, J. C., 1978. *Preliminary Map Geothermal Energy Resources of Arizona.* Arizona Bur. of Geology and Mineral Technology.

Hamilton, R. M. and Muffler, L. J. P., 1972. Microearthquakes at the Geysers, California Geothermal Area (abs. only) *EOS Trans. Amer. Geophys. Un.,* v. 52, p. 862.

Harthill, N., 1976. Time-Domain Electromagnetic Sounding, *IEEE Trans.,* GE-14, p. 256-260.

Hermance, J. F. and Pedersen, J., 1976. Assessing the Geothermal Resource Base of the Southwestern United States, Status Report of a Regional Geomagnetic Traverse (abs.). *46th Ann. Internat. Meet. SEG,* Houston, Texas.

Hill, D. P., 1976. Structure of Long Valley caldera, California, from a Seismic Refraction Experiment. *J. Geophys. Res.,* 81: 745-753.

Hockstein, M. P., 1976. Geophysical Exploration of Kawah Kamojang Geothermal Field, West Java. *Proc., Second U.N. Symposium on the Development and Use of Geothermal Resources,* San Francisco, CA, 2: 1049-1058.

Hoover, D. B., Frischknect, F. C. and Typpens, C. L., 1976. Audiomagneto-Telluric Sounding as a Reconnaissance Exploration Technique in Long Valley, California. *J. Geophys. Res.,* 81: 801-809.

Isherwood, W. F., 1976. Gravity and Magnetic Studies of the Geysers-Clear Lake Geothermal Region, California, U.S.A. *Proc., Second U.N. Symposium on the Development and Use of Geothermal Resources,* 2: 1065-1074.

Iyer, H. M. and Hitchcock, T., 1976. Seismic Noise Survey in Long Valley, California, *Jour. Geophys. Res.,* 81: 821-840.

Jackson, D. B. and Keller, G. V., 1972. An Electromagnetic Sounding Survey of the Summit of Kilauea Volcano, Hawaii. *J. Geophys. Res.,* 77: 4957.

Jiracek, G. R., Reddy, I. K., Phillips, R. J. and Whitcomb, J. H., 1977. Magnetotelluric Soundings of the Rio Grande Rift, COCORP Seismic profile in New Mexico, Abs. *47th SEG Conf.,* Calgary, Alberta, Canada.

Keller, G. V. and Rapolla, A., 1974. Electrical Prospecting Methods in Volcanic Areas. In: *Physical Volcanology,* L. Civetta et al. (eds.), Elsevier Scientific, Amsterdam, 133 pp.

Lachenbruch, A. H., Sass, J. H., Monroe, R. J., and Moses, T. H., 1976. Geothermal Setting and Simple Heat Conduction Models for the Long Valley, California. *J. of Geophys. Res.,* 81: 769-784.

Laughlin, A. W. and Eddy, A. C., 1977. Petrography and Geochemistry of Precambrian Rocks from GT-2 and EE-1, Los Alamos Scientific Laboratory Informal Report, LA-6930-MS, 50 pp.

Matlick, J. D., III and Buseck, P. R., 1976. Exploration for Geothermal Areas Using Mercury: A New Geochemical Technique, *Proc. Second U.N. Symposium on the Development and Use of Geothermal Resources,* San Francisco, CA, 1: 785-793.

Muffler, L. J. P., 1976. Summary of Section II — Geology, Hydrology and Geothermal Systems. *Proc. Second U.N. Symposium on the Development and Use of Geothermal Resources,* San

Francisco, CA, 20-29 May, 1975. 1: xlv-li. Three volumes, 2466 pp. (Available from the U.S. Gov't. Printing Office, Washington, D.C., 20402.)

Muffler, L. J. P., 1976. Tectonic and Hydrologic Control of the Nature and Distribution of Geothermal Resources. *Proc., Second U.N. Symposium on the Development and Use of Geothermal Resources,* San Francisco, CA, 20-29 May, 1975. 1: 499-507. Three volumes, 2466 pp. (Available from the U. S. Gov't. Printing Office, Washington, D.C. 20402.)

Naeser, C. W. and Forbes, R. B., 1976. Variation of Fission Track Ages with Depth in Two Deep Drillholes (Abs.). *Trans. Am. Geophys. Un.,* 57:353

Pederson, J. and Hermance, J. F., 1978. Evidence for Molten Material at Shallow to Intermediate Crustal Levels Beneath the Rio Grande Rift at Santa Fe (Abs.). *Trans Am. Geophys. Un.,* 59: 390.

Peters, W. C., 1978. *Exploration and Mining Geology,* John Wiley and Sons, New York, 696 pp.

Petracco, C. and Squarci, P., 1976. Hydrological Balance of Larderello geothermal region. *Proc. Second U. N. Symposium on the Development and Use of Geothermal Resources,* 1: 521-530.

Phelps, D. W. and Buseck, P. R., 1978. Natural Concentrations of Hg in the Yellowstone and Coso Geothermal Fields: In Geothermal Energy, a Novelty Becomes a Resource, *Transactions,* Volume 2, Section 2, Geothermal Resources Council, pp. 521-522.

Reiter, M., Edwards, C. L., Hartman, H. and Weidman, C., 1975. Terrestrial Heat Flow Along the Rio Grande Rift, New Mexico and Southern Colorado. *Geol. Soc. Am. Bull.,* 86: 811-818.

Reiter, M., Shearer, C. and Edwards, C. L., 1978. Geothermal Anamolies Along the Rio Grande Rift, New Mexico. *Geology,* 6: 85-88.

Simmons, G. and Eddy, A. C., 1976. Microcracks in GT-2 Core, (Abs. only) *EOS Trans. Amer. Geophys. Un.,* 57: p. 353.

Slemmons, D. B., 1975. Fault Seismicity in the Los Alamos Scientific Laboratory's Test Site, Jemez Mts. Los Alamos Report LA-5911-MS.

Smith, R. L. and Shaw, H. R., 1975. Igneous-Related Geothermal Systems. *U.S.G.S. Circ.* 726, pp 58-83. In: *Assessment of Geothermal Resources of the U.S.-1975,* D. E. White and P. L. William (eds).

Stanley, W. D., Jackson, D. B., and Zohdy, A. A. R., 1976. Deep Electrical Investigations in the Long Valley Geothermal Area, California. *Journal of Geophysical Research,* 81: 810-820.

Steeples, D. W. and Iyer, H. M., 1976. Low-velocity Zone under Long Valley Caldera as Determined from Teleseismic Events, *J. Geophys.* Res., 81: 849-860.

Steeples, D. W. and Pitt, A. M., 1976. Microearthquakes in and near Long Valley, California, *Jour. Geophys. Res.,* 81: 841-847.

Sugimura, A. and Uyeda, S., 1973. *Island Arcs: Japan and Its Environs.* Elsevier, Amsterdam, 247 pp.

Swanberg, C. A., 1976. The Mesa Geothermal Anomaly, Imperial Valley, California: A Comparison and Evaluation of Results Obtained from Surface Geophysics and Deep Drilling, *Proc. Second U. N. Symposium on the Development and Use of Geothermal Resources,* San Francisco, CA, 2: 1217-1230.

Swanberg, C. A., Morgan, P., Stoyer, C. H. and Witcher, J. C., 1977. Regions of High Geothermal Potential, an Appraisal Study of the Geothermal Resources of Arizona and Adjacent Areas in New Mexico and Utah and Their Value for Desalination and Other Uses.

Truesdell, 1976. Summary of Section III, Geochemical Techniques in Exploration. *Proc. Second U. N. Symposium on Development and Use of Geothermal Resources,* San Francisco, CA, 1:liii-lxiii.

Turner, D. L. and Forbes, R. B., 1976. K-Ar Studies in Two Deep Basement Drillholes: A New Geologic Estimate of Argon Blocking Temperature for Biotite (Abs.). *Trans. Am. Geophys. Un.,* 57:353.

Ward, P. L., 1972. Microearthquakes: Prospecting Tool and Possible Hazard in the Development of Geothermal Resources. *Geothermics,* 1: 3-12.

West, F. G. and Laughlin, A. W., 1979. Aquarius Mountains Area, Arizona: A Possible HDR Prospect. Los Alamos Scientific Report LA-7804-MS.

White, D. E., 1965. Geothermal Energy. *U. S. Geol. Surv. Circ. 519,* p.1-17.

White, D. E., 1967. Some Principles of Geyser Activity, Mainly from Steamboat Springs, Nevada. *Am. J. Sci.,* 265: 641-684.

Williams, P. L., Mabey, D. R., Zohdy, A. A. R., Ackermann, H., Hoover, D. B., Pierce, K. L. and Oriel, S. S., 1976. Geology and Geophysics of the Southern Raft River Geothermal Area, Idaho, U. S. A. *Proc., Second U. N. Symposium on the Development and Use of Geothermal Resources,* San Francisco, CA, 2: 1273-1282.

Williston, McNeil and Associates, 1979. A Time Domain Survey of the Los Alamos Region, New Mexico, Los Alamos Scientific Laboratory Informal Report, LA-7657-MS, 32 pp.

Zohdy, A. A. R., Anderson, L. A. and Muffler, L. J. P., 1973. Resistivity, Self-potential and Induced Polarization Surveys of Vapor-Dominated Geothermal System. *Geophysics,* 38: 1130-1144.

Ken Greene, Dresser Security (retired)
Lee Goodman

5

Geothermal Well
Drilling and Completion

Introduction

The drilling of geothermal wells is very similar to drilling oil and gas wells in that the basic laws of physics and rock destruction apply in both instances. The basic differences between drilling geothermal and hydrocarbon wells are threefold:

1. Excluding the geopressured geothermal testing now underway in the Gulf Coast area of U.S., nearly all other geothermal drilling is performed at low pressures.
2. Most geothermal wells are of relatively shallow depth having high formation temperatures, with the exception of U.S. Gulf Coast area.
3. Except for certain sedimentary basins, the rocks being drilled are igneous and metamorphic.

This chapter presents a review of drilling rigs, rig equipment, casing design, bit and hydraulics programs, muffler systems, use of drilling fluids and drilling case histories.

Drilling Rigs and Rig Equipment

Generally, for most geothermal drilling operations, rigs in the range of 300 to 1,000 hp are adequate depending upon the depth of the reservoir and the casing and pipe loads required. In some areas, where high rotary torque is encountered, an extra heavy, independently-powered rotary table is used.

The drilling rig should be designed for the particular area of drilling. For example, in the Imperial Valley of California a typical rig might be one that is trailer-mounted and highly portable, because drilling operations are fairly uncomplicated and can be performed in a relatively short time. (Figure 5-1 shows a type used in New Zealand.) In contrast, a typical rig for the Geysers Area, U.S.A. may require at least 1,000 hp to meet the hoisting requirements of wells in that area. In addition, the rotating system must be extra heavy-duty to tolerate excess torque. The rotary clutch, shaft, and chain are very susceptible to damage and, therefore, must be properly designed. To help preserve the surface equipment, a direct drive in the power train to the rotary table is avoided. This is accomplished by the independent drive already mentioned or by using fluid couplings on the compound.

The *mud pumps* on the rig should be of sufficient size to complement the hydraulics program. One 500-hp pump may be sufficient for operations in the Imperial Valley, California, whereas, two 600-hp pumps or their equivalent are needed for drilling a 17½-in. surface hole at the Geysers area, California.

The *drill string* also will vary depending upon the location. A typical string in the Imperial Valley may consist of 3½-in. drill pipe with 6-in. collars. The drill string at the Geysers area, however, is subjected to very severe conditions. Consequently, the drill string should be properly designed for the job and inspected regularly to ensure that its condition is excellent.

The *blowout prevention equipment,* vital to the drilling operation, also varies in different areas. A typical blowout stack arrangement in the Imperial Valley, California, consists of pipe and blind rams, hydril, spacers, spools, and a rotating head (Figure 5-2). A recommended stack for the Geysers area consists of two ram preventers (pipe and blind) below the banjo box to shut-in the well if it becomes necessary to repair the banjo box or blooie line. A modified ram preventer (banjo box), to which the blooie line is attached, is in the middle of the stack with a double set of ram preventers on top to permit the well to be shut-in during testing or circulating through the blooie line. A rotating head is also used for drilling with accompanying spools and spacers (Figure 5-3). In some Geyser area operations, water is injected under the rotating head to help preserve the ram rubbers. A drilling muffler (discussed later) is installed at the end of the blooie line to control dust and noise.

Casing Design

A very brief review of the maximum load approach is presented to illustrate the rationale behind the procedure used in designing the various casing strings for geothermal wells.

Inasmuch as high pressures, which are characteristic of oil and gas drilling, are absent in a typical geothermal well (excluding geopressured geothermal wells), casing design focuses some attention on burst and collapse loading, with special attention given to the type of threading used. A typical casing program for the Geyser area, California is shown in Figure 5-4.

Figure 5-1. Typical geothermal drilling rig on location in New Zealand.

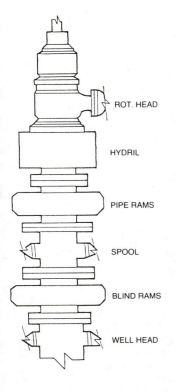

ROT. HEAD

HYDRIL

PIPE RAMS

SPOOL

BLIND RAMS

WELL HEAD

Figure 5-2. Typical blowout preventer stack, Imperial Valley, California.

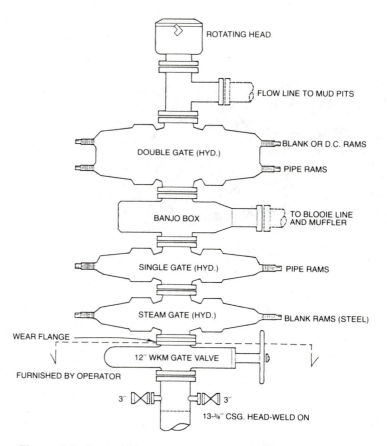

ROTATING HEAD

FLOW LINE TO MUD PITS

BLANK OR D.C. RAMS

DOUBLE GATE (HYD.)

PIPE RAMS

BANJO BOX

TO BLOOIE LINE
AND MUFFLER

SINGLE GATE (HYD.)

PIPE RAMS

STEAM GATE (HYD.)

BLANK RAMS (STEEL)

WEAR FLANGE

12" WKM GATE VALVE

FURNISHED BY OPERATOR

3"

3"

13-3/8" CSG. HEAD-WELD ON

Figure 5-3. Typical blowout preventer stack, Geysers, California.

Burst Load

A typical burst load would be based upon the fracture gradient anticipated below the string of casing after it is set. At the bottom of the string, the burst load is influenced by the column of fluid, which backs up the actual load behind the casing:

$$BL_B = (FG - MG) \times D \qquad (5\text{-}1)$$

where: BL_B = burst load at the bottom of the casing, psi
FG = fracture gradient below casing, psi/ft
MG = mud gradient behind casing, psi/ft
D = depth of casing string, ft

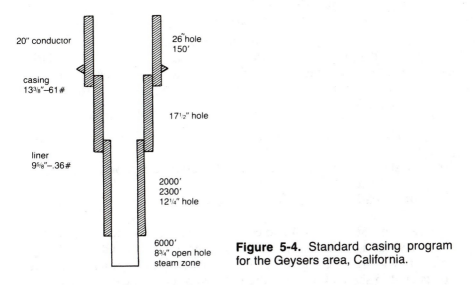

Figure 5-4. Standard casing program for the Geysers area, California.

20" conductor

casing
$13\frac{3}{8}$"–61#

liner
$9\frac{5}{8}$"–.36#

26" hole
150'

$17\frac{1}{2}$" hole

2000'
2300'
$12\frac{1}{4}$" hole

6000'
$8\frac{3}{4}$" open hole
steam zone

Either water or the actual drilling mud density can be used in calculating MG. The choice would depend upon the designer and the assumptions made.

At the top of the string, the burst load is influenced by the fracture gradient, with no backup fluid. One could assume that a column of some fluid always exists inside the casing and should be considered in calculating the maximum burst pressures that the casing would experience:

$$BL_T = (FG - WG) \times D \qquad (5\text{-}2)$$

where: WG = water gradient inside the casing; in the Geysers area, this could be considered the gradient of steam, i.e., practically zero.

BL_T = burst load at the top of the casing, psi

A graphic example of a typical burst load versus depth is presented in Figure 5-5. The safety factors used in calculating the burst load vary and are dependent upon the company which is designing. Typical burst safety factors (SF_B) range from 1.1 to 1.25. Actual safety factors (burst) are calculated by using the following equation:

$$SF_B = \text{rated burst (psi)/burst load (psi)} \qquad (5\text{-}3)$$

Collapse Load

Collapse loads (CL) are usually calculated by using the formula for hydrostatic pressure:

$$CL = WT \times D \times .052 \text{ (psi)} \tag{5-4}$$

where: WT = density of the mud in which casing is run

Typical collapse loads are calculated assuming that the casing will be empty (Figure 5-6). Typical design safety factors in calculating collapse loads vary from 0.8 to 1.25. Actual safety factors are calculated by using the following equation:

$$SF_c = \text{adjusted collapse rating (psi)/collapse load (psi)} \tag{5-5}$$

The adjusted collapse rating accounts for tension effect on collapse.

Type of Threads

Because of the expansion and contraction of casing experienced in geothermal wells, as a result of varying high temperatures, standard API eight round threads (ST&C and LT&C) are not used as a general rule. These joints tend to slip and override the threads as the casing expands. Consequently, the buttress thread is more widely used. In many instances, threads with even a higher efficiency are used, such as the Hydril Triple Seal or better.

Stresses Due to Thermal Expansion

The added stresses imposed on casing strings and resulting failure have been described. It should be emphasized that casing strings should be cemented such that the entire annulus around the casing is filled with cement. This is further discussed in the cementing section of this chapter. The following explains some modes of failure when the above is not true.

When water (or mud) is trapped in a void in the annulus area under cool conditions, the increase in temperature during production increases the pressure in the annulus. This pressure may exceed the collapse rating of the inner casing or the burst rating of the outer casing. Ideally, a good design ensures the latter (Dench, 1970).

Axial compression stresses may exceed the strength of the weakest part of the string (Dench, 1970). In the case of API 8 round threads, this would be the joint. The use of buttress threads or better has been discussed. Axial tensile stresses due to cooling may also exceed the strength of the joint (Dench, 1970).

Thermal expansion of well materials is of major concern in geothermal operations. As a consequence, it appears desirable that pipe and casing incorporate materials and threads, such as the buttress type, to resist the stresses that develop (Whiting, 1974).

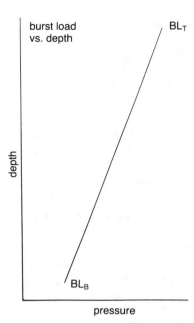

Figure 5-5. Burst load versus depth graph.

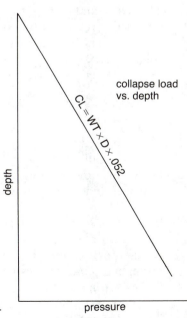

Figure 5-6. Collapse load versus depth graph.

Cementing and Casing

Because of the higher temperatures encountered in geothermal wells, cement design is similar to that for deep gas or oil wells having high bottomhole temperatures (see Chapter 6 for details). Typical cement designs in geothermal wells can be summarized as follows:

1. *Conductor casing* — A cement slurry for conductor casing could be Class G cement with some calcium chloride for accelerated setting time.
2. *Surface casing* — A surface casing slurry might consist of a 2:1 ratio of Class G and Litepoz cements with 30% silica flour for retarders. In many cases, most silica flour is used with some gel (bentonite). Friction reducers may also be used. Cement for surface casing should be pumped back to the surface to aid in preventing the casing from parting. As pointed out by Combs and Muffler (1973), because casing and cement coefficients of thermal expansion are nearly identical, cementing and casing will expand together. In the case of limited water availability and/or to reduce the pumping time, cementing of surface casing may be done through the drill pipe using a stab-in type float collar.
3. *Intermediate casing* — A typical cement slurry for intermediate casing string could consist of API Class G cement and perlite (1:1 ratio), with 40% silica flour and appropriate retarders, water loss reducing agents, and friction reducers. In some areas (e.g., Imperial Valley, California) it is necessary to stage-cement this string by using DV cementer port collars. Also, it is a common practice to pump cool water ahead of the cement for cooling purposes, followed by a mud flush and then the cement.

Drilling Bits

Selection of drilling bits for a geothermal drilling operation does not differ from that of any other drilling operation. Information needed to design the bit program includes the following:

1. Bit records from offset wells.
2. Geological cross section.
3. Rock types, cementation, etc.
4. Formation temperatures and pressures.
5. Casing setting depths.
6. The depth at which steam will be encountered in areas of vapor-dominated reservoirs.

A typical bit program design for a 6,500-ft well at East Mesa in the Imperial Valley, California, is as follows:

Surface hole to 2,000 ft
    number of bits — 1 bit
    type of bit — very soft formation, jet
    energy levels — 20#M lb/100 rpm
Intermediate hole to 5,500 ft
    number of bits — 5-6 bits
    type of bit — soft formation, jet
    energy levels — 30#M lb/90-100 rpm
Production interval to 6,500 ft
    number of bits — 4-5 bits
    type of bit — soft formation, jet
    energy levels — 20#M lb/70-80 rpm

In this particular area very soft formations are encountered with little or no drilling problems from a bit standpoint. The wells can be drilled in a relatively short time to completion depths.

In contrast, each well drilled in the Geysers area is essentially a wildcat because of the presence of severe faulting, folding, and tectonics. A general bit program for that area is included in Table 5-1.

Because of the hard rock drilling at the Geysers area, nearly all of the bits used are of the insert type. In the portion of the hole where mud is used, drilling rates may range from 10 to 20 ft/hr. On the other hand, in the bottom part of the borehole where air is the circulating medium, penetration rates may reach 75 ft/hr (Anderson, 1972).

Table 5-1
Typical Well, Geysers, California

Bit diameter	17½ ins.	12¼ ins.	8¾ ins.
Depth out: (Average) (ft)	1575	4040	6950
Range of depth Out (ft)	1400 to 2400	2200 to 6300	5400 to 10,000
Range of bits used	1 to 9	2 to 19	7 to 29
Avg. no. bits per interval	3.2	9.4	12.5
Weight on bit (1000#)	30/70	10/50	10/50
Rpm	60/100	60/100	45/80 (60)
Bit type	Tooth & Insert	Insert	Air Insert

Because the production interval at the Geysers, California is drilled with air, the air circulation insert bit is used. Emphasis on these bits includes the wear on the exterior of the cones (gauge wear). Tungsten carbide is used in this area to prevent this. These bits are usually 8¾ ins. in diameter. The shock absorber has been used advantageously in many areas of "hard rock" drilling. This device helps eliminate the shock of the drill string by absorbing it just above the bit (Figure 5-7). A sufficient number of drill collars should be run for the maximum bit weight anticipated. In calculating the length of collars, some operators use the following formula:

$$L_C = \frac{B_W}{C_W \times BF} \qquad (5\text{-}6)$$

where: L_C = length of collars, ft
B_W = bit weight, lb
C_W = collar weight, lb/ft
BF = buoyancy factor, dimensionless

The buoyancy factor, BF can be calculated by using the following formulas:

$$BF = \frac{65.4 - D_1}{65.4} \qquad (5\text{-}7)$$

$$BF = \frac{489.2 - D_2}{489.2} \qquad (5\text{-}8)$$

$$BF = \frac{7.85 - D_3}{7.85} \qquad (5\text{-}9)$$

where: D_1 = mud specific weight, lb/gal
D_2 = mud specific weight, lb/cu ft
D_3 = mud specific weight in terms of specific gravity

Other operators use the following formula for calculating the length of collars:

$$L_C = \frac{(1\text{-}BF)\,(P_W \times TD) + BW}{C_W + (1 - BF)\,(P_W - C_W)} \qquad (5\text{-}10)$$

where: P_W = drill pipe weight, lb/ft
TD = total depth ft

Equation 5-10, which gives a longer length of collars, is designed primarily to keep the entire drilling string in tension to lessen the wear effect on the pipe.

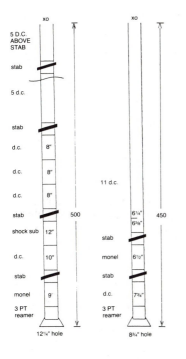

Figure 5-7. Examples of bottomhole assemblies used in drilling a geothermal well in the Geysers area, California.

Muffler Systems

In areas of steam-dominated reservoirs, such as the Geysers area, air drilling is employed to drill into the producing zone. Increasing numbers of operators are employing muffler systems attached to the end of the blooie line to reduce the noise. In addition, this system provides a collection system for cuttings and condensates, and aids in the reduction and abatement process of H_2S produced in the steam.

A typical muffler system consists of a large diameter pipe, e.g., 144 ins., with a smaller vent pipe, e.g., 90 ins. in diameter, as a vent at the top. (Figure 5-8). During drilling, the returns (air, steam, cuttings) enter the larger pipe tangentially. The centrifugal force resulting from a tangential entry, forces the solids to the side while the steam and air move out through the vent pipe. A conical bottom on the larger pipe causes the unit to act similar to a desander or desilter as the solids escape through the cone at the bottom. Boiler plate is currently being used inside the muffler to resist the abrasiveness of the high-velocity cuttings. Another typical design that is used is shown in Figure 5-9. It employs the same principal but has a maximum pipe

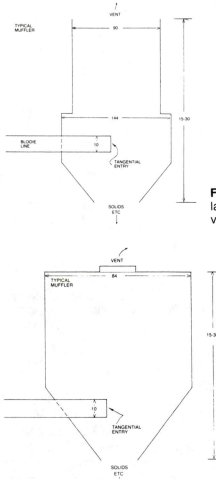

Figure 5-8. Muffler system consisting of a large diameter pipe (144 ins.) and a smaller vent pipe (90 ins.)

Figure 5-9. Another typical muffler design, but has a smaller maximum pipe diameter (84 ins.)

diameter of 84 ins. These muffler systems vary in height from 15 to 30 ft depending upon design.

The amount of H_2S produced with the steam at the Geysers area varies (20-200 ppm range) with each well and is dependent upon the amount of steam being produced. The concentration of H_2S is sufficient to give rise to an offensive odor causing environmentalists to place more restrictions on the quality of emissions released to the atmosphere.

Various chemicals have been injected into the steam-air mixture in an attempt to reduce the H_2S concentration. Ammonium hydroxide is one of the chemicals that has been used. According to Budd and Chester (1973):

Various quantities of ammonium hydroxide were mixed with a solution of water and Unisteam (a corrosion inhibitor). This solution was then mixed with the air and injected down the drill pipe at high pressures. Hydrogen sulfide reacts with ammonium hydroxide as follows:

$$2NH_4OH + H_2S \rightarrow (NH_4)S + H_2O \qquad (5\text{-}11)$$

Budd and Chester (1973) also pointed out that the combined effect of reduced flow rate injection of ammonium hydroxide with air and water injection would be in the range of 49-73% reduction in sulfide emission. This reduction in emission of H_2S may not be adequate to meet air pollution requirements for wells to be drilled near populated areas. The ammonium sulfide of Reaction 5-11 will regenerate hydrogen sulfide if the pH of the disposal pond is allowed to drop below 9.3.

A more recent development for use in treating H_2S involved hydrogen peroxide in combination with sodium hydroxide. According to Budd and Chester (1973), the combined effect of injecting NaOH and H_2O_2 into the blooie line during normal drilling operations (drilling with air and water injection) would be in the range of 68-98% abatement or a reduction from 34 lbs/hr to a range of 11 to 0.7 lb/hr of H_2S.

The reactions among NaOH, H_2O_2, and H_2S can be presented as follows:

$$H_2S + NaOH \rightarrow NaHS + H_2O \qquad (5\text{-}12)$$

$$NaHS + 4H_2O_2 \rightarrow NaHSO_4 + 4H_2O \qquad (5\text{-}13)$$

$$H_2S + 2NaOH \rightarrow Na_2S + 2H_2O \qquad (5\text{-}14)$$

$$Na_2S + 4H_2O_2 \rightarrow Na_2SO_4 + 4H_2O \qquad (5\text{-}15)$$

The products of the reaction, Na_2SO_4 and $NaHSO_4$, will not physically revert to hydrogen sulfide (Budd and Chester, 1973, p. 129). There are precautions that should be observed in the handling of compounds used for H_2S treatment. Special surface equipment is necessary to prevent accidental contact by rig personnel.

Drilling Fluids

In a geothermal drilling operation, the type of drilling fluid and its proper control is very important to the success of the drilling of the well. Generally, producing geothermal reservoirs may be classified into two types: (1) vapor-dominated systems where dry steam is being produced, and (2) hot-water systems where hot water being produced "flashes" into steam near the surface as pressures are reduced. Although both systems are found in geothermal areas throughout the world, the hot-water systems are more prevalent. Prior to drilling into the production zones of both systems, the drilling fluid types may be similar in nature. In drilling into a vapor-dominated (dry steam) system, however, the zone is normally drilled with air in order not to kill the producing zone by hydrostatic column of fluid. A hot-water system is normally drilled with conventional drilling fluids (muds) to completion.

Problems of Geothermal Drilling Fluids

Problems associated with drilling fluids in geothermal drilling in many instances are similar to those which occur with drilling for hydrocarbons. Some of these problems, however, may occur more frequently and may be magnified in a geothermal drilling operation. Common to all geothermal drilling are relatively high bottom hole temperatures as the well is drilled into the producing zone. Temperatures may range from 300°F to 800°F depending on the area that the well is being drilled. High-temperature gelation of the drilling fluid may occur if the drilling fluid is not properly treated. The drilling fluid may be chemically treated to withstand these very high temperatures. In some cases, the hot drilling fluid, which is being pumped out of the well bore, is circulated through a cooling tower as the well is being drilled. There is a great degree of contact of the fluid with the cool air at the surface. As a result, the drilling fluid temperature is reduced and mud circulating temperatures are kept at a minimum.

Obviously, high-temperature stability of drilling fluids is the one problem common to all categories of geothermal exploration. Many types of drilling muds are stable up to 150°C (302°F).

With temperatures above 150°C (302°F), mud stability becomes a major problem, and an intractable one at the temperatures above 200°C (392°F). It is necessary, therefore, to discuss in some detail the effect of high temperature on drilling muds.

Water-base muds start to degenerate significantly at about 150°C (302°F) and the rate of degeneration increases rapidly with increase in temperature. Fundamentally, the problem arises because the clay colloids flocculate giving rise to high shear and gel strengths (>20 lbs/100 ft²) (Figure 5-10). The gel strengths can be cut by the addition of lignosulfonates, but unfortunately, these start to degrade at about the same temperature, and are virtually useless at temperatures above 190°C (374°F). At these temperatures lignosulfonates degrade and give off H_2S, which causes severe corrosion problems. The flocculation of the clay also makes it difficult to control filtration rates, and filtration control agents, such as carboxy methyl cellulose, also start to degrade rapidly at 150°C (302°F). The higher the proportion of solids in the mud, particularly the clay fraction (Figure 5-11), the more sensitive the mud becomes to high-temperature flocculation. The presence of salts, even in small quantities, causes further sharp increases in shear and gel strengths.

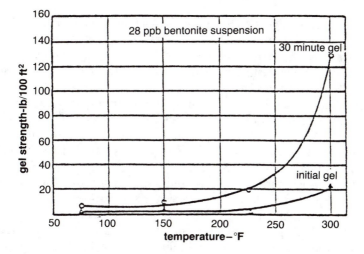

Figure 5-10. Effects of temperature on initial and 30-minute gel strength.

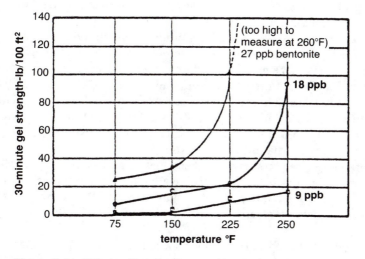

Figure 5-11. Effects of temperature and bentonite concentration on 30-minute gel strength.

With unweighted muds, the problem can be handled to a limited extent by watering back and replacing the degraded treating agents with fresh materials. However, this is expensive. Also of benefit is the use of mechanical separators (such as fine screens and cyclone separators) to remove drill cuttings and dispersed particles to keep the solids content as low as possible.

With *weighted muds* the problems are much more severe. The barite itself greatly increases the solids content of the mud, making rheological control much more difficult. In order to remove drilled solids without removing the barite, a power driven centrifuge is necessary. Centrifuges are expensive and only handle a small fraction of the mud stream. Watering back is also very expensive because of the fresh additions of barite required to restore the density. Worst of all, if the mud becomes contaminated, for example, by a salt water flow, the resulting high gel strengths or thick filter cakes on permeable strata, may cause the drill pipe to stick. This will entail an expensive fishing job, or even loss of the well.

Because of these difficulties, the limiting temperatures for water-base muds is generally considered to be about 190°C (374°F), although some products, which are discussed later in this report, now being developed show promise of extending the range further.

Oil-base muds are more stable, and some of them withstand temperatures up to 250°C (475°F), but oil muds are not considered suitable for geothermal drilling because contamination of the aquifer by oil would decrease productivity. Furthermore, most operators are hesitant to use oil-base muds because of the possibility of environmental damage resulting from spills, separator discharges, or blowouts.

It is important to note that the limiting temperatures previously mentioned are mud temperatures, which are well below the maximum formation temperature while the mud is circulating. This is because the mud cools the formations at the bottom of the hole, and transfers the heat to the cooler formations above and to the atmosphere at the surface. When circulation is stopped the temperature of the mud rises towards equilibrium with the formation, but does not have time to reach the undisturbed geothermal gradient during normal drilling operations.

Temperatures can be very high at shallow depths. Geothermal well temperatures as high at 350°C (662°F) have been encountered at a depth of 1,500m (4,920 ft). Boiling of the mud may be prevented by three methods, or any combination thereof:

1. Cooling the mud at the surface so that temperatures throughout the whole circulating system are lowered sufficiently.
2. Holding a back pressure at the surface, and flashing the mud in a separator.
3. Increasing the density of the mud.

Obviously, method (1) is to be preferred because it has the additional advantage of making the control of the mud properties much easier. However, it is inoperative during round trips to change bits, etc., and there is the danger of a blow-out when circulation is resumed, and the hot mud from the bottom is carried up the hole. It therefore must be used in combination with method (2). Method (3) is the least desirable, and should be avoided if at all possible because it increases the difficulty and expense of controlling the other properties of the mud.

Because of this gap between the mud and the formation it has been found possible to maintain adequate mud properties with formation temperatures of 350°C (660°F). In the Salton Sea area, for example, formation temperatures of this order are encountered, but bottom hole mud temperatures have been found to be only about 250°C (450°F) after an eight-hour shutdown. Nevertheless, mud costs are high and there is always the danger that the mud will break down completely should circulation be stopped for a prolonged period, such as during a fishing job or during completion operations.

Lost Circulation

In many geothermal areas, the geology consists of highly fractured and faulted sequences which, when drilled, can give rise to full or partial loss of circulation of the drilling fluid. To combat this problem, conventional methods of regaining circulation are used. Coarse-sized lost circulation material may be introduced (spotted) or, if the problem is severe, a cement squeeze job to cement off the thief zone is often implemented. In some cases, the hole may be drilled without returns (drilling blind) until the next casing point is reached.

Stuck Drill Pipe

Sticking of the drill string is a problem as common to geothermal drilling as it is in the drilling for hydrocarbons, and the methods to free the stuck drill pipe are essentially the same. Hydrostatic pressures should be maintained as low as possible to prevent differential sticking.

Corrosion

Corrosion of tubular goods (tubing, casing, etc.) can be a serious problem in geothermal operations. In hot-water systems, particularly if the water being produced is a brine, a highly corrosive environment exists. Likewise, if air is being used to drill the well, conditions are favorable for corrosion to occur. Proper procedures for corrosion monitoring and treatment with inhibitors, which are added to the drilling fluid system, may be implemented to reduce corrosion rates.

Environmental Problems

Protection of the environment is as important an aspect in geothermal drilling as it is in all types of operations. In some geothermal areas H_2S gas is encountered. To protect the environment of its toxic effects as well as to reduce the possibility of corrosion of the drill string, a sulfide scavenger is to be added to the drilling fluid.

Drilling Fluid Systems Used In Geothermal Wells

Various types of drilling fluid systems are used in geothermal drilling, and the type used depends on where the well is being drilled. In terms of fluid properties, however, the muds from one area may be quite similar to another:

1. Mud specific weights usually range from 8.8 to 10.0 lbs/gal (65.8-74.8 lbs/ft³).
2. API fluid loss is normally controlled at 10-20 cc/30 min.
3. Funnel viscosities may range from 35 to 45 s/qt.
4. pH normally is controlled in the 9.0-10.0 range.

The advantages and disadvantages of drilling fluid systems prevalent in geothermal drilling can be summarized as follows:

Bentonite-water system. This system is primarily composed of bentonite for viscosity and fluid loss control with additions of caustic soda for adjusting pH.

Advantages:
1. Can be an economical system.
2. "Simple" system in that very few products are added to the drilling fluid.

Disadvantages:
1. Not readily tolerant to contaminants, e.g., cement and salts, and solids.
2. Little or no heat stability which, at high temperatures, may result in high-temperature gelation and excessively high viscosities.
3. Fluid properties may be difficult to control.

Bentonite-lignite - caustic soda system. This system differs from the previous system in that lignite is incorporated in the fluid for greater thermal stability and better viscosity and fluid loss control.

Advantages:
1. A relatively economical mud system.
2. More stable than bentonite-water system.

Disadvantages:
1. Although more tolerant to contaminants than bentonite-water systems, not completely inhibitive.
2. May be subject to gelation if bottomhole temperatures are excessive.

Chrome lignite — chrome lignosulfonate system. Chrome lignite and chrome lignosulfonate are added to the drilling fluid to impart greater overall stability.

Advantages:
1. Greater inhibition is provided.
2. High-temperature stability.
Disadvantages:
1. More expensive mud system than preceding types.
2. Due to local environmental regulations, it can not be used in some areas.

Polymer systems. Polymer muds are predominantly composed of VAMA type polymers and poloyacrylate polymers. The use of these polymers results in bentonite extension and flocculation of drill solids thereby creating a low solids mud system.

Advantages:
1. Low solids system with consequent increased penetration rates.
2. Greater hole cleaning ability due to good rheological properties (lower N value).
Disadvantages:
1. Not tolerant of contaminants.
2. System may become too thin if overtreated.
3. Requires maximum utilization of solids control equipment.

Sepiolite system. Sepiolite mud contains sepiolite clay for viscosity control, modified polymers for fluid loss reduction, and caustic soda for adjusting pH. Sepiolite clay is substituted for bentonite because it does not flocculate at high temperatures.

Advantages:
1. Not susceptible to high-temperature gelation and flocculation.
2. Tolerant of contaminants.
3. Economical due to the lower maintenance costs, because this mud is relatively unaffected by contaminants and high temperature.
Disadvantages:
1. High-shearing mixing equipment is required to impart viscosity by the sepiolite.
2. Efficient solids control equipment is required.

Case Histories

Two case histories are presented to illustrate two previously discussed techniques and problems.

Dry Hot Rock — Typical Well

At the present time only one successful well has been drilled in a dry hot rock area (Caldera, N. M.) and many questions remain unanswered. These wells are about 10,000 feet deep. Basically the holes are drilled into hot granite formations, which are fractured hydraulically by pumping water under pressure into the formation.

This fractured zone is then intersected with the second hole. The temperature of the rock in this first experiment is approximately 204°C (400°F). It is believed that greater temperatures can be obtained by drilling deeper, 260°C (500°F) at 13,000 ft, although in some areas this has not been the case — the temperature has decreased. Water is pumped down one hole and circulates through the fractured rock, rises through the second hole, and flashes to steam as it reaches the surface. This is a closed-loop system.

One of the most important unsolved problems is "What is the life expectancy of such a reservoir?" If the life of such a reservoir is of sufficient duration this could be an alternate source of energy.

The most successful endeavor in harnessing geothermal energy is the utilizing of dry steam from a geothermal reservoir. As of this writing there are only three known areas that have a dry steam field with sufficient quantity of steam to economically produce electrical energy. These dry steam areas are the Geysers, California; Larderello, Italy; and Matsukawa, Japan.

Geysers — Typical Well

There is *no* typical well at The Geysers. The wells vary from 5,400 to 10,000 ft as can be seen from Table 5-1, although some wells are as shallow as 600 ft.

The equipment used in drilling a steam well in The Geysers is basically the same as used in drilling a semi-deep well. Some modifications, such as the blowout preventer stack and a muffler at the end of the blooie line to reduce the noise level when steam is encountered, have been discussed.

A typical well program proceeds as follows: Drill a 17½-in. hole to 150-200 ft and open up with a 26-in. hole opener. Set 20-in. conductor casing and nipple up with a 20-in. Hydril and flow line.

Drill a 17½-in. hole to approximately 2,500 ft and set a 13⅜-in. casing. (See Figure 5-3 diagram for typical blowout preventors.)

Drill a 12¼-in. hole to approximately 6,000 ft and set a 9⅝-in. liner. This can be a very critical point in a drilling program. There are three basic formations in The Geysers between the 13⅜-in. and 9⅝-in. casings. These are the Serpentine, the Greenstone, and the Graywacke. The Greenstone is one of the most difficult formations to identify and can be one of the most difficult problems in drilling a successful steam well. This is a sloughing formation that will infiltrate the hole. The Graywacke is the formation sought by geologists for a sufficient casing seat for the 9⅝-in. casing. Above this the hole must be drilled with mud as the drilling medium. However, air drilling must be initiated below the 9⅝-in. casing. This is the reason the casing seat is so critical. After this casing is set, drilling proceeds using air until sufficient production is encountered.

Conclusion

Geothermal energy is still in its infancy. It has, however, made great strides in helping to supply the consumer with the energy needed. In the future, this viable energy source will continue to add its part to the energy picture. Drilling in the outlying areas will necessitate some changes in the methods and equipment now needed to drill for steam or hot water. These challenges will be met with newer concepts and more efficient drilling processes and equipment.

References

Anderson, D.N., 1972. Geothermal Development in California, paper SPE 4180 presented at the SPE-AIME 43rd Ann. Calif. Reg. Meet., Bakersfield, CA, Nov. 8-10.

Budd, Jr., C.F., 1973. Steam Production at The Geysers Geothermal Field. *Geothermal Energy,* 129-144.

Combs, J. and Muffler, L.J.P., 1973. Exploration for Geothermal Resources. *Geothermal Energy,* 95-128.

Cromling, J., 1972. Geothermal Drilling Technology in California, paper SPE 4177 presented at the SPE-AIME 43rd Ann. Calif. Reg. Meet., Bakersfield, CA, Nov. 8-10.

Cromling, J., 1973. How Geothermal Wells Are Drilled and Completed. *World Oil,* Dec. vol. 177 (no. 7): 41-45.

Dench, N.D., 1970. Casing String Design for Geothermal Wells, U.N. Symposium on the Development and Utilization of Geothermal Resources, Piza, 1970. vol. 2, part 2.

Gardner, R., 1973. A report on geothermal drilling fluids. Dresser Industries Interoffice Correspondence, Sept.

Hamshire, L.R., Woertz, B.B. and Castrantas, H.M., 1977. Hydrogen Sulfide Emissions Reduction and Abatement While Drilling A Geothermal Well. Report, Union Oil of Calif., Union Geothermal Division, July, 12-16.

Harder, P., 1964. Steam: Hot New Target for U.S. Drilling Rigs. *Drilling,* Apr. vol. 25 (no. 6): 78-81.

Krause, H., 1977. A Report on Muds for Geysers. Dresser Industries Interoffice Correspondence, Oct.

White, D.E., 1973. Characteristics of Geothermal Resources. *Geothermal Energy,* 69-94.

Whiting, Robert L., 1974. Drilling and Production Techniques in Geothermal Reservoirs. SPE paper 5169, presented at the Deep Drilling and Production Symposium of the Society of Petroleum Engineers, AIME Amarillo, Texas, Sept. 8-10.

Herman H. Rieke III, TRW, Inc.
George V. Chilingar, University of Southern California

6

Casing and Tubular Design Concepts

Introduction

The cost of drilling and completing geothermal wells hinders development of such resources because it is currently two to four times that of conventional oil and gas wells of comparable depths. These higher costs are due to high temperatures, types of formations, and erosion and corrosion effects that are found during drilling in geothermal reservoirs. The concepts and guidelines presented here are of value in extending oil/gas well casing design principles to geothermal wells.

This chapter presents an overview of the state-of-the-art in concentric casing-tubular technology for geothermal and geopressured-geothermal wells. One of the most important considerations in drilling any type of geothermal energy recovery well is to provide adequate control through proper planning. The proper selection of casing or tubing size and grade is imperative to the success of completing and producing the well. Inasmuch as drilling in geothermal areas is accompanied by the problems of lost circulation, abnormal formation pressures, differential sticking of drill pipe, and crooked holes, the need for properly designed casing string configurations is well recognized. The attainable optimum condition is to design a casing-tubular program for a geothermal well that will provide the means to handle the following problems at minimum cost. This does not mean, however, that all mechanical constraints will be satisfied at minimum cost.

1. Strong fluid flows create cement washouts
2. High-velocity debris causes erosion and mechanical damage
3. Corrosive fluids
4. Scaling fluids
5. Pitting caused by acid gases and hot brines

6. Thermal stresses
7. Metal fatigue
8. Entrapped fluid expansion
9. Formation loading

Geothermal well completions pose a special problem in casing design, because of high flow rates of large volumes of superheated brines or steam. Wells are usually drilled in four types of geothermal reservoirs: vapor-dominated, liquid-dominated, hot dry rock, and geopressured-geothermal. The lithologies involved may be sedimentary, metamorphic, or igneous, and include abrasive, highly-fractured and vugular formations. Temperatures range up to 350°C and pressures are typically hydrostatic, except for the geopressured wells. In the producing zone, pressures are either greater or less than hydrostatic. Well depths can be as great as 4 km in hard rocks and up to 7 km in sedimentary basins. Formation fluids are usually corrosive and may contain H_2S and CO_2. Total dissolved solids may be as high as 200,000 ppm (Varnado, 1979).

The maximum load casing design concept is presented. By choosing the least expensive weight and grades of casing that will satisfy the burst loading, and upgrading the design using the proper sequence can result in the most inexpensive possible configuration. This configuration will fulfill the maximum load requirements for the geothermal and/or geopressured well. Design considerations must include more than the conventional concepts of burst, collapse and tension. In the geothermal environment, buckling, corrosion, and reduced yield strengths also have to be considered in designing a casing string.

Production of geothermally-heated fluids from these unfamiliar types of reservoirs, having unusual pore pressure conditions, demand unique solutions in completion technology.

Geothermal Well Completion Technology

Snyder (1978) pointed out that geothermal wells present a different set of problems to completion engineers with petroleum backgrounds. These difficulties are best discussed after a review of conventional completion practices used in the petroleum industry.

There are three basic types of completions used in oil and gas wells, which have also been used in geothermal wells:

1. Open-hole completion — The Geysers, California (Capuano, 1979)
2. Liner completion — Wairakei, New Zealand (Craig, 1961), and U.S. Gulf Coast
3. Perforated casing completion — Roosevelt, Utah (Rudisill, 1978)

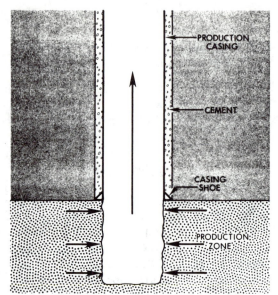

Figure 6-1. Well configuration of a standard open-hole (bear-foot) completion.

Open-Hole Completion

In open-hole completions, the production casing is set above the geothermal producing zone prior to drilling into the zone (Figure 6-1). Thus, the well is completed with the producing interval being open to the wellbore. Main advantages are minimum cost for completing the well and easy conversion to a liner configuration. Such a completion may lead to excess casing wear during cleanout operations because of debris created by high flow rates of steam.

Liner Completions

There are two types of liner assemblies: (1) the screen or slotted liner assembly type and the downhole perforated liner type (see Figures 6.2a and 6.2b). The screen or liner completion technique involves setting an uncemented screen or liner assembly below the production casing. This prevents fill-up of the wellbore with debris. In the downhole perforated liner completion method, the liner is cemented in place and then perforated or notched for production. This liner configuration gives better control on the influx of sand-size particles into the well; however, it is expensive because of the costs of in-place perforation, gravel-sand pack, or additional cementing required. Additional rig-time costs may also be considerable.

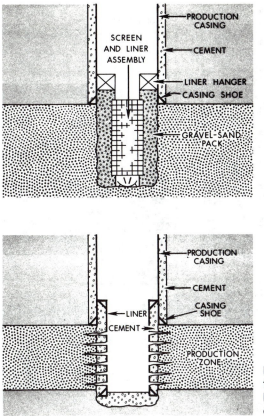

Figure 6-2. Well configuration for (A) screen and liner assembly and (B) perforated liner completions.

Perforated Casing Completions

In the perforated casing completion method, the production casing is cemented through the producing zone and the zone is selectively perforated (Figure 6.3).

Completion Problems

For geothermal wells, the general procedure used in implementing completion is as follows:

1. Locate the top of the zone producing hot water or steam.
2. Drill enough vertical section to permit commercial flow at existing permeability.

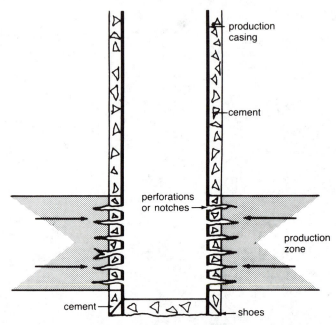

Figure 6-3. Well configuration for a standard downhole perforated or notched casing completion.

3. Isolate the commercial production from cooler waters using the specified casing and cement.
4. Remove any drilling fluid damage affecting the permeability of the producing formation.
5. Complete the open hole if possible in competent rock for least cost. If the producing formation is unconsolidated or incompetent, use a slotted liner or cemented casing.
6. Perforate or notch cemented casing with adequate holes if necessary.
7. Install the appropriate tubular goods.

Difficulties encountered in completion practices are thought to be best handled by implementing the following concepts.

Wellbore cleanup. Stimulation of the production zones by temperature-resistant acids, pressure washes, or water injection has been successful in removing shallow formation damage in the critical flow area immediately around the wellbore. Cased and perforated completions allow greater control for selectively testing and stimulating individual production horizons.

Table 6-1
Selection Guide for Optimum Geothermal Well OD Casing Sizes

Steam flow, (tons/hr)	Open hole (ins.)	Casing size O.D. (ins.)	Casing classification
10 to 25	17	13⅜	Surface
	12¼	9⅝	Intermediate
	8⅝	7	Production
	6¼	4½	Slotted liner
25 to 50	18	16	Surface
	14¾	11¾	Intermediate
	10⅝	8⅝	Production
	7⅝	6⅝	Slotted liner
50 to 80	22	18	Surface
	17	13⅜	Intermediate
	12¼	9⅝	Production
	8⅝	7	Slotted liner

After Matsuo, 1973

Casing size selection. The primary consideration in casing design is to assure presence of an adequately large diameter to permit commercial flow from a given depth with a given formation productivity index. It is also necessary to keep friction losses through wellheads, flow lines, and valves to a minimum to assure maximum flow. Deep-well centrifugal pumps may be needed to increase hot water flow rates. Wellheads are usually simple, consisting of a large valve and a cross or tee to the flowline (Varnado, 1979). Complex wellheads are required for the geopressured-geothermal wells.

Matsuo (1973) provided a guide for estimating the optimum well diameter for open-hole and casing (Table 6-1). A diameter that is too small creates a high resistance to upward fluid flow, whereas too large a diameter could result in the well being unproductive. Although procedures for calculating the optimum diameter are not lacking, they cannot be applied in a meaningful manner to wildcat wells or even to some in-fill wells.

Plugging. Completion problems can occur in any phase of the producing system. Productivity loss between formation and wellbore can be caused by:

1. Plugging of producing formations
2. Scale buildup within liner slots or perforations
3. Swelling or dispersion of clay particles in sandstones
4. Compaction around perforations

Perforations may be ineffective due to limited performance of temperature sensitive charges.

Corrosion. Completion problems can arise in wells that produce wet steam, because it can erode casing as the high-speed fluid strips water film from the casing surface. Thermal and corrosion effects can cause failure of liners. Higher quality casing with flush joints must be used to reduce the problem. When hot and acidic brines or H_2S are present, downhole equipment is subject to embrittlement and corrosion. Existence of these conditions also requires special procedures and training for the crews. In addition, more expensive, corrosion-resistant materials must be used, and additives may be required in the drilling fluid to help control corrosion.

Scale buildup and diameter reduction of casing can be significant, particularly in portions of the string where obstructions, such as packers, hangers, and pump casings, cause turbulence and pressure instability. Mixing of waters from different zones can cause scaling. Although single-diameter strings are desirable, they require special methods to cement effectively above a slotted casing section without contaminating productive zones. Cemented and perforated casing would be the most reliable single-diameter system to install.

Wellbore stability. Borehole stability in less competent sands, such as those found in the Imperial Valley, California, may require positive support from cemented casing or gravel-packed liners. Compaction within the zone and subsidence of strata above the zone as a result of fluid withdrawal can affect borehole stability, cause casing and/or cement failure, or create surface problems.

Temperature effects. The high-temperature environment encountered in geothermal wells produces adverse effects on drilling fluids, casing, cement, bits, and elastomeric materials (Varnado and Stoller, 1978).

Drilling-fluid-related difficulties form the single most frequently cited category of geothermal drilling problems. Lost circulation is a pervasive difficulty in the geothermal field because of the highly fractured formations in which geothermal resources are found. Fluids that remain stable at high temperature often do not have adequate filtration characteristics for control of lost circulation. Furthermore, conventional muds at elevated temperatures tend to gel when circulation is stopped for tripping, logging, and running casing. This can lead to induced problems, such as stuck tools and failures of the drillstring due to the differential sticking. Expensive procedures are required to correct these failures.

Drilling fluids can also affect the quality of the cementing job, because they tend to contaminate the cement. In addition, muds that have gelled and thickened leave excessive filter cake on the casing and formation, which inhibits good cement bonding. Also, a thickened mud can cause the cement to channel behind the casing and can give rise to large uncemented regions. The water in water-based mud can vaporize at high temperatures. The pressures that build up as a result have the potential for causing collapse of the casing.

Thomas (1967), Karlsson (1978), and Snyder (1979) discussed in detail the decrease in yield and/or maximum ultimate tensile strength of casing caused by high temperature in the well. Various casing grades used in geothermal wells, such as N-80, P-105, C-75, and P-110 up to 900°F, when compared to strength at 100°F, show reduction in their rated yield strengths starting at 600°F (Snyder, 1979). The reduction in ultimate tensile strength is more marked than the drop-off in yield strength. Thomas (1967) reported that above 650°F the yield/tensile ratio tends to increase as the temperature increases. Fatigue failure can occur with temperature cycling and repeated contraction and elongation. High-strength casing is inherently more sensitive to work hardening and notch failure (Snyder, 1979).

Currently the best available elastomers fail at temperatures of 175-225°C depending upon use and environment. Inasmuch as these materials cease to be elastomeric when subjected to high temperatures and pressures, their value in geothermal applications is limited. Unfortunately, elastomers are important to nearly every aspect of drilling.

Well control. The master valve is attached to the production casing, and a blowout preventer (BOP) is mounted on this valve to allow drilling into the producing zone. When the well is completed, the valve provides control after the BOP is removed. Geothermal wellheads are subject to severe erosion, corrosion, and mechanical damage. The single master valve and wide-open casing system would directly expose the producing zone to the surface in case of valve failure (leakage and cracking). Thus, protection of wellhead components from weakening or damage is very important. Well control in deep geopressured-geothermal wells is much more demanding than in simple shallow hydrothermal wells.

Casing Design Procedure

Geothermal wells are typically shallow, being less than 9,000 ft in depth; however, geopressured-geothermal wells located in the Tertiary Gulf Coast basin can range in depth from 13,000 ft to >20,000 ft. Snyder (1979) stated that the factors limiting casing diameter are cost of drilling and cementing in large-diameter holes and collapse rating limitations. In order to properly evaluate the maximum loads imposed on different casing types, each one has to be considered separately. In geothermal well completions there are typically five types of casing used: conductor casing, surface casing, intermediate casing, production casing, production liner, and tieback casing.

An example of geothermal well completion is illustrated in Figure 6-4. This configuration is used in The Geysers, California, where competent metamorphosed sediments allow air-drilled open-hole completions (Snyder, 1979, p. 81). Casing heads and master valves frequently are attached directly to the tieback string by killing the well with clean water and cutting off surface pipe after the blowout preventers (BOPs) are removed.

Acording to Snyder (1979), a common variation of the design illustrated in Figure 6-4 is found in the Imperial Valley, California, with production through surface pipe and liner without a tieback. Larger-diameter pipe allows large-diameter centrifugal pumps to be set at a depth of 1,000 ft (or deeper). Slotted liners over long uncemented intervals give near-open hole productivity with control of sloughing sands and shales (Snyder, 1979, p. 82).

Cigni et al. (1976) pointed out that in order to operate economically, the producing wells must have high flow rates, as the enthalpy of the endogenous fluids is relatively low (650 to 700 cal/kg) even under the best circumstances.

For medium-depth geothermal wells the most common casings are API J55 grade steel, with special buttress joints and having a greater thickness than calculated theoretically. The reason for this is that they must withstand as much as possible the stresses caused by thermal phenomena, which are very difficult to control. The J55 steel seems to be particularly suitable for use in corrosive environments.

The most common production casing used in Italy is 13⅜-in. API grade casing in a 16-in. hole. In the exploratory or relatively deeper boreholes, this casing is run as an intermediate column for safety reasons. Production casing then measures 9⅝ ins. (API) in the 12¼-in. hole. Drilling is then completed with an open hole having a diameter of 12¼ ins. or 8½ ins., respectively, as no liners are needed. Figure 6-5 shows two typical medium-depth geothermal well completion profiles.

Each one of the above mentioned casing types is discussed separately. By initially selecting the least expensive weights and grades of casing, which satisfy thermal stress and burst loading, and then upgrading the casing using the prescribed sequence outlined below, the resulting design will provide adequate protection for the well's geothermal reserves. The least expensive possible design, which fulfills all maximum loading requirements for the steam, hot water, or hot brine geopressured wells, is selected. Generally, geothermal wells fall into the "conventional" completion category in which the production casing size is larger than 4½ ins. OD (Table 6-1).

The procedure for casing design can be summarized as follows:

1. Select the appropriate diameter for the hot-water or steam well casing.
2. Design for maximum load at minimum cost.

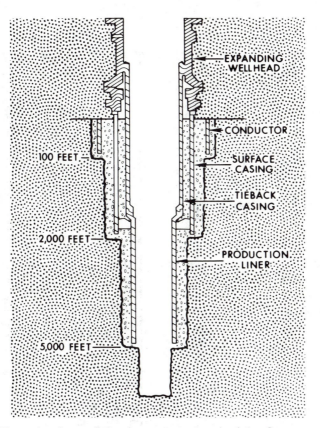

Figure 6-4. Example of completion in competent rock of the Geysers reservoir, which allows open-hole completion in geothermal wells and air drilling. Imperial Valley completions in California exclude the tieback, but have long slotted liners opposite softer sands. (After Snyder, 1979)

3. Use low-to moderate-strength steels for maximum resistance to corrosion, work hardening, and possible H_2S stress corrosion cracking.
4. Use the API buttress type couplings (or other premium shouldered, recessed, free-bore connections) to prevent thermally induced stress failures. In the latter case, eliminating coupling recesses provides better protection against corrosion.

Different applications of casing strings in geothermal wells dictate different loading configurations. The burst loading should be considered first in the design, inasmuch as it dictates a majority of the maximum load on the

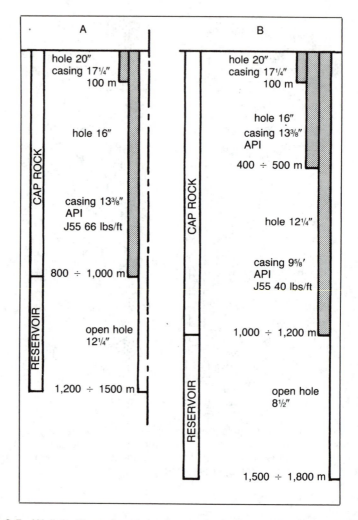

Figure 6-5. Well profile and casing program for Italian geothermal well. Standard casing is shown at A, whereas casing with intermediate string is shown at B. (Cigni et al., 1976; courtesy of United Nations.)

strings. The concept for designing consists of an iterative determination of the weights, grades, and section lengths for prevention of the burst failure. The collapse load is evaluated next and the previously selected casing string sections are upgraded if necessary. The tension load is evaluated after the weights, grades, and section lengths have been found to satisfy the requirements for withstanding both burst and collapse loadings. The tubular goods are upgraded, if necessary, and the buttress type couplings are selected. Final step in the design procedure is to check for any biaxial reductions in the collapse, burst, and tension loads and correct for thermal stress. If these reductions show that the strength of any part of the section is less than the potential load, the casing sections in question are upgraded to withstand this loading.

Design Procedure for Casing Strings

The selection of number and type of casing string, size, setting depth, and the determination of the downhole environment are important requirements in properly planning a geothermal well. Thomas' (1979) early work on the high-temperature tensile properties of casing and tubing provided an initial rationale for developing a geothermal casing string design procedure. Experience from drilling in high-temperature geothermal areas of Iceland indicated that expected bottomhole temperatures and pressures may be assumed to follow the boiling curve based on the assumption that water is at boiling condition for any depth (Karlsson, 1978). Dench (1970) also used the saturation pressure and temperature values based on the boiling water column in designing casing strings for the geothermal field in New Zealand. The design values were modified if pressure and/or temperature values exceeded those defined by the boiling water column with respect to depth. In some New Zealand fields pressure approached a hydrostatic pressure of a cold water column in the well (Dench, 1970).

The acceptable design of casing string for dry hot rock wells follows the basic principles set forth in the petroleum literature. Current oil field practice utilizes the primary design aids from the American Petroleum Institute (API) specifications (API Spec. 5A, 5AC, 5AX) and bulletins (API Bulletins 5C2 and 5C3) for manufacturing, testing, minimum performance properties: and applicable API standards for service for oil field tubular products, such as drill pipe, casing and tubing. Data on non-API specified tubulars are obtainable from the manufacturers. Reasonable failure modes are assumed to be only axial tension or compression within the wall or joints of the casing, bending within the wall, and/or a combination of the two. Collapse or burst failure should not occur because excessive pressures or radical changes in fluid densities within or external to the casing are not to be expected in dry hot rocks drilling areas. A universal procedure to extend the range of design coverage for geothermal wells is available.

The design context for the five types of casing strings used in these wells is discussed here employing the maximum load design procedure as developed by Prentice (1970). This procedure is especially applicable to deep wells such as those projected to test and produce the geopressured-geothermal resource in the Gulf Coast, U.S.A.

Definitions of the various casing strings are presented to provide an understanding of the different purposes the various strings serve.

Conductor Pipe

The first of the outermost casing strings is the conductor pipe, which is known sometimes in other countries as surface casing (Figure 6-4). This string can vary from 40 to 80 ft in length and from 18 to 30 ins. in outside diameter. Commonly, an 18-in. OD string is used in shallow wells, whereas a 24-to 30-in. OD string is employed in deep wells. The conductor casing is commonly cemented back to the surface but, in some cases, it may be driven. This pipe lends structural support to the well and wellhead.

Conductor casing in a geothermal well is not designed for burst, collapse, or tension loads. Design consideration, however, must be given to the compressive loads supplied by all subsequent string weights which may be transmitted to the conductor pipe. When surface casing is not strong enough to support loads imposed by subsequent strings, there are means of transferring that load to the conductor pipe available. Onshore, it is possible to provide a bearing plate that will support the wellhead and transfer the casing load to the surrounding ground and the conductor pipe.

In marshy and offshore areas, the load can be transferred in a similar manner to conductor or drive pipe. In this case, drive pipe is preferred. When pipe is driven with 250 blows/ft with a diesel hammer, there is little possibility of subsidence with even the heaviest casing strings. Additional support given to the well site can be also supplied by grouting the area around the conductor hole, which is practiced at Wairakei, New Zealand.

Goins et al. (1966) provided a procedure for calculating the maximum allowable compressive loads on the conductor string.

Surface Casing

The next string after conductor pipe is the surface casing or anchor casing (Figure 6-4), which can vary in length from 100 to 400 ft. This string serves the multiple purpose of providing a sound foundation, if not provided by the conductor string, and protection of fresh-water aquifers from contamination as required by law in the U.S.A. The size of most commonly used surface casing in geothermal wells is 13⅜ ins. OD (Table 6-1).

Surface casing strength is increased if the subsequent casing strings are cemented to the surface or otherwise fastened. Even if the cement job between the conductor casing and the surface casing is not perfect, the bonding strength of the commonly used cements is such that only a relatively few feet of good cement are required to provide ample vertical support. Sufficient lateral support can be provided by even a poor cement job.

Cement must be in position around the surface casing to the top and inside the conductor pipe. The cement serves to transfer casing load to the outer strings. Farris (1946) showed that cement having an 8-psi tensile strength and 100-psi compressive strength would support a load of 20,000 lbs per lineal bonded foot. Thus, 100 ft of cement around the top of a surface casing string would support approximately 2,000,000 lbs of casing. Even assuming that the cement around the top joints of the surface string is not continuously bonded, 100 ft of cement for 500,000-lbs casing load and 50 ft of cement for shallow wells seem to be adequate. All cements used in geothermal wells have strengths much higher than 100 psi, i.e., compressive strength of cement used by Farris. At the present time, 500 psi is a practical minimum compressive strength of cements.

Another factor that may impose a significant limitation on surface casing loading is the load that may be safely set on a wellhead. Casing joints are about equally as strong in compression as they are in tension (Goins et al., 1966). The load set on the wellhead would be equal to the rated tensile strength of the surface casing when the casing alone supports the load. Designers can safely use the API or manufacturer's rated tensile strength when designing surface casing for compressive loads.

Utilization of API buttress type or premium shouldered connections in geothermal wells greatly increases the load-carrying capacity of the strings. API round thread couplings cannot tolerate extreme compressive and tensile loading (Varnado, 1979). Most pipe used for surface casing in oil and gas wells is equipped with these threads. Such threads are about 50% efficient, i.e., the string is approximately twice as strong as the threads (Goins et al., 1966). Several joints with buttress type couplings at the top of the surface string will almost double its load carrying capacity. Premium couplings (non-API) offer metal-to-metal sealing and shouldering features, which make them ideal for deep high-temperature steam wells (Varnado, 1979). While premium couplings are more expensive, they are often more cost effective.

If there is unsupported casing for some appreciable distance below the wellhead, a design approach used to calculate the allowable load by Casner (1963) can be employed. The casing string is considered to be a long, slender column. The column buckling of the surface casing is determined using equations presented in Table 6-2. These equations can help to ensure that in deep geothermal wells ($\sim$ 9,000 ft) the weight of subsequent pipe strings does not cause columnar buckling of the surface casing. This depends not only on the grade and size of the casing, but also on the slenderness ratio (length/radius), which usually ranges from 0 to 400. Local hole conditions may cause greater eccentricity or crookedness than the equations in Table 6-2 provide for. These equations, however, provide an excellent guide for solving the columnar-strength problem in surface casing. Table 6-3 provides the values of the radius of gyration for common sizes of oil-well casing.

Table 6-2
Equations for Column Buckling of Surface Casing

Casing grade (API)	Ranges of slenderness ratio, L/r	Average working stress
F-25	0-202	$P/A = 15{,}838 - 0.1890 \, (L/r)^2$
	202-400	$P/A = \dfrac{32{,}200}{1 + (1/12{,}800) \, (L/r)^2}$
H-40	0-160	$P/A = 24{,}615 - 0.4838 \, (L/r)^2$
	160-400	$P/A = \dfrac{55{,}300}{1 + (1/7{,}290) \, (L/r)^2}$
J-55	0-136	$P/A = 33{,}846 - 0.9147 \, (L/r)^2$
	136-400	$P/A = \dfrac{78{,}300}{1 + (1/5{,}100) \, (L/r)^2}$
N-80	0-113	$P/A = 49{,}231 - 1.935 \, (L/r)^2$
	113-400	$P/A = \dfrac{117{,}000}{1 + (1/3{,}400) \, (L/r)^2}$
P-110	0-96	$P/A = 67{,}692 - 3.659 \, (L/r)^2$
	96-400	$P/A = \dfrac{163{,}000}{1 + (1/2{,}430) \, (L/r)^2}$

After Goins et al, 1966, table 2, p. 23
L = length, in.
r = radius of gyration, in.
P = working stress, lb
A = area of the pipe, in.^{2}

Example Design Problem

What is the greatest weight of casing that can be safely landed on the wellhead without column failure? The geothermal well is to be drilled in an area where surface conditions are competent.

An example of casing string design is presented for the Raft River geothermal production well No. 5 (RRGP-5), Raft River Valley, Idaho (Miller and Prestwich, 1979).

Table 6-3
Radius of Gyration for Some Common Sizes of Oil Well Casing Used in Geothermal Wells.*

Grade	Casing O.D. (ins.)	Nominal weight (lbs/ft)	Wall thickness (in.)	Radius of gyration (ins.)
J-55, K-55	4½	11.60	0.250	1.50
P-110	7	29.00	0.408	2.33
J-55, K-55	7⅝	26.40	0.328	2.58
J-55, K-55	8⅝	36.00	0.400	2.91
P-110	9⅝	53.50	0.545	3.22
J-55, K-55	9⅝	40.00	0.395	3.27
J-55, K-55	10¾	51.00	0.450	3.65
J-55, K-55	11¾	60.00	0.489	3.82
J-55, K-55	13⅜	54.50	0.380	4.33
J-55	16	62.53	0.375	5.53
	18	70.59	0.375	6.23
H-40	20	78.60	0.375	6.94
	24	94.62	0.375	8.35

*Essentially a linear relationship exists between casing OD and the radius of gyration (r) for a given grade. Thielsch (1973) presented the following equation for calculating r:

$$r = 0.25\sqrt{D^2 + d^2}$$

where D is the OD of the casing, and d is the ID of the casing

Surface Casing: 20-in. H-40 ST&C (94 lbs/ft); length of 212.93 ft weighed 20,015.4 lbs

Intermediate Casing: 13⅜-in. K-55 ST&C (54.5 lbs/ft); length of 1510.3 ft weighed 82,329.3 lbs

Production Casing: 9⅝-in. K-55 Butt (36 lbs/ft); length of 2,121.04 ft weighed 76,357.4 lbs

Total weight: 178,702.1 lbs

1. Given:
   a. Surface casing — 20 in. OD
   b. Grade — H-40
   c. Weight — 94 lbs/ft

    d. Length of unsupported joint — 42.90 ft

    e. Unsupported interval — 20 ft above the cellar floor

2. Method of calculating the weight:

    a. First calculate the slenderness ratio for the unsupported length of surface casing. From Table 6-3, the radius of gyration for 20-in. casing is 6.94 ins. Column length is 20 ft (= 240 ins.) of 20-in. OD casing and the L/r ratio is equal to 240 ins. divided by 6.94 ins., or 34.58.

    b. Using Table 6-2, H-40 casing with a slenderness ratio of 34.58 has an average working stress (P) of 647.062 lbs. Using equations:

$$P/A = 24{,}615 - 0.4838 \, (L/r)^2$$

where A is area of the pipe in ins.^{2} (use manufacturer's casing tables),

$$P/26.92 = 24{,}615 - 0.4838 \, (34.58)^2$$

and

$$P = 647{,}062 \text{ lbs}$$

    c. The surface casing can safely support a column load of a casing and tubing weight not to exceed 647,082 lbs, which is sufficient in the Raft River wells.

It is anticipated that this method will be needed only in rare cases, because in most geothermal wells cement, which is used at the surface, adequately supports the casing. Casing is considered as a short column in which the casing tensile strength and the bearing strength of the ground control the design.

Burst and collapse can occur in surface casing. Collapse can occur when a lost circulation zone is encountered during drilling below the surface casing, and the drilling fluid is lost. The design approach concept is discussed in the section on intermediate casing below.

Intermediate Casing

The intermediate casing string necessitates the most complex set of considerations. The approach used by Prentice (1970) is to first evaluate the burst loading by establishing values for the surface and bottomhole burst limits. Bursts are considered only for geothermal wells in reservoirs having pressures greater than the hydrostatic, such as those present in the Gulf Coast geopressured-geothermal wells. Reservoirs with subnormal pressures do not require burst design calculations.

Burst calculations

These calculations can be summarized as follows:

1. The surface burst pressure limit is normally set equal to the related working pressure of the wellhead with BOP's.
2. Burst limitation at the bottom of the intermediate casing string is calculated as a pressure equal to the predicted fracture gradient (expressed as mud weight in pounds per gallon, ppg) of the rock immediately below the casing shoe. A design factor of 1.0 ppg (mud weight equivalent) is assumed. Maximum load will occur under a gas kick condition. Kick loading characteristically occurs when two or more fluids are present in the well while drilling below the intermediate string. Inasmuch as design is made for maximum loads, the fluids considered are the heaviest drilling fluid projected for use below the casing string and gas, as the single influx fluid. Such a condition could occur in the geothermal wells being drilled into high fluid pressure zones of the Gulf Coast area U.S.A. According to Prentice (1970), the position of these fluids in the wellbore is important. Figure 6-6 illustrates this concept. In the case of line 2, where the heaviest mud is on top and gas is on bottom, there is greater burst loading on the casing than in the case of line 1. Thus, line 2 is the proper maximum load design line.
3. The lengths of the drilling fluid and gas columns are determined by assuming a gas gradient value appropriate for the area (0.115 psi/ft for Gulf Coast). In the case of water and steam see Table 1-10. Column lengths are calculated by solving the following simultaneous equations (Prentice, 1970):

$$x + y = D \qquad\qquad (6\text{-}1)$$

$$p_s + G_m + yG_g = 0.052 \, (G_f + SF)D \qquad\qquad (6\text{-}2)$$

where: x = length of drilling fluid column, ft, in.
y = length of gas (steam) column ft, in.
D = setting depth of the casing, ft, in.
p_s = surface pressure, psi, kPa.
G_m = gradient of heaviest mud weight to be used, psi/ft, kPa/m.
G_g = gas gradient (assumed to be 0.115 psi/ft for the Gulf Coast area).
G_f = fracture gradient, ppg.
SF = design factor — 1.0 ppg
0.052 = conversion factor (psi × gal/lb × ft)

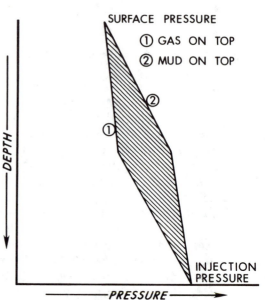

Figure 6.6. Position of the load line relative to the positions of the fluids in the well. (After Prentice, 1970; courtesy of Society of Petroleum Engineers.)

4. Next the values for the surface pressure, drilling fluid-gas boundary point, and the pressure at the bottom of the intermediate casing string can be plotted as the load line shown in Figure 6-7. The burst load line shows the burst pressure values at every increment of depth which act on the intermediate casing.

5. A load resisting this burst is normally calculated based on the fluid known as the backup fluid, if any, occupying the annular space behind the casing string. For Gulf Coast wells, the backup fluid is assumed to have the density of salt water (gradient = 0.465 psi/ft). The burst resisting backup load is plotted and subtracted from the burst load line (Figure 6-7). The plotted line is termed the resultant burst load line. Sometimes a design factor of 1.1 is applied to allow for wear in crooked or deviated holes. This will shift the resultant burst load line to the right. If no design factor is necessary, then the resultant burst load line is the design line.

6. Starting at the top of the final burst design line, the published values of minimum yield strength for the least expensive weight and grade of casing, which exceeds the calculated design load, are plotted (Figure 6-8). The length of the casing section is determined by the intersection with the design line. The minimum yield strength of the next applicable casing weight or grade is plotted to intersect the abscissa (pressure axis). This procedure is repeated until the string is completely designed for burst load. The grades, weights, and section lengths of the casing are then set aside pending evaluation of the collapse loading.

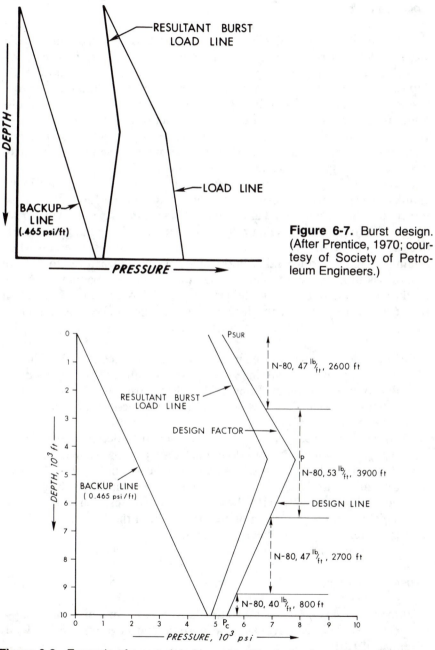

Figure 6-7. Burst design. (After Prentice, 1970; courtesy of Society of Petroleum Engineers.)

Figure 6-8. Example of a completed burst loading design for a 10,000-ft well. An economic decision must be made whether to complete the last 800 ft with 40 lbs/ft, N-80 casing, or to extend 47 lbs/ft, N-80 casing to total depth (TD).

Collapse Calculations

Collapse calculations are of importance in geothermal well design. Snyder (1979, p. 82) described the possible fracture modes leading to casing collapse. One basic limitation in geothermal wells is the inability to consistently and reliably cement casing strings solidly from bottom to top due to problems such as lost circulation, hole gage control, mud gellation, erosion-dissolution of cements by hot water, and tool failures.

Principles of point loading and casing collapse by downhole fault movement or sloughing-flowing salt or shale formations are documented in petroleum engineering literature (Suman and Ellis, 1977).

The collapse load is imposed by the fluid in the annular space. Fluid in the annular space between the intermediate casing and the formations is assumed to be the heaviest mud in which the casing string is expected to be run. The maximum collapse loading will occur when the drilling fluid level inside the casing drops due to the loss of circulation, which is common when drilling in highly-porous and/or permeable volcanic or fractured rocks with subnormal fluid (pore) pressures. Maximum collapse loading in a well results when minimum backup fluid is present. This occurs when circulation is lost while drilling with the heaviest drilling fluid below the emplaced string. The following procedure can be followed:

1. Construct the backup line and subtract it from the collapse load line to obtain a resultant load line (Figure 6-9). Prentice (1970) pointed out that only the lower sections of the casing string will be affected by collapse considerations; therefore, using a full column of fluid as the backup fluid is valid.
2. Apply the appropriate design factor, if any. If no design factor is used, then the resultant collapse load line is the collapse design line.
3. Plot and check the collapse resistances of the sections dictated by burst loading (Figure 6-10). If the collapse resistances fall below the collapse design line, the section is to be upgraded for collapse. Thus the grades, weights, and section lengths that satisfy both the burst and collapse maximum loads have been determined.

Tension Calculations

With the weights, grades, and section lengths based on burst and collapse design known, one can evaluate the positive and negative tension loads (Prentice, 1970).

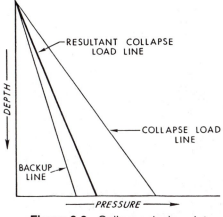

Figure 6-9. Collapse design plot.

Buoyancy cannot be overlooked in a maximum load design because of the way in which the buoyant force is applied to a casing string. The burst and collapse resistances are altered by the effect of biaxial stresses. Prentice (1970) discussed the effects of buoyancy in detail.

Buoyancy, or reduction in string weight, as noted on the surface is actually the result of forces acting on all the exposed, horizontally-oriented areas of the casing string. The forces are equal to the hydrostatic pressure at each depth times all the exposed areas. These forces are defined as negative if they are acting upward. The areas referred to are the tube-end areas, the shoulders at points of changing casing weights, and, technically speaking, to a small degree, the shoulders on collars.

Prentice (1970) presented a diagram explaining the forces acting on the exposed areas of a casing string (Figure 6-11). According to Prentice (1970), "buoyancy" can act upward, downward, and the summation of forces may be equal to zero.

In the opinion of the writers, however, the buoyant force on the casing is equal to the weight of displaced drilling fluid and it always acts upwards. This opinion is based on some preliminary experiments.

Reduction in load, as observed at the surface, is equal to the weight of the displaced fluid or can be determined by using the "buoyancy factor" method. Tension loadings, however, can differ greatly.

According to Prentice (1970), once the magnitude and location of the forces are determined, the tension load line may be constructed graphically (Figure 6-12). It is noteworthy that more than one section of the casing string may be loaded in compression.

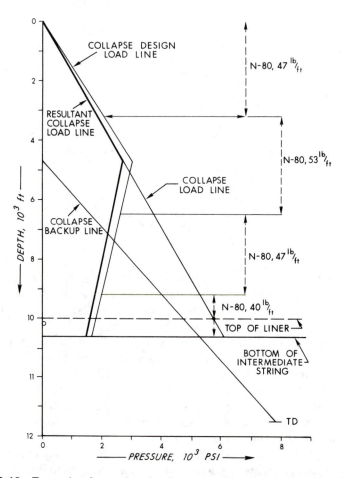

Figure 6-10. Example of completed collapse loading design data (see Figure 6-8). Both the burst and collapse conditions are now satisfied.

Obtaining a design line for tension, it is recommended by Prentice (1970) that a design factor be used with a conditional minimum overpull value included. Casing capacity loads are based on ultimate material strengths. The 1.6 safety factor is normally associated with yield strength which in turn is associated with tubing design. Prentice (1970) has recommended values for the safety factors: 1.6 for the design factor and/or 100,000 lbs of overpull, whichever is greater. Snyder (1979) recommended the following design factors: 1.10 for internal yield; 1.125 for collapse; and 1.80 for casing, based on ultimate tensile strength (a catch-all factor). This is a compromise with the "Marginal Loading" concept of Goins et al. (1966), and allows for safely pulling on stuck casing to some definite predetermined value such as 100,000 lbs in the above case.

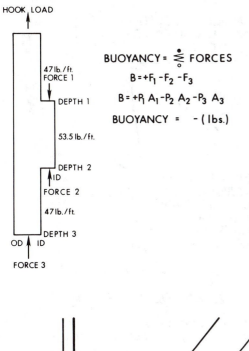

BUOYANCY = $\overset{\bullet}{\underset{\circ}{\lessgtr}}$ FORCES

$B = +F_1 - F_2 - F_3$

$B = +P_1 A_1 - P_2 A_2 - P_3 A_3$

BUOYANCY = $\quad - (lbs.)$

Figure 6-11. Effect of buoyancy. The forces are acting at each exposed area of a casing string, with the resultant loading indicated as negative tension (compression). Forces acting on the areas of collar shoulders are for practical purposes negligible in casing design. (After Prentice, 1970; courtesy of Society of Petroleum Engineers.)

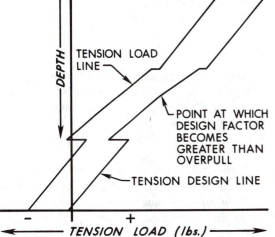

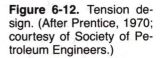

Figure 6-12. Tension design. (After Prentice, 1970; courtesy of Society of Petroleum Engineers.)

The graphical representation of this combination of design factors, which is shown in Figure 6-12, is labeled the "tension design line."

The weakest part of a joint of casing in tension is normally the coupling; therefore, the tension design line is used primarily to determine the type of coupling to be used (Prentice 1970). The least expensive coupling strengths that satisfy the design can be plotted and the proper couplings that must be used can be selected (Figure 6-13).

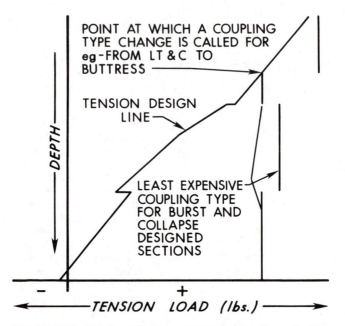

Figure 6-13. Relation of tension design to coupling selection. (After Prentice, 1970, courtesy of *Journal of Petroleum Technology, SPE.*)

Thus, the entire string is designed for burst, collapse, and tension, and the weights, grades, section lengths, and coupling types are known. At this point, it is necessary to check the reduction in burst resistance and collapse resistance caused by biaxial loading (Prentice, 1970).

Biaxial Load Calculations

The tension load line, which shows the relationship between tension loading and depth, is used to evaluate the effect of biaxial loading. One should not use the design line. By noting the magnitude of tension plus or minus compression loads at the top and bottom of each section, the strength reductions can be calculated using the biaxial yield stress ellipse (Figure 6-14) (Holmquist and Nadai, 1939). With the reduced values known at the end of each section, a new strength line can be constructed by connecting the end points with a straight line. Should the reduced values indicate an under-design, the section should be upgraded. One should be aware of the following facts:

1. Tension increases burst resistance
2. Tension decreases collapse resistance
3. Compression increases collapse resistance
4. Compression decreases burst resistance

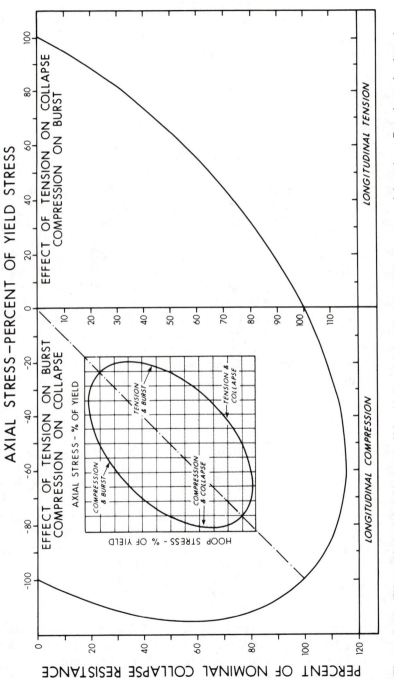

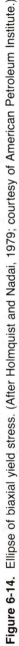

Figure 6-14. Ellipse of biaxial yield stress. (After Holmquist and Nadai, 1979; courtesy of American Petroleum Institute.)

Some service companies do not recommend the use of the biaxial increase in burst and collapse resistance due to an axial load increase or decrease. This assures the conservative approach, which should be required in a geothermal-geopressured well design.

Production Casing

Burst Calculations

Surface pressure is maximum when the well is shut-in. The surface tubing pressure is added as a burst load over the entire length of the production casing. Thus, the worst possible well condition is to have a gas leak in the tubing near the surface, which results in the introduction of the bottomhole pressure minus the gas pressure over the entire interval of the production casing.

Another assumption is that the density of the packer fluid is equal to the weight of the mud (the mud the string was run in) remaining in the annular space behind the casing (Prentice, 1970). This is not consistent with the previous discussion, but makes a good case for using low density packer fluids. According to Prentice (1970), the result of this assumption is that the effects of the load and of the backup fluids cancel each other and, consequently, the casing has no burst or backup loads.

Fracture gradient and pressure at bottom of casing are not used in designing production casing (Figure 6-15).

Collapse Calculations

Owing to the possibility of tubing leaks, presence of artificial lift, and plugged perforations, the collapse design for production casing incorporates no consideration for backup fluid, and the string is designed assuming that it is dry inside. The collapse load is determined by the hydrostatic pressure of the heaviest mud the string is to be run in, and the design factor is applied directly to this load (Prentice, 1970). The resulting design line is used to check and upgrade the burst design as necessary. The tension and biaxial reductions are evaluated as previously discussed (Prentice, 1970).

Tubing Design

Tubing design procedure is essentially the same as that used for casing design. Tapered tubing strings are commonly used in deep geopressured wells. Because of the frequency with which tubing must be pulled, uniform strings are more desirable than tapered strings. It is difficult to keep a tapered string in the proper sequence of weight and grade. Goins et al.

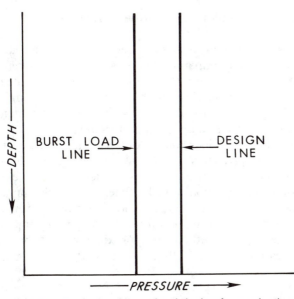

Figure 6-15. Example of burst load design for production casing.

Table 6-4
Uniform Weight Upset Tubing String Setting Depths in Air

| Type | Safety Factor | | |
	1.75	1.60	1.50
J-55	8,000	9,600	10,200
N-80	12,700	13,900	14,800
P-105	16,700	18,300	19,500

After Goins et al, 1962

(1962) provided data on safe tensile limit for uniform weight upset tubing set in air (Table 6-4). For most cases a tension design factor of 1.60 is recommended by Goins et al. (1962) for tubing. Under no circumstances should the tubing be subjected to pressures higher than its rated pressure divided by 1.1 without testing. Burst is a more serious factor in the case of tubing than it is in the case of casing.

Stress analysis of tubular strings in thermal wells is based on the knowledge of the equilibrium temperature distribution and the effects of mechanical buckling. The design techniques discussed below must be applied to the maximum load design results, especially in designing deep, hot, geopressured wells.

Design for Thermal Stress

Additional design techniques must be used to prevent mechanical problems associated with high-temperature effects on casing, tubing, and associated downhole equipment, after the string is designed using the maximum load approach. The following three problems are of utmost importance in the case of tubular goods: *compression — tension* failure in cemented pipe; *radial expansion and cement damage;* and *buckling.*

Leutwyler and Bigelow (1964) pointed out that one of the keys to successful solution of these problems is accurate knowledge of the downhole environment. Ramey's (1962) early work in determining wellbore heat transmission for steam injection wells has been effectively applied to geothermal conditions in order to obtain data about the downhole environment. Thermal stress analysis of casing requires information on the temperature distribution in the well. A method for approximating distribution of temperature in the casing and for analyzing mechanical buckling criterion is presented below.

Effect of Heat on Casing and Tubing

Inasmuch as geothermal wells in certain areas are hot enough to affect strength properties of the steel, weakening of strength should be considered in design.

Compression — tension failure in cemented casing, was first noticed in steam injection wells. The rate of linear expansion of tubular goods is directly proportional to the rate of change in the average temperature. Wellheads on steam injection wells may rise above ground level as a result.

Considerable changes in stress level of the elastically unstable downhole pipe occur if thermal elongations are purposely or accidentally prevented (Leutwyler and Bigelow, 1964). In order to evaluate the change in stress conditions and to prevent plastic failures, it is necessary to determine (approximate) temperature changes and to analyze the nature of these changes.

Temperature measurements can be made either inside the casing using temperature logging tools or thermocouples, or outside the casing by thermocouples attached as the well is completed. The results of these expensive measurements are not always conclusive. Ramey (1962) presented an expression for the liquid temperature as a function of depth and time within a well. The following is an adaptation of Ramey's equation:

$$T_1(z, t) = az + b - aA + [T_0(t) + aA - b] e^{-z/A} \qquad (6\text{-}3)$$

where: T_1 = surface temperature, °F
z = height above the producing horizon, ft
t = producing time, days
a = geothermal gradient, °F/ft
b = ambient surface temperature, °F
A = time function (see Equation 6-6)
T_0 = bottomhole temperature , °F

A schematic diagram illustrating the parameters important in calculating the geothermal wellbore temperature distribution is presented in Figure 6-16.

The time function $A(t)$ depends upon conditions specified for heat conduction. Ramey (1962) presented a means of calculating the approximate values of $A(t)$, which provide engineering accuracy. Figure 6-17 presents the time function for several different internal boundary conditions. The convergence at long times ($>$one week) is analogous to pressure buildup theory. The time function may be estimated from solutions for radial heat conduction from an infinitely long cylinder. The equation for $f(t)$, which considers the line source for long times, is:

$$f(t) = -ln \frac{r'_2}{2\sqrt{\alpha t}} - 0.290 + 0(r_2'^2/4\alpha t) \tag{6-4}$$

where: r'_2 = outside radius of casing, ft
α = thermal diffusivity of the formation, ft²/day
0 = "on the order of"

The evaluation of the overall heat-transfer coefficient is the most difficult step involved in solving wellbore heat-transmission problems (Ramey, 1962). Assuming that heat is transferred radially away from the wellbore and that there are no phase changes, then:

$$A = \frac{Wc[k + r_1 \, Uf(t)]}{2\pi r_1 \, UK} \tag{6-5}$$

where: W = liquid production rate, bbl/day, lb/day
c = specific heat at constant pressure of fluid, Btu/lb-°F
r_1 = inside radius of tubing, ft
U = overall heat-transfer coefficient between inside of tubing and outside of casing based on r_1, Btu/day-ft²-°F
K = thermal conductivity of the formation, Btu/day-ft-°F

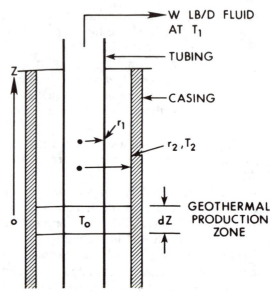

Figure 6-16. Generalized schematic diagram illustrating the parameter's importance in calculating the wellbore temperature distribution. (After Ramey, 1962, courtesy of SPE)

If thermal resistance in the wellbore is assumed to be negligible, then the overall heat-transfer coefficient can be assumed to be infinite and Equation 6-5 becomes:

$$A(t) = \frac{Wcf(t)}{2\pi k}, U = \infty \tag{6-6}$$

Heat transfer from the production of hot gases (steam) leads to the following modification of Equation 6-3 as suggested by Ramey (1962):

$$T_1(Z,t) = aZ + b - A(a \pm \frac{1}{778c})$$
$$+ [T_0 - b + A(a \pm \frac{1}{778c})]\, e^{-z/A} \tag{6-7}$$

The negative sign on the potential-energy term in Equation 6-7 is used for flow up the well with depth increasing (positive) upward from the geothermal producing interval. For flow down a well positive sign is used for the potential-energy term with depth increasing (positive) from the surface. This is applied to thermal injection wells used to recover heavy crudes.

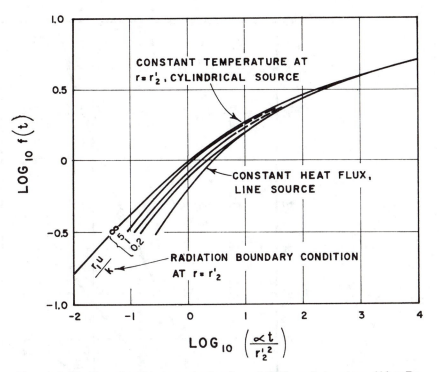

Figure 6-17. Transient heat conduction in an infinite radial system. (After Ramey, 1962; courtesy of Society of Petroleum Engineers.)

The elastic and plastic properties of the casing string are important. The magnitude of the thermal stress can be obtained on using the following equation:

$$S_t = \beta \Delta t E \qquad (6\text{-}8)$$

where: S_t = thermal stress, psi
β = linear thermal expansion coefficient, 6.9×10^{-6} (in./in.-°F)
Δt = temperature rise, °F
E = modulus of elasticity for steel, 29×10^6 psi

The thermal elongation can be calculated from the following relation:

$$\Delta L = L\beta\Delta t \qquad (6\text{-}9)$$

where: ΔL = thermal elongation, in.
L = casing length, in.

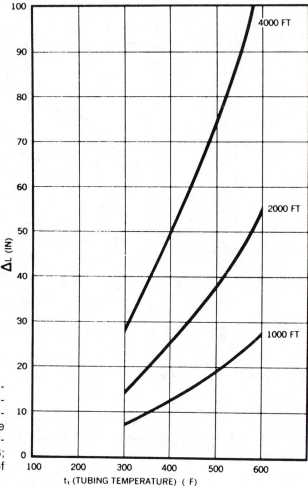

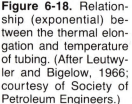

Figure 6-18. Relationship (exponential) between the thermal elongation and temperature of tubing. (After Leutwyler and Bigelow, 1966; courtesy of Society of Petroleum Engineers.)

Figure 6-18 illustrates the magnitude of the thermal elongation that can occur. Thermal elongation is shown in this figure as a function of producing temperatures and some selected cases of casing and tubing combinations (Leutwyler and Bigelow, 1966). This casing movement is restrained by the shear bond strength of the cement and the resistance of collars embedded in cement.

Stress behavior of casing upon heating and cooling is very complicated. As soon as a geothermal well is placed on production, the temperature rise creates stresses on the casing. Ductile deformation takes place if the yield point of the steel is exceeded. This plastic failure occurs in compression.

During shut-in periods the casing will cool down. Upon certain amount of cooling, tensional stresses are created that can cause failure of the casing or coupling. Buttress or premium threads give increased coupling strength to minimize such tensile failures.

Casing buckling due to thermal expansion in uncemented intervals poses a serious problem. Support may be lost due to the presence of lost circulation zones, or where long channels are left as a result of poor displacement of thickened mud by cement (Figure 6-19). Stage cementing tool failures caused by heat may force operators to perforate and squeeze cement. Thus, there is strong possibility that some intervals will be left uncemented. Slotted liners may also buckle, if movement is restrained, by sloughing of the producing interval.

The elongation of the casing string creates an end force at the bottom end, which governs the behavior of the casing. In such cases, one can use the following equations (Leutwyler and Bigelow, 1966):

For conditions where $F \leqslant w L$,

$$\Delta L = \frac{-L}{AE}(F) - \frac{r^2}{8EIw}(F^2) \tag{6-10}$$

where: F = end force, lb
ΔL = elongation, in.
A = cross-sectional casing area, in.^{2}
E = modulus of elasticity for steel usually assumed to be 29×10^6 psi
L = casing length, in.
I = moment of inertia of casing cross-section, in^4
r = radial clearance between casing and open hole, in.
w = weight per unit length of string, lb/in.

For conditions where $F > w L$,

$$\Delta L = \frac{L}{E} \left[\left(\frac{1}{A} + \frac{r^2}{4I} \right) F - \frac{r^2 Lw}{8I} \right] \tag{6-11}$$

These equations are used to determine the end forces in order to evaluate the stress distribution along the string. Helical deformation of the string can occur even at moderate geothermal temperatures. Enlarged hole sections, lost circulation zones, and other reasons that prevent a good cement bond may cause buckling of the unsupported casing. Figure 6-19 shows how unsupported pipe tends to elongate upon heating (Snyder, 1979).

There is little danger of helical buckling of the casing in enlarged hole sections, except where the shear energy yield criterion is approached.

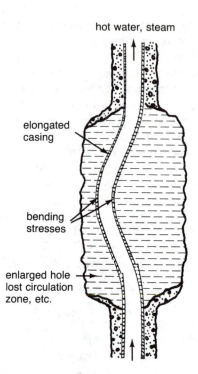

Figure 6-19. Schematic diagram showing how enlarged hole sections, lost circulation zones, and other problems that prevent a good cement job may result in casing buckling as the unsupported pipe attempts to elongate with heating. (After Snyder, 1979; courtesy of Gulf Publishing Company.)

Lubinski et al. (1962) and Timoshenko and MacCullough (1959) showed that the outer stress distribution can be determined by using the following equation :

$$S = \frac{F/w - L + Z\,(F)}{F/w} \left[\frac{1}{A} + \frac{Dr}{rI} \right] \qquad (6\text{-}12)$$

where: S = reduced stress in the outer layer, psi
Z = depth, ft
D = OD of casing, in.

The pitch of the helix at any depth (Figure 6-20) can be calculated from the following equation:

$$P_z = \pi \sqrt{\frac{8EIn}{(n - L + z)F}}$$

where: P_z = pitch of the helix at any depth z
n = F/w

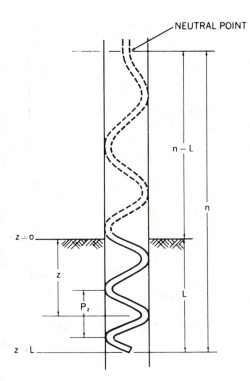

Figure 6-20. Geometrical relationship between neutral point, length L and depth z for a casing or a tubing string. (After Leutwyler and Bigelow, 1966; courtesy of Society of Petroleum Engineers.)

Thomas (1967), Holliday (1969), Karlsson (1968), and Snyder (1981) have discussed the yield strength behavior of various casing grades above 700°F. Recent test data indicate that there is about a five-ten percent yield strength loss in J-55 and K-55 grade pipe over a temperature range of 623°F. (77°F − 700°F) and a twenty percent or greater loss in N-80 grade casing over this same temperature range. Snyder (1981) reported that the tensile strength of P-110 is reduced by twenty-five percent at 800°F and the ductility is significantly increased. The recommended design curves for tubing and casing materials are presented in Figure 6-21. The plots of yield strength ratios for the various casing grades are shown up to 900°F and are compared to the strengths at 100°F.

The above casing string design has ignored the imposed load when large cement volumes are used on strings set in relatively low-weight drilling fluids. This gives use to a maximum load condition. The weight of the string is maximum when the cement reaches the casing shoe or when the top cement plug is released. Increase in weight can approach the allowable pull remaining in the string and, therefore, has to be considered.

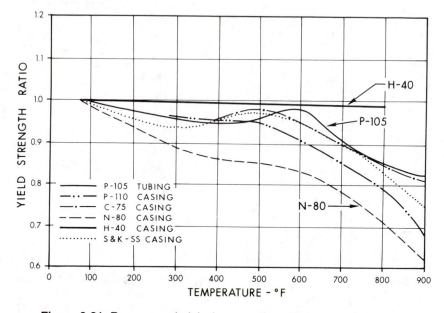

Figure 6-21. Recommended design curve for tubing and casing material.

Special Applications

There is potential failure from H_2S stress corrosion cracking and chloride stress cracking. Varying amounts of H_2S are present in most geothermal reservoirs. Snyder (1969) reported that lower-grade steels diminish this problem, and that downhole temperatures are hotter than those at which H_2S stress cracking is most severe, i.e., below 150-175°F.

Chloride stress cracking of austenitic stainless steels at elevated temperatures is a real possibility owing to the ready source of chlorides in brines and the popularity of stainless steel because of its corrosion resistance. Contrary to the behavior of H_2S, chloride stress cracking increases at temperatures above 150-160°F.

Angles (1979) utilized fiberglass casing in low-temperature corrosive geothermal wells in France. Specifications of this special fiberglass casing run in the development wells to a depth of 6,600 ft were as follows:

1. OD = 7 ins. (177.8 mm)
2. ID = 6 ins. (152.4 mm)
3. OD joints = 8 ins. (203.2 mm)
4. Threads = API-LTC
5. Length of joint = 30 ft (9 m)

Maximum operating specifications for this particular casing at 150°F (66°C) were:

1. Burst = 2,500 psi (175 bars)
2. Collapse = 3,000 psi (210 bars)
3. Joint tension = 120,000 lb (55 t)

No problems were encountered in cementing the fiberglass casing; however, the results from the cement bond log were not conclusive. The cement used was Class C without an accelerator. Laboratory testing indicated that cement adherence to the fiberglass pipe was higher than to the steel casing.

Tubing

The length of the tubing string must always be accurately estimated before it is landed in the well. Movement of the tubing caused by temperature effects and pressure changes creates serious and expensive problems. Thus, it is necessary to provide a proper slack-off weight (Equation 6-10) for the tubing string so that it can properly move in the well.

Inasmuch as the thermal stress levels encountered can easily exceed those of conventional oil field steel grades, tubing expansion joints can be used, although they leak and are unreliable (Figure 6-22) (Leutwyler and Bigelow, 1966). The expansion joints can be installed at the surface and downhole. Tubing elongation can be predicted by using Equation 6-9.

In order to improve the columnar characteristics of tubing and to avoid friction locks between tubing and casing due to elastic instability, centralizing guides must be placed on the string, especially in the case of deeper wells. Flexible connections between tubing and surface steam lines are required for such installations, utilizing high-temperature hoses or swivel joints.

Inasmuch as the tubing expansion joint has been found to be the most applicable because it is designed as an integral part of a retrievable packer. Other expansion joints, which are installed in the string several joints above the pack-off device, are self-contained units. Installation of these expansion joints several joints high has the advantage of keeping them out of the heavy residue, which often accumulates above the packer after the annulus fluid is boiled off.

Seals

Snyder (1979) noted that the elastomer seals cannot tolerate continual movement in deep, hot wells. The following downhole seal points should be considered (Leutwyler and Bigelow, 1966):

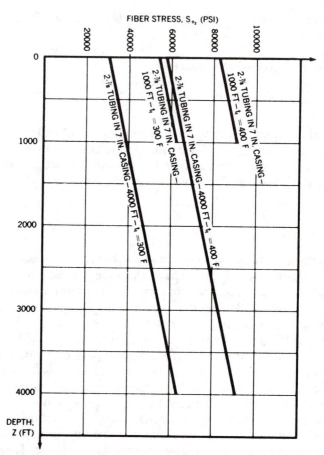

Figure 6-22. Even moderate injection temperatures subject tubing to excessive fiber stresses. (Leutwyler and Bigelow, 1966; courtesy of Society of Professional Engineers.)

1. Casing to open hole
2. Tubing to casing annulus
3. Packer to casing
4. Tubing to packer

Casing-to-open-hole and tubing-to-casing annulus seals necessarily require thread connections. Thread sealing compounds are needed to prevent leaks through the helical leak path between crest and root. Conventional sealing compounds for casing and tubing threads cannot be used above 300°F, because the physical properties of their fillers become inadequate under the influence of higher temperatures. Special high-temperature sealing compounds with non-metallic fillers have been developed, however, which are used as thread sealants and lubricants in steam-injection systems at temperatures up to 600°F. Metal to metal primary seals can eliminate the need for pipe dope to be used as a sealing element.

Packer-to-casing seal systems cannot depend on the usual elastomer compounds, because their thermal limit is about 400°F (204°C). These temperatures are too high for standard oil well packer equipment. Some limited success was obtained with inflatable type formation packers using Viton™ bladders. Although elastomeric pack-off systems have been tried with varying success up to 475°F under laboratory and field conditions, they were found to perform unreliably. Thus, they were replaced with inorganic sealing materials, such as those used in high-temperature steam glands. Upton and Spriggs (1979) reported the development and the laboratory testing of hydraulically actuated treating packers up to 570°F and 5,000 psi with no leakage or deterioration of parts. This high-temperature packer technology consists of using an old-style hydraulic formation packer design and incorporating a spiral wrapped asbestos sealing element, or standard casing packer design with formed rope asbestos seal elements and valves (Figure 6-23).

Blowout Preventers (BOP's)

In the case of deep geothermal-geopressured wells, it is necessary to provide specifications and procedures for surface blowout prevention. These specifications and procedures were developed by industry and have been adopted by IADC and API.

Containment equipment for shallow geothermal wells has not been specified as clearly as those for the deep wells drilled in the Gulf Coast. A model for geothermal wells could be those wells being drilled in Idaho for DOE, with the following sequence of installation steps:

1. Construct a 8 × 10 × 8 ft reinforced cellar to accommodate the BOP stack.
2. Install a single-gate blowout preventer between the casing head and drilling nipple for drilling the surface hole.
3. After setting the surface casing, a containment stack is used during drilling to total depth (TD) (Figure 6-24).
4. Install the permanent wellhead. Example of a wellhead used in the DOE Idaho geothermal operations is shown in Figure 6-25.

Standard operating procedure in equipping Geyser steam wells with BOP stacks has been presented by Capreano (1979). Geyser well BOP stack consists of blowdown lines, a double-ram preventer, and rotating head.

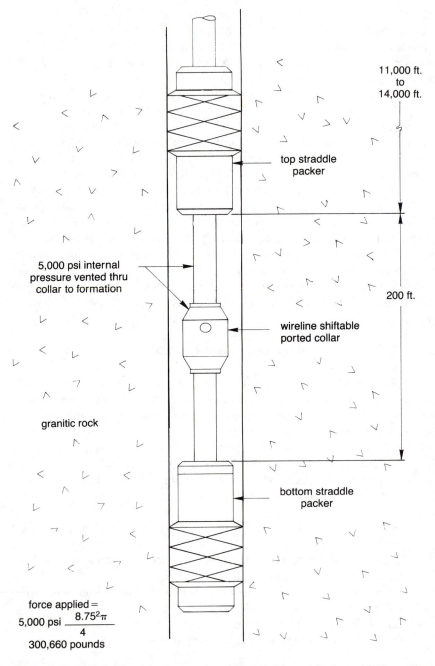

Figure 6-23. Schematic of the test packer. (After Upton and Spriggs, 1979; courtesy of American Society of Mechanical Engineers.)

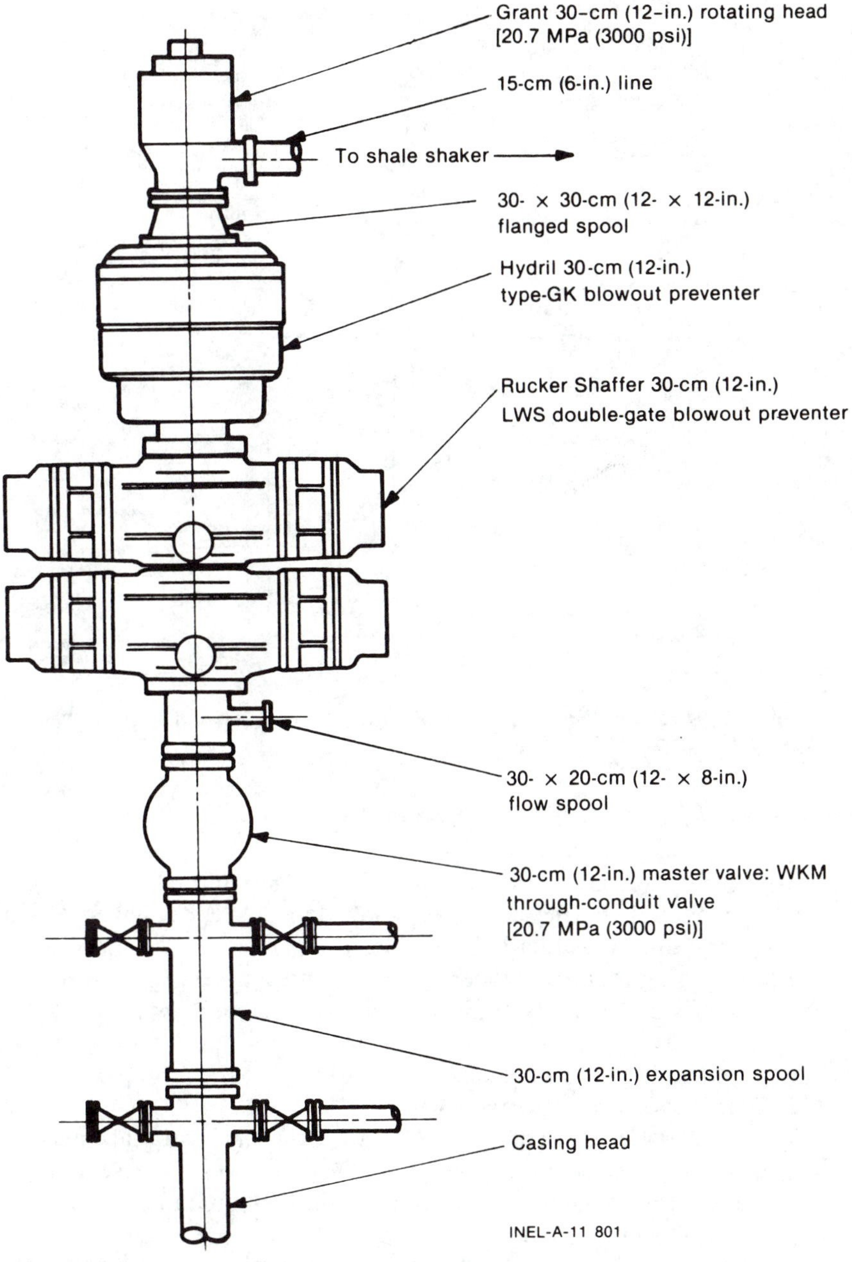

Grant 30–cm (12–in.) rotating head
[20.7 MPa (3000 psi)]

15-cm (6-in.) line

To shale shaker ———⟶

30- × 30-cm (12- × 12-in.)
flanged spool

Hydril 30-cm (12-in.)
type-GK blowout preventer

Rucker Shaffer 30-cm (12-in.)
LWS double-gate blowout preventer

30- × 20-cm (12- × 8-in.)
flow spool

30-cm (12-in.) master valve: WKM
through-conduit valve
[20.7 MPa (3000 psi)]

30-cm (12-in.) expansion spool

Casing head

INEL-A-11 801

Figure 6-24. Blowout preventer for drilling below the 34-cm (13-3/8-in.) casing shoe. (After Miller and Prestwich, 1979.)

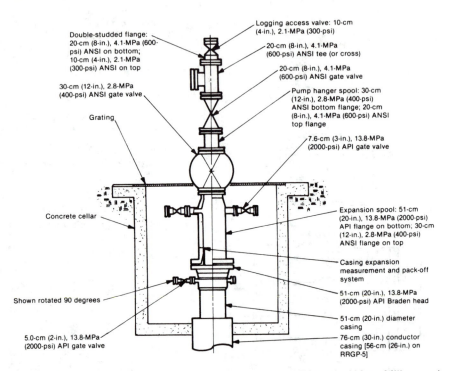

Double-studded flange:
20-cm (8-in.), 4.1-MPa (600-psi) ANSI on bottom; 10-cm (4-in.), 2.1-MPa (300-psi) ANSI on top

30-cm (12-in.), 2.8-MPa (400-psi) ANSI gate valve

Grating

Concrete cellar

Shown rotated 90 degrees

5.0-cm (2-in.), 13.8-MPa (2000-psi) API gate valve

Logging access valve: 10-cm (4-in.), 2.1-MPa (300-psi)

20-cm (8-in.), 4.1-MPa (600-psi) ANSI tee (or cross)

20-cm (8-in.), 4.1-MPa (600-psi) ANSI gate valve

Pump hanger spool: 30-cm (12-in.), 2.8-MPa (400-psi) ANSI bottom flange; 20-cm (8-in.), 4.1-MPa (600-psi) ANSI top flange

7.6-cm (3-in.), 13.8-MPa (2000-psi) API gate valve

Expansion spool: 51-cm (20-in.), 13.8-MPa (2000-psi) API flange on bottom; 30-cm (12-in.), 2.8-MPa (400-psi) ANSI flange on top

Casing expansion measurement and pack-off system

51-cm (20-in.), 13.8-MPa (2000-psi) API Braden head

51-cm (20-in.) diameter casing

76-cm (30-in.) conductor casing [56-cm (26-in.) on RRGP-5]

Figure 6-25. Schematic diagram of completed wellhead. (After Miller and Prestwich, 1979.)

Cementing

Excellent research work on cementing of geothermal wells, which is treated in detail in Chapter 7, is being done by Brookhaven National Laboratory and by Sandia Laboratories.

The following required properties of cements for use in geothermal wells have been listed by the Brookhaven National Laboratory (1978, p. 1):

1. Bond strength to steel casing = >10 psi.
2. Permeability to water = <0.1 md.
3. Compressive strength = >1,000 psi, 24 hours after placement.
4. Stability = no significant reduction in strength or increase in permeability after prolonged exposure at 750°F (~400°C) to 25% brine solutions, flashing brine, or dry steam.
5. Placement ability = capable of 3-4 hr retardation at expected placement temperatures.

In addition, cement must be compatible with drilling fluid and should be noncorrosive to steel well casing.

As pointed out by Sandia Laboratories (1979), if an adequate cement sheath is not emplaced initially, the casing will tend to buckle upon thermal loading because it is not uniformly supported. In addition, the hot, bare metal of casing exposed to formation fluids will corrode rapidly.

If thermal shocks (cyclic temperature changes) of several hundred degrees are frequent, failure of the cement and casing can be expected after 2 to 3 years. Because of the large volume of brines present and deteriorated state of the casing and remaining cement, squeeze cementing of such damaged areas would probably have only a very limited success.

Testing methods, therefore, must be developed to evaluate cements under dynamic conditions involving changing pipe stresses, cyclic temperatures, and leaching solutions having different chemical compositions.

The long-term cement problems are caused by cracking caused by casing expansion (ballooning), longitudinal movement from elongation and shrinkage, thermal shocks, and leaching by hot water, which results in cement deterioration.

Extensive research work is being conducted on identifying, evaluating and selecting high-temperature well cementing materials, e.g., PC containing crosslinked mixtures of styrene (St), acrylonitrile (ACN), and acrylamide (Aa); organosiloxanes; and portland cement combined with silica sand and vinyl-type monomers, such as methacrylate (MMA) and St to form PC.

Research on high-temperature *inorganic* cements is in progress by G.L. Kalousek at the Colorado School of Mines; T.J. Rockett at the University of Rhode Island; B.E. Simpson at the Dowell Division of Dow Chemical; R.S. Kalyoncu at the Battelle's Columbus Laboratories; D.M. Roy at the Pennsylvania State University; D.K. Curtice at the Southwest Research Institute; and others.

In explaining properties of the cements and in the phase identification, one can use X-ray diffraction, differential scanning calorimenter (DSC), and infrared analysis (IR).

In the Brookhaven National Laboratory Progress Report No. 8 (1978), the following test was described for measuring the cement-steel bond strength: The specimen assemblage for bond strength consists of a 3/4-in.-diameter by 2-in. mild steel rod, which is centered in a 2-in.-diameter by 1-in. high container. The cement slurry is placed around the rod and cured for 1 day at 150°C. After an additional curing at 250°C for 7 days, the steel rod is loaded in a compression tester to failure of the bond. The suitable binders include gyrolite + scawtite, xonotlite, and tobermorite.

References

Angles, P.J., 1979. Fiberglass Casing Used in Corrosive Geothermal Wells. *O&G Jour.*, 77(42): 131-132.

Brookhaven National Laboratory, 1977-1979. Cementing of Geothermal Wells. Progress Reports Nos. 7 through 12. United States Dept. Energy, Brookhaven Natl. Lab., Upton, N. Y. 11973.

Capuano, L.E., Jr., 1979. How Geysers Steam Wells Are Drilled and Equipped. *World Oil*, 188(2): 69-72.

Casner, J.A., 1963. New Equations Help Prevent Surface-Casing Overload. *O&G Jour.*, 61(50): 84-87.

Cigni, V., Fabbri, F. and Giovannoni, A., 1976. Advancement in Cementation Techniques in the Italian Geothermal Wells. *Proc. 2nd UN Symp. Dev. and Use Geoth. Resc.*, vol. 2, pp 1471-1481.

Craig, S.B., 1961. Geothermal Drilling Practices at Wairakei, New Zealand. *Proc. U.N. Conf. on New Source of Energy*, Rome, Italy, 3(II): 121-133.

Farris, R.F., 1946. Method for Determining Minimum Waiting-on-Cement Time. *Trans. AIME*, 165: 175-188.

Gallus, J.P. and Pyle, D.E., 1978. Performance of Oil-Well Cementing Compositions in Geothermal Wells. 53d Annu. Fall Tech. Conf., Soc. Petrol. Engrs., Houston, Texas, SPE Paper No. 7591:6pp.

Goins, W.C. and Sheffield, R., 1981. *Blowout Prevention*. Houston: Gulf Publishing Co., 300pp.

Goins, W.C., 1980. Better Understanding Prevents Tubular Buckling Problems. *World Oil*, 189(2): 35-40.

Goins, W.C., Collings, B.J. and O'Brien, T.B., 1966. A New Approach to Tubular String Design. *World Oil*, 162(1): 79-84.

Holmquist, J.L. and Nadai, A., 1939. Collapse of Deep Well Casing API Drilling and Production Practices.

Kalousek, G.L. and Chaw, S.Y., 1976. Research on Cements for Geothermal and Deep Oil Wells. *Soc. Petr. Engrs. Jour.*, 16(6): 307-309.

Leutwyler, K. and Bigelow, H.L., 1966. Temperature Effects on Subsurface Equipment in Steam Injection Systems. *Jour. Petr. Tech.*, 17(1): 93-101.

Lubinski, A., Althouse, W.S. and Logan, J.L., 1962. Helical Buckling of Tubular Goods. *Jour. Petr. Tech.*, 14(6): 665-670.

Matsuo, K., 1973. Drilling for Geothermal Steam and Hot Water. Geothermal Energy (Review of Research and Development), UNESCO, Paris, France, 186pp.

Miller, L.G. and Prestwich, S.M., 1979a. Completion Report: Raft River Geothermal Production Well Five (RRGP-5). DOE, Idaho Operations Office, IDO-10082, Idaho Falls, Idaho, 19pp.

Miller, L.G. and Prestwich, S.M., 1979b. Completion Report: Raft River Geothermal Injection Well Seven (RRGI-7). DOE, Idaho Operations Office, IDO-10084, Idaho Falls, Idaho, 10pp.

Prentice, C.M., 1970. "Maximum load" Casing Design. *Jour. Petrol. Tech.* 22(7): 805-811.

Ramey, H.J., Jr., 1962. Wellbore Heat Transmission. *Jour. Petr. Tech.*, 14(4): 427-432.

Roy, D.M., 1956. Subsolidus Data for the Join Ca_2SiO_4-$CaMgSiO_4$ and the Stability of Merwinite. *Min. Mag.* 31(233): 187-195.

Rudisill, J.M., 1978. A Case History: The Completion of a Shallow, Over-pressured Geothermal Well. *Geoth. Res. Council Trans.* 2:587-590.

Slaughter, J., Kerrick, D.M. and Wall, V.J., 1975. Experimental and Thermodynamic Study of Equilibria in the System CaO-MgO-SiO_2-H_2O-CO_2. *Am. J. Sci.*, 275:143-162.

Snyder, R.E., 1978. Geothermal Well Completions: State of the Art. *Geoth. Res. Council Trans.*, 2:601-603.

Snyder, R.E., 1979. How Geothermal Wells Are Completed and Produced. *World Oil,* Oct.: 81-87.

Snyder, R.E., 1979. Geothermal Well Completions: A Critical Review of Downhole Problems and Specialized Technology Needs. SPE Paper 8211, 54th Ann. SPE Meeting, Las Vegas, NV, 6pp.

Suman, G.O., Jr. and Ellis, R.C., 1977. *World Oil's Cementing Handbook, Including Casing Handling Procedures.* Gulf Publ. Co., Houston, TX, 73pp.

Timoshenko, S. and MacCullough, G.H., 1959. Elements of Strengths of Materials. 3rd Edition, New York: D. Van Nostrand Co., Inc., 374pp.

Upton, T.E. and Spriggs, D.M., 1979. Hydraulically Actuated Treating Packers for Dry Rock Geothermal Applications. ASME, Pressure Vessels and Piping Conf., San Francisco, CA, 79-PVP-23, 4pp.

Varnado, S.G., 1979. Report of the Workshop on Advanced Geothermal Drilling and Completion Systems. Sandia Lab., Rept. No. 79-1195, Albuquerque, NM, 88pp.

Varnado, S.G., and Stoller, H.M., 1978, Geothermal Drilling and Completion Technology Development. *Geoth. Res. Council Trans.*, 2:675-678.

S.H. Shryock, Halliburton Services

7

Geothermal Cementing

Introduction

Wells drilled in known geothermal areas for producing hot water or steam to be used primarily for generating electricity are similar in many aspects to wells drilled for the production of hydrocarbons. Successful completion of these wells requires utilization of temperature-stabilized drilling fluids, high-grade steel casing strings with special threaded couplings, and temperature-stabilized portland cementing compositions. Because of hostile environmental conditions, special planning is necessary to ensure the well's integrity.

Drilling techniques and drilling fluids selection are covered in other chapters of this book, whereas this chapter covers the current state of the art of cementing operations and casing design.

Well Design

A typical geothermal well might have a 20-in. casing set at a depth of 300 ft in a 26-in. hole, a 13⅜-in. casing set at 2,000 ft in 17½-in. hole, and a 9⅝-in. liner run in 12¼-in. hole to a depth of 5,000 ft, with its top being 150-200 ft above the shoe of the 13⅜-in. casing. Each string must be properly cemented in such a manner that the casing-hole annulus is totally filled with uncontaminated temperature-stabilized cements (see Figure 7-1).

Casing centralizers designed to position the casing in the center of the hole should be used in sufficient quantities to allow the cement slurry to flow uniformly around the full length of the casing. Other casing accessory equipment usually include a float collar and shoe, which enable filling and guiding of the casing as it is run into the well; and positive back-pressure seals, which keep the cement slurry in the casing-hole annulus following the cementing job.

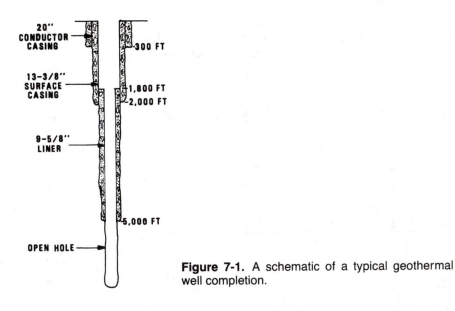

20″
CONDUCTOR
CASING

300 FT

13-3/8″
SURFACE
CASING

1,800 FT
2,000 FT

9-5/8″
LINER

5,000 FT

OPEN HOLE

Figure 7-1. A schematic of a typical geothermal well completion.

Importance of Obtaining Complete Casing Cementing

The reasons for what might be called total or complete cementing are threefold. First, the hydrated cement slurry must have sufficient strength to support and anchor the casing. The steel casing being subjected to the high temperatures of the geothermal well tend to elongate due to temperature changes. These temperature changes will occur probably at three different times: (a) The first change occurs as the casing is run in the mud-filled hole. If the atmospheric temperature is equal to 60°F and the mud temperature is 160°F (temperature change of 100°F), and elongation of 8.28 ins./1,000 ft will occur. (b) The second change occurs during the hydration period of the cement as it solidifies. In this case, the temperature change is also around 100°F, resulting in an elongation of 8.28 ins./1,000 ft of casing (Table 7-1). Although the first two periods have less significance than the third one, they must be considered in order for the casing to remain centered in the hole by maintaining tension strain at the surface. (c) Upon completion of the well and initiation of production, the temperature change (difference between temperature after setting of cement and production temperature) could be as much as 200°F. If any part of the casing is free and the produced fluid or vapor temperature is 450°F, the casing will tend to elongate another 15-16 ins./1,000 ft, and casing buckling will probably occur.

Table 7-1
Elongation of Tubing and Casing Caused
by Temperature Changes

Length of Pipe Feet	Temperature (°F)									
	50°	100°	150°	200°	250°	300°	350°	400°	450°	500°
	Inches									
500	2.07	4.14	6.21	8.28	10.35	12.42	14.49	16.56	18.63	20.70
1000	4.14	8.28	12.42	16.56	20.70	24.84	28.98	33.12	37.26	41.40
1500	6.21	12.42	18.63	24.84	31.05	37.26	43.47	49.68	55.89	62.10
2000	8.28	16.56	24.84	33.12	41.40	49.68	57.96	66.24	74.52	82.80
2500	10.35	20.70	31.05	41.40	51.75	62.10	72.45	82.80	93.15	103.50
3000	12.42	24.84	37.26	49.68	62.10	74.52	86.94	99.36	111.78	124.20
3500	14.49	28.98	43.47	57.96	72.45	86.94	101.43	115.92	130.41	144.90
4000	16.56	33.12	49.68	66.24	82.80	99.36	115.92	132.48	159.04	165.60

The second reason for complete casing cementing is the fact that it requires the total displacement of the drilling fluids, mud spacers, and preflushes by the cement slurry. Most drilling fluids used during the drilling of geothermal wells are water-based. Mud spacers and preflushes are also water-base fluids. Although each one of these fluids has an important role in obtaining a successful completion, subsequently they must be removed from the borehole. Drilling fluid remaining in the casing-hole annulus usually contaminates the cement slurry resulting in an increase of the slurry viscosity, which may require significantly higher pump pressures to maintain its movement in the annulus. Higher pump pressures result in increased pressures imposed on the exposed formation, and often cause fracturing and subsequent lost circulation. Sufficient quantities of mud-contaminated cement are also thought to be responsible for channeling of the cement slurry in some parts of the annulus, leaving portions of the casing only partially cemented (Figure 7-2).

When the casing-hole annulus is not totally filled with a cement of good quality, it is possible to leave tightly-seated pockets of water from the drilling fluid, preflush fluid, or mud spacer. This water is heated as the well temperature reverts back to static conditions. Upon initiation of production, higher temperatures from the produced fluid or vapor increase the temperature of this trapped water. The expansion of water caused by increasing temperatures generates forces of the order of 45-50 psi/°F, which cause failures. In several instances of casing collapse in geothermal wells, the only reason for

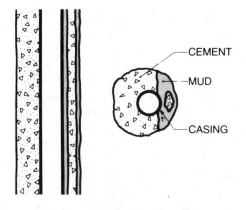

Figure 7-2. Presence of channel behind the casing on a primary cementing job.

the collapse is thought to have been the expansion of water trapped in these pockets.

The third reason for complete casing cementing is to protect the casing from corrosive fluids that are found in many formations. A protective cement sheath properly placed around all of the casing eliminates the corrosion of the outside of the casing. Other remedies, however, are necessary for the control of corrosion on the inside of the casing.

Cementing Compositions

Portland cement loses a major portion of its compressive strength when cured at temperatures above 250°F. This may not present a major problem, except while undergoing "strength retrogression" it is becoming permeable (often as much as 10 md). Inasmuch as portland cement remains the primary product for casing cementing, although considerable research has been done in an attempt to find other products, it is necessary to eliminate this strength retrogression and permeability increase.

Cements exhibiting strength retrogression contain two hydration products: *calcium hydroxide* and *dicalcium silicate alpha-hydrate*. These products appear together and sometimes singularly, depending on the temperature at which the cementing compounds had been cured and the length of time elapsed prior to testing. In the presence of either or both of these hydration products, there is a loss in compressive strength of the set cement.

To eliminate this retrogressive characteristic, a finely-divided silica is added to portland cement. Quantities of 30-60% by weight of the cement achieve high-temperature stability. The "silica flour" initially reacts with the calcium hydroxide to form dicalcium silicate alpha-hydrate, which converts to *tobermorite* at elevated temperatures. The tobermorite group of calcium silicate hydrates provides the temperature stability necessary to maintain high compressive strength and low cement permeability at high temperatures existing in geothermal steam wells.

It is of prime importance to place cement in the casing-hole annulus all the way to the top, i.e., the full length of the casing. Because geothermal boreholes encounter not only high temperatures, but also often penetrate zones containing natural fractures and gravel beds having high permeabilities, other cement additives are required.

Lost circulation, i.e., when the displaced fluid and/or cement slurry fails to return to the surface, is a common problem encountered during cementing operations in geothermal wells. Addition of bridging additives, such as a perlite, to the portland cement-plus-silica flour mixtures has proven to greatly assist in maintaining full circulation throughout the cementing jobs. Perlites weigh 8-10 lbs/ft³ (dry), occur in particle sizes of -4 to 40 mesh (U.S. Series), and are cellular, which allows them to absorb water under pressure. Because perlite is light in its dry state, it is necessary to use 2-4% bentonite (by weight of the total cement) in order to keep perlite particles evenly dispersed throughout the slurry. The most commonly used cementing composition is: 1 ft³ (94 lbs) of cement plus 1 ft³ (8 lbs) of perlite, with 40% silica flour and 2% bentonite (gel), both based on the weight of the cement. In special applications, the perlite content has been increased to as much as 3 ft³/sack cement, but this is not commonly recommended because of the substantial increase in the required volume of mixing water. (The normal amount of mixing water with 1 ft³ of perlite per sack of cement is equal to 1.46 ft³, resulting in an in-hole slurry density of 104 lbs/ft³ and a slurry volume of 2.28 ft³/sack of cement.) Some prefer to use a tail-in cement composition, which provides high early strength: portland cement plus 40% silica flour mixed with 0.91 ft³ of water (specific weight = 116 lbs/ft³).

Long-term testing of geothermal well cements indicates that those compositions having low water/solids ratios are the most temperature stable. Because cement slurry is a non-Newtonian fluid, with viscosity being a function of shear rate, and the method of its placement requires the viscosity to remain low enough to be pumpable throughout the entire operation, there is a limitation of minimum W/S ratios. Special chemical dispersants can be added to the cement in concentrations of 0.5-1.0 lb/sack of cement, in order to increase wettability of particles and, thus, allow the use of reduced mixing water volumes without increasing the viscosity. Reducing the volume of mixing water increases the slurry density, so it is important to know if the hole conditions will tolerate increased hydrostatic pressures.

High-density slurries are mostly used for plug-back cementing for "kick-off plugs." If it is necessary to redrill a part of the hole or drill around a "fish" (part of the drilling assembly), an extremely hard, high-density cement plug will greatly aid in side-tracking or changing direction with the drilling assembly (Figure 7-3).

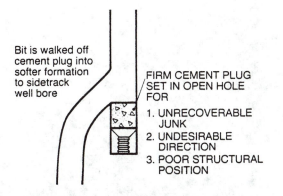

Bit is walked off
cement plug into
softer formation
to sidetrack
well bore

FIRM CEMENT PLUG
SET IN OPEN HOLE
FOR

1. UNRECOVERABLE
   JUNK
2. UNDESIRABLE
   DIRECTION
3. POOR STRUCTURAL
   POSITION

Figure 7-3. Schematic diagram of sidetracking.

In all cases (casing, liner, plugback or squeeze cementing), the cement slurry must remain pumpable until the placement is complete. An instrument which is used to determine the pumpability of cement slurries under downhole conditions of high temperature and pressure is called a *thickening time tester* or *consistometer* (Figure 7-4). It is calibrated in such a manner that the units of consistency of the slurry in the cup can be recorded by a voltmeter. Thus, one can obtain a permanent record of the slurry viscosity or consistency throughout the duration of each test. Once the value of consistency units reaches a designated value, such as 100, the test is terminated and the time is referred to as the "thickening time."

The thickening time test schedule depends on the type of job to be performed. Test schedules are formulated prior to starting the test in such a manner that the rates of temperature and pressure increase for the test slurry are as close as possible to those for the actual slurry to be used for the specific well. Normally, thickening times for casing cementing jobs are 2½ to 3½ hours. It is a common practice to compute the time to do the job and then add an additional hour for safety. Because the thickening time test schedule for a squeeze or plugback cementing job is different than that for a casing job, one should not use test results interchangeably.

Thickening time tests are performed at temperatures referred to as the bottomhole circulating temperatures (BHCT), which are always less than the bottomhole static temperatures (BHST), and represent actual measurements in wells (Figure 7-5). Because geothermal wells are generally rather shallow, i.e., 4,000-10,000 ft, the BHCT is seldom more than one-half the BHST. In order to obtain the desired thickening time for a specific type of job, chemical retarders are usually added to cement slurries. Retarders extend the thickening time when added at concentrations of 0.1-1.5% by weight of the cement, with the most common concentration being 0.2-0.4%.

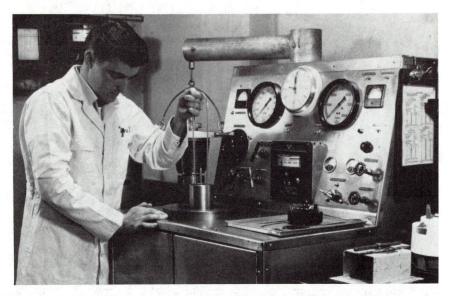

Figure 7-4. API pressure-temperature thickening time tester.

It might seem that using sufficient retarder to obtain 5-6 hours thickening time would be advisable. Unfortunately, however, extending the thickening time also increases the setting time, and one may have to wait twice as long as desired by overtreatment.

Cement Mixing Methods

The most common method of mixing cement is with a jet mixer, as shown in Figure 7-6. The jet mixer consists of a funnel-shaped hopper, a mixer bowl, discharge line, mixing tub, and water supply lines. A stream of water is forced through a jet, across the bowl into a discharge line, and then into a sump tub from which the cement slurry is removed by the cementing pumps (Figure 7-7).

Because of the importance of having no free water, new mixing methods have been divised utilizing mechanical batch mixing equipment. By placing all of the ingredients into the batch mixer, it is possible to prepare a specific volume of slurry having exact physical requirements before it is displaced downhole (Figure 7-8).

Often, job success depends largely on a proper density control and proper amounts of all the additives used. Based on experience, high temperatures drastically affect cement slurries during the first few minutes of exposure. To

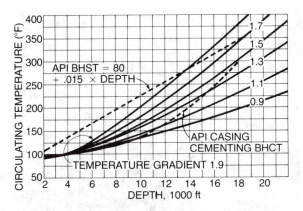

Figure 7-5. Relationship between circulating temperature and depth for various temperature gradients.

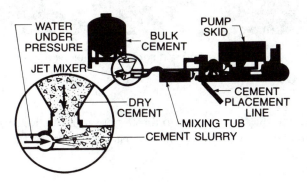

Figure 7-6. Typical cement-mixing operation.

control these effects, a surface mixing time of 30 minutes in batch mixers prior to exposure to high temperatures has been found to be necessary. This is more critical in the plugback or squeeze cementing than on casing cementation jobs.

Plugback cementing jobs to control lost circulation during drilling operations require thorough planning; however, selection of materials and placement procedure are of greater concern than temperature stability and mixing technique. Because drilling mud is usually lost into natural fractures or intervals having very high permeability, successful lost circulation control is seldom obtained with cement alone. For the effective lost circulation control, a viscous gel containing bridging agents, such as gilsonite, can be

Figure 7-7. Field set-up using jet mixer.

introduced into the system prior to the cement introduction. Plugback jobs for lost circulation control are time consuming because they involve the following:

1. The drilling assembly has to be removed, which involves a trip out of the hole with the drillpipe.
2. The drillpipe is then run back in the hole to lay in the cement plug.
3. Subsequently, the waiting-on-cement time is usually equal to 8-16 hours, during which the drillpipe can be tripped to pick up the bit and associated tools.
4. If cement fills part of the hole, drilling through it delays progress, and even upon completion, lost circulation may occur again.

Thus, products other than cement are being investigated in order to reduce the required time.

Material Handling Aspects

Once the cement composition is chosen and a cementing service company is hired, it is necessary to work out the logistics.

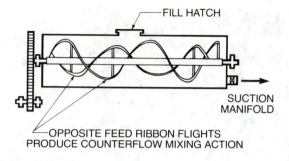

Figure 7-8. Schematic diagram of ribbon-type batch mixer.

An adequate space must be provided near the drilling rig for the cement mixing and pumping equipment. A readily available source of mixing water and easy access to displacing fluid (mud or water) is necessary. The casing should be positioned in such a manner that the plug holding head can be installed from the rig floor leaving room to connect the discharge lines from the cement pumps.

Nearly all cementing materials are handled in bulk. Pneumatic transfer is utilized to move the cement from pressurized bulk cement tanks to the surge bin or batch mixers. If the volume required for the job requires several bulk material trucks, it is advisable to use field storage bins that hold 1,000-1,300 ft³ of dry materials compared to 300-500 ft³ capacity in the bulk trucks (Figure 7-9). Locating bulk materials nearby to avoid shutting down operation to move bulk material equipment greatly reduces the chance that air and/or water pockets, caused by the slurry falling in the pipe while moving this equipment, will form.

All of the materials used in the preparation of cement should be dry blended (e.g., at the cementing service company facility) to ensure a uniform blend having correct proportions of each component (Figure 7-10). Some additives, such as retarders and dispersants, are used in such small concentrations that uniform blending is very difficult if not done at plants with the facilities to do the job. The total job is ruined if only a very small portion of the total volume of the cement slurry sets prematurely. Casings that are left partially or totally filled with solidified cement have no value, and are very expensive to clean out.

Cement Placement Techniques

The most common cementing job is the primary cement job whereby cement slurry is pumped down the casing and up the annulus. In certain cases, hole conditions require modifications of the basic *normal displace-*

Figure 7-9. Field storage bins for bulk cement.

Figure 7-10. Land-based cement blending station.

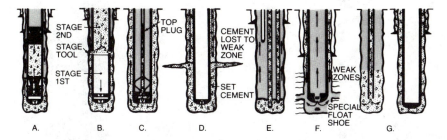

Figure 7-11. Various techniques of placing cement downhole. A—normal displacement method. B—two-stage cementing. C—inner-string cementing. D—outside cementing. E—multiple-string cementing. F—reverse-circulating cementing. G—delayed-set cementing.

ment technique. Seven different techniques, with the first four (A-D) being most commonly used in geothermal wells, are shown in Figure 7-11.

Good results are obtained with the normal displacement method, when casings below the conductor pipe are being moved (reciprocation or rotation) during the job. Casing movement has proven to be a very significant factor in obtaining effective mud displacement, which is required for achieving a good quality primary casing cementing job. Casing movement should be slowed down as the cement reaches bottom and increased when the cement is present in the annulus.

The normal displacement method using top and bottom displacement plugs is presented in Figure 7-12. These displacement plugs separate the fluids in the pipe; and provide a positive indication when the casing has been displaced and the top plug reaches, or bumps on, the float collar or baffle, which is usually placed one to three joints above the casing shoe. Use of the first or bottom plug, which wipes the drilling mud from the inside of the casing, is not recommended when high concentrations of bridging agents are used in the cement. The absence of a bottom plug increases the possibility of mud contamination in that part of the cement which is left in the casing below the top or final displacement plug (Figure 7-13).

Upon displacement of cement in the casing and "bumping" of the plug, the positive seal in the float collar and shoe keeps the cement in place. Valves on the casing head should not be closed during the waiting-on-cement period, because the temperature of the fluid left in the casing will increase and, thus, could cause pressure increase, which will expand the casing. Upon pressure release, the casing will contract leaving space between the cement and the casing (outside diameter). This space is called micro-annulus and could allow fluids to move (up or down) outside the casing.

Figure 7-12. Normal cement displacement method.

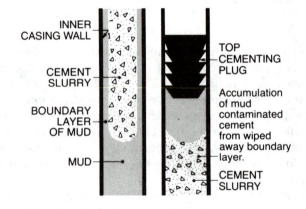

INNER
CASING WALL

CEMENT
SLURRY

BOUNDARY
LAYER
OF MUD

MUD

TOP
CEMENTING
PLUG

Accumulation
of mud
contaminated
cement
from wiped
away boundary
layer.

CEMENT
SLURRY

Figure 7-13. Mud removal through use of bottom cementing plug.

When the hydrostatic pressure gradient of the cement slurry is higher than the formation fracture gradient, it is common to perform the so-called *stage cement job* (Figure 7-14). This allows the placement of the cement slurry in the annulus from the casing shoe to just above the weak zone during the first stage. A combination of plugs and baffles enable the stage collar to be manipulated hydraulically, shutting off access to the casing below this collar and opening ports in the collar so that fluid can be circulated from this point to the surface. After allowing sufficient time for the first-stage cement to set, cement circulation is performed through these ports back to the next stage or to the surface. Two-stage cementing is the most widely used method, although a three-stage method can be employed if necessary.

Because of the necessity for full-string cementing in geothermal wells, cementing the larger-diameter casings (9⅝ ins. and larger) can best be achieved by using a technique called *inner string cementing*. This technique

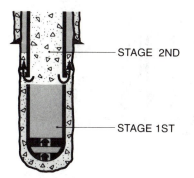

Figure 7-14. Two-stage cementing.

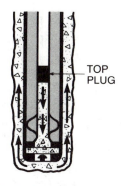

Figure 7-15. Inner-string cementing.

utilizes a modified float collar together with sealing adapters on the drill pipe (Figure 7-15). The cementing is done through the drill pipe, which has been "stabbed" into this modified float collar. Cement mixing is continued until cement returns to the surface. When the drill pipe plug is released, only the volume of cement present in the drill pipe is displaced to the sump. This ensures full-string cementing and reduces the amount of cement returns.

Liner cementing is similar to casing cementing method except for the technique used to get the cement in place. Once the liner is in place, the cement slurry is circulated down the drill pipe out of the liner, and, then, up the outside of the liner. Plugs are used to prevent mixing of drilling fluid and cement in the liner just as they are in the casing string. The liner plug, which is installed prior to running the liner in the hole, has an opening that allows passage of the cement slurry. At the proper time, a drill pipe plug is released from the surface. It displaces the cement slurry from the drill pipe and upon seating in the liner plug forces the latter down the liner displacing the slurry into the annulus. Seldom is the liner in a geothermal well moved, although there are tools available to achieve this.

Upon the completion of the cementing job, the drill pipe is picked up 2-3 stands above the liner top and the excess cement is reversed out of the hole. The remaining portion of the cement on the liner top is later drilled out.

Because a tight seal is necessary in the liner lap, in the event such a seal did not form, a squeeze cementing job is required. *Squeeze cementing* in geo-thermal wells involves pumping cement through the drill pipe, which is attached to a squeeze packer set in the casing above the liner lap. Cement is squeezed into the voids and forms a seal that was not achieved during the primary casing or liner job (Figure 7-16).

Waiting-on-cement time requires considerable patience, the lack of which is the major cause for numerous failures. Cement slurries used in most

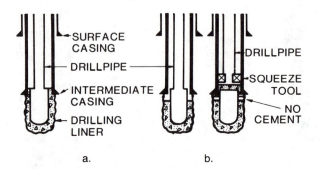

Figure 7-16. Liner-cementing in high-pressure zones. A—single-stage operation in which cement is circulated around liner (top to bottom) in one continuous operation. B—two-stage operation in which cement is first circulated around 70% of the annulus volume; cement squeeze job is then performed above the liner top.

geothermal wells contain retarders, which are required for successful placement. Retarders affect the setting time and compressive strength development as well as the thickening time, and it is seldom advisable to resume operations until the cement has been in place for at least 16 hours. Waiting-on-cement time can be reduced only if laboratory testing of that particular slurry at the conditions existing in the well has been performed. Most often, it is more advisable to add 6 to 8 hours to the waiting-on-cement time rather than to reduce it by 3 to 4 hours. It is recommended that the compressive strength of cement should not be less than 500 psi before drill-out operation, following a primary casing or liner job, and should be greater than 4,000 psi for a kick-off plug.

Summary

All cementing operations (casing, liner, plugback or squeeze) require proper planning. Preparation for these jobs is essential and attention must be given even to the smallest details. Selection of proper cementing materials, equipping the casing with adequate numbers of centralizers, and installation of the floating-guiding equipment should be done well in advance of the actual operation. Proper planning, coordinating, and executing of a cement job in a geothermal well will prevent failure to obtain a successful cementing job which can result in costly problems sometimes during the life of the well.

References

Budd, C. F., Jr., 1973. Steam Production at the Geysers Geothermal Field. In: *Energy Resources, Production, and Stimulation,* P. Kruge and C. Otte (eds.), pp. 129-137.

Carter, G. and Shryock, S. H., 1966. Thermal Well Completions Can Be Improved. *World Oil,* Oct.: 146-149.

Patchen, F. D., 1960. Reaction and Properties of Silica-Portland Cement Mixtures Cured at Elevated Temperatures. *Trans. Am. Inst. Min. Metall. Engrs.,* 219: 281-287.

Smith, D. K., 1976. *Halliburton Energy Institute Manual.* Chapter 6: 1-97.

Walter Fertl, Dresser Industries
Harold Overton, Consultant

8

Formation Evaluation

Objectives of Geothermal Well Logging

Very high-temperature, hostile borehole conditions, unusual geologic environments (fractured, igneous, metamorphic), and specific, often novel, key reservoir parameter objectives frequently restrict the application of existing well logging instrumentation and related interpretive techniques for exploration, resource assessment, and production of geothermal systems. An overview of the National Geothermal Exploration Technology Program has been presented by Ball et al. (1979) recently, whereas a nine-year geothermal logging instrumentation development program plan was outlined by Veneruso et al. (1978).

Well logging in the petroleum industry has developed over the last fifty years into a mature industry, whereas geothermal well logging is a relatively new enterprise. Well logging instrumentation standard for petroleum industry applications was used in the early stages of geothermal resource evaluation. More recently, several well logging companies have developed logging instrumentation applicable in geothermal environments. This includes temperature-upgrading of existing equipment and the design of new devices. At the same time, attention is being focused on the development of reliable interpretation techniques.

Furthermore, geothermal logging operations, such as in the Geysers steam field, often require special precautions. In other areas, wellbores frequently need to be cooled down. This is accomplished by extensive well conditioning (circulating or even icing-down); however, in air-drilled steam wells such procedures are not applicable.

With the exception of geothermal-geopressure resources in clastic sequences, fundamental differences also occur in the geologic environments and key objectives of both geothermal and petroleum-related logging applications (Table 8-1). Geothermal reservoirs frequently occur in fractured igneous or metamorphic rocks, which contain hot water or steam at temperatures exceeding 150°C (Sanyal et al., 1979). See Table 8-2.

Based on the Geothermal Measurement Workshop (Baker et al., 1975) several key parameters needed to evaluate geothermal resources have been identified. Whereas importance and application of individual parameters vary with the type of geothermal resource, geologic environment, and state of resource development, the following specific factors have been stressed:

1. *Temperature* — wellbore and static formation temperature.
2. *Pressure* — hydrostatic wellbore pressure, formation and overburden pressure, and/or pressure gradient.
3. *Flow rate* — low and high volume rates on test or production.
4. *Permeability and associated parameters* — fracture systems, reservoir shaliness, fluid velocity, interconnected porosity, pore geometry, size of pores, and compressibility.
5. *Hydrogeochemical data:* such as total dissolved and undissolved solids and gases, pH, Eh, fluid density, chemical composition, and compressibility.
6. *Porosity:* interconnected, isolated, effective, primary versus secondary (fractures, vugs, solution, dolomitization).
7. *Casing:* Corrosion, scaling problems, and design.
8. *Cementation:* fluid isolation, channeling, cement properties, and micro-annulus.
9. *Geologic properties:* such as reservoir thickness, rock type, stratigraphy, orientation, mineral composition, and discontinuities.
10. *Heat flow and thermal conductivity.*

A priority list of parameters desired for open hole exploration and reservoir assessment, summarized by Ball et al. (1979), is shown in Table 8-3, whereas the technical impact and economical significance of improved geothermal logging techniques and associated interpretive techniques, as reviewed by Rigby and Reardon (1979), are shown in Table 8-4.

Evaluation of clastic sediments and carbonate formations is well established and documented in extensive logging literature. Despite several recent studies and important developments, well log analysis in igneous and metamorphic rocks however, is still in its infancy (Ritch, 1975; Nelson and Glenn, 1976; West et al., 1975; Ehring et al., 1978; Sanyal et al., 1979; Keys, 1979; Sethi and Fertl, 1979). Additional research will have to focus on basic petrophysical models for the response of conventional, temperature-upgraded, and/or novel logging devices in igneous and metamorphic forma-

Table 8-1
Typical Energy Resource Characteristics

Properties	Hydrocarbon Resources	Geothermal Resources
Reservoir	Sedimentary rocks	Igneous, metamorphic, well-indurated sedimentary rocks
Porosity	Mainly integranular	Mainly secondary (fractures, solution channels, vugs, etc.)
Fluids	Fresh to saline	Corrosive brines or steam
Fluid distribution	Hydrocarbons plus water (i.e., less than 100% saturated with formation water)	Formation brines (100%) or steam
Static formation temperature	Less than 400°F	350°F to 500°F (and higher)
Energy resource	Hydrocarbons: easily stored or transported at surface	Heat: not easily stored or transported at surface

Table 8-2
Distribution of U.S. Geothermal Reservoirs

Lithology	Igneous (crystalline and glassy)	31.1%
	Sedimentary	22.2%
	Volcanic ash and associated formations	20.0%
	Hydrothermally altered	20.0%
	Metamorphic	6.7%
Pore geometry	Fractures	50.0%
	Integranular	31.3%
	Vuggy or vesicular	18.7%
Fluid type and temperature regime	Water in excess of 400°F	33.3%
	Water at 300°F to 400°F	33.3%
	Water below 300°F	22.3%
	Steam	7.4%
	Dry rock	3.7%
Fluid salinity	Below 5,000 ppm total solids	52.3%
	5,000 ppm to 35,000 ppm	30.4%
	35,000 ppm to 100,000 ppm	8.7%
	Exceeding 100,000 ppm	4.4%
	Dry	4.4%

After Sanyal et al., 1979

Table 8-3
Borehole Parameter Priorities

- Formation temperature
- Formation pressure
- Flow rate
- Hole geometry (may be critical in log interpretation)
- Fracture system (location, orientation, permeability, etc.)
- Fluid composition (pH and types and amounts of dissolved solids and gases)
- Permeability

- Porosity (interconnected, effective, and isolated)
- Formation depth and thickness
- Thermal conductivity
- Electrical conductivity or resistivity
- Heat capacity
- Lithology and mineralogy
- Acoustic wave velocity
- Formation density

After Ball et al., 1979

Table 8-4
Impact of Improved Interpretation and Application
of Logging Information

Impact	Activity	Requirements
Improved drilling success rate	Exploration development program	Better interaction with survey data, new techniques; well siting from reservoir model.
Reduced drilling costs	Development program Reservoir modeling	Preplanning wells; improved or more rapid completion; better selection of total depth.
Reduced testing time	Assessment	Reduced flow testing of production wells based on greater comparability and relation of logged parameters to production characteristics.
Improved reinjection program	Environmental protection	Justification for less than 100% reinjection; justification for shallow reinjection.
	Reservoir management	Better control of reinjection well clogging.
Improved well pump placement	Development program	Better location of flashing level and scale characteristics.
Reduced permitting	Environmental protection	Better control of subsidence and seismicity; ability to monitor reservoir.

(Table continued next page)

Table 8-4 continued

Impact	Activity	Requirements
Accelerated development	Reservoir modeling	Earlier utility commitment.
Facilitation of unitization agreement or other regulations controlling development	Development program	Earlier agreement, more equitable allotment of revenues or drilling rights.
Plant design	Plant operation	Easier or earlier determination of brine chemistry.
Improved well performance	Reservoir management	Development of techniques of well stimulation.
Extended field life	Reservoir management	Reservoir monitoring, improved modelling, and reinjection control.
Location of new resources or resources otherwise overlooked	Exploration	New techniques and interaction with surface data. Special significance for direct use.

After Rigby and Reardon, 1979

tions, specific tool calibrations tailored to these geologic environments, and various fracture detection concepts. A compilation of the latter is presented in Table 8-5. Furthermore, a great need exists for extensive correlations of core and logging data in many types of igneous and metamorphic rocks to properly evaluate well log responses.

In the past, geothermal test wells have not been available to calibrate specific log responses. Recently, geothermal test wells have been made available at Roosevelt Hot Springs KGRA in Southwestern Utah at 440°F (225°C) and East Mesa, California, at 330°F (165°C). Several well logging service companies have already run various logging instrumentation in these calibration wells. Unfortunately, no core data is available in either of the two wells.

The test well in Utah is partially cased (5½ in.) down to 4,200 ft (1,280 m), with open hole (8½ in.) completion from 4,200 ft to total depth of 6,885 ft (2,098 m). It penetrates igneous rocks including diorite, granodiorite, and granite.

The East Mesa calibration well in California is 6,175 ft (1,880 m) deep. It is cased to TD with 7⅞-in. pipe and penetrates Plio-Pleistocene deltaic

Table 8-5
Detection of Fractures Intersecting Wellbores

Methods	Comments
Impression packer	Physical imprint of fracture on rubber sleeve; fracture orientation capability (FOC).
Straddle packer	Locates fractured interval; no fracture orientation capability (NFOC).
Flow spinners	Fluid entry points into wellbore defined by mechanical wireline device; NFOC.
Temperature log	"Cooling" anomaly due to drilling fluid invading open fracture system; NFOC.
Temperature scans	Conceptional; FOC.
Downhole TV camera	Requires clean borehole fluid or empty (gas, steam filled) wellbore; stationary NFOC or rotating FOC.
Caliper	Mechanical multi-arm (6+) devices; FOC (?).
Spontaneous potential curve	Erratic curve shape, streaming potentials; NFOC.
Resistivity logs	Anomalies due to different response characteristics of induction and micro-resistivity devices; NFOC.
Diplogs (dipmeter)	Multi-arm (4) micro-resistivity devices locate vertical extent of fractured interval, also FOC.
Radioactive tracers	Artificially induced radioactivity is monitored, NFOC.
Spectral gamma ray logging	Uranium salts precipitated over geologic time recorded, NFOC.
Magnetic induction technique	In uncased wellbores to map fractures and thickness within several meters of wellbore.
Log-inject-log	Based on properly engineered changes of mud salinities for resistivity logs and/or boron-spiked borehole fluids for pulsed neutron logs, NFOC.
Density logging pad devices	Correction curve ($\Delta\rho$) affected by mud in open fractures in gauge hole or shows "clipping" effect (i.e., washouts) over fractured interval, NFOC. "Photoelectric effect" — curve may locate barite (from weighted mud system) penetrated into fracture system, NFOC. For fractures to be seen they have to be on side of borehole where logging pad is in contact with formation.
Acoustic techniques	Conventional and/or sidewall acoustic logs, full acoustic wave train recording (variable density, signature,

(table continued on next page)

Table 8-5 continued

	micro-seismogram, etc.); fracture detection, but NFOC. Acoustic Seis (Tele-) viewer: 360° scanning, FOC. Circumferential acoustic: multi-set of transducers and receivers circumferentially located at the sonde records the shear wave perturbations, FOC.
Secondary porosity index (SPI)	Comparative porosity log responses. Total porosity ϕ_T (primary + vugs and fractures) from density and neutron logs, whereas primary porosity ϕ_p is recorded by acoustic logs. Hence, SPI $= \phi_T - \phi_p$; NFOC.
Cementation m exponent	Apparent m exponent calculated from porosity and resistivity logs; $m\!<\!<\!2.0$ in fractured interval; NFOC.

sandstones, siltstones, clays and shales. Temperature, pressure, caliper, density, neutron, gamma ray, and cement bond logs run by four service companies have been studied on a comparative basis by Mathews et al. (1979).

Geophysical Well Logs

Generally, well logging denotes any operation wherein some characteristic data of the formation penetrated by a borehole are recorded in terms of depth. Such a record is called a log. The log of a well, for example, may simply be a chart on which abridged descriptions of cores are written opposite the depths from which cores were taken. A log may also be a graphic plot with respect to depth of various characteristics of these cores, including porosity, horizontal and vertical permeability, and water hydrocarbon saturations.

In geophysical well logging, a probe is lowered in the well at the end of an insulated cable, and physical measurements are performed and recorded in graphical form as functions of depth. These records are called geophysical well logs, well logs, or simply logs. Often, when there is no ambiguity, geophysical well-logging operations are referred to shortly as well logging or logging.

Various types of measuring devices can be lowered on cables in the borehole for the sole purpose of measuring (logging) borehole and *in-situ* formation properties. These logging tools, or logging sondes, contain sensors that measure the desired downhole properties, whether thermal, magnetic, electric, radioactive, or acoustic. Insulated conductive cables not only suspend these sondes in the borehole, but also provide power to the sondes and transmit recorded signals (data) to the surface, where the latter are recorded as a log. Hence, geophysical well-logging methods provide a detailed and economical evaluation of the entire length of drilled hole.

Most well-logging methods have been initially designed and developed to assist in answering some major questions associated with exploration, evaluation, and production of oil and gas. Today, however, many of these logging instrumentations and associated interpretation concepts have been successfully applied in exploring and evaluating metallic deposits, non-metallic minerals, and in mine design. They are being increasingly used in the evaluation of geothermal resources.

Most logging is performed with the sonde moving uphole, but due to the hostile environment, geothermal wells are frequently logged with the sonde going down into the borehole.

Several geophysical well logs can be used to evaluate geothermal resources, as far as lithology variations, thickness, porosity, fracturing, formation pressures, temperature, salinity, entry points, and quality of steam, etc., are concerned. Before discussing properly selected logging programs and associated interpretation methods, geophysical well logs applicable for geothermal reservoir evaluation will be reviewed briefly. Detailed and extensive information on each instrument type is available from the logging service companies.

Caliper Log

This type of log measures and records the size (diameter) of the borehole. Basically, flexible springs or arms, which ride against the wall of the borehole, are mounted on the body of a sonde. Owing to changes in hole size, the movement of these springs or arms generates voltages, which are transmitted to the surface and recorded in terms of hole diameter.

For example, the 4-arm caliper measures two diameters at right angles to one another and is used in hole size and hole shape determinations, fracture indications, location of parting casing, borehole profile for cement volume calculations, etc. Through-tubing calipers, such as a 1⅜-in., 3-arm caliper, are useful in casing or tubing strings where internal corrosion, paraffin deposits, or mineral scaling is a problem.

Calipers can be run in combination with virtually all openhole logging instruments and work equally well in boreholes filled with fresh-water or salt-water drilling muds or completion fluids, oil-base muds, or gas (e.g., air).

In geothermal resource evaluation, the caliper log assists in washout corrections on porosity log responses and, particularly, in fracture detection.

For example, 3-arm calipers were run in wells No. EE-1 and No. GT-2 for the Los Alamos Hot Dry Rock Project at Fenton Hill, New Mexico. Results from the U.S.G.S. borehole televiewer scans and the oriented caliper log agree quite well with the fractures observed and fracture directions measured on oriented cores. Figure 8-1 shows the 3-independent-arm oriented caliper log in well GT-2b at a depth of 8,920 ft (2,719 m). The correlation between these caliper anomalies and the conceptual fracture model is illustrated in Figure 8-2 (Brown et al., 1979).

Nonfocused Electric Log

This electric log, which is an older-type logging instrument, measures the resistivity of materials surrounding the logging sonde. Frequently, several different electrode configurations have been applied in combination to measure several resistivity parameters exhibiting different depths of investigation. Quality of measurements depends on the type of borehole fluid, wellbore size, formation resistivity, reservoir thickness, mud filtrate invasion, etc.

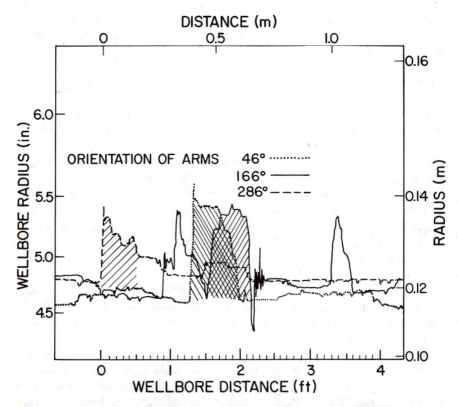

Figure 8-1. Three-independent-arm oriented caliper log in GT-2B well at a depth of 2,719 m (8,920 ft). (After Brown et al., 1979.)

The short normal curve is very useful for qualitative detection and quantitative overpressure evaluation, which are prerequisite for geothermal-geopressure resource evaluation. Electrical resistivity measurements are mostly off-scale in the massive igneous and/or metamorphic sections.

Electric logs cannot be run in air (gas)-filled wellbores, oil-base muds, or in cased boreholes.

Focused Conductivity (Induction) Type Logs

This log measures the electric conductivity of formations traversed by the borehole to obtain an estimate of the true formation resistivity. Formations are energized by electromagnetic induction and the resulting electromagnetic forces (eddy currents) are detected by receiver coils and transmitted to the surface, where formation conductivity and its reciprocal resistivity are recorded.

A *dual induction-focused log* provides information from four measurements performed simultaneously, i.e., deep-investigation induction, medium-investigation induction, shallow-investigation guard-type focused curve, and a lithology curve (gamma ray or spontaneous potential (SP) curve). This log is most effective in medium-porosity to low-porosity formations drilled with freshwater-base muds, where mud filtrate invasion is common. From this log true formation resistivity and the depth of filtrate invasion can be determined.

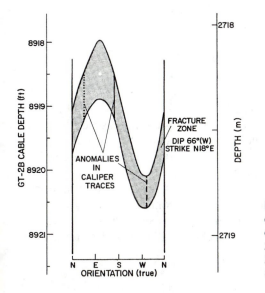

Figure 8-2. Correlation between oriented three-independent-arm caliper anomalies and fracture zones in GT-2B well at a depth of 2,719 m (8,920 ft). (After Brown et al., 1979.)

The conductivity curve is frequently used in quantitative overpressure evaluation techniques and it also assists in evaluating some fractured igneous and metamorphic intervals.

With the exception of very saline borehole fluids (particularly in large or severely washed-out wellbores) the induction-electrolog works equally well in gas-filled holes and oil-base or fresh-water muds particularly in formations having medium to high porosity.

Focused Resistivity Log

Several types of this log are routinely used for measuring the apparent resistivity of a thin segment of the formation perpendicular to the borehole. These logging instruments perform well in the saltwater muds and in detecting highly resistive and relatively thin zones, but are ineffective in gas-filled boreholes and in oil-base muds.

Formation resistivity is measured by recording the voltage at a central electrode from which a constant current is so focused that the latter goes laterally a certain distance into the formation and then fans out vertically. This focusing is accomplished by proper arrangement of electrodes, situated above and below the measuring electrode.

Dual laterologs simultaneously measure deep and shallow formation resistivities. In addition, a gamma ray curve is most frequently run as a lithology curve (if so desired, the spontaneous potential curve can replace the gamma ray).

Short-spaced resistivity logs assist in determining net pay thickness, define the amount of movable hydrocarbons present in potential reservoir rocks, and locate porous and permeable zones. In geothermal resource evaluation, however, they are of limited use.

Gamma Ray Log

This is used as a lithology log and to determine shaliness and/or radioactive heavy minerals in zones investigated. It basically measures the intensity of the natural gamma rays of the formations penetrated by the borehole. This natural radiation intensity is measured by proper downhole detector, amplified, and then transmitted to the surface.

Applications include depth correlation with other logs, determination of stratigraphic profiles, differentiation of shales from nonshales, determination of rock types, estimation of reservoir shaliness, etc. The gamma ray log can be run in any liquid-or air-filled borehole, either cased or uncased.

Spectralog (Gamma Ray Spectral Logging)

Basically, the vacuum flask assembly in the pressure housing of the Spectralog sonde contains a high-resolution gamma spectrometer. The spectrometer consists of the thallium-activated sodium iodide crystal (2 in. by 12 in.) optically coupled to a photomultiplier (PM) tube. A downhole, high-quality electronic amplifier assures voltage amplitude proportionality when transmitting the pulses through either a single conductor or multi-conductor logging cable to the surface panels. These include an electronic amplifier, a multichannel analyzer, a digital panel, and a conventional logging camera.

Pulse signals reaching the surface first pass through the electronic amplifier (its gain set during calibration). Amplified pulses are then filtered by the multichannel analyzer, which not only displays the entire spectrum, but also selects pulses within preselected energy windows. Next, the digital panel computes background corrected radiation count rates from the raw logging data by means of a mathematical spectrum stripping technique.

Four count rate meters (CRM) are available to tally the total number of gamma rays measured (total count rates, counts/minute) and the background corrected count rates for potassium, uranium, and thorium in each of the energy windows. The outputs from each CRM are displayed on the camera and recorded on film as a function of depth.

Hence, in addition to total gamma ray counts, the Spectralog measures and records the gamma rays emitted by:

1. Potassium-40 (K^{40}) at 1.46 MeV
2. Uranium series, nuclide bismuth-214 (Bi^{214}) at 1.764 MeV
3. Thorium series, nuclide thallium-208 (TI^{208}) at 2.614 MeV

Over the last few years, extensive field experience with the Spectralog in open hole and cased wellbores has clearly shown its wide and versatile potential for oil, mineral, and geothermal resource evaluation.

Application of such gamma ray spectral data may be made either qualitatively, such as in detailed stratigraphic correlation, recognition of rock types in argillaceous, clastic, carbonate, igneous, and metamorphic formations, fracture and high-permeability zone identification, location of watered-out zones in oil reservoirs, etc.; or quantitatively, to determine reservoir shaliness, cation exchange capacity of clays in potential reservoir rocks, clay diagenesis, source rock potential of argillaceous formations, potash concentration, ash content of coal seams, evaluation of uranium and other mineral deposits, and geothermal resource assessment.

Quantitative Spectralog applications include the determination of rock types, detecting fracture zones, and examining mobility of the elements uranium, potassium, and thorium. Distribution of potassium, uranium, and thorium in several rocks and minerals is described in Table 8-6 (Fertl, 1979) and Table 8-7 (Frondel et al., 1956).

Gamma-Gamma Density Log

This log basically measures the electronic density (i.e., number of electrons per cm) of downhole formations. This electron density is directly related to the true bulk density which, in turn, is a function of the composition of rock matrix material, formation porosity, and the density of the fluids and/or gases filling the pore space.

A gamma ray source emits medium-energy gamma rays into the formation where they collide with electrons in the rock. This collision causes a backscattering and a decrease in gamma ray energy (Compton effect). As the scattered gamma rays reach two detectors (on the logging instrument) located at a fixed distance from the radioactive source, they are digitally corrected and recorded as a corrected formation bulk density. With the appropriate formation matrix density, these bulk density values can then be used to estimate formation porosity.

Density logging of geothermal wells assists in porosity determination, crossplots for determining rock types, fracture location, steam entry, and steam quality estimates. It also has application in quantitative evaluation of subsurface overpressure regimes.

Density logs can be run equally well in all types of borehole fluids, including empty wellbores.

Acoustic (Sonic) Log

An acoustic log records the time required for a compressional sound wave to traverse one foot of formation (microseconds per foot). A transducer creates elastic wave pulses that travel a given distance through the formation and are then picked up at receivers in the logging instrument.

Travel time, total acoustic wave trains, signal amplitudes, etc., can be recorded in open and cased wellbores. Acoustic logs can be run in liquid-filled boreholes but not in air or gas-filled holes.

The recorded interval transit time is the reciprocal of the velocity of the compressional sound wave and is affected by formation lithology, porosity, and type and amount of fluids in the pore space.

Long-spaced borehole-compensated acoustic logs are available for acoustic measurements that are even less affected than the standard borehole-compensated acoustic logging information. Furthermore, acoustic "televiewers" (seisviewers) are available, which scan the borehole walls

Table 8-6

Potassium (K), Uranium (U), and Thorium (Th) Distribution in Several Rocks and Minerals

	(K) (%)	U (ppm)	Th (ppm)
Accessory Minerals			
Allanite		30-700	500-5000
Apatite		5-150	20-150
Epidote		20-50	50-500
Monazite		500-3000	2.5×10^4-20×10^4
Sphene		100-700	100-600
Xenotime		500-$3,4 \times 10^4$	Low
Zircon		300-3000	100-2500
Andesite (average)	1.7	0.8	1.9
A., Oregon	2.9	2.0	2.0
Basalt			
Alkali basalt	0.61	0.99	4.6
Plateau basalt	0.61	0.53	1.96
Alkali olivine basalt	<1.4	<1.4	3.9
Tholeiites (orogene)	<0.6	<0.25	<0.05
(non orogene)	<1.3	<0.50	<2.0
Basalt in Oregon	1.7	1.7	6.8
Carbonates			
Range (average)	0.0-2.0(0.3)	0.1-9.0(2.2)	0.1-7.0(1.7)
Calcite, chalk, limestone, dolomite (all pure)	<0.1	<1.0	<0.5
Dolomite, West Texas (clean)	0.1-0.3	1.5-10	<2.0
Limestone (clean)			
Florida	<0.4	2.0	1.5
Cretaceous Trend, Texas	<0.3	1.5-15	<2.0
Hunton Lime, Okla.	<0.2	<1.0	<1.5
West Texas	<0.3	<1.5	<1.5
Clay Minerals			
Bauxite		3-30	10-130
Glauconite	5.08-5.30		
Bentonite	<0.5	1-20	6-50
Montmorillonite	0.16	2-5	14-24
Kaolinite	0.42	1.5-3	6-19
Illite	4.5	1.5	
Mica			
Biotite	6.7-8.3		<0.01
Muscovite	7.9-9.8		<0.01
Diabase, Va.	<1.0	<1.0	2.4
Diorite, quartzodiorite	1.1	2.0	8.5
Dunite	<0.02	<0.01	<0.01
Feldspars			
Plagioclase	0.54		<0.01
Orthoclase	11.8-14.0		<0.01
Microcline	10.9		<0.01

Table 8-6 continued

	(K) (%)	U (ppm)	Th (ppm)
Gabbro (mafic igneous)	0.46-0.58	.84-.9	2.7-3.85
Granite (silicic igneous)	2.75-4.26	3.6-4.7	19-20
Rhode Island	4.5-5	4.2	25-52
New Hampshire	3.5-5	12-16	50-62
Precambrian (Okla., Minnesota, Col., Tex.)	2-6	3.2-4.6	14-27
Granodiorite	2-2.5	2.6	9.3-11
Colorado, Idaho	5.5	2.-2.5	11.0-12.1
Oil shales, Colorado	<4.0	up to 500	1-30
Peridodite	0.2	0.01	0.05
Phosphates		100-350	1-5
Rhyolite	4.2	5	
Sandstones, range (average)	0.7-3.8(1.1)	0.2-0.6(0.5)	0.7-2.0(1.7)
Silica, quartz, quartzite (pure)	<0.15	<0.4	<0.2
Beach sands, Gulf Coast	<1.2	0.84	2.8
Atlantic Coast (Flo., N.C.)	0.37	3.97	11.27
Atlantic Coast (NJ Mass.)	0.3	0.8	2.07
Shales			
"Common" shales (range & av.)	1.6-4.2(2.7)	1.5-5.5(3.7)	8-18(12.0)
Shales (200 samples)	2.0	6.0	12.0
Schist (biotite)		2.4-4.7	13-25
Syenite	2.7	2500	1300
Tuff (feldspathic)	2.04	5.96	1.56

After Fertl, 1979

across 360 degrees and thus provide information on lithology and fractures intersecting the wellbore. Acoustic-type logs

1. Determine intergranular reservoir porosity in liquid-filled holes (provided lithology is known) and secondary (fracture, solution channels, and vugs) porosity.
2. Identify lithology.
3. Locate fractured intervals.
4. Assist in interpreting seismic data (formation velocity information).
5. Detect and quantitatively evaluate subsurface pressure regimes.
6. In combination with bulk density data, using the acoustic-derived shear wave information, define the four important elastic rock constants (Young's modulus, bulk modulus, shear modulus, Poisson's ratio).

Table 8-7
Thorium and Thorium-Bearing Minerals

Name	Composition	ThO$_2$ content, %
Thorium minerals		
Cheralite	(Th,Ca,Ce)(PO$_4$SiO$_4$)	30, variable
Huttonite	ThSiO$_4$	81.5 (ideal)
Pilbarite	ThO$_2$·UO$_3$·PbO·2SiO$_2$·4H$_2$O	31, variable
Thorianite	ThO$_2$	Isomorphous series to UO$_2$
Thorite**	ThSiO$_4$	25 to 63-81.5 (ideal)
Thorogummite**	Th(SiO$_4$)$_{1-x}$(OH)$_{4-x}$;x<0.25	24 to 58 or more
Thorium-bearing minerals		
Allanite	(Ca,Ce,Th)$_2$(Al,Fe,Mg)$_3$Si$_3$O$_{12}$(OH)	0 to about 3
Bastnaesite	(Ce,La)Co$_3$F	Less than 1
Betafite	About (U,Ca)(Nb,Ta,Ti)$_3$O$_9$·nH$_2$O	0 to about 1
Brannerite	About (U,Ca,Fe,Th,Y)$_3$Ti$_5$O$_{16}$	0 to 12
Euxenite	(Y,Ca,Ce,U,Th)(Nb,Ta,Ti)$_2$O$_5$	0 to about 5
Eschynite	(Ce,Ca,Fe,Th)(Ti,Nb)$_2$O$_6$	0 to 17
Fergusonite	(Y,Er,Ce,U,Th)(Nb,Ta,Ti)O$_4$	0 to about 5
Monazite*	(Ce,Y,La,Th)PO$_4$	0 to about 30; usually 4 to 12
Samarskite	(Y,Er,Ce,U,Fe,Th)(Nb,Ta)$_2$O$_6$	0 to about 4
Thucholite	Hydrocarbon mixture containing U, Th, rare earth elements	
Uraninite	UO$_2$ (ideally) with Ce, Y, Pb, Th, etc.	0 to 14
Yttrocrasite	About (Y,Th,U,Ca)$_2$(Ti,Fe,W)$_4$O$_{11}$	7 to 9
Zircon	ZrSiO$_4$	Usually less than 1

After Frondel et al., 1956
*Most important commercial ore of thorium. Deposits are found in Brazil, India, USSR, Scandinavia, South Africa, and U.S.A.
**Potential thorium ore minerals.

7. Allow estimation of degree of sand consolidation and possible sand production.
8. Describe quality of cement bond behind pipe, etc.

Neutron Log

This log measures the abundance of hydrogen atoms in downhole formations or the relative abundance of epithermal neutrons arriving at the detector.

High-energy, fast neutrons are continuously emitted from a radioactive source, which is mounted in the sonde. Owing to collision with nuclei of the rock, the neutrons will be slowed down until finally they are captured by the nuclei of atoms, such as chlorine, hydrogen, and silicon. The capturing nucleus emits a high-energy gamma ray, and depending on the type of neutron logging instrumentation, either these capture gamma rays or the neutrons (thermal, epithermal) themselves are counted by detectors in the sonde. For many applications, the standard neutron log and the sidewall neutron devices have been superceded by the compensated neutron log. Instrumentation for the latter utilizes a source and two detectors, one near and one distant. These two measurements then allow the determination of apparent limestone porosity.

Neutron logs can be run equally well in uncased and cased wellbores, independent of, but affected by, the type of fluid in the borehole. These logs assist in porosity and lithology determination and secondary porosity identification. In combination with bulk density measurements, they locate gas-bearing intervals and evaluate entry zones and quality of steam in geothermal wells.

Pulsed neutron logging. The *dual detector neutron lifetime log* (DNLL) has its optimum application in high-porosity reservoirs containing high-salinity formation brines. It can be run equally well in fluid or gas-filled wellbores. The DNLL measures the macroscopic thermal neutron absorption cross-section (Σ) of the bulk formation. The value of Σ is primarily a function of porosity, formation water salinity, type and quantity of hydrocarbons in pore space and the type of rock matrix. The DNLL logs:

1. Can be correlated with other open and cased-hole logs.
2. Determine stratigraphy.
3. Establish semiquantitative values for porosity.
4. Evaluate hydrocarbon resources by determining fluid saturation distribution in reservoir pore space and locating gas and gas/liquid contacts.

This log can be run through drill pipe and can quantitatively evaluate subsurface overpressure regimes in open or cased wellbores. In fresh water environments, the DNLL provides reliable shaliness estimates, especially in clastic reservoir rocks.

The *carbon/oxygen log* (C/O), which is another pulsed neutron device, determines the quantity of hydrocarbons behind the casing. Its response is unaffected by fresh, mixed, or varying formation water salinities. Lithologies are identified by simultaneous calcium and silicon measurements. Whereas DNLL instrumentation is rated up to 350°F, the even more sophisticated C/O instrumentation is presently rated to 275°F only.

High Resolution 4-Arm Diplog (Dipmeter)

The focused, high-resolution 4-arm diplog is a borehole wall contact logging instrument that is designed to detect changes in formation resistivity. Data is obtained from short-spaced, focused electrode systems imbedded in insulating pads designed to maintain a uniform contact with the borehole wall.

Simultaneous measurements are recorded at each pad electrode system. Each recording is correlated in order to establish dip angles across the borehole. A hole directional survey and a hole caliper survey are recorded simultaneously in order to correct the dip across the borehole to true formation dip. These data are recorded on film and/or magnetic tape for the necessary interpretive digital processing.

Analysis of this computed data provides information on the presence or absence of structural changes — identification of faults, unconformities, crossbedding, sand bars, reefs, channels, deformation around salt domes, and other structural anomalies. Application of diplogs is very much limited to clastic sequences and of not much use in igneous and metamorphic rock sequences.

Formation Multi-Tester (FMT)

This tester has a major advantage over a standard wireline formation tester because it can record an unlimited number of pressure tests on a single trip into the borehole, and recover two independent formation fluid samples.

Basically, the FMT instrumentation utilizes a sidewall sealing pad to isolate formation fluids from the fluids in the wellbore. Accuracy of the pressure measurements using the proper field calibration and temperature correction is $\pm$ 0.13% with a recorded pressure resolution of $\pm$ 1.0 psi.

Accurate depth control is ensured by correlation with the spontaneous potential curve. Subsurface measurements of hydrostatic and formation pressures, combined with the surface evaluation of the recorded fluids, provide valuable information for reliably predicting reservoir characteristics, including formation permeability determined from pressure drawdown and/or buildup.

Geothermal application of the FMT is limited to clastic hot water wells and is not practical in igneous or metamorphic rocks. Generally, an FMT should not be used when the downhole temperature exceeds 350°F.

Temperature Log

Most temperature logs provide a continuous measurement of two parameters, the absolute borehole fluid temperature and a continuous differential temperature. Over the last few years, several types of temperature logs have been run in geothermal wells on a comparative basis, sometimes with conflicting results.

Today, special high-temperature logs that have been successfully field tested in several KGRA's are available to the geothermal industry.

Flowmeter (Spinner Surveys)

Continuous flowmeter surveys can be used for metering fluid flow rates in cased or open hole wells. Velocity of fluid movement in the borehole, toward the surface in producing wells and toward the bottom in injection wells, can be determined. Units of measurements are revolutions per second, barrels per day, and percentage of full flow.

Applications include production or injection profiles to indicate fluid movement, detection of zones of lost circulation in open hole, production loss due to crossflow thief zones, detection of packer leaks, and location of fractured, high-flow capacity intervals in geothermal wells.

High-temperature flowmeter instrumentation has been successfully applied in several geothermal areas, including the Geysers dry steam field in California.

Production Logging Instrumentations

Besides the temperature log and flowmeter, many other production logging devices are available to the oil industry to monitor reservoir behavior around the wellbore, check the integrity of casing and entire well completion, during production or injection, and to obtain the necessary information for properly engineered well workover and recompletion efforts.

It is outside the scope of this brief review to discuss any of these production logging services. It should be noted however, that high-temperature production logging will drastically increase in capability and importance with increasing maturity of the geothermal industry.

Subsurface Pressure Concepts

Overburden pressure originates from the combined weight of the formation matrix (i.e., rock) and the fluids (i.e., water, oil, and gas) in the pore space overlying the formation of interest.

Generally, it is assumed that overburden pressure increases uniformly with depth. For example, average Tertiary formations on the U.S. Gulf Coast and elsewhere exert an overburden pressure gradient of 1.0 psi/ft of depth ($0.231 \, \mathrm{kgcm^{-2}m^{-1}}$). This corresponds to a force exerted by a formation with an average bulk density of 2.31 g/cm^3. Worldwide experience also indicates that the probable maximum overburden gradient in clastic rocks may be as high as 1.35 psi/ft ($0.312 \text{ kgcm}^{-2}\text{m}^{-1}$). Furthermore, field observations over the last few years have resulted in the concept of a varying (i.e., not constant) overburden gradient for fracture gradient predictions used in drilling and completion operations.

Hydrostatic pressure is caused by the weight of interstitial fluids and is equal to the vertical height of a fluid column times the specific weight of fluid. Size and shape of this fluid column have no effect on the magnitude of this pressure.

The hydrostatic pressure gradient is affected by the concentration of dissolved solids (i.e., salts) and gases in the fluid column and the magnitude of varying temperature gradients. An increase in dissolved solids (i.e., higher salt concentration) tends to increase the pressure gradient, whereas increasing amounts of gases in solution and higher temperatures would decrease the hydrostatic pressure gradient. For example, a pressure gradient of 0.465 psi/ft ($0.1073 \text{ kgcm}^{-2}\text{m}^{-1}$) assumes a water salinity of 80,000 parts per million (ppm) NaCl at a temperature of 77°F (25°C).

Typical average hydrostatic gradients encountered during drilling for oil and gas range from 0.433 psi/ft ($0.10 \text{ kgcm}^{-2}\text{m}^{-1}$) in rocks containing fresh and brackish water to 0.465 psi/ft ($0.1074 \text{ kgcm}^{-2}\text{m}^{-1}$) in formations containing saltwater.

Formation pressure is the pressure acting upon the fluids (i.e., formation water, oil, and gas) in the pore space of the formation. Normal formation pressures in any geologic setting will equal the hydrostatic head (i.e., hydrostatic pressure) of water extending from the surface to the subsurface formation. Abnormal formation pressures, by definition, are then characterized by any departure from the normal trend line. Formation pressures exceeding hydrostatic pressure are defined as abnormally high formation pressures (surpressures), whereas formation pressures less than hydrostatic are called subnormal (subpressures). See Figure 8-3.

Abnormally high pore fluid pressures are encountered worldwide in formations ranging in age from Pleistocene to as old as the Cambrian age. They may occur as shallow as a few hundred feet below the surface or at depths exceeding 20,000 ft and can be present in shale-sand sequences and/or massive carbonate-evaporite sections.

Detection and quantitative evaluation of overpressured formations are critical to exploration, drilling, and production operations involving hydrocarbon resources. Worldwide experience indicates a significant correlation between the presence and magnitude of formation pressures and the shale/sand ratio of sedimentary sections. Distribution of oil and gas is related to regional and local subsurface pressure and temperature environments. Knowledge of the expected pore pressure and fracture gradients is the basis for (a) efficiently drilling wells with correct mud weights; (b) properly

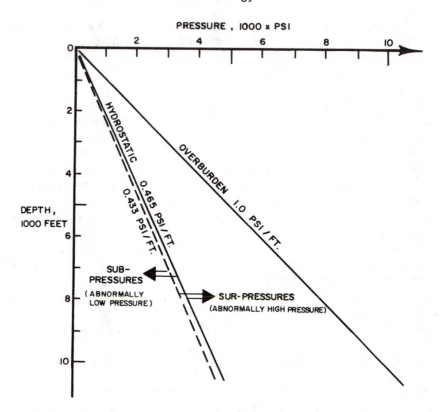

Figure 8-3. Schematic diagram of subsurface formation pressure environment concepts. (After Fertl and Chilingarian, 1976, courtesy of SPE.)

engineered casing programs; and (c) proper completions, which must be effective, safe, and allow for killing of the well without excessive formation damage. In reservoir engineering, formation pressures influence compressibility and the failure of reservoir rocks, and can be responsible for water influx from adjacent overpressured shale sections as an additional driving mechanism in hydrocarbon production.

Many factors can cause abnormal formation pressures. In some areas, a combination of these factors prevails. To place the possible causes of abnormal formation pressures in proper perspective, one must understand the importance of petrophysical and geochemical parameters and their relationship to the stratigraphic, structural, and tectonic history of a given area or basin (Fertl, 1976).

Worldwide experience indicates that costly and hazardous misinterpretations are best avoided by using and studying a combination of several such recorded pressure indicators. These indicators are numerous and can be obtained prior to spudding a well, while drilling, or from wireline logging and formation tester operations. The state of the art of this technology is summarized in Table 8-8 (Fertl and Ilavia, 1977).

Not all of these indicators will always be available, usable, or necessarily needed in any given situation. All data available, however, must be interpreted as an interrelated group by an interdisciplinary team of professionals. Considered individually without relation to other pressure indicators, erroneous interpretation and incorrect decisions could easily be made because numerous parameters may affect formation pressure evaluation techniques. See Table 8-9 (Fertl and Chilingarian, 1976).

The professional team has to do a thorough job of well design, which is safe, well engineered, and economical, by incorporating all known information, "educated guesses" about the unknown, and the flexibility to deviate from the plan when necessary.

Formation Pressure Determination from Well Logs

Wireline logging methods and their evaluation are "after-the-fact" techniques, because they are used after penetration of the geologic interval by the drill bit. Nevertheless, these methods significantly aid engineering planning even though short intervals sometimes have to be logged to monitor pore pressure variations continually. Several types of logs investigating electric, acoustic, and nuclear properties of formations can be used:

1. Formation parameters from geophysical well logs — Electrical surveys: resistivity, conductivity, salinity, shale formation factor. Acoustic (sonic) surveys: interval transit time and wave train presentations (VDL, signature log, etc.). Bulk density surveys, density log, downhole gravity meter. Hydrogen index (neutron-type logs). Thermal neutron capture cross-section (pulsed neutron logging). Nuclear magnetic resonance. Gamma ray spectral analysis logs.
2. Formation multi-tester (FMT) — Unlimited pressure tests in permeable formations. Normalization of normal pressure trend lines.

Except for spontaneous potential (SP) curve data and FMT tests, all parameters recorded in shale formations, rather than in sands, are plotted versus depth. Trend lines are then established for normal compaction. Interpreting such plots depends on their departure from the normal trend (Figure 8-3). Quantitative evaluation can be carried out for a specific formation, area, or region; or the equivalent depth method can be used (Fertl, 1976).

One must be aware of possible pitfalls in order to utilize fully the potential of well logs, particularly the acoustic and electrical curves, as one of the oil

(text continued on page 350)

Table 8-8
Possible Overpressure Indicators

Type	Conditions	Comments
Seismic operations	Subsurface information prior to spudding well.	Presence, depth, and magnitude of overpressures. "Prediction log." High frequency seismic methods for shallow depths.
Gravity data	From surface prior to spudding or in borehole.	Delineating overpressure zones at shallow depth.
Drilling operations	Drilling parameters, instantaneous.	Drilling rate, torque, drag, hole fill (reaming), modified d-exponent, analytical drilling model concepts.
	Logging while drilling, instantaneous or semi-instantaneous.	Electrical and acoustic-type transmission concepts using cables, drill string, mud system analysis of drill string vibrations, downhole gas detection tool, etc.
	Drilling mud parameters while drilling, but delayed by circulation lag-time.	Mud density, gas content, temperature, flow rate, hole fill-up, pit level and flow sensors, salinity (resistivity, conductivity), well kicks.
	Drill cutting analysis while drilling, but delayed by time required for sample return.	Cutting density, shape, size, volume over shale shaker, color, moisture content, "litho-function" plots, shale factor (cation exchange capacity). Cutting slurry and/or filtrate: resistivity color, redox and pH-potential, bicarbonate content, special anion and cation concentrations, filtration rate.
Well logging parameters	Formation parameters from well logs.	Electrical surveys, resistivity conductivity, salinity, shale formation factor. Acoustic surveys, interval transit time and wave train presentations. (VDL, signature log, etc.). Bulk density surveys, density log, downhole gravity meter. Hydrogen index (neutron-type logs). Thermal neutron capture cross section (pulsed neutron logging). Nuclear

Table 8-8 continued

Type	Conditions	Comments
		magnetic resonance. Gamma ray spectral analysis logs.
Well testing	Direct formation pressure measurements.	Formation pressure tests on wireline or drill string, pressure buildup tests, production tests.

After Fertl and Ilavia, 1977

Table 8-9
Parameters Affecting Practical Formation Pressure
Evaluation Techniques

Geologic factors	Geologic age changes. Compaction effects on regional (basin edge vs basin center) and local scale and differential compaction across structures. Sand-shale ratio in clastic sediments. Lithology effects: pure shales (soft, hard); limey and silty shales; bentonitic markers; gas-bearing ("shale gas") organic-rich, bituminous shales; types and amounts of clay minerals in shales (depending on depositional environment and/or diagenesis). Heavy minerals (siderite, pyrite, mica, etc.). Drastic formation water salinity variations in subsurface. High geothermal gradients. Stratigraphic and tectonic features (acting as overpressure continuities or barriers), including unconformities, pinchouts, and faults; proximity to large salt masses, mud volanoes, geothermal "hot" spots, etc. Steep, thin, overturned beds. Pore pressure gradients within single thick shale interval.
Borehole environment	Borehole size, shape, and deviation. Shale alteration and hydration (exposure time of open hole to drilling mud). Type of drilling mud (freshwater, saltwater, oil-base). Type and amount of weighting material (barite, etc.) and/or lost circulation material (mica, etc.) Degree of "gas-cutting" in mud.
Drilling conditions	Hole size, shape, and deviation. Mud programs and mud hydraulics (circulation rate). Rotary speed. Bit type (button, diamond, insert, etc.) Bit weight-to-bit diameter ratio. Bit wear (sharp, new bits vs dull, old bits). Degree of overbalance. Floater ("heave" action) vs fixed onshore or offshore rig.

(table continued on next page)

(Table 8-9 continued)

Sample selection	Type and size of sample (avoid sand, cavings, recirculated shales). Sampling technique. Sampling frequency. Analysis methods (for example, cutting density-variable density column, multiple density solution technique (float and sink method), mercury pump technique, mud balance technique. Proper calibration is of utmost importance.
Geophysical well logging	Different basic measuring principles (and shales are anisotropic). Sonde spacing. Depth of tool investigations. Temperature ratings. Proper tool calibration. Tool malfunction (overlapping repeats or reruns).
Parameter plotting techniques	Interval (or sample) selection. Sampling frequency. Linear, logarithmic plots. Plot comparable data (not compatible are: bulk density from logs vs cuttings short normal vs induction log resistivities). Proper selection of "normal" compaction trendlines (discrepancies become enhanced with increasing depth of wells). Use all information available. Experienced and properly trained personnel.

After Fertl and Chilingarian, 1976

(text continued from page 347)

industry's best methods to locate and evaluate abnormal formation pressures. Complications may arise from unrecognized logging tool problems, severe hole conditions, and unusual formation characteristics (Table 8-9).

Hydrocarbon Distribution in Clastic Overpressured Environments

Important findings are shown in crossplots of formation pressure gradient and temperature (Figure 8-4, a and b). A temperature range of 215-290°F (102-143°C) coincides with the range of highest pressure gradients in hydrocarbon zones. This is particularly noteworthy because this temperature range lies in the bulk part of second-stage clay dehydration as proposed by Burst (1969), i.e., in the zone of maximum fluid distribution. At the same time, extremely high pressure gradients are encountered in high-temperature aquifers and zones with noncommercial oil and gas shows. Burst's region of "aquifers with gas in solution" conforms with the region of dissolved methane in brines investigated presently by the geothermal -geopressured resource projects sponsored by DOE.

A generalized correlation has been developed between a "typical" U.S. Gulf Coast shale resistivity profile (Figure 8-5) and the distribution of oil and gas fields in the area (Timko and Fertl, 1970). Application of such findings indicates whether it is possible for commercial production to exist below the depth to which the well has already been drilled and logged, and whether it is economically attractive to continue drilling a borehole below a given depth in shale-sand sequences.

A shale resistivity ratio parameter, which is a function of thermodynamic and geochemical effects, is often used. Initially developed for the U.S. Gulf Coast area, additional experience has shown the model to hold true in California and several other Tertiary basins with shale-sand sequences throughout the world. Correlations similar to that present in the U.S. Gulf Coast area, however, have to be established (Fertl, 1976).

Formation Temperature

The subsurface formation temperature available from open hole well logs listed on the log heading is always lower than the true, or static, formation temperature. Because of the cooling of formations by circulating drilling mud (conditioning a well prior to logging, for example), the recorded bottomhole temperature (BHT) may be 20°F to 80°F lower than the actual formation temperature.

Inasmuch as true or static formation temperature is an important parameter in exploration, drilling, logging, well completion, and reservoir engineering, there is a method that permits the determination of static formation

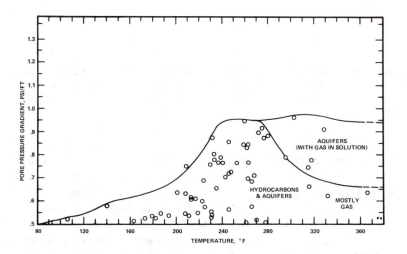

Figure 8-4a. Formation pressure gradient versus formation temperature in 60 overpressured U.S. Gulf Coast wells. (After Timko and Fertl, 1970.)

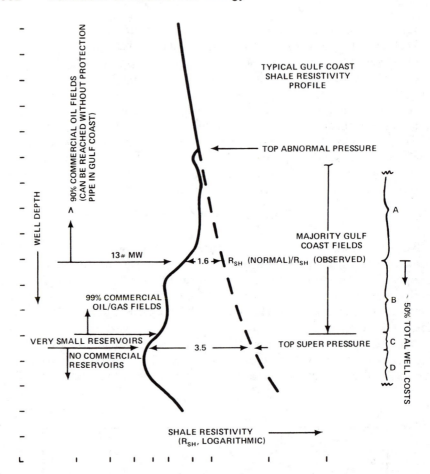

Figure 8-4b. Typical Gulf Coast shale resistivity profile based on the short normal curve correlated to distribution of gas-oil reservoirs. Profile is based on hundreds of commercially productive wells. (After Timko and Fertl, 1970.)

temperature from maximum recording thermometer (BHT) data recorded during all routine logging operations. The recommended technique requires the use of BHT data on each logging run, including information as to mud circulating time and time that the logging device was last present at the bottom of the wellbore (Fertl and Timko, 1972; Dowdle and Cobb, 1975).

The basic concept is the straight-line relationship on semilogarithmic paper of BHT in °F (from well log heading) versus the ratio of $(\Delta T/(T + \Delta T))$, where $\Delta T =$ time in hours after stopping of circulation and $T =$ circulating time in hours for well conditioning. Then, extrapolation of this

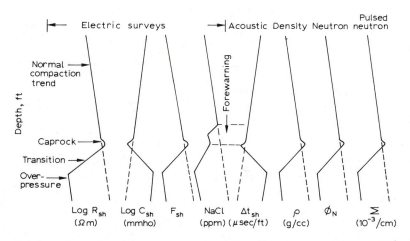

Figure 8-5. Schematic response of well-logging parameters to normal and overpressured environments. (After Fertl, 1976.)

straight line to a ratio of $(\Delta T/(T+\Delta T)) = 1.0$ determines the true static formation temperature. Figure 8-6 illustrates the extrapolation technique for true static formation temperature in two wells, which have been drilled in quite different geothermal regimes. Well number 1 is a high-temperature well located in the South China Sea and well number 2 is a deep hole drilled onshore in Texas (Fertl, 1976).

Application of this method to a geothermal well in the Roosevelt Hot Springs field, Utah, is illustrated in Figure 8-7 (Davis and Sanyal, 1979).

Numerous other empirical methods have been proposed. For example, based on statistical data, Kehle (1971) suggested a general depth-dependent relationship to estimate equilibrium (static) formation temperature (T_f, °F) from the well depth (D, ft) and the measured bottomhole temperature (BHT, °F):

$$T_f(°F) = BHT - 8.819 \times 10^{-12}D^3 - 2.143 \times 10^{-8}D^2 + 4.375 \times 10^{-3}D - 1.018$$

$$(8\text{-}1)$$

Temperature, which is one of the key parameters in geothermal wells,

1. Reflects lithology variations, overpressures, and hot water and steam quality.
2. Defines fluid-steam entries into wellbore.
3. Characterizes methane solubility.
4. Defines application limits of other logging devices.
5. Affects rock properties and drilling and completion practices.

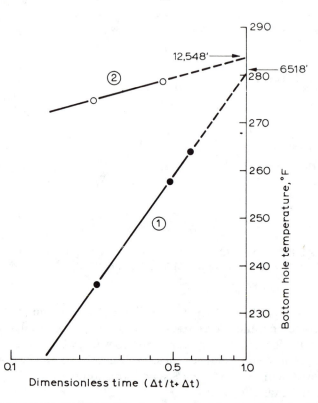

Figure 8-6. Extrapolation technique for true formation temperature. 1—High-temperature well. Four logs were run to 6,518 ft (1,987 m). First log, 4 hours after stopping of mud circulation recorded 237°F (114°C). Straight line extrapolation to infinite time (log 1.0) indicates a BHT of 281°F (138°C). 2—Deep Texas onshore well. Three logs were run to 12,548 ft (3,826 m). First-recorded temperature was 272°F (133°C), whereas actual stablized BHT is 284°F (140°C). t = circulating time (hrs); Δt = time after stopping of circulation (hrs). (After Fertl, 1976.)

Furthermore, at sufficient depths, rock hot enough to be a potentially useful energy source exists everywhere, a concept already successfully field-tested at Fenton Hill, New Mexico. The temperature variation in geothermal test hole GT-2 reached a temperature of 387°F (197°C) at 9,610 ft (2,929 m) as illustrated in Figure 8-8, which also shows the types of rocks penetrated by this test well.

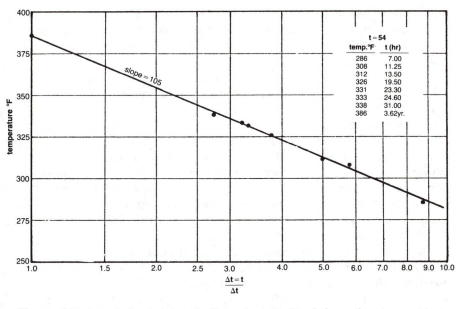

Figure 8-7. Log-derived determination of static (true) formation temperature, Roosevelt Hot Springs, Utah. (After Davis and Sanyal, 1979.)

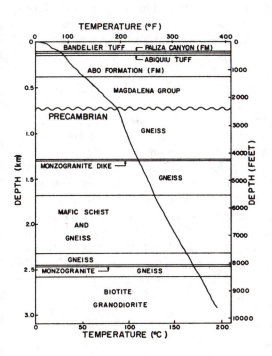

Figure 8-8. Temperature and lithology variations in geothermal test hole GT-2, Hot Dry Rock Project, Fenton Hill, New Mexico. (After Mortensen, 1977.)

Recently developed concepts to evaluate thermal rock conductivities and "true" formation temperature methods (Goss and Combs, 1976; Veneruso and Coquat, 1979; Vagelatos et al., 1979, etc.) are not discussed in this chapter.

Measurements of Na/K atomic ratios and silica content of geothermal waters are also used by geochemists to determine reservoir temperatures (White, 1970).

In a pressure transition zone, the formation pressure increases at a rate above the normal increase with depth. The same appears to hold true with temperature. A problem exists in determining the formation temperature, however, because the only accurate way to measure static temperature would be in a temperature-stabilized environment after drilling. In drilling operations, interest lies in before-the-fact temperature-pressure related measurements. This can be done by monitoring the flowline temperature of the mud stream.

Inasmuch as heat conductivity varies with rock and fluid components of subsurface formations, overpressured, high-porosity shales act as "thermal barriers," thereby locally increasing the geothermal gradient as has been observed in the field previously (Jones, 1968; Fertl and Timko, 1970; Lewis and Rose, 1970). Changes in flowline temperature gradients of up to 10°F/100 ft (18.2°C/100 m) have been observed prior to and/or when entering overpressured intervals (Figure 8-9).

This pressure indicator, however, is also affected by lithology, circulation and penetration rate, tripping the drill string for bit change, etc. Thus, certain precautions and refinements in use and interpretation of such data have to be considered (Wilson and Bush, 1973). Recommendations for plotting the flowline temperature include using temperature end points of each drill bit run, and replotting segments end-to-end without regard to actual temperature values.

Furthermore, worldwide experience also indicates that steeper than normal temperature gradients do occur in overpressure environments. This is illustrated in the field case from offshore Louisiana (Figure 8-10).

Exploration-type application of log-derived temperature data can be used successfully to locate and define areas with the potential for containing geopressured-geothermal fluids in economic quantities, so-called "geothermal fairways." The latter, for example occur in the Wilcox Group along the U.S. Gulf Coast region, where the gulfward-dipping sandstone-shale wedge thickens abruptly across a complex growth-fault system. A recent study by Bebout et al. (1978) utilized well log data to identify these areas where the Wilcox Group contains significant thicknesses of sandstone reservoir rocks with a subsurface fluid temperature higher than 300°F.

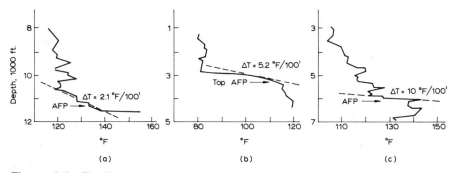

Figure 8-9. Flowline temperature data in normal and overpressured environments in (A) South Texas well, (B) North Sea well, and (C) South China Sea well. (After Wilson and Bush, 1973). AFP = abnormal formation pressure.

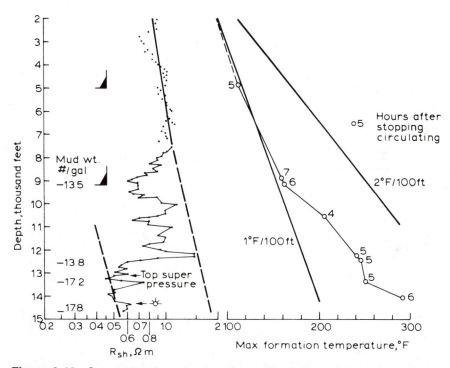

Figure 8-10. Conventional short normal shale plot indicates abnormal pressure below about 8,000 ft (2,439 m) and superpressures below 13,000 ft (3,963 m), as characterized by a very high shale/sand ratio and > 17 lb/gal (2.04 kg/dm³) mud weight. Plot of log-derived formation temperature suggests an "elbow" region which corresponds closely to the high pressure. (After Timko and Fertl, 1972.)

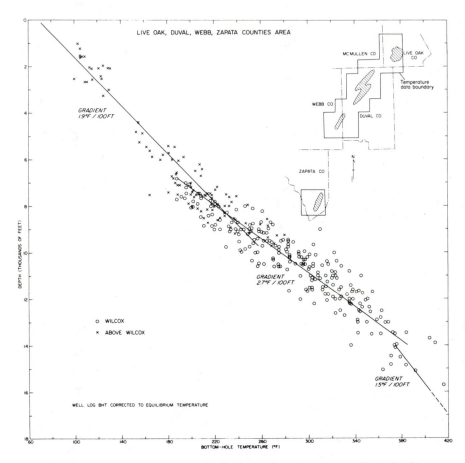

Figure 8-11. Temperature-versus-depth plot and geothermal gradients for the area including Live Oak, McMullen, Duval, Webb, and Zapata Counties, South Texas. (After Bebout et al., 1978.)

Whereas Figure 8-11 shows the change of formation temperature with depth and geothermal gradients for data obtained in five counties, Figure 8-12 illustrates the location of eight geothermal fairways (thick sandstones with temperatures >300°F) within the Wilcox group as derived from well log data.

Whereas temperature logs provide the only direct measurement of the geothermal resource, it should be recalled that numerous factors may affect such measurements. It is quite useful therefore, to make a series of repeat logging runs over a properly selected time period to obtain the maximum amount of information for better interpretative application.

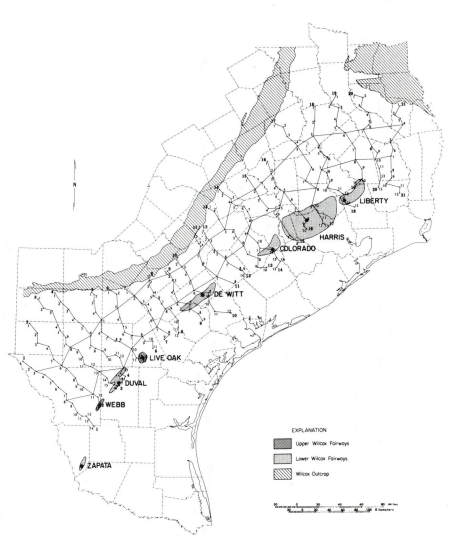

Figure 8-12. Wilcox geothermal fairways, South Texas. Dots show locations of representative well logs. (After Bebout et al., 1978.)

For example, several of the temperature logs run in well GGEH-1, Coso Hot Springs KGRA, California are shown in Figure 8-13 (Galbraith, 1978). Figure 8-14 presents a more detailed temperature distribution in the upper part of the wellbore. Two major temperature changes occur at depths of 500 ft and about 900 ft. In combination with other temperature logs run previously in the wellbore, several interpretive conclusions can be reached

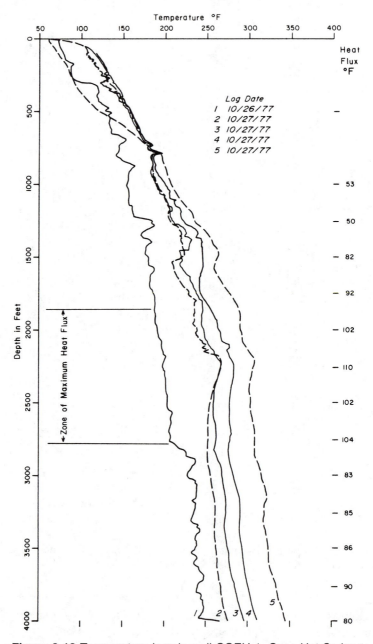

Figure 8-13. Temperature logs in well CGEH-1, Coso Hot Springs KGRA, California. (After Galbraith, 1978.)

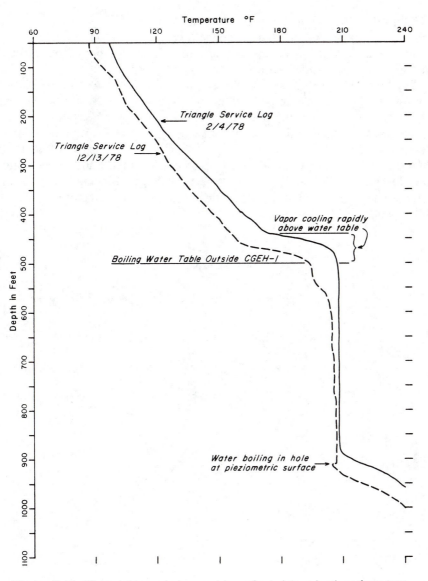

Figure 8-14. Water table and piezometric surface determinations from temperature logs in well CGEH-1, Coso Hot Springs KGRA, California. (After Galbraith, 1978.)

(Galbraith, 1978). The temperature change at a depth of about 900 ft, observed in the cased portion of the well, corresponds to the fluid level in casing. The temperature change at a depth of 500 ft corresponds to the top of the water table outside casing.

Furthermore, comparative studies of temperature logs in several wells not only give information as to vertical temperature variations, but also provide the basic data for real temperature maps and correlations with localized structural features (such as deep faults, etc.). Figure 8-15 shows several temperature logs in the East Mesa area, which have been successfully used in such studies (Kassoy and Goyal, 1979). Higher temperatures are encountered in wells 44-7 and 48-7 to the south (drilled by Magma Power Company), when compared to the others drilled in the northern part of the East Mesa field.

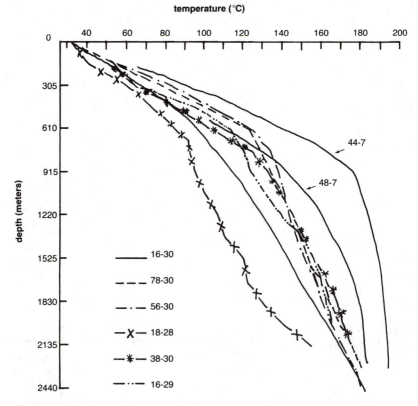

Figure 8-15. Temperature profiles in several geothermal wells at Mesa geothermal anomaly, Imperial Valley, California. (After Kassoy and Goyal, 1979.)

Steam Quality

The Geysers geothermal area, which contains one of the world's few producing vapor-dominated (dry-steam) fields, is located 80 miles north of San Francisco in the Mayacmas Mountains along Big Sulfur Creek in the northern coast ranges in California.

The water-phase diagram describes the state of water in its liquid, vapor and/or mixed phases as a function of pressure and temperature. (See Figure 1-5.) Both of these latter parameters can be obtained from well logs.

From such a phase-diagram one can also calculate the quality of the vapor + liquid mixture. For saturated vapor or a mixture of vapor and liquid, the pressure is a function of the temperature only, and the volume of the mixture depends upon the temperature and quality of the mixture. For the vapor in the superheated state, the volume depends on both temperature and pressure which may be varied independently.

According to Garrison (1972), the vapor within the Geyser geothermal system is derived from boil-off from a deep water table, which is postulated to lie at depths greater than 9,000 ft. Within the vapor range, temperature and pressure behavior is basically that of a continuous steam phase. The most probable temperature and pressure ranges for a vapor-dominated system is 236°C to 246°C and 38 kg/cm^2 (455 psi to 483 psi). Pressure and temperature, however, may exceed these limits because of the effect of dissolved solids and the pressure of other gases (Figure 8-16).

In a recent study of log-derived information for the Geysers geothermal area it has been demonstrated by Ehring et al. (1978) that it is possible to calculate the fluid density (ρ_f) from the density-neutron combination. From this fluid density, then, the specific volume, V_i($V_i = 1/\rho_f$) and, therefore, steam quality can be computed from a standard set of steam tables for any given temperature or pressure. This is easily seen by referring to the pressure-volume (water-phase) diagram shown in Figure 8-17. Steam quality is defined only for points between the saturated liquid line and the saturated vapor line. For points to the right of the saturated vapor line and to the left of the saturated liquid line or above the critical point, ρ_f and V_i can still be determined, but both the temperature and pressure need to be specified.

Formation Water Salinities

Characterization of the physical and chemical properties of geothermal brines in liquid-dominated geothermal resource areas is very important. Unfortunately, however, it is complicated due to the hostile downhole environment, high temperature, corrosivity, flow rates, the diminishing value of measurements on stagnant fluids collected for study, and other problems.

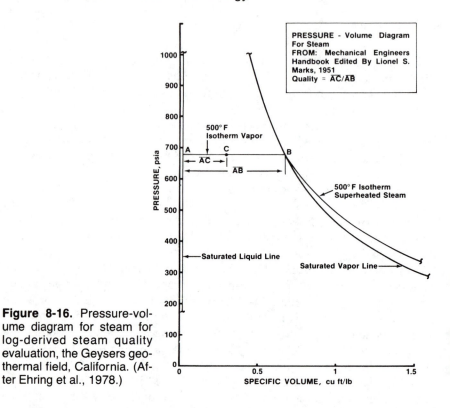

Figure 8-16. Pressure-volume diagram for steam for log-derived steam quality evaluation, the Geysers geothermal field, California. (After Ehring et al., 1978.)

Often, geothermal brines in the U.S.A. contain a wide variety of chemical species that tend to precipitate within the wellbore, causing scaling, corrosion, fouling, reduced flow rates, gas formation, and other undesirable effects.

The following phases can occur within the total mass flow inside a wellbore: liquid (brines of various concentrations), gas (steam, other gases), and solids (precipitates and particles from the rock formation).

Analyses of geothermal brines have been compiled by Cosner and Apps (1977) and Reeber (1977), and observed concentration ranges are listed in Table 8-10.

Such data greatly differs not only with geographic location and the geologic subsurface environment, but also with the type of analytical analysis. For example, if the brine is flashed prior to analysis, its composition may change. This complicates any detailed well-to-well comparisons.

Similar to salinity variations encountered in exploration of oil and gas, geothermal brines greatly vary in concentration from as low as approximately 500 ppm total dissolved solids encountered in geothermal test holes in Idaho to salinities exceeding 300,000 ppm in the Salton Sea area, California. Whereas wells in Idaho and in the Imperial Valley in California primar-

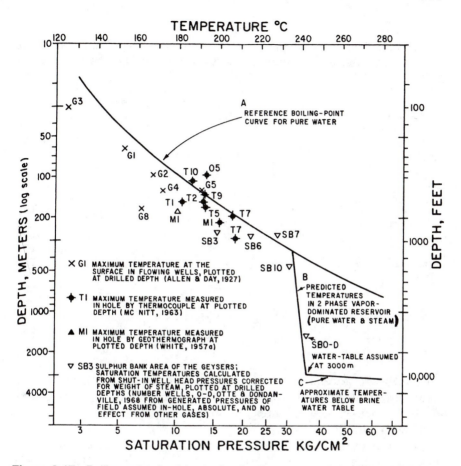

Figure 8-17. Boiling point-depth curve for the Geysers geothermal field, California. (After Garrison, 1972.)

ily contain sodium and calcium chlorides, with small amounts of silica, Salton Sea brines exhibit high concentrations of many metals in addition to sulfur, boron, and arsenic.

For example, among the elements reported in Salton Sea wells are up to 5,000 ppm rubidium, 400 ppm lithium, 50 ppm cesium, and 2,700 ppm arsenic. Other area wells were reported to contain up to 4,800 ppm strontium, 3,000 ppm barium and 3,400 ppm calcium as major constituents. Another detailed investigation of the chemical composition of these subsurface waters has been undertaken by Bailey (1977).

The oldest rocks in the Coso Hot Springs KGRA in Inyo County, California consist of intermediate to mafic metamorphic rocks occurring as roof pendants and xenoliths in granitic rocks. The basement rocks are partially

Table 8-10
Probable Median Ranges and Maximum Reported Concentrations
of Important Chemical Constituents of Geothermal Brines

	Concentration, ppm	
Species	**Range**	**Maximum**
Total dissolved solids	1,000-10,000	360,000
Chloride	100-1,000	260,000
Sodium	100-1,000	87,000
Sulfate	50-500	84,000
Calcium	10-100	65,000
Magnesium	1-10	40,000
Potassium	50-140	30,000
Aluminum	0.5-5	7,200
Iron	1-10	4,600
Silica	50-500	1,060
Ammonium	0.5-5	1,050
Nitrate	Not estimated	1,020
Carbon dioxide	0.5-5	500
Lead	0.5-5	110
Hydrogen sulfide	Not estimated	75
Silver	Not estimated	2

covered by Pleistocene rhyolite domes, ash fall tuffs, and basalt cinder cones and flows (Hulen, 1978). Water chemistry data from this area is listed in Table 8-11 and clearly indicates the presence of a chloride-rich hot water system (Galbraith, 1978).

Chemical water analyses data for the Cerro Prieto geothermal field, located in Baja California, Mexico are listed in Table 8-12 (Reed, 1975).

Water chemistry data for the dilute sodium chloride brines in the Roosevelt Hot Springs geothermal area, located on the western margin of the Mineral Mountains, Utah, are presented in Table 8-13 (Ward et al., 1978).

In geopressured-geothermal resource evaluation, the geochemistry of subsurface brines in Tertiary clastic sediments, such as the ones encountered along the U.S. Gulf Coast area, also becomes very important. Table 8-14 lists the average composition and concentration ratios of several constituents in the Tertiary U.S. Gulf Coast brines (Collins, 1970). Additional information on formation water salinities and formation pressures are given in Table 8-15 (Jones, 1968). Importance and application of these data will become apparent in the discussion of the salinity principle later in this section.

Log-derived salinity estimates are possible, for example, from the spontaneous potential (SP) curve. Basically, the SP-curve is a millivolt reading (as a

Table 8-11
Water Chemistry Data for Wells CGEH-1 and Coso-1, Coso CGRA Hot Water System, Inyo County, California

	Make up water	10:45am	11:00am	11:15am	11:30am	11:45am			
SiO$_2$	63.0	710.0*	710.0*	710.0*	710.0*	710.0*	50.0	27.0	154.0
Ca	100.0	110.0	98.0	99.0	93.0	98.0	72.8	359.0	74.4
Mg	29.8	3.0	2.5	2.3	2.7	2.5	0.5	0.6	1.0
Na	100.0	1,600.0	1,600.0	1,580.0	1,590.0	1,590.0	1,764.0	2,808.0	1,632.0
K	0.5	122.0	123.0	125.0	126.0	126.0	154.0	172.0	244.0
Li	1.0	9.6	9.7	10.0	10.3	10.0			
Rb	<0.02	0.103	0.105	0.106	0.112	0.118			
Cs	<0.01	<0.01	<0.01	<0.01	<0.01	<0.01			
HCO$_3$	307.0	150.0	286.0	273.0	297.0	279.0	134.2	CO$_3$**50.4? / 0.0 / OH**76.2?	77.4? / 0.0 / 1.7?
SO$_4$	234.0	314.0	268.0	266.0	257.0	245.0	38.0	216.0	52.8
Cl	81.0	2,330.0	2,360.0	2,420.0	2,460.0	2,480.0	2,790.0	3,681.0	3,042.0
F	0.1	3.8	3.8	3.8	3.8	4.2	3.7	1.6	2.2
B	0.92	54.0	53.0	54.0	56.0	58.0	48.0	57.4	71.6
pH	7.67	8.14	7.74	8.15	8.14	8.22	8.9	9.8	8.5
TDS	918.0	5,410.0	5,518.0	5,547.0	5,610.0	5,606.0	5,744.0	6,894.0	5,228.0

After Galbraith, 1978
Data source: Fournier et al. (1978) and Austin and Pringle (1970)
Sample Interval = 3,488 − 4,824 ft (12/1/77)
*Silica content includes possibly colloidal clay dispersed in water
**Impossible combination. Transcription error (?)

Table 8-12
Calculated Average Fluid Composition and Reservoir Conditions for Producing Wells — Cerro Prieto Field, Baja California, Mexico

Well	Li	Na	K	Mg	Ca	HCO$_3$	Cl	SO$_4$	SiO$_2$	B	CO$_2$	H$_2$S	Average production depth, m	Average pressure, bar	Average temperature, °C	pH
M-5	14.2	5,250	1,290	0.5	330	27	9,810	<3	630	13	1,920	481	1,200	103	289	5.26
M-8	10.9	4,730	1,180	0.2	272	38	9,040	9	590	12	2,580	624	1,220	105	305	5.43
M-9	10.7	4,730	750	1.5	358	56	8,530	27	430	11	823	200	1,070	97	228	5.52
M-11	13.7	5,600	1,230	0.8	369	27	11,400	27	610	13	1,250	310	1,150	129	261	5.34
M-20	10.0	4,590	1,050	0.9	329	37	8,270	<3	520	11	2,450	463	1,100	96	281	5.50
M-25	14.9	5,610	1,300	0.4	380	29	11,000	5	580	13			1,240	107	280	5.4†
M-26	12.8	5,630	1,370	0.6	522	25	10,400	<3	620	12			1,240	106	292	5.3†
M-29	12.7	5,450	1,010	3.1	406	46	10,200	13	420	15	1,580	229	870	81	235	5.44
M-30	14.6	5,650	1,320	0.6	389	24	10,900	11	630	13			1,290	114	273	5.4†
M-31	11.1	4,410	1,110	0.1	287	28	8,820	3	490	11	2,210	593	1,160	111	313	5.35
M-34	13.5	5,330	900	2.3	484	36	9,840	30	450	12	1,270	260	1,310	117	231	5.52
M-39	10.1	4,400	780	1.4	328	44	8,150	34	470	13			1,300	113	246	5.4†

Chemical constituents, mg

After Reed, 1975

* At reservoir conditions these constituents are distributed among many species.

† Estimated from the pH of well fluids with similar composition and production characteristics.

Table 8-13
Selected Roosevelt Hot Springs Water Analyses (mg/l)

	Sample 1*	Sample 2*	Sample 3*	Sample 4*
Na	1,840	1,800	2,072	2,500
Ca	122	107	31	22
K	274	280	403	488
SiO_2	173	107	639	313
Mg	25	24	.26	0
Cl	3,210	3,200	3,532	4,240
SO_4	120	70	48	73
HCO_3	298	300	25	156
Al			1.86	.04
Fe			.016	
Total dissolved solids	6,063	5,888	6,752	7,792
Temperature	25°C	28°C	92°C	55°C
pH	6.5	6.43	5.0	7.9
Na-K-Ca temperature	241°C	239°C	274°C	283°C
SiO_2 temperature	170°C	140°C	283°C	213°C

After Ward et al., 1978

*Sample (1): Roosevelt Seep, University of Utah, June, 1975. Sample (2), Roosevelt Seep: Phillips Petroleum Co., August, 1975. Sample (3), Thermal Power Company well 72-16. University of Utah, Jan., 1977. Surface leakage. Sample (4): Roosevelt Hot Springs, Sept., 1957, U.S.G.S., Mundorff (1970).

function of depth) of naturally occurring potential differences between the fixed potential of a surface electrode and a moveable electrode, i.e., the logging sonde, in a borehole filled with conductive fluid. Main SP-components are the electrochemical potential (shale membrane potential), the liquid junction potential (diffusion potential), and the electro-kinetic (streaming) potential.

In clean, essentially shale-free, permeable formations the total electrochemical potential is the sum of the shale membrane and liquid-junction potential, which can be expressed as:

$$E_c = -K \log (a_w/a_{mf}) \qquad (8\text{-}2)$$

where K, the electrochemical coefficient, is directly proportional to absolute temperature, and a_w and a_{mf} are the chemical activities of the formation water and the mud filtrate at formation temperature, respectively. For pure so-

Table 8-14
Concentration Ratios of Some Constituents in the Brines
Taken from Tertiary Age Rocks

Constituents	Average composition, mg/l		Concentration ratio*	Excess factor**
	sea water	Tertiary brines		
Lithium	0.2	3	15	12.5
Sodium	10,600	37,539	3.5	2.9
Potassium	380	226	.6	.5
Calcium	400	2,077	5.2	4.3
Magnesium	1,300	686	.5	.4
Strontium	8	148	18.6	15.5
Barium	.03	73	2,439	2,033
Boron	4.8	20	4.1	3.4
Chloride	19,000	63,992	3.4	2.8
Bromide	65	79	1.2	1
Iodide	.05	21	426	355
Sulfate	2,690	104	.03	.03
Mg†	1,543	1,947	1.3	1.1

After Collins, 1970
*Amount in brine/amount in sea water ratio
**Concentration ratio of a given constituent/concentration of bromide
†Magnesium equivalent of calcium plus magnesium in brine.

dium chloride solutions that are not too concentrated, resistivity values are inversely proportional to activities. This inverse proportionality, however, does not hold exactly at high ionic concentrations or low-salinity brines characterized by a prevalence of significant quantities of multivalent ions, such as Mg^{2+}, Ca^{2+}, and SO_4^{2-}.

Under normal circumstances, the SP curve mainly depends on the salinity (resistivity) contrast of drilling mud filtrate and formation water. In general, therefore, use of freshwater muds gives the best results. Small or non-existent salinity contrast, such as in wells drilled with saltwater mud, results in minor if any SP development and the SP curve becomes rather featureless.

The spontaneous potential (SP) cannot be recorded in boreholes filled with nonconductive muds (oil-base, air, gas) or in cased holes. The SP-curve also loses character in high-resistivity formations and is affected mainly by invasion of mud filtrate into permeable formations, presence of multivalent ions, saltwater flows, circulation loss, stray currents causing spurious potentials, and streaming potentials under overbalanced drilling conditions.

Table 8-15
Geostatic Ratios and Formation Water Composition in Wells Located in the Texas and Louisiana Gulf Coast Area

Depth, ft	Original pressure, psi	Geostatic ratio[2]	Dissolved solids, mg/l	Sodium, mg/l	Potassium, mg/l	Calcium, mg/l	Magnesium, mg/l	Bicarbonate, mg/l	Sulfate, mg/l	Halides as chloride, mg/l
12,992	11,960	0.92	92,000	38,000	520	4,300	580	700	260	52,300
10,400	10,400	0.87	98,000	38,000	200	1,200	200	1,100	26	56,700
12,500	10,500	0.85	99,000	36,000	200	1,100	200	1,200	27	56,600
11,106	9,030	0.81	111,000	40,000	320	2,700	360	500	—	—
10,082	8,000	0.79	55,000	19,000	130	1,100	200	3,100	82	30,200
10,875	8,110	0.74	41,000	15,000	100	340	60	2,900	700	21,600
10,401	7,700	0.74	146,000	46,000	320	6,700	600	—	100	76,000
10,870	7,900	0.73	136,000	43,000	500	5,400	700	400	53	73,000
12,552	8,690	0.69	175,000	63,000	540	6,100	700	300	—	98,000
16,064	9,600	0.60	93,000	30,000	480	4,200	300	200	10	53,500
9,051	5,000	0.55	14,000	4,800	30	290	60	600	60	7,700
11,000	5,950	0.53	66,000	24,000	100	1,300	300	1,400	25	38,200
12,200	6,500	0.54	97,000	36,000	220	1,300	300	1,000	—	55,500
11,200	5,830	0.52	72,000	26,000	210	1,100	200	100	12	40,900
10,500	5,440	0.52	100,000	35,000	210	1,600	400	800	—	57,000
13,000	6,600	0.51	45,000	16,000	100	560	100	600	—	26,000
Gulf of Mexico water			35,800	10,970	429	423	1,324	147	2,750	19,770

After Jones, 1968

*Ratio of aquifer fluid pressure to overburden pressure.

Principal uses of the SP-curve include:

1. Detection and correlation of permeable formations.
2. Determination of formation water resistivity and salinity (appropriate nomographs and computer programs for handheld calculators are published in the extensive logging literature and/or available on request from most well logging service companies).
3. Clay content in permeable reservoir rocks.
4. Indication of tectonic stress for use in abnormal formation pressure detection.
5. Fracture identification.

Concerning the last use, for example, in the LASL Hot Dry Rock geothermal project at Fenton Hill, New Mexico, an artificial fracture was formed immediately below a depth of 6,420 ft (1,957 m) in well EE-1. Several SP-logs were run over this interval with pressure increasing and then again decreasing, and with the fracture completely depressurized (Kintzinger et al., 1977).

These comparative test data, shown in Figure 8-18, suggest that during times of fluid changes in fractured intervals the SP-curve may assist in locating the position of fractures. However, after the fracture system has come to equilibrium with fluid parameters (salinity, temperature, pressure, etc.), the SP response to fractures may no longer be evident.

Log-derived salinity profiles illustrate concentration variations as a function of depth. Salinity information of geothermal waters provides an indication of its quality, anticipated log responses, and methane content of subsurface brines, and relates to possible problems of scaling, plugging, corrosion, and disposal. It can also be used to predict abnormal formation pressures.

For example, Littleton and Burnett (1977) computed salinity profiles from SP and dual induction logs for 10 East Mesa wells, California. Salinity reversals were observed frequently and the calculated salinities, particularly from resistivity logs, were found to agree well with those determined from chemical analyses of produced waters. Presence of fresh formation waters and significant concentrations of multivalent ions make SP-derived salinity values less reliable. Figure 8-19 shows salinity variations as a function of depth for East Mesa test well No. 6-1 (Kassoy and Goyal, 1979).

Major salinity reversals in several East Mesa wells are important in the interpretation of geothermal reservoir characteristics and will be equally applicable to environmental impact studies. Black (1975) utilized the salinity profiles calculated for the East Mesa wells to establish cross-sectional salinity distribution maps for this geothermal field. Interpretation of the data suggested that hot, saline subsurface waters probably ascend near wells

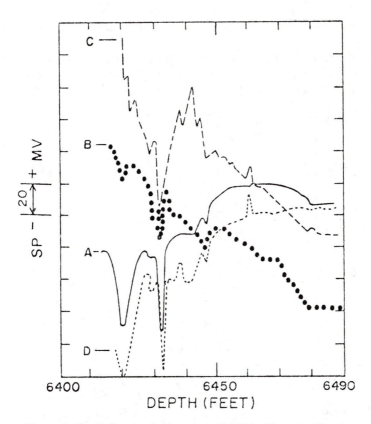

Figure 8-18. Self-potential logs in well EE-1 with and without pressure, Fenton Hill, New Mexico. A—hydrostatic pressure; B—pressure buildup (1,675 psi); C—venting (1,750 psi); D—hydrostatic pressure. Main hydraulic fracture was produced at 6,432 ft during stage B; minor fracture formed at 6,432 ft created by initial pressurization and reinflated at stage B. (After Kintzinger et al., 1977.)

Mesa No. 6-1 and No. 6-2, and then spread laterally to merge with the regional water flow into the NW direction. These waters again descend in the vicinity of wells No. 8-1, No. 5-1 and No. 31-1 as a result of the density effect of concentrated brines.

The Salinity Principle

As early as 1927, Soviet investigators have cited field observations of a marked freshening or decrease of formation water salinities with increasing depth. Similar observations are well documented for the U.S. Gulf Coast area, the U.S. mid-continent region, and many other Tertiary basins around the world.

Formation water salinity and associated formation pressures for the U.S. Gulf Coast area have been already presented in Table 8-15. The low salinity values were observed in many of abnormally high-pressured formations.

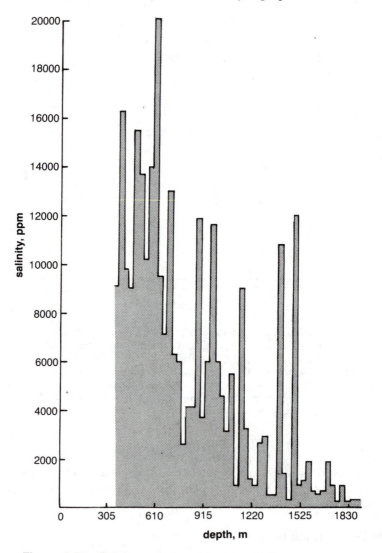

Figure 8-19. Salinity variation with depth for East Mesa test well 6-1, Mesa geothermal anomaly, Imperial Valley, California. (After Kassoy and Goyal, 1979.)

Overton and Timko (1969) proposed to plot salinity variations as calculated from the SP-curve for abnormal pressure detection work. A remarkably simple relationship, the so-called salinity principle, exists between salinity (C_w) of clean sands and porosity of adjacent shales (ϕ_{sh}):

$$C_w \cdot \phi_{sh} = \text{constant} \tag{8-3}$$

Formation water salinity in assumed equilibrium between sand and shales, therefore, will vary inversely with the porosity in adjacent shales. Under normal compaction, shale porosity decreases with increasing depth, whereas water salinity in associated sands tends to increase (Rieke and Chilingarian, 1974). Abnormal formation pressure environments, however, cause divergence from such normal trends which in turn suggests a decrease in formation water salinity, provided the salinity principle is applicable. Decrease in the formation water salinity, as indicated by the SP-curve, should then indicate proximity of overpressure. (This may be attributed to the influx of fresher water from the associated thick sequences of undercompacted shales.) Numerous case studies in several Tertiary basins worldwide have confirmed this concept (Fertl and Timko, 1970). Figures 8-20 and 8-21 show SP-curve derived salinity profiles in overpressured wells located in Matagorda County, Texas, and offshore Louisiana, respectively. It was also noted, however, that throughout these geologic sections the pore waters appear to be less saline in the shales than in the associated sandstones (Figure 8-22) (Fertl and Timko, 1970).

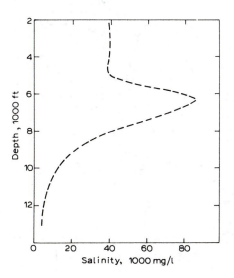

Figure 8-20. Salinity trend in Matagorda County, Texas. Decrease in the formation water salinity, as indicated by the SP curve, can indicate approach or proximity to abnormal-pressure environments. (After Myers and Van Siclen, 1964.)

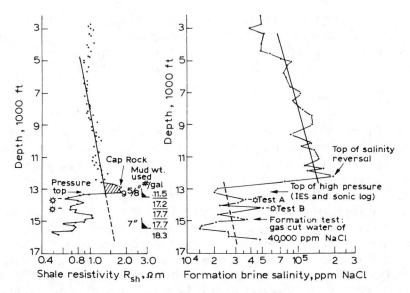

Figure 8-21. Sudden salinity divergence from the normal trend at a depth of about 800 ft (244 m) above overpressure top exhibited in the Louisiana wildcat well. Overpressure top is defined from logs (such as shale resistivity plot on left) and drilling data. Casing points, mud weights, and formation tests are shown. (After Fertl and Timko, 1971.)

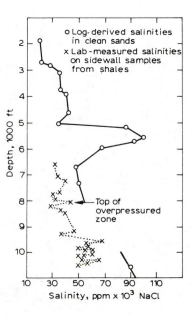

Figure 8-22. Salinity trend in an offshore Louisiana well. Comparison of log-derived salinities in clean sands and laboratory-measured salinities of sidewall samples from shale zones. (After Fertl and Timko, 1970.)

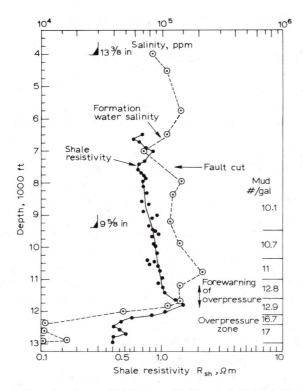

Figure 8-23. Comparison of shale resistivity and formation water salinity plots in a Gulf Coast well, Iberia County, Louisiana. (After Fertl and Timko, 1970.)

The concept of using the SP-curve as a tectonic stress indicator in sand-shale sequences has in many regions proven to be quite successful. Frequently, freshening of formation water has been observed prior to the drill bit naturally penetrating overpressures, providing a forewarning of impending downhole formation pressure changes (Figure 8-23).

In some areas in the U.S.A., the Himalayan foothills, the South China Sea region, and elsewhere, observed salinity reduction in overpressure environments at greater depth become so drastic that SP reversals occur, i.e., the recorded SP-value changes polarity and the SP-deflections in over-pressured sands become positive.

Straightforward application of the salinity principle is not always possible because of the problems inherent to SP measurements and interpretation and/or the fact that formation water salinities can vary quite markedly due to the presence of faulting, proximity to salt domes and unconformities, and other local or regional geologic factors.

Formation Water Salinity and Methane Content

Over the last decade serious consideration has been given to the recovery of geothermal heat, of overpressured fluids, and the methane content dissolved in these fluids. Presence of high methane concentration in over-pressured zones is evidenced by gas and water flows, "kicks" encountered while drilling, and analysis of fluid samples from normally pressured and overpressured aquifers.

Published data on the solubility of methane (Figure 8-24) in water ranging from 70°F to 680°F (21°C to 360°C) and from 600 psi to 16,000 psi have been used by Bonham (1978) to construct depth-versus-solubility curves for given geothermal and geopressure gradients. Figure 8-25 is an example of the practical application of such curves to estimate the maximum amount of methane that could be dissolved in subsurface waters in environments similar to those existing in the U.S. Gulf Coast region. These curves are based on solubility of methane in pure water. Methane solubility decreases with salinity. For example, in formation waters having total salinity of about 100,000 ppm, methane solubility decreases by 30 to 40% (Brill and Beggs, 1975).

Estimates of the total geothermal-geopressured resources in trillion cubic feet of gas, located along the U.S. Gulf Coast area, onshore and offshore in Texas and Louisiana are listed in Table 8-16.

Methane solubility in subsurface waters is governed by pressure, temperature, and salinity. Methane content increases with increasing pressure and temperature above 200°F, but decreases with higher salinities. It should be noted that these three key parameters — formation pressure, temperature, and salinity — can be determined from geophysical well logging techniques.

Geothermal Well Log Analysis

Interpretive Concepts in Clastic and Carbonate Reservoir Rocks

A multitude of crossplot concepts for well logging parameters is available to the formation evaluation specialist. Raw and/or calculated logging parameters can be crossplotted on linear, semilogarithmic, or other exponential scales. Input data include resistivity, acoustic, and nuclear logging measurements and, if available, core, test, and production data.

Applications of specific crossplot concepts allow recognition of log calibration problems, normalization of basic log measurements, and determination of lithologic reservoir characteristics of clastic rocks and those having complex mineralogy. Additional information includes primary and total porosity, formation water salinity, distinction between oil and gas, estimation of reservoir rock grain size, distribution prediction of irreducible water

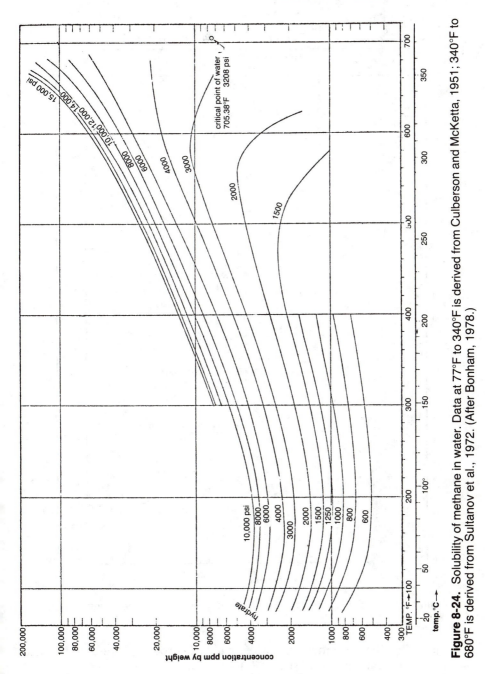

Figure 8-24. Solubility of methane in water. Data at 77°F to 340°F is derived from Culberson and McKetta, 1951; 340°F to 680°F is derived from Sultanov et al., 1972. (After Bonham, 1978.)

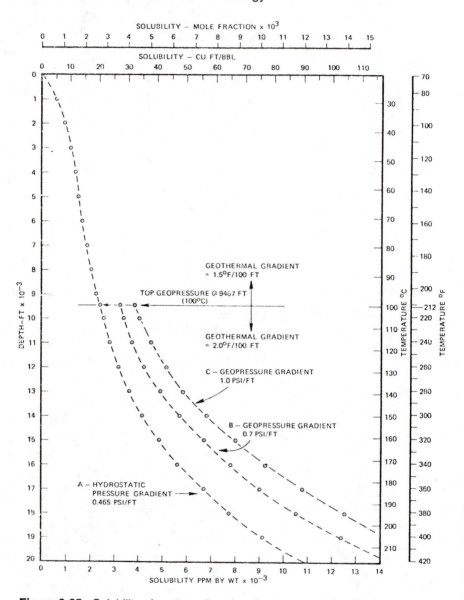

Figure 8-25. Solubility of methane in subsurface waters. (After Bonham, 1978.)

saturation (hence anticipated water-cut in hydrocarbon reservoirs), and heavy mineral analysis (Fertl, 1978).

Crossplotting of porosity logging data has been utilized since the early 1960s. Today, an extremely large variety of two-dimensional and three-dimensional crossplots (z-plots) is available. Whereas computer processing

Table 8-16
Estimates of Total Geothermal-Geopressured Resources
Along U.S. Gulf Coast

Source	Resources, Trillion cubic feet			
	Thermal	**Mechanical**	**Gas**	**Total**
Brown-Hudson	—	—	60,000	60,000
			100,000	100,000
Dorfman-Texas	—	—	5,735	5,735
Hawkins-LSU				
Louisiana only*	19.5	1.2	13.6	34.3
Jones-LSU				
sands and shales	—	—	100,000	100,000
sands only	—	—	49,000	49,000
Papadopulos-U.S.G.S. Circ. 726				
sands and shales assessed	43,331	198.0	23,927	67,456
onshore only				
unassessed only	One and one-half to two and one-half times the assessed quantities			
Wallace-U.S.G.S. Circ. 790				
assessed onshore	5,800	—	3,220	9,100
sandstone only				
assessed offshore				
sandstone only	5,200	—	2,800	8,000

*Only recoverable amounts are calculated.

allows quick and easy data handling, hand plotting still provides an effective check for the experienced log analyst.

Monomineral Porosity Model

If lithology of a reservoir rock is known, and it is clean and/or shale-corrected, then porosity (ϕ) can be expressed as follows (for mathematical concepts see Tixier, 1968; and Gaymard and Poupon, 1968):

$$\phi_D, \phi_N = f \text{ (matrix type, bulk density, hydrogen index, and amount and type of fluid in pore space, i.e., mud filtrate, oil, and gas)}$$

$$\phi_{AC} = f \text{ (matrix type, porosity, amount and type of fluid in pore space, degree of compaction, and secondary porosity)}$$

Binary (Two-Mineral) Porosity Model

Complex reservoir rocks. If two basic mineral constituents are present, the mathematical solution requires, in addition to the unity equation, two porosity response functions listed below. Ambiguity exists in certain crossplots due to nonlinear dolomite response of neutron-type logs.

ρ_b, ϕ_N = f (type and bulk volume percentage of matrix components, porosity, types and amounts of various fluids in pore space and type and amount of clay content)

Δt = f (type and bulk volume percentage of matrix components, porosity, types and amounts of various fluids in pore space, and type and amount of clay content, and secondary porosity)

1.0 = Porosity $+\Sigma$ (bulk volume matrix components) = unity equation

A brief review of the three possible porosity log crossplot combinations leads to the following conclusions:

Density (Y) versus acoustic travel time (X) (ρ_b vs. Δt) crossplots. These provide good shaliness definition; reliable evaporite determination (anhydrite, halite, polyhalite; gypsum in clean, non-gas-bearing zones); poor porosity resolution; and definition of ρ_{sh}, Δ_{sh}, $(M_{sh})_{cal}$. Light hydrocarbons, gas, and gypsum shift the points towards NW; shales and compaction effects shift points towards E; secondary porosity shifts trend towards W; borehole rugosity and washouts shift trend towards N.

As early as 1963, a crossplot technique of density-acoustic travel time data was proposed by Alger et al. (1963) based on the graphical solution (Figure 8-26) of the two linear response functions:

$$\rho_b = f(\phi_{SD}, \rho_{ma}, \rho_f, \rho_{SH}, p) \qquad (8\text{-}4)$$
$$\Delta_t = f(\phi_{SD}, \Delta t_{ma}, \Delta t_f, \Delta t_{SH}, p) \qquad (8\text{-}5)$$

where p = shale fraction of formation based on the laminated sand-shale model (Poupon et al., 1954); and ϕ_{SD} = sand laminae porosity.

In Figure 8-26, the lines parallel to QC, such as 12, define constant shaliness (p). Lines parallel to QC, such as 34, describe constant effective porosity (ϕ_e), whereas lines originating on QF and converging in the clay point, C, define constant values of the sand-laminae porosity (ϕ_{SD}).

In such a crossplot, the proper adjustments have to be made for compaction and light hydrocarbon (gas) effects. In the past, this crossplot has been used without considering the type and distribution of clay present.

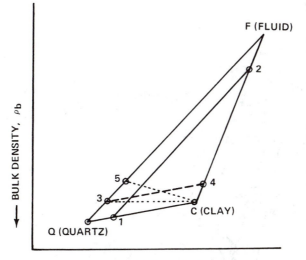

ACOUSTIC TRAVEL TIME, Δt ⟶

Figure 8-26. Basic crossplot concept of bulk density versus acoustic travel time in shaly clastic sediments based on laminated shale model. (After Alger et al., 1963.)

An interesting empirical crossplot correlating (Figure 8-27) log-derived parameters, such as the q-factor ($q = (\phi_{AC} - \phi_{DEN})/\phi_{AC}$), effective and total porosity, and production test data has been developed for several geological provinces (Fertl, 1976).

The *DEN-AC* crossplot has also been proposed for refined, log-derived overpressure evaluation in the North Sea area (Campbell, 1974).

Neutron porosity (Y) versus acoustic travel time (X) (ϕ_N vs. Δt) crossplots. These enable good porosity and lithology (quartz, calcite, and dolomite) resolution in clean, water-bearing intervals; and resolution of gypsum in clean, non-gas-bearing zones. Presence of shales (and definition of Δt_{sh} and ϕN_{sh}) shifts point towards NE (easily seen on z-plots); compaction shifts point towards E; light hydrocarbons and gas shift points towards S; gypsum, hole rugosity, and washouts shift points towards N; and secondary porosity shifts points toward W.

Density (Y) versus neutron (X) (ρ_b vs. ϕ_N) crossplots. Important and quite frequently used, these provide satisfactory resolution of porosity and definition of ρ_{sh}, ϕN_{sh}, and (N_{sh}) cal, and good lithology resolution for quartz, calcite, and dolomite. There is basic insensitivity to lithology with simple

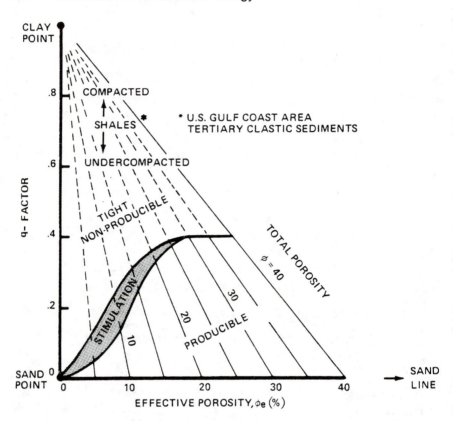

Figure 8-27. Empirical correlation among *q*-factor, effective and total porosity, and production characteristics. (After Fertl, 1976.)

approximation of $\phi = 0.5\,(\phi_{DEN} + \phi_{NEU})$, both in limestone units. The secondary porosity effect is absent because both logs measure total porosity. Shales and gypsum points trend towards E, NE, whereas light hydrocarbon and gas result in trend towards NW. Gypsum determination is made in clean (or shale-corrected) reservoirs devoid of gas. Dolomite response of neutron log is nonlinear.

The basic porosity log responses to sand-clay models, which have been mathematically described in the logging literature, may be summarized as follows:

$$\phi_{DEN} = \text{effective porosity} \pm \text{clay effect} + \text{light hydrocarbon}$$
$$\text{(gas) effect} \qquad\qquad (8\text{-}6)$$

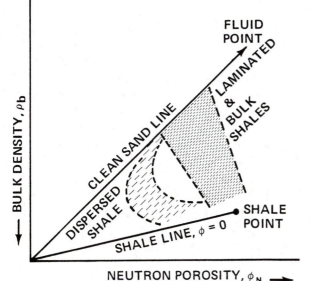

Figure 8-28. Basic crossplot of density versus neutron-type data and clay distribution patterns.

ϕ_{NEU} = effective porosity + clay effect (light hydrocarbon (gas) effect or + heavy oil effect) (8-7)

ϕ_{AC} = effective porosity + clay effect + compaction (+ light hydrocarbon, gas) (8-8)

At the present time, ρ_b versus ϕ_N is the most widely used crossplot (Figure 8-28) for porosity evaluation in shaly pay sands. The mathematical solution provides an effective porosity value, which has been corrected for clay and light hydrocarbon (gas) effects. Attempts have also been made to differentiate between several types of clay distribution (such as laminated, dispersed, and structural) and to estimate the silt content of reservoir rocks. Numerous sandstone evaluation models, published over many years by oil and/or service company personnel, heavily rely on such *DEN* vs. *NEU* crossplot information.

Generally, in this crossplot the gas effect moves points towards a NW direction, whereas shaliness effects push the data towards SE.

Acoustic travel time versus neutron porosity (Δt vs. ϕ_N) crossplots. These are similar to the limited application of the *AC* vs. *NEU* crossplots in shaly sand analysis.

Ternary (Three-Mineral) Porosity Model

M(Y) vs. N(X) or litho-porosity crossplot. This crossplot is a two-dimensional display of all three-porosity log responses in complex reservoir rocks. M and N are lithology-dependent parameters, which are essentially independent of primary porosity (Burke et al., 1969). Nonlinearity of neutron response in low-porosity dolomite requires three pseudo-matrix points for dolomite. All matrix point locations also vary with the type of neutron logs and fluid in the borehole. As stated by Frost (1978), however, the scaling of constituents in ternary mineral models is more complex than for binary mineral combinations, because M and N values are defined by the tangents of the matrix-to-fluid-point lines. Basically, however:

$$M = f(\Delta t_f, \Delta t, \rho_b, \rho_f) \cdot 0.01 = \text{slope in AC vs. DEN crossplot} \qquad (8\text{-}9)$$

$$N = f(\phi_{NF}, \phi_N, \rho_f) = \text{slope in NEU vs. DEN crossplot} \qquad (8\text{-}10)$$

A graphical example of an M vs. N crossplot is shown in Figure 8-29. It can be used for lithology determination, lithology trends, gas detection, and clay mineral classification. Each mineral has a unique set of M and N values and the concept is applicable to ternary and binary mineral models. Additional superimposed effects include gas shifts points in the NE direction, secondary porosity shifts points to N, gypsum shifts points to NNW, and shales shift points to SSW.

Additional information is found in the classic paper on litho-porosity by Burke et al. (1969).

MN-product. Application of the MN-product (Heslop, 1971) to the evaluation of complex reservoir rocks assists in computer manipulations of voluminous data in the case of unknown lithology. As in the case of the litho-porosity plot, the MN-product is porosity-independent and, for a given mineral combination, will give a constant value.

Furthermore, in a plot of MN vs. ρ_{ma}, all four major reservoir rock constituents (quartz, calcite, dolomite, and anhydrite) plot on a straight line (Figure 8-30). Knowledge of ρ_{ma} and ρ_b enables calculation of the effective porosity. MN vs. ρ_b crossplots also provide valuable information as applied to log calibration checks.

Similar to other crossplot concepts, several restrictions exist, e.g., shale effect moves points towards SW, and the presence of light hydrocarbons, gas, and/or secondary porosity causes the MN-product values to be too high.

MID plot $((\rho_{ma})_a$ vs. $(\Delta t_{ma})_a)$. For field interpretation, the less tedious *MID* plot (Clavier and Rust, 1976) approach has been proposed to replace the

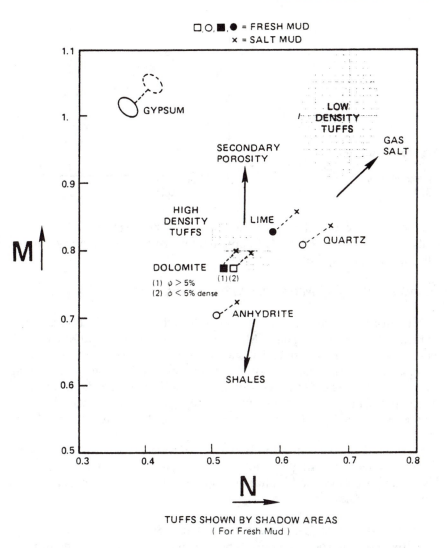

Figure 8-29. M versus N-crossplot model. (After Burke et al., 1969.)

litho-porosity technique. Again, two porosity-independent parameters are utilized, namely apparent density $(\rho_{ma})_a$ and apparent travel time $(\Delta t_{ma})_a$. Both parameters are derived from specially scaled *DEN-NEU* and *AC-NEU* crossplots, taking into account differences in neutron-type responses and the type of fluids in the borehole (fresh or salty). The proper *MID* plot grid is entered with appropriate values of $(\rho_{ma})_a$ on the y-axis and $(\Delta t_{ma})_a$ on the x-axis.

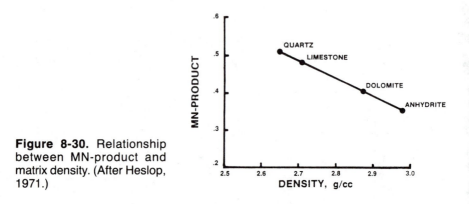

Figure 8-30. Relationship between MN-product and matrix density. (After Heslop, 1971.)

Several well-established log responses affect the *MID* plot, as schematically indicated in Figure 8-31. Supplementary z-plots (gamma ray, caliper, etc.) are frequently helpful.

Standardization and Normalization of Logging Data

Present-day digital data handling facilitates the standardization and normalization of well logging data (Miyari et al., 1976; Neinast and Knox, 1973) based on crossplotting concepts, such as frequency plots, two- and three-dimensional histograms, etc., for a given instrument, interval, well, or field. Correlation coefficients then classify the comparison of such data. Raw and/or computed parameters can be investigated ($\rho_b, \Delta t, \phi_N, R_t, \ldots, \rho_{ma}, \Delta t_{ma}$) and their crossplots studied (ρ_b *vs.* Δt, $\rho_b(\phi_N, \Delta t)$ *vs.* ϕ_e, *MN* vs. ρ, *M* vs. $\phi_N (\Delta t, \rho_b,$ etc.).

Close study of such information will be required because no unique decision based on a single histogram, for example, is possible. Logging instrument, wellbore, and formation response characteristics must be sorted out prior to making any final decision.

Clastic sediments (sand-shale sequences). Significant clay content of potential reservoir rocks not only imposes drilling, completion, and production problems, but to a varying degree, affects all electric, radioactive, and nuclear logging techniques. To obtain values of effective porosity and of water saturation, the influence of varying clay content on wireline log data must be considered.

As reviewed and discussed in extensive logging literature, clay content estimates can be obtained from various well logs (Fertl, 1979; Frost and Fertl, 1979). Each one of these methods exhibits advantages and inherent limitations. Normally, these shaliness indicators give a useable approximation or an upper limit of the clay content.

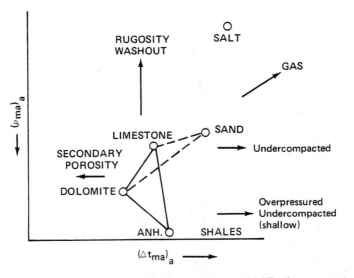

Figure 8-31. Schematic diagram showing MID-plot concept and several possible effects. (After Clavier and Rust, 1976.)

Understandably, both porosity and calculated water saturation must be corrected for the effects of the varying shale or clay content. Table 8-17 compiles the more common shaliness indicators (Fertl, 1979). The corresponding remarks section in the table is particularly important in applying the listed equations. Furthermore, in order to accurately utilize these equations, compensation must be made for the effects due to the presence of hydrocarbons.

All of the listed equations are based on the assumption that the response characteristics of the adjacent shales will also apply for the interval of interest. Clay minerals in the adjacent shales, however, are frequently not identical to the clay minerals present in the sand sequences. Hence, this limiting condition provides an additional uncertainty in these log-derived shaliness estimates.

Interpretive Concepts in Igneous and Metamorphic Rocks

Igneous and metamorphic rocks greatly differ from sedimentary formations in several petrophysical characteristics. At the present time, only limited information is available in the well logging literature, despite the fact that a major part of geothermal resources is encountered in these rock types.

(text continued on page 394)

Table 8-17
Log Derived Clay Content Indicators

Logging Curve	Mathematical relationship	Favorable conditions	Unfavorable conditions
Spontaneous potential (SP-curve)	$V_{cl} = 1.0 - (PSP/SSP) = 1.0 - \alpha$	Waterbearing, laminated shaly sands ($<R_t$).	R_{mf}/R_w approaches 1.0 Thin. $>>R$ zones. Hydrocarbon bearing. Large electro-kinetic and/or invasion effects.
	$V_{cl} = (PSP - SP_{min})/(SSP - SP_{min})$	c < 1.0 as function of clay type.	
	$V_{cl} = 1.0 - C \cdot \alpha$ $1.0 - \alpha = \log A/\log \{(A - V_{cl} \cdot B)/(1 - V_{cl} \cdot B)\}$ where $A = R_t/R_{xo}$, $B = R_t/R_{cl}$	Knowledge of several parameters required, including d, R_t, R_{xo}, R_{cl}. Similar limitations as for straight forward SP-equations.	
	$1.0 - \alpha = (K V_{cl} W)/(K V_{cl} W + \phi S_{xo})$	K = log derived coefficient. W = clay porosity from bulk and matrix ρ_{cl}, S_{xo} = flushed zone water saturation; laboratory-derived, too many requirements.	
Gamma ray	$V_{cl} = (GR - GR_{min})/(GR_{max} - GR_{min})$	Only clay minerals are radioactive.	Radioactive minerals other than clays (mica, feldspar, silt)
	$V_{cl} = C(GR - GR_{min})/(GR_{max} - GR_{min})$	C < 1.0, frequently approximately 0.5 when $V_{cl} < 40\%$	Only potassium-deficient kaolinite present. Uranium enrichment in permeable fractured zones.
	$V_{cl} = (GR - W)/Z$	W, Z = geologic area coefficient.	Radiobarite scales on casing. Severe washouts ($<<GR$).

Table 8-17 continued

Logging Curve	Mathematical relationship	Favorable conditions	Unfavorable conditions
Gamma ray	$V_{cl} = 0.33(2^{2VCL} - 1.0)$	Highly consolidated and Mesozoic rocks.	Younger, unconsolidated rocks
	$V_{cl} = 0.083(2^{3.7VCL} - 1.0)$ where $VCL = (GR - GR_{min})/(GR_{max} - GR_{min})$	Tertiary, clastics.	Older, consolidated rocks.
Spectralog Gamma ray spectral logging provides individual measurements of potassium (K, %) and thorium (Th ppm) content	$V_{cl} = (A - A_{min})/(A_{max} - A_{min})$ $V_{cl} = C(A - A_{min})/(A_{max} - A_{min})$ $V_{cl} = 0.33(2^{2VCL} - 1.0)$ $V_{cl} = 0.083(2^{3.7VCL} - 1.0)$ where $VCL = (A - A_{min})/(A_{max} - A_{min})$	Conditions similar to gamma ray discussion. A = Spectralog readings (K in %. Th in ppm). A_{min} = minimum value (K or Th) in clean zones. A_{max} = maximum values (K, Th) in essentially pure shales.	Similar to gamma ray discussion. However, uranium enrichment in permeable, fractured zones and radiobarite build up are no limitations. If Th-curve is used, localized bentonite streaks should be ignored.
Resistivity If several resistivity logs are available. Use the one which exhibits highest resistivity values in subject well.	$V_{cl} = (R_{cl}/R_t)^{1/b}$ where b = 1.0 b = 2.0	Low porosity zones (carbonate, marls), pay zones with low $(S_w - S_{wir})$. R_{cl}/R_t from 0.5 to 1.0 R_{cl} approaches R_t	High porosity water sand, high R_{cl} - values.

Table 8-17 continued

Logging Curve	Mathematical relationship	Favorable conditions	Unfavorable conditions
	$V_{cl} = \left\{ R_{cl}(R_{max} - R_t)/(R_t(R_{max} - R_{cl})) \right\}^{1/b}$ $V_{cl} =$ same as above where $(1/b) = 1.0$ when $R_{cl}/R_t < 0.5$ $(1/b) = 0.5/(1 - R_{cl}/R_t)$ when $R_{cl} R_t < 0.5$	In clean hydrocarbon-bearing zones one calculates $V_{cl} = 0$	
Neutron	$V_{cl} = \phi_N/\phi_{Ncl}$ $V_{cl} = (\phi_N - \phi_{MIN})/(\phi_{Ncl} - \phi_{MIN})$	High gas saturation or very low reservoir porosity. ϕ_{min} can be varied	ϕ_{Ncl} is low
Pulsed neutron	$V_{cl} = (\Sigma - \Sigma_{min})/(\Sigma_{max} - \Sigma_{min})$ $V_{cl} = (\Sigma_{cl}/\Sigma)(\Sigma - \Sigma_{min})/(\Sigma_{max} - \Sigma_{min})$	Fresh water environment low porosity and gas bearing zones. V_{cl} calculates zero in clean zones.	
Density-neutron	$V_{cl} = \dfrac{\rho_B(\phi_{Nma} - 1.0) - \phi_N(\rho_{ma} - \rho_f) - \rho_f\,\phi_{Nma} + \rho_{ma}}{(\rho_{sh} \cdot \rho_f)(\phi_{Nma} - 1.0) - (\phi_{Nsh} - 1.0)(\rho_{ma} - \rho_f)}$		Too low V_{cl} in prolific gas zones. Don't use with severe hole conditions Lithology affected

Table 8-17 continued

Logging Curve	Mathematical relationship	Favorable conditions	Unfavorable conditions
Density-acoustic	$$V_{cl} = \frac{\rho_B \cdot (\Delta t_{ma} - \Delta t_f) - \Delta t \cdot (\rho_{ma} - \rho_f) - \rho_f \cdot \Delta t_{ma} - \rho_{ma} - \Delta t_f}{(\Delta t_{ma} - \Delta t_f)(\rho_{sh} - \rho_f) - (\rho_{ma} - \rho_f)(\Delta t_{sh} - \Delta t_f)}$$	Less dependent on lithology and fluid conditions than DEN-NEU crossplot. Use in gauge boreholes	Badly washed out wellbores Highly undercompacted formations (shallow overpressures)
Neutron-acoustic	$$V_{cl} = \frac{\phi_N \cdot (\Delta t_{ma} - \Delta t_f) - \Delta t - (\phi_{Nma} - 1.0) \cdot \Delta t_{ma} + \phi_{Nma} - \Delta t_f}{(\Delta t_{ma} - \Delta t_f)(\phi N_{sh} - 1.0) - (\phi N_{ma} - 1.0)(\Delta t_{sh} - \Delta t_f)}$$	Use only in gas-bearing zones with low S_w.	Similar effects due to shaliness on both logs.

After Fertl, 1979

(text continued from page 389)

Generally, interpretive log analysis objectives in igneous and metamorphic rocks focus on the identification of rock types, recognition of hydrothermally altered intervals, determination of formation porosity and, particularly, location and evaluation of natural fracture systems.

Igneous rocks can be grouped into four major, distinctly different groups (Wyllie, 1971). These include basalt (originating from the liquid fraction of mantle pyrolite but without volatile enrichment); peridotite (the residual fraction of pyrolite), which is extremely poor in uranium and thorium content; agpaitic alkalic rocks (phonolites), and miaskitic alkalic rocks (kimberlite and carbonatites).

A simplified classification of several more common igneous rock types and some associated log-derived parameters is presented in Table 8-18 (Keys, 1979).

Classification of rock properties and related well log responses of metamorphic rocks, unfortunately, are even more complex.

Whereas porosity in vesicular basalt and volcanic detritus, such as tuff, can reach values up to 65%, primary porosity in igneous rocks is generally quite low (less than 2-5%). Evaluation of tuffites and tuffaceous formations in Argentina has been discussed by Khatchikian and Lesta (1973) based on

Table 8-18
Simple Classification of Igneous Rocks with
Well Log Response Parameters

| Fine-medium crystalline | Deep-seated intrusives | | | | | |
	Syenite	Granite	Monzonite	Diorite	Gabbro	Peridotite
Major feldspar	alkali	←equal amounts→			soda-lime	none
Silica		increasing →				
Aluminum		← increasing				
Fe and Mg		increasing			→	
Density		increasing			→	
Radio-activity		← — — increasing — — — —				
Very fine crystalline	Trachyte	Rhyolite	Latite	Andesite	Basalt	
	Shallow intrusives and surface flows					

After Keys, 1979

open-hole logs, whereas Sacco (1979) presented a cased-hole evaluation technique for such tuffaceous, hydrocarbon-bearing reservoirs based on the carbon/oxygen log and spectral gamma ray logging information. Further-more, light and heavy tuffs are easily recognized and distinguished based on their locations in the *M-N* crossplot (Figure 8-29). These techniques have been already successfully applied in Argentina, Alaska, Nevada (Sethi and Fertl, 1979), and elsewhere (Sanyal et al., 1979).

In tuffs, most of the uranium and thorium remaining after consolidation are locked in glass, which is relatively insoluble. Greater concentration of uranium in rhyolite and phonolitic tuffs has been noted (Gabelman, 1977).

Potassium, uranium, and thorium distribution in basalts and associated rocks is shown in Figure 8-32 whereas data for several Italian lavas are presented in Figure 8-33. Additional log responses in basaltic rocks have been discussed by Keller (1974), for a well drilled at the flank of Kilauea Volcano, Hawaii. For example, based on extensive core analysis data, the following relationship between formation resistivity factor, F and porosity, ϕ, was established: $F = 18 \ \phi^{-1.05}$. Several log responses in rhyolite tuffs, basalt, phyllites, chlorite schist with hornfels, and granite have been investi-gated by Sethi and Fertl (1979) in a geothermal well located in Nevada.

Metamorphic rocks in an interval from 3,900 ft to 19,595 ft in a deep well in Hill County, Texas have been analyzed by Rick (1975). Four major rock types were encountered; quartzite (a metamorphosed sandstone), phyllite (metamorphosed fine-grained sediments), calcite marble (metamorphosed limestones). Low-porosity, natural fracture systems, and heavy minerals (micas) complicated the log analysis in the subject well. The positions of the four metamorphic rock types on the *M-N* crossplot, however, were found to be quite similar to those of their sedimentary counterparts. Table 8-19 lists the average log responses of these four metamorphic rock types.

Average concentrations of potassium, uranium, and thorium for several groups of granite rocks are listed in Table 8-20. In granite, the potassium concentration varies from 2 to 6%. Uranium and thorium concentrations in igneous rocks vary with silica content. Granites are rich in uranium and thorium contents, whereas ultramafic rocks exhibit low potassium, ura-nium, and thorium concentrations.

In granites, a well-defined linear relationship is frequently observed be-tween uranium and thorium contents (Figure 8-34). Deviations may result from weathering effects or late magmatic intrusions of basic dikes (Figure 8-35). Generally, uranium concentrations of subsurface rocks may not be representative due to the relative mobilization or uranium. Furthermore, accessory minerals (e.g., zircon, sphene, rutile, and cassiterite) can exhibit a significant effect on the natural radioactivity of igneous rocks (Tables 8-6 and 8-7). Figure 8-36 shows the relationship between uranium and thorium contents and the potassium feldspar/plagioclase ratio of granite rocks.

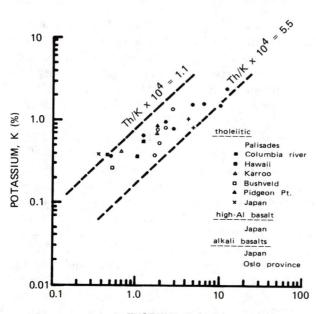

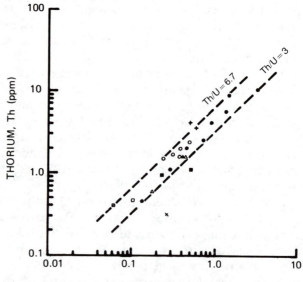

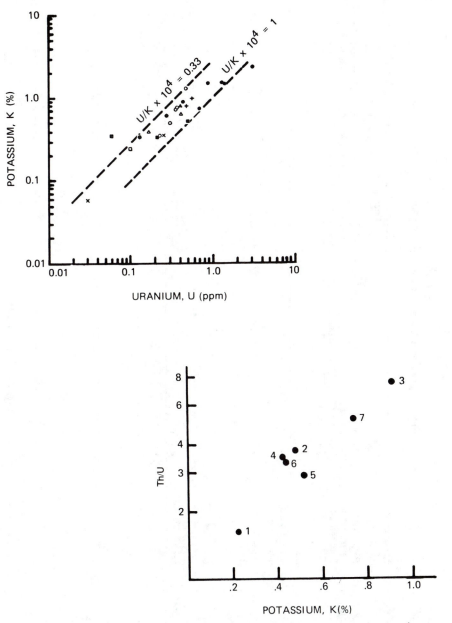

Figure 8-32. Relationship between potassium and thorium, potassium and uranium, and thorium and uranium. Lower right: potassium versus Th/U ratio in primary basic magma. 1—Japan, tholeiite; 2—Japan, high-alumnium basalt; 3—Japan, alkali olivine basalt; 4—Duluth, layered gabbro; 5—South California batholith, gabbro; 6—Columbia River basalt; 7—Palisades sill. (After Heier et al., 1963.)

Table 8-19
Average Log Response in Metamorphic Rocks; Well E.W. Barett #1*

Petrophysical parameter measured	Log type yielding measured petrophysical parameter	Rock type			
		quartzite	phyllite	calcite marble	dolomite marble
Resistivity, Ω-m	IES, DI or LL-3	50-100+	3-35	600+	600+
Interval transit time, μ s/ft	BHC-SLC	52-55	55-62	49-50	43-44
Velocity, ft/sec	BHC-SLC	19,231-18,182	18,182-16,129	20,408-20,000	23,256-22,727
Porosity, Lime liquid. Matrix. %	SNP	-1 to -3	6 to 10	0	0
Bulk Density, g/cc	FDC density	2.68 - 2.72	2.77 - 2.81	2.76 - 2.78	2.80 - 2.85
Bulk density, g/cc and % ϕ	Core and Cuttings	Cuttings 2.68 - 2.73	Cuttings 2.75 - 2.80	Cores 2.73 & 0.4%	Cores 2.84 & 0.8%
Natural radioactivity, API units	Gamma ray in water-filled hole	40-60	100-120	10-30	10-30
S.P. development, Δ mv. with Rmf = 0.64 @ 75°F	IES	50-130	0-30	80-110	80-100
			(Corrected to 77°F)		

After Ritch, 1979
*Interval = 3,000 to 19,650 ft, Hill County, north central Texas

Table 8-20
Gamma Ray Spectral Average Data for Various Groups
of Granitic Rocks

Location	K (%)	U (ppm)	Th (ppm)	Th/U ratio	Average Th Average U
Rocky Mountains	3.3	2.9	15.7	6.9	5.4
Canadian Shield	2.5	1.8	11.0	5.9	6.1
Pre-Cambrian	3.2	2.6	14.0	5.6	5.4
Post-Cambrian	2.2	3.0	8.6	3.3	2.9
U.S. West Coast	1.8	3.1	8.5	2.9	2.7

After Whitefall et al., 1959

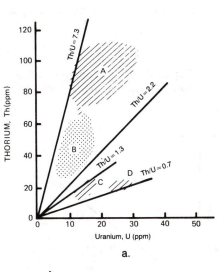

Figure 8-33a. Relationship between thorium (Th) and uranium (U) contents in Italian Ischian lavas: A—phonolites, alkali trachytes; B—trachytes, latites, Somma-Vesuvia; C—trachytic lavas; D—leucite tephrites and tephrites leucitites. (After Gasparini, 1963; Luongo and Rapolla, 1964.)

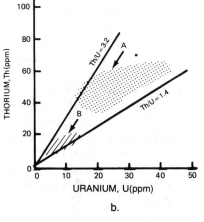

Figure 8-33b. Relationship between thorium (Th) and uranium (U) contents in Italian Roccamonfina lavas: A— leucite bearing rocks; B—trachytes to basalts. (After Civetta et al., 1966.)

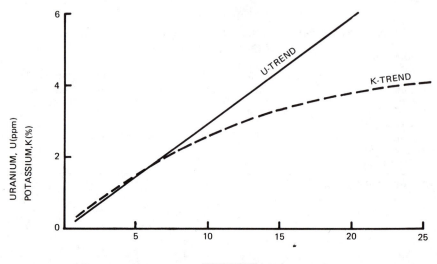

Figure 8-34. Uranium (U) and potassium (K) contents versus thorium (Th) content in granite rocks. (After Whitfield et al., 1959.)

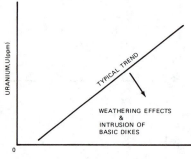

Figure 8-35. Relationship between uranium (U) and thorium (Th) contents in granites. "Typical" trend and possible effects due to weathering or intrusion of basic dikes are presented. (After Fertl, 1979.)

Recent applications of spectral gamma ray logging in crystalline basement rocks in boreholes of the Los Alamos Scientific Laboratories' Dry Hot Rock Geothermal Project clearly indicated their usefulness for determination of rock types, detection of fracture zones, and examination of the heat-producing elements uranium, thorium, and potassium. Figures 8-37 and 8-38 illustrate the Spectralog response in open and closed fracture systems in igneous basement rocks on Fenton Hill, Valles Caldera, New Mexico (West and Laughlin, 1976).

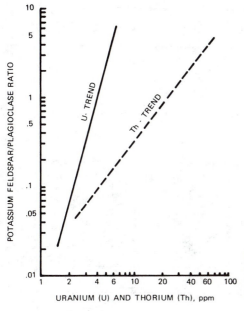

Figure 8-36. Uranium (U) and thorium (Th) contents versus potassium feldspar/plagioclase ratio of granites. (After Whitfield et al., 1959.)

Furthermore, Stuckless and Nkomo (1978) suggested that fracturing was a key factor in migration and precipitation of uranium in the Granite Mountains, Wyoming. Similar observations were made in hot-water transmitting fractures in hydrothermally altered zones at Raft River, Idaho (Keys, 1979).

Spectral gamma ray data are also useful to define the extent of hydrothermally altered zones. The latter appear to be reflected in decreased potassium content and precipitation of uranium, with potassium again being reprecipitated near the margins of such zones (Figure 8-39).

In the first demonstration test of the Los Alamos Hot Dry Rock Geothermal Energy Project, volcanics, red beds, shales, and limestones were penetrated before encountering the Precambrian target zone consisting of gneiss, monzogranite dikes, mafic schist, biotite, and granodiorite. Temperature in test hole FT-2 reached 387°F at a depth of 9,610 ft and in well EE-1 402°F at 10,053 ft, with a distance of 235 ft separating two wells. Besides several novel experimental devices, various electrical (normal, induction, lateral, microdevices, diplog), gamma ray, spectral gamma ray, acoustic, density, neutron, and caliper logs were run in the subject wells and their responses correlated to drill cutting and core analysis data (West et al., 1975). As mentioned previously, special emphasis was placed on interpretive applications of

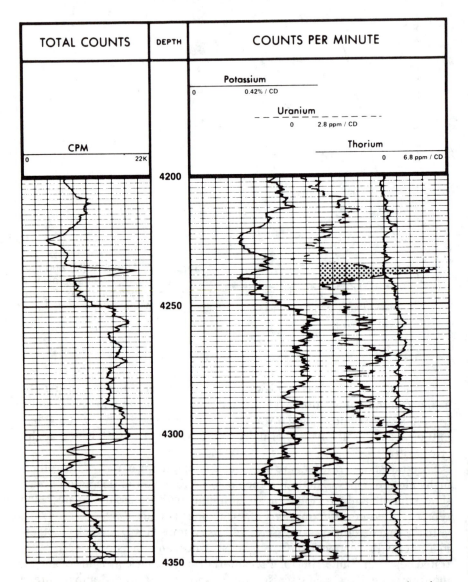

Figure 8-37. Coincidence of high U-peaks with natural fracture zones has been observed in test wells GT-2 and EE-1, located on the western flank of the Valles Caldera, New Mexico. Presence of an open, unhealed natural fracture at a depth of approximately 4,237 ft coincides with an excessively high U-peak and was also confirmed by full wave acoustic data and electric logging information. The Spectralog response between 4,250 to 4,305 ft is characteristic of a monzogranite dike. (After West et al., 1976.)

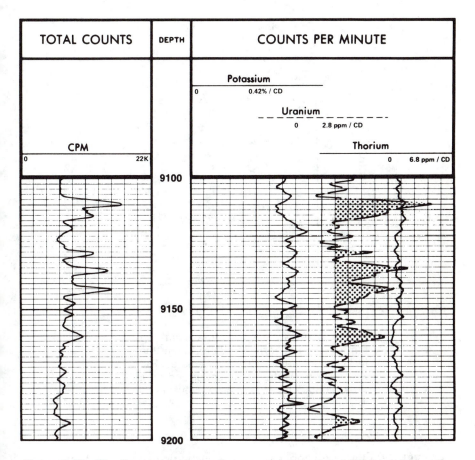

Figure 8-38. The Spectralog shows the response in test well GT-2 located on the western flank of the Valles Caldera, about 32 km west of Los Alamos, New Mexico. In the interval shown the homogeneous biotite granodiorite is cut by a sealed fracture system that is enriched in uranium. Note that from the Spectralog alone one cannot distinguish between open and sealed natural fracture of fissure systems. (After West et al., 1976.)

spectral gamma ray logging (West and Laughlin, 1976). The extent of the logging suites run in test well EE-1 is illustrated in Figure 8-40 a, b and c.

Possible effects on neutron log response by bound water present in the igneous rock matrix, particularly in hydrothermally altered zones, has been investigated by Nelson and Glenn (1975). Based on the log response in low-porosity diabase and andesite-tatite formations, modified relationships for the neutron-density and neutron-acoustic crossplot were proposed.

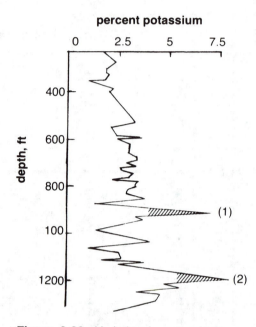

Figure 8-39. Variation in potassium content over hydrothermally altered zone that includes hot-water producing fractures, Raft River KGRA, Idaho. (1) is upper and (2) is lower contact of altered zone. (After Keys, 1979.)

Keys (1979) illustrated application of acoustic-neutron, density-neutron, and gamma ray-neutron crossplots in igneous rocks to distinguish granite, syenite, and hydrothermally altered zones. For example, syenite and hydrothermally altered zones can be distinguished by using the gamma ray-neutron crossplot, whereas the density-acoustic crossplot cannot be used in this case. Furthermore, crystal size appears to affect acoustic measurements. Several fracture detection techniques, including application of a high-temperature acoustic televiewer, has been reviewed by Keys (1979) in some detail (Figure 8-41).

Several investigators have compiled useful data on basic logging parameters for various minerals and rocks (Edmundson and Raymer, 1979; Fertl, 1979; Sanyal et al., 1979). As mentioned previously, interpretation of well logs in igneous intervals is complicated by the various rock types encountered. Nevertheless, quartz ($>\rho, <Ra$) and quartz monzonite ($<\rho, >>Ra$) can be distinguished by density, ρ, and natural radioactivity, Ra. Mica schist can be isolated on the porosity-independent M versus N crossplot, which

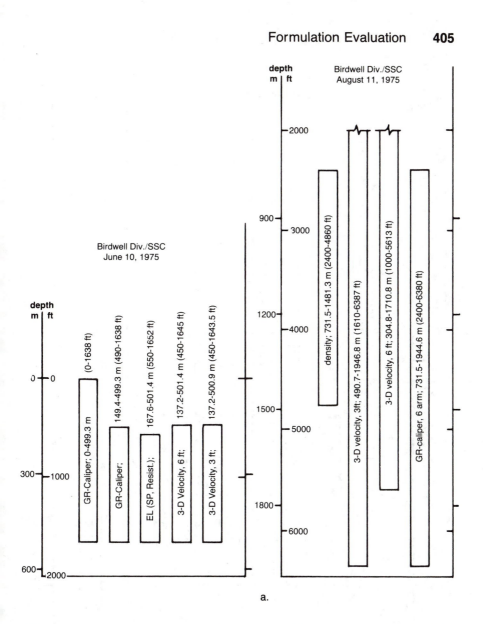

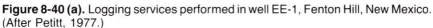

Figure 8-40 (a). Logging services performed in well EE-1, Fenton Hill, New Mexico. (After Petitt, 1977.)

(figure continued next page)

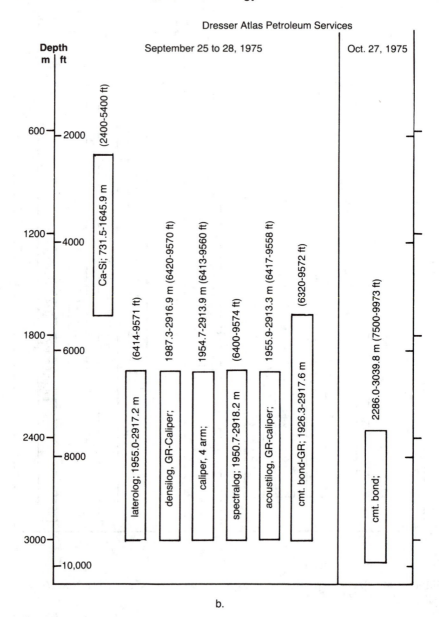

b.

Figure 8-40 (b). Logging services performed in well EE-1, Fenton Hill, New Mexico. (After Petitt, 1977.)

(figure continued next page)

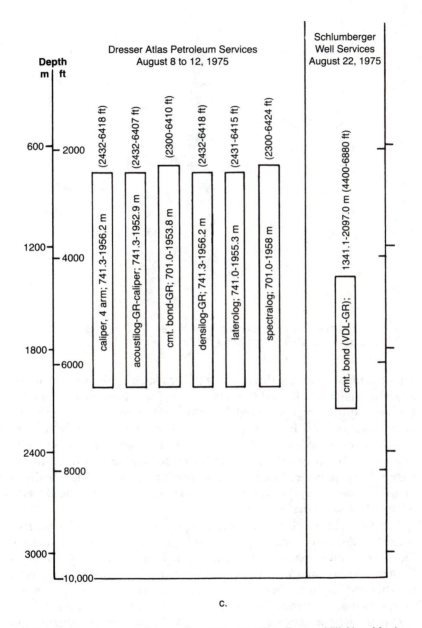

Figure 8-40 (c). Logging services performed in well EE-1, Fenton Hill, New Mexico. (After Petitt, 1977.)

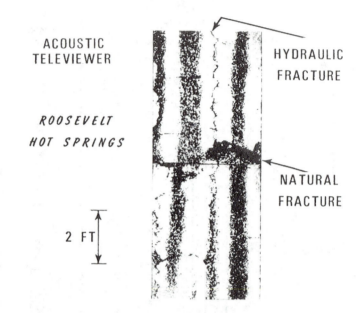

Figure 8-41. Acoustic televiewer log, Roosevelt Hot Springs, Utah. (After Keys, 1979.)

requires input from all three porosity logs such as density, neutron, and acoustic (Sanyal et al., 1979). Furthermore, identification and characterization of hydrothermally altered formations is important in geothermal resource assessments. For example, biotite can be hydrothermally altered to chlorite as observed in the schist zone at Raft River (Keys, 1979). Whereas density and neutron logs show similar response, the biotite schist exhibits significantly higher radioactivity. In addition, crossplotting density versus neutron data clearly differentiates between chlorite schist and the adjacent adamelite and quartzite.

The Geysers geothermal field, one of the few vapor-dominated (dry steam) fields in the world, is producing from natural fracture systems within the Franciscan graywacke from a depth interval of 2,000 to 10,000 ft. Franciscan rocks may be generally characterized as consisting of graywacke, interbedded argillite, chert, greenstone, and schist. These rocks are metamorphosed and, in some areas, exhibit geothermal alteration.

Due to high formation temperature (500°F) and large producing volumes (135,000 lbs steam/hour), these steam wells had not been evaluated across the Franciscan producing zone with geophysical wireline logs prior to 1977. In 1977, the McKinley No. 1 well was successfully logged at 500°F over a number of producing intervals utilizing a logging suite that included gamma ray, a Compensated Densilog®, and a neutron-neutron log (Ehring et al.,

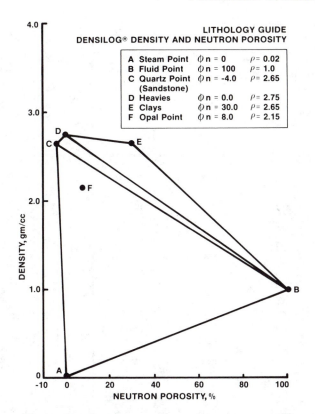

Figure 8-42. Density versus neutron porosity crossplot with basic lithology components as used in the Geysers area, California. (After Ehring et al., 1978.)

1978). More recently, high-temperature, four-arm caliper, flow meter, and a temperature log were added to this special geothermal logging package.

Based on this logging information, a geothermal Epilog® analysis was successfully developed to define several important formation parameters, including lithology for geologic subsurface control, reservoir porosity, steam entry intervals and steam quality estimates. Figure 8-42 shows the basic lithology component triangles used in this analysis, whereas the computed Epilog® is presented in Figure 8-43.

Whenever possible, such log-derived evaluation incorporates mud logging and drill cutting analysis. Inasmuch as air drilling causes fractionated cuttings (ranging from small chips to powder), which may be offset in depth from 20 to 30 ft, this effect has to be taken into account.

Cochran (1979) reported results of formation evaluation on eight wells drilled down to a depth of about 4,000 ft. Conventional wireline logs

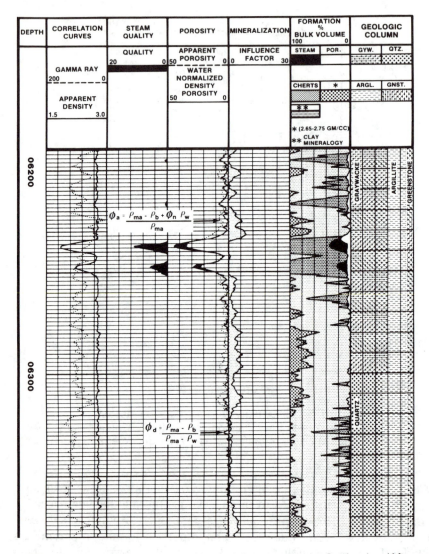

Figure 8-43. Geothermal Epilog®, the Geysers area, California. (After Ehring et al., 1978.)

(gamma ray, density, and neutron), mud logging, extrapolated temperature, and steam entry data were studied. Unfortunately, wireline logs were only run in the overlying or nonproductive reservoir sections. Nevertheless, integration of digital wireline and mud log data have successfully identified five basic lithologies, i.e., chert, graywacke, argillite, greenstone (layered intrusive and extrusive igneous rock), and ultramafic.

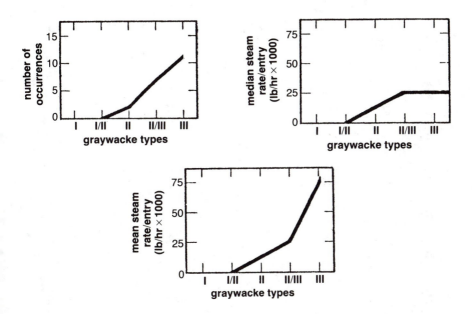

Figure 8-44. Steam distribution versus lithology variations, the Geysers steam field, California. (After Cochran, 1979.)

The graywacke target interval was divided into three textural zones and two transitional intervals. Zone I is unmetamorphosed, Zone II is cataclastic, whereas Zone III is completely recrystallized to quartz-mica schist. A significant correlation was found between the degree of metamorphism of graywacke and steam productivity of a section. Higher-grade metamorphic graywacke sections (Zones I/II and III) correspond to the most productive intervals, whereas low-grade graywacke (Zones I and II) has little steam potential (Figure 8-44).

In the hot-water geothermal resource area of the northern Hot Springs Mountains, Churchill Co., Nevada, the geothermal well Desert Peak No. B-23-1 was logged in Spring 1979 to a total depth of 9,642 ft, with a temperature of 408°F observed on the thermometer run with one of the logging tools. An empirical, statistical log-derived analysis, based on various crossplot concepts and histograms of gamma ray, spectral gamma ray, resistivity, acoustic, density, and neutron logging data, yielded a semi-quantitative evaluation of this complex lithological sequence, which was collaborated by drill cutting data (Sethi and Fertl, 1979). Specific log responses resolved various lithological units in the rock sequence, which consisted in the upper section of volcanic rocks (primarily dacite and rhyol-

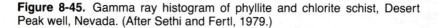

Figure 8-45. Gamma ray histogram of phyllite and chlorite schist, Desert Peak well, Nevada. (After Sethi and Fertl, 1979.)

lite tuffs with a few basalt dikes underlain by phyllite); whereas the deeper part of the borehole penetrated chlorite schist plus hornfels interspersed with granite dikes, with massive fresh granite at the bottom.

Some of several characteristic log responses observed include significantly lower resistivity and negative SP deflections for phyllite. Gamma ray response (Figure 8-45) does not distinguish between phyllite and chlorite schist, whereas density and resistivity log responses (Figures 8-46 and 8-47) clearly do. Granite and chlorite schist both exhibit high density and resistiv-

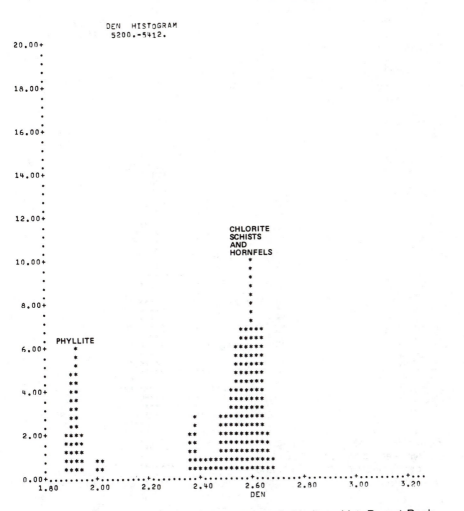

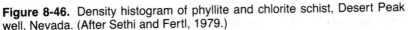

Figure 8-46. Density histogram of phyllite and chlorite schist, Desert Peak well, Nevada. (After Sethi and Fertl, 1979.)

ity values. They are easily distinguished by the gamma ray response because granite is much more radioactive than granite schist.

Bentonitic clays and rhyolite tuffs are differentiated by the potassium content as recorded on the spectral gamma ray log (Figure 8-48), gamma ray log, formation resistivity and density histograms (Figure 8-49). The neutron log is unable to resolve these two lithologies (Figure 8-50).

Sethi and Fertl (1979) concluded that important lithologic parameters can be determined in these volcanic, metamorphic, and igneous rocks with

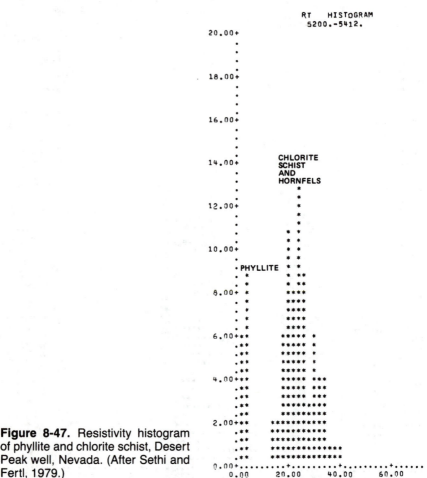

Figure 8-47. Resistivity histogram of phyllite and chlorite schist, Desert Peak well, Nevada. (After Sethi and Fertl, 1979.)

proper planning of geothermal logging suites and log calibrations. Correlation of geologic, drilling, and logging information is also imperative.

Elastic Rock Properties

Mechanical properties of rocks, such as Young's modulus, bulk modulus, shear modulus, and Poisson's ratio, are important engineering parameters for the design of well stimulation (fracture treatment), sand control methods in producing hydrocarbon and/or geothermal wells, casing programs, drilling programs, and perforations. They are also important in determining fluid injection rates, analyzing subsidence, solving sand stability problems, and mine design and operations.

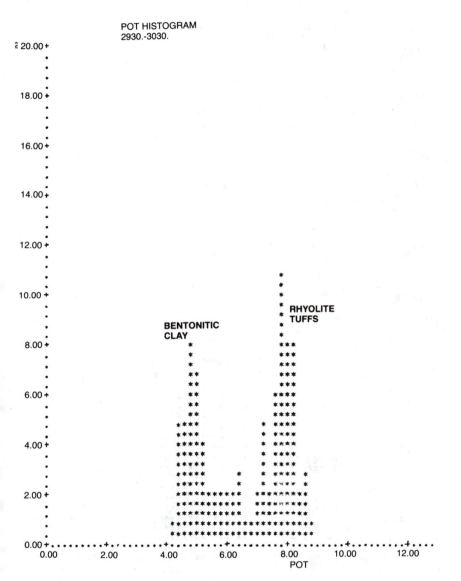

Figure 8-48. Potassium-40 histogram of bentonitic clay and rhyolite tuffs, Desert Peak well, Nevada. (After Sethi and Fertl, 1979.)

These elastic rock properties can be determined from well logs, such as acoustic and density devices. The results provide the dynamic properties, whereas similar measurements on cores under standard triaxial tests will describe the corresponding static rock properties. This has to be kept in mind when comparing log-derived and core-derived data.

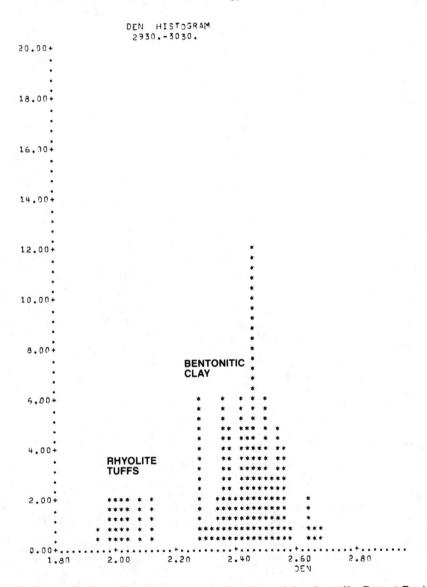

Figure 8-49. Density histogram of bentonitic clay and rhyolite tuffs, Desert Peak well, Nevada. (After Sethi and Fertl, 1979.)

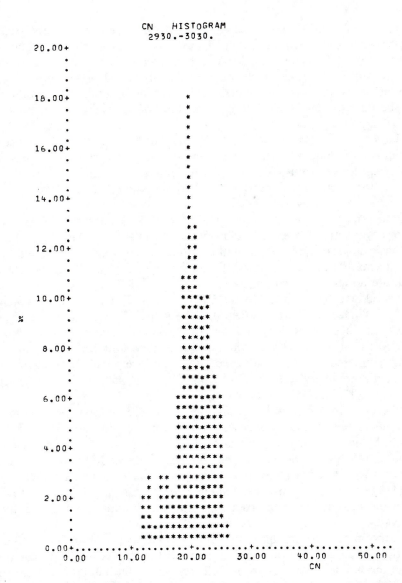

Figure 8-50. Compensated neutron histogram of bentonitic clay and rhyolite tuffs, Desert Peak well, Nevada. (After Sethi and Fertl, 1979.)

Bond et al. (1971) described the strength of formations, with special emphasis on coal mining, using the dynamic elastic properties. Modulus of deformation (EMD) can be calculated from a combination of acoustic and density logs. Experience has shown, however, that this EMD parameter will not define the strength characteristics of a formation under all conditions, but should give an indication of the upper limit of bed competence (Bond et al., 1971).

Mathematical relationships for the Young's modulus, bulk modulus, shear modulus, and Poisson's ratio have been established to relate these properties to log-derived values from formation density, and acoustic information, such as the travel times for shear and compressional waves (Myung and Helander, 1972; Kowalski, 1975).

Unfortunately, quantitative interpretation of the shear wave information is not always straightforward. Table 8-21 defines all four elastic constants, the basic equations, and their mathematical transformations into well logging terms. These elastic moduli can be determined along the entire wellbore under *in-situ* conditions.

Several investigations (Stein and Hilchie, 1971; Tixier et al., 1973; Stein, 1975; Kohlhaas, 1976) have illustrated the value of using log-derived bulk density and acoustic information for correlation with anticipated sand production in oil wells. These techniques are important for evaluating the friable, undercompacted sand reservoirs, which are possible candidates for the geothermal-geopressured DOE test projects that will be carried out under very high producing flow rates.

Whereas some of the above techniques are based on sophisticated digital handling of density and acoustic log information, a rapid manual estimate of the probability of a clastic reservoir rock producing sand may be obtained from the raw logging data using the nomograph shown in Figure 8-51. This nomograph is based on a ratio of shear modulus (μ) to the bulk compressibility ($1/K$) of 0.8×10^{12} psi. For example, in the Tertiary U.S. Gulf Coast sands, no sanding has been observed if this ratio is equal to or greater than this value.

Furthermore, the fracturability of a formation is controlled by these elastic moduli and the principal stresses within the earth. By calculating these moduli and stresses from well logs in a continuous fasion in the borehole along the interval of interest, the tendency for a fracture to propagate through various strata can be predicted (Rosepiler, 1979).

A rock classification method, based on log-derived acoustic density data, using Young's modulus (E) and bulk density (D) has been proposed by Myung and Helander (1972) in Table 8-22. An interesting, interrelated correlation of several rock properties, including dynamic elastic moduli, shear velocity, porosity, and acoustic travel time, has been observed for the volcanic rocks at the Nevada Test Site, and is shown in Figure 8-52.

Table 8-21
Definition of Elastic Constants

Elastic constants	Basic equations	Interrelation of equations	Equation in well logging terms
Young's Modulus*	$E = \dfrac{9K\rho V_s^2}{3K + \mu}$	$E = \dfrac{3K\mu}{3K + \mu} = 2\mu(1+\sigma) = 3K(1-2\sigma)$	$E = \left(\dfrac{\rho}{\Delta t_s^2}\right)\left(\dfrac{3\Delta t_s^2 - 4\Delta t_c^2}{\Delta t_s^2 - \Delta t_c^2}\right) \times 1.34 \times 10^{10}$
Bulk Modulus**	$K = \rho\left(V_c^2 - \dfrac{4}{3}V_s^2\right)$	$K = \dfrac{E\mu}{3(3\mu - E)} = \mu\dfrac{2(1+\sigma)}{3(1-2\sigma)} = \dfrac{E}{3(1-2\sigma)}$	$K = \rho\left(\dfrac{3\Delta t_s^2 - 4\Delta t_c^2}{3\Delta t_s^2\Delta t_c^2}\right) \times 1.34 \times 10^{10}$
Shear Modulus†	$\mu = \rho V_s^2$	$N = \dfrac{3KE}{9K-E} = 3K\dfrac{1-2\sigma}{2+2\sigma} = \dfrac{E}{2+2\sigma}$	$\mu = \left(\dfrac{\rho}{\Delta t_s^2}\right) \times 1.34 \times 10^{10}$
Poisson's Ratio††	$\sigma = \dfrac{1}{2}\,\dfrac{\left(\dfrac{V_c^2}{V_s^2}\right) - 2}{\left(\dfrac{V_c^2}{V_s^2}\right) - 1}$	$\sigma = \dfrac{3K-2\mu}{2(3K+\mu)} \quad \left(\dfrac{E}{2\mu} - 1\right) \quad \dfrac{3K-E}{6K}$	$\sigma = \dfrac{1}{2}\left(\dfrac{\Delta t_s^2 - 2\Delta t_c^2}{\Delta t_s^2 - \Delta t_c^2}\right)$

After Kowalski and Fertl, 1976

ρ = bulk density g/cc
V_s = shear velocity, ft/s
Δt_s = shear travel time, μs/ft

V_c = compressional velocity, ft/s
Δt_c = compressional travel time, μs/ft
1.34×10^{10} = conversion factor

*Young's Modulus (E) measures opposition of a substance to extensional stress, $E = \dfrac{F/A}{\Delta 1/1}$

**Bulk Modulus (K) is the coefficient of incompressibility and measures opposition of a substance to compressional stress, $K = \dfrac{F/A}{\Delta V/V}$

†Shear Modulus μ, also called rigidity modulus, measures the opposition of a substance to shear stresses.

†Finite values for solids, zero values for fluid, $\mu = \dfrac{F/A}{\tan S}$

††Poisson's Ratio (σ) is the ratio of relative decrease in diameter to relative elongation, $\sigma = \dfrac{\Delta d/d}{\Delta 1/1}$

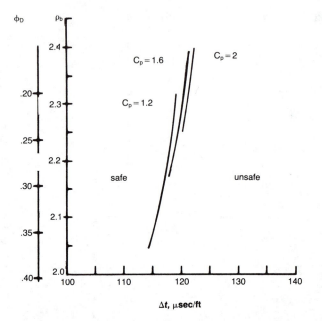

Figure 8-51. Estimation of formation strength and "sanding" from acoustic (Δt) and density (ρ) log data, as computed for ratio of shear modulus (μ) to bulk compressibility ($1/K$) of $0.8*10^{12}$ psi. Compaction correction factor for acoustic log is $Cp = \Delta t$ shale/100. Bulk density ρ and porosity (ϕ_D) are derived from density log. (Modified from Tixier et al., 1973.)

Table 8-22
Rock Classification Method Based on Log Derived Acoustic and Density Data

Rock Index No.	Description	Condition
1	Very hard	$E^*>10\times10^6$ $D^{**}>2.76$
2	Hard	$7\times10^6<E<11\times10^6$ $2.60<D<2.76$
3	Medium	$5\times10^6<E<8\times10^6$ $2.46<D<2.59$
4	Soft	$3\times10^6<E<6\times10^6$ $2.30<D<2.45$
5	Very soft	$E<3.0\times10^6$ $D<2.29$

After Myung and Helander, 1972
*E = Young's modulus
**D = density

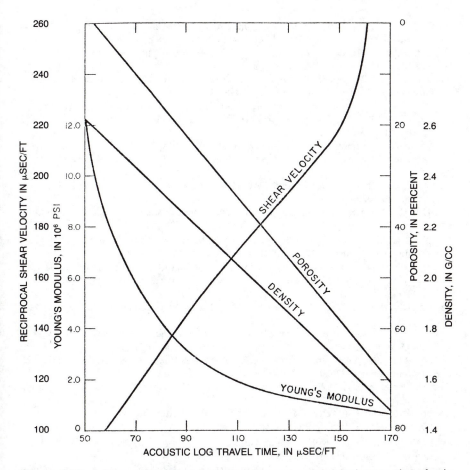

Figure 8-52. Interpretation chart for estimating various physical properties of volcanic rocks at the Nevada Test Site based on the acoustic (sonic) log readings. (After Carrol, 1968.)

References

Alger, R. P., Raymer, L. L., Hoyle, W. R. and Tixier, M. P., 1963. Formation Density Log Applications in Liquid-Filled Holes. *J. Pet. Tech.,* 15 (3); 321-332.

Bailey, T. P., 1977. *A Hydrogeological and Subsurface Study of Imperial Valley Geothermal Anomalies, Imperial Valley, California.* Geological Sciences, Univ. Colorado, Unpubl. Report: 101 pp.

Baker, L. E., Baker, R. P. and Hughen, R. L., 1975. *Report of the Geophysical Measurements in Geothermal Wells Workshop.* Sandia Laboratories, Report SAND 75-0608, Dec.: 76 pp.

Ball, L., Salisbury, J. W., Kintzinger, T. R. and Veneruso, A. F., 1979. The National Geothermal Exploration Technology Program. *Geophysics,* 44 (10): 1721-1737.

Bebout, D. G., Gavenda, V. J. and Gregory, A. R., 1978. *Geothermal Resources, Wilcox Group, Texas Gulf Coast,* DOE Contract EY-76-S-05-4891, Report ORO/4891-3: 82 pp.

Black, H. T., 1975. *A Subsurface Study of the Mesa Geothermal Anomaly, Imperial Valley, California.* Report CUMER 75-5, Univ. Colorado, May: 58 pp.

Bond, L. O., Alger, R. P. And Schmidt, A. W., 1971. Well Log Application in Coal Mining and Rock Mechanics. *Trans. SME,* 20:355-362.

Bonham, L. C., 1978. Solubility of Methane in Water at Elevated Temperatures and Pressures. *Bull. Am. Assoc. Petrol. Geol.,* 62 (12): 2478-2481.

Brill, J. P. and Beggs, H. D., 1975. *Two-phase Flow in Pipe.* Univ. Tulsa, Tulsa, Okla., 250 pp.

Brown, W. M., 1976. *100,000 Quads of Natural Gas?,* Hudson Inst. Res. Memorandum, No. 31, HI-2451/2P, July: 4 pp.

Brown, M. C., Duffield, R. B., Siciliano, C.L.B. and Smith, M. C., 1978. *Hot Dry Rock Geothermal Energy Development Program.* Ann. Report, LA-7807-HDR, UC-66a, Apr.: 53-54.

Burke, J., Campbell, R. and Schmidt, A., 1969. The Litho-porosity Crossplot. *The Log Analyst,* 10(6): 25-43.

Burst, J. F., 1969. Diagenesis of Gulf Coast Clayey Sediments and Its Possible Relation to Petroleum Migration. *Bull. Am. Assoc. Petrol. Geol.,* 53(1): 73-93.

Campbell, R., 1974. *Abnormal Pressure Detection Using Log Data.* Presentation at Well Evaluation Conference North Sea, Schlumberger, London. June: 171 pp.

Carrol, R. D., 1968. Application of In-hole Geophysical Logs in Volcanic Rocks, Nevada Test Site, pp. 125-134. In: E. B. Eckel (Editor), *Nevada Test Site,* Mem. 110, Geol. Soc. Am., Inc.: 290 pp.

Calavier, C. and Rust, D. H., 1976. MID-Plot: A New Lithology Technique. *The Log Analyst:* 18 (6): 16-24.

Cochran, L. E., 1979. *Formation Evaluation in the Geothermal Environment, The Geysers Steam Field, California.* SPE 8452, 54th Ann. Fall Mtg., Las Vegas, Nevada, Sept., 23-26.

Collins. A. G., 1970. Geochemistry of Some Petroleum-Associated Waters from Louisiana. *RI Bur. Mines Report* 7326, Jan.: 31 pp.

Cosner, S. and Apps, J., 1977. *Geothermal Water Data File.* Lawrence Berkeley Laboratory, Feb.: 68 pp.

Culberson, O. L. and McKetta, J. J., 1951. The Solubility of Methane in Water at Pressures to 10,000 psia. *AIME Petrol. Trans.,* 192:223-226.

Davis, D. G. and Sanyal, S. K., 1979. *Case History Report on East Mesa and Cerro Prieto Geothermal Fields.* Report LA-7889-MS, UC-66a, June: 171 pp.

Dorfman, M. and Kahle, R. O., 1974. Potential Geothermal Resources of Texas. *Texas Univ. Bur. Econ. Geol.,* Circ. 74-4: 45 pp.

Ehring, T. W., Lusk, L. A., Grubb, J. M., Johnson, R. B., Devries, M. R. and Fertl, W. H., 1978. Formation Evaluation Concepts for Geothermal Resources, paper FF. *Trans. SPWLA,* 19: 14 pp.

Fertl, W. H. and Timko, D. J., 1970. Occurrence and Significance of Abnormal Pressure Formations. *Oil Gas J.,* 68 (1): 97-108.

Fertl, W. H. and Timko, D. J., 1970. Association of Salinity Variations and Geopressures in Soft and Hard Rocks, paper J. *Trans. SPWLA,* 11: 24 pp.

Fertl, W. H. and Timko, D. J., 1972. How Downhole Temperature, Pressures Affect Drilling. *World Oil,* 175 (1): 45-50.

Fertl, W. H. and Timko, D. J., 1972-73. How Downhole Temperature, Pressures Affect Drilling, 10-part series. *World Oil,* June 1972 through Mar. 1973.

Fertl, W. H., 1976. *Abnormal Formation Pressures — Implications to Exploration, Drilling, and Production of Oil and Gas Resources.* Elsevier Publ. Co., Amsterdam-New York, 382 pp.

Fertl, W. H., 1976. Simplified Shaly Sand Analysis in Development Wells. *Dresser Atlas Tech. Mem.,* 7(4): 1-10.

Fertl, W. H. and Chilingar, G. V., 1976. *Importance of Abnormal Formation Pressures to the Oil Industry.* SPE 5946, SPE-European Spring Mtg., Amsterdam, The Netherlands, Apr. 7-9.

Fertl, W. H. and Ilavia, P. E., 1977. Detection and Evaluation of Abnormally High Formation Pressures Using the Team Approach. In: *6th Form. Eval. Symp., Canad. Well Logging Soc.,* Calgary, Canada, Oct. 24-26.

Fertl, W. H., 1978. *Open Hole Crossplot Concepts — A Powerful Technique in Well Log Analysis.* EU 90, SPE European Offshore Petrol. Conf., London, England, Oct.

Fertl, W. H. And Frost, E., 1979. *Evaluation of Shaly Clastic Reservoir Rocks.* SPE 8450, 54th Ann. Fall Mtg., Las Vegas, Nevada, Sept. 23-26.

Fertl, W. H., 1979. Basics in Shaly Sandstone Reservoirs. In: *Dresser Atlas, California, Formation Evaluation Symp.,* Bakersfield, Calif., 68 pp.

Frondel, C., 1956. In: L. R. Page, H. E., Stocking and H. B. Smith, *U.S. Geol. Surv. Prof. Papers,* No. 300: 55 pp.

Frost, E. and Fertl, W. H., 1979. Integrated Core and Log Analysis Concepts in Shaly Clastic Reservoirs, paper C. *7th Form. Eval. Symp., Canad. Well Logging Soc.,* Calgary, Oct. 21-24.

Gabelman, J. W., 1977. Migration of Uranium and Thorium Exploration Significance. *Studies in Geology No. 3, Am. Assoc. Petrol. Geol.,* Tulsa, Okla.: 165 pp.

Galbraith, R. M., 1978. *Geological and Geophysical Analysis of Coso Geothermal Exploration Hole No. 1 (CGEH-1), Coso Hot Springs KGRA, California.* Report IDO-1701-2, DOE Contract EG-78-C-07-1701, 85 pp.

Garrison, L. E., 1972. Geothermal Steam in the Geysers — Clear Lake Region, California. *Geol. Soc. Am. Bull.,* 84(5): 1449-1468.

Gaymard, R. and Poupon, A., 1968. Response of Neutron and Density Logs in Hydrocarbon-Bearing Formations. *The Log Analyst,* 9(5): 3-12.

Goss, R. and Combs, J., 1976. *Thermal Conductivity Measurement and Prediction from Geophysical Well Log Parameters with Borehole Application.* Natl. Tech. Information Service Report PG-262-372: 31 pp.

Hawkins, M. F., 1977. *Final Report — Investigation of the Geopressure Energy Resource of Southern Louisiana.* Louisiana State Univ., Dept. Petrol. Eng., ERDA — Contract No. EY-76-S-05-4889, Apr. 15: 110 pp.

Heslop, A., 1971. Mixed Lithology Using MN-Product. *J. Canad. Well Logging Soc.,* 4(1): 85-95.

Hulen, J., 1978. *Geology and Alteration of the Coso Geothermal Area, Inyo Co., California.* UURI-ESL Report, DOE Contract EG-78-C-07-1701, 65 pp.

Jones, P. H., 1968. *Hydrodynamics of Geopressures in the Northern Gulf of Mexico Basin.* SPE 2207, 43rd AIME Fall Mtg., Houston Texas; also *J. Pet. Tech.,* 1969. 21:802-810.

Khatchikian, A. and Lesta, P., 1973. Log Evaluation of Tuffites and Tuffaceous Sandstone in Southern Argentina, paper K, *Trans. SPWLA,* 14: 24 pp.

Kassoy, D. R. and Goyal, K. P., 1979. *Modeling Heat and Mass Transfer at the Mesa Geothermal Anomaly, Imperial Valley, California.* Report LBL-8784-GREMP-3, UC-66a, DOE Contract W-7405-ENG-48, Feb.: 152 pp.

Kintzinger, P.R., West, F. G. and Aamodt, R. L., 1977. *Downhole Electrical Detection of Hydraulic Fractures in GT-2 and EE-1.* Report LA-6890-145, UC-66b, July: 13 pp.

Kohlhaas, C. A., 1976. *Evaluation of Well Completions in Southeast Asia for Sand Control.* SPE Offshore Southeast Asia Conf., Singapore, Feb. 17-21.

Kowalski, J. J., 1975. Formation Strength Parameters from Well Logs, paper N. *Trans, SPWLA,* 16: 19 pp.

Kowalski, J. J. and Fertl, W. H., 1976. Application of Geophysical Well Logging to Coal Mining Operations. *Erdol-Erdgas Zeitschrift,* 92(9): 301-305 (in German); also *Energy Sources,* 1977, 3(2): 133-147 (in English).

Lewis, C. R. and Rose, S. C., 1970. A Theory Relating High Temperatures and Overpressures. *J. Pet. Tech.*, 22: 11-16.

Littleton, R.T. and Burnett, E. E., 1977. Chemical Profile of the East Mesa Field, Imperial County, California. In: *2nd Workshop on Sampling and Analysis of Geothermal Effluents*, sponsored by U.S. Environ. Protection Agency, Las Vegas, Nevada, 45 pp.

Mathews. M., Arney, B. and Sayer, S., 1979. Log Comparison from Geothermal Calibration/Test Well CT-1, paper RR. *Trans. SPWLA*, 20: 25 pp.

Miyairi, M., Itoh, T. and Okabe, F., 1976. Water Saturation in Shaly Sands; Logging Parameters from Log-Derived Values, paper G. *Trans. SPWLA*, 17: 22 pp.

Mortensen, J. H., 1977. *The LASL Hot Dry Rock Project Geothermal Energy Development Project, LASL-Mini Review*, 77-8: 4 pp.

Myung, J. I. and Helander, D. P., 1972. Correlation of Elastic Moduli Dynamically Measured by *In-situ* and Laboratory Techniques, paper H. *Trans. SPWLA*, 13: 16 pp.

National Academy of Sciences, Washington, D.C., 1977. Assessment of the Characterization (in-situ downhole) of Geothermal Brines, National Materials Advisory Board Publication. NMAB-344: 115 pp.

Neinast, G. S. and Knox, C. C., 1973. Normalization of Well Log Data, paper I. *Trans. SPWLA*, 14: 18 pp.

Nelson, P. H. and Glenn, W. E., 1975. Influence of Bound Water on the Neutron Log in Mineralized Igneous Rock. *Trans. SPWLA*, 16: 8 pp.

Overton, H. L. and Timko, D. J., 1969. The Salinity Principle — Tectonic Stress Indicator in Marine Sands, paper M, *Trans, SPWLA*, 10: 10 pp.

Papadoupulos, S. S., Wallace, R. H., Wesselman, J. B. and Taylor, R. E., 1975. Assessment of Onshore Geopressured Geothermal Resources in the Northern Gulf of Mexico Basin. In: D. E. White and D. L. Williams (eds.), *Assessment of Geothermal Resources of the U.S. — 1975. U.S. Geol. Surv.* 726: 35 pp.

Pettit, R. A., 1977. *Planning, Drilling, Logging, and Testing of Energy Extraction Hole EE-1, Phases I and II*. Report LA-6906-MS, Aug., 75 pp.

Poupon, A., Loy, M. E. and Tixier, M. P., 1954. A Contribution to Electrical Interpretation in Shaly Sands. *Trans. AIME*, 201: 138-145.

Reeber, R. R., 1977. *Report to the NMAB Committee on Assessment of the Characterization (In-Situ Downhole) of Geothermal Brines*. Woods Hole, Ma., Aug. 25: 88 pp.

Rigby, F. A., and Reardon, P., 1979. *Benefit/Cost Analysis for Research in Geothermal Log Interpretation*. Report LASL L68-1908E-1, May: 120 pp.

Ritch, H. J., 1975. An Open Hole Logging Evaluation in Metamorphic Rocks, paper V. *Trans. SPWLA*, 16: 11 pp.

Rosepiler, M. J., 1979. *Determination of Principal Stresses and Confinement of Hydraulic Fractures in Cotton Valley*. SPE 8405, 54th AIME Fall Mtg., Las Vegas, Nevada, Sept. 23-26.

Sacco, E. L., 1978. Carbon/Oxygen Log Application in Shaly Sand Formations Contaminated with Tuffite Minerals, paper HH. *Trans. SPWLA*, 19: 10 pp.

Sanyal, S. K., Wells, L. E. and Bickham. R. E., 1979. *Geothermal Well Log Interpretation*. Mid-term Report LA-7693-MS Informal Report, Feb.: 178 pp.

Sanyal, S. K., Juprasert, S. and Jusbache, S. K., 1979. An Evaluation of a Rhyolite-Basalt-Volcanic Ash Sequence from Well Logs, paper TT. *Trans. SWPLA*, 20: 14 pp.

Sethi, D. K. and Fertl, W. H., 1979. *Geophysical Well Logging Operations and Log Analysis in Geothermal Well Desert Peak No. B-23-1*. DOE-LASL: 69 pp.

Stein, N. and Hilchie, D. W., 1972. Estimation of Maximum Production Rates Possible from Friable Sandstones Without Using Sand Control. *J. Pet. Tech.*, 24(11): 1157-1161.

Stein, N., 1976. Mechanical Properties of Friable Sands from Conventional Log Data. *J. Pet Tech.*, 28(7): 757-764.

Sultanov, R. G., Skripka, V. G. and Namiot, A. Y., 1972. Solubility of Methane in Water at High Temperatures and Pressures. *Gazovaya Promyshlennost'*, 17(5): 6-7.

Timko, D. J. and Fertl, W. H., 1970. Hydrocarbon Accumulation and Geopressure Relationship and Prediction of Well Economics with Log-Calculated Geopressures. *J. Pet. Tech.*, 23: 923-933.

Tixier, M. P., 1968. Log Evaluation of Low-Resistivity Pay Sands in the Gulf Coast. *Trans. SPWLA*, 19: 18 pp.

Tixier, M. P., Loveless, G. W. and Anderson, R. A., 1973. *Estimation of Formation Strength from the Mechanical Properties Log.* SPE 4532, 48th Ann. Fall Mtg., Las Vegas, Nevada, Sept. 30-Oct. 3.

Vagelatos, N., Steinman, D. K. and John, J., 1979. True Formation Temperature Zone, paper LL. *Trans. SPWLA*

Veneruso, A. F., Polito, J. and Heckman, R. C., 1978. *Geothermal Logging Instrumentation Development Program Plan.* Sandia Laboratories, SAND 78-0316, Aug.: 72 pp.

Veneruso, A. F., and Coguat, J. A., 1979. Technology Development for High Temperature Logging Tools, paper KK. *Trans. SPWLA,* 20: 13 pp.

Wallace, R. H., Kraemer, T. F., Taylor, R. E. and Wesselman, J. B., 1978. Assessment of Geopressured Resources in the Northern Gulf of Mexico Basin. In: L.J.P. Muffler (ed.), *Assessment of Geopressured Resources of the U.S., U.S. Geol. Sur. Cir.* 790: 28 pp.

Ward, S. H., Parry, W. T., Nash, W. P., Sill W. R., Cook, K. L., Smith, R. B., Chapman, D. S., Brown, F. H., Whelan, J. A. and Bowman, J. R., 1978. A Summary of the Geology, Geochemistry, and Geophysics of the Roosevelt Hot Springs Thermal Area, Utah. *Geophysics,* 43 (7 Dec.): 1515-1542.

West, F. G. and Laughlin, A. W., 1976. Spectral Gamma Logging in Crystalline Basement Rocks. *Geology,* (4 Oct.): 617-618.

West, F. G., Kintzinger, P. R. and Laughlin, A. W., 1975. *Geophysical Well Logging in Los Alamos Scientific Laboratory Geothermal Test Hole No. 2.* LA-6112-MS Informal Report, Univ. California, Los Alamos, New Mexico: pp. 1-11.

Wilson, G. J. and Bush, R. E., 1973. Pressure Prediction with Flow Line Temperature Gradient. *J. Pet. Tech.*, 25: 135-142.

White, D. E., 1970. Geochemistry Applied to the Discovery, Evaluation, and Exploitation of Geothermal Energy Resources. In: *UN Symp. on the Development and Utilization of Geothermal Resources,* Pisa, Proc., Geothermics, Spec. Issue 2: 1-58.

Wyllie, P. J., 1971. *The Dynamic Earth,* John Wiley and Sons, New York, 416 pp.

Fernando Samaniego V., Instituto de Investigaciones Electricas
Heber Cinco-Ley, Stanford University

9

Reservoir Engineering Concepts

Introduction

Geothermal reservoir engineering is related in many aspects to oil and gas reservoir engineering, which began receiving attention in the 1930s (Richardson, 1973). However, it wasn't until the end of the '60s, with the classic studies on the Wairakei field by Whiting and Ramey (1969) and the Geysers field by Ramey (1970) that geothermal reservoir engineering began receiving special attention. Geothermal reservoir engineering includes activities that begin with locating the wells, well logging, drilling and flow measurements, identification of the production mechanism, and performance prediction of reservoir behavior to find the optimum production conditions that would lead to maximum economic heat recovery.

The basic tasks of the reservoir engineer are the prediction of the long-term behavior of the well and the reservoir (Ramey, 1977). There are a few important questions that he or she must answer:

1. What is the optimum development plan for the reservoir?
2. How many wells and what kind of pattern will be required for optimum development of the reservoir?
3. What will be the rate of production for the wells?
4. How much heat will be recovered?
5. What will be the variation of temperature versus time?
6. Would it be feasible to implement an enhanced recovery process to recover additional heat?

To answer all these and other possible questions, the engineer must undertake a continuous and very careful work since the beginning of the life

426

of the reservoir. As production from the reservoir increases and more data becomes available, the reservoir engineer has more data for history matching purposes, allowing him or her to update previous studies and, consequently, to make better predictions of the reservoir behavior. Unfortunately, all data concerning the reservoir becomes available only when it is fully depleted.

It is important that the reservoir engineer quantitatively appreciates the physical processes that occur within the geothermal system, because this permits optimum exploitation of the reservoir. This "appreciation" can be divided into three main steps: first, the physical processes associated with the particular geothermal system under study must be identified and used to develop a conceptual model of the reservoir; second, a careful assessment of the physical and thermal properties of the rock and fluids must be made (these data will be extremely useful for simulation studies purposes); and third, a mathematical or physical model of the reservoir is developed, using the previously determined information about the reservoir. This model should include the properly identified initial and boundary conditions for the system. Once the model is at hand, it can be updated and refined as new production data becomes available. The reservoir response under production should be carefully matched with the model. This technique of matching the observed production history data by means of a suitable model and using the model to predict future performance is fundamental to the subject of reservoir engineering.

About the middle of the 1960s geothermal reservoir engineering relied heavily on the theory developed for oil and gas reservoir engineering. In cases where the characteristics of the geothermal and petroleum systems are alike, the oil and gas reservoir engineering techniques can be properly employed, if inherent differences in the systems are considered. Examples of successful applications of these techniques to geothermal reservoirs are available (Whiting and Ramey, 1969; Ramey, 1970; Atkinson et al., 1978). However, there are several factors that make geothermal reservoir engineering a unique subject in itself, such as very high reservoir temperatures; formations containing the fluid in many cases are highly fractured and of volcanic type; chemical deposition of solids during flow of fluids in the reservoir; and steam flashing. Due to the interest in geothermal energy, geothermal reservoir engineering has shown great advancements in the last fifteen years (Takahashi et al., 1975).

This chapter presents a review of the basic principles of geothermal reservoir engineering. The presentation includes a discussion of some of the practical aspects of reservoir engineering, such as well test analysis; and mathematical reservoir simulation.

Types of Geothermal Systems

Geothermal systems can be classified in four main types: vapor-dominated (dry steam); liquid-dominated (hot water); geopressured (hot) accumulations; and dry (hot rock) formations. Although each one of these systems has potential for exploitation, the vapor-dominated system offers the optimum conditions for electricity production. According to White (1973), technology currently available for geothermal electricity generation makes use of steam. Based on present knowledge, there are some requirements that have to be met for electricity generation from geothermal steam: (a) temperature of reservoir should be high (at least 180°C, and preferably above 200°C); (b) reservoir depth must be less than 3 km; (c) reservoir volume must be adequate; (d) the reservoir should contain natural fluids for transferring the heat to surface and power plants; (e) permeability of the formation should be adequate to ensure sustained delivery of fluids to wells at high enough rates to meet power production needs; (f) there must be no major unsolved technology problems. Unfortunately, these requirements are not easily met in the earth's crust.

Vapor-Dominated Systems

These systems are the rarest found in nature and the most desirable (Burnham, 1973), because thay provide a clean and environmentally safe energy source with minor production problems. There are a few vapor-dominated systems, among which the two most important ones are the Lardarello fields of Italy and The Geysers of California, which produce dry or superheated steam with no associated liquid (White, 1973). This is the reason why they are commonly known as "dry-steam" systems. White et al., (1971) have concluded, however, that liquid water and vapor usually coexist in the reservoir, with vapor being the continuous, pressure-controlling phase.

Vapor systems contain far less heat than the hot-water systems (Ramey et al., 1973), but, as mentioned earlier, problems related with their utilization are minor. The production mechanism of steam reservoirs is similar to that of natural-gas reservoirs, because pressure is reduced by the expansion of the gas in-place. The fractional mass production of original mass in place is usually high, being around 85-90%. The situation for energy recovery, however, is quite different, being quite low. Ramey et al. (1973) presented a table in which they compared the recovery for a hot-water and a steam reservoir. For the latter, a fractional energy recovery is 5.6%, because most of the heat is stored in the rock and, since the steam flow process in the reservoir is essentially isothermal, almost none of the heat contained in the rock is recovered. To recover heat stored in the rock, the production process should reduce the rock temperature. For these systems, additional energy can be recovered by means of an enhanced recovery process, using liquid-reinjection to the reservoir.

Hot Water Systems

In these systems, water is the continuous, pressure-controlling fluid phase. Such a system may contain some vapor, found as discrete bubbles in the shallow, low-pressure zones. Among geothermal systems discovered to date, hot liquid-dominated systems are far more common than vapor-dominated systems (White, 1973). Water in these reservoirs is a dilute aqueous solution containing sodium, potassium, lithium, calcium, chloride, bicarbonate, sulfate, borate, and high percent of silica. Some of the most important liquid-dominated systems are those located in the Imperial Valley in California, Wairakei in New Zealand, and Cerro Prieto in Mexico.

As mentioned previously, the energy content of these systems is higher than that for the vapor-dominated systems. Ramey et al. (1973) have shown that the fractional energy recovery for these kinds of systems is higher than for the vapor-dominated systems. This is caused by flashing of the water in the reservoir or by reinjection of water, which are the key factors in the recovery of geothermal energy from these systems. The problems encountered in the utilization of these systems are more difficult than those found for the vapor-dominated systems; however, current elaborate research programs may overcome these difficulties in the future.

Geopressured (Hot) Accumulations

Geopressured systems are composed of rocks that contain fluids at pressures far greater than normal (hydrostatic). Knapp et al. (1977) presented an explanation for the existence of these high-pressure zones. Briefly, these zones have a highly impermeable overlying formation that prevents the migration of fluids out of the zone and, thus, causes the fluids saturating this zone to partially bear the overburden load. This results in an increase in the fluid pressure. Pressure gradients in a geopressured zone usually approach lithostatic pressure, i.e., 1 psi/ft. The geopressured sediments have a low thermal conductivity and high heat capacity, causing the system's temperature to be high. The most important geopressured zone known to date is the Gulf Coast region of the United States, which extends on land and offshore from Texas to the Mississippi estuaries.

Fluids contained in geopressured systems are commonly saturated with methane. At the pressures and temperatures encountered, the quantity of methane associated with water saturating the system can be important. This is an additional benefit that can be obtained from the exploitation of these accumulations. Another factor that augments the solution of methane in the water of these reservoirs is water salinity, which is lower under the geopressured conditions than at normal conditions (Burst, 1969). Low-salinity waters can hold more methane in solution than high-salinity waters.

Dry (Hot-Rock) Formations

Hot dry formations are systems that do not contain water to act as heat transport medium. The incidence of this type of geothermal resources is far greater than that of the fluid-saturated geothermal sites (Burnham and Stewart, 1973). These resources can be an important energy supply, provided that means can be found to extract and use such heat economically. The most simple, practical, and economical way to recover heat from these systems is introducing water into the formation, permitting it to circulate until it has been heated to a sufficiently high temperature, and then recovering the fluid as steam or hot water (Smith et al., 1973 and 1975).

Frequently, dry formations have very low permeability, and the problems of containing and recovering the injected water can be overcome by means of creating flow passages of sufficient surface area, through which water can flow for economically long periods of time. This would allow heat recovery through contact of the water with the surface area of the flow passages. The basic heat extraction system tested to date consists of an injector and a producer well, intercommunicated by a hydraulic fracture.

Geothermal systems saturated with fluids (convective systems) can be classified on the basis of the location of their initial conditions on a pressure-temperature diagram. Such a classification has been presented by Whiting and Ramey (1969) and by Martin (1975). Figure 9-1 presents a pressure-temperature diagram for pure water. The solid line is the boiling curve. Similar diagrams for geothermal brines should be modified to include the effect of dissolved salts. It is helpful to examine events subsequent to producing a reservoir at different assumed initial conditions.

Several points in Figure 9-1 must be considered. Point A represents the initial conditions for a single-phase (steam) reservoir that initially existed entirely within the vapor region. There is no formation of hot water because reservoir conditions prohibit it; therefore, the flow is isothermal and the boiling curve is not crossed. At the end of production, the temperature of the reservoir would be high. Thus, additional energy can be recovered by water injection. Point B represents the initial conditions for a reservoir; these conditions fall on the vapor pressure curve and, consequently, there are two phases (hot water and steam) originally present in the reservoir. The analog of this reservoir in petroleum reservoir engineering is the gas-cap reservoir. According to Whiting and Ramey (1969), for this system production is a mixture of hot water and steam that could range from saturated

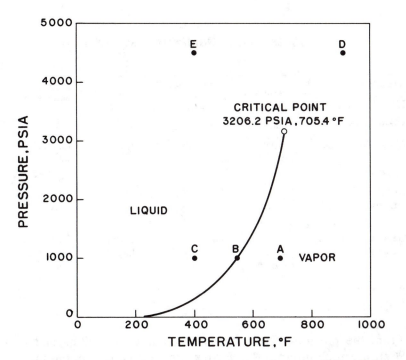

Figure 9-1.   Pressure-temperature diagram for water (After Whiting and Ramey, 1969, Fig. 1, p. 894, by permission of the S.P.E. of A.I.M.E.)

liquid to saturated steam. Point C corresponds to a reservoir that originally had only hot water. The production mechanism is such that eventually, as reservoir pressure declines, the conditions of the vapor-pressure curve will be reached. From then on, production is similar to that of reservoir B previously discussed. Whereas the reservoir fluid is hot water, the flow in the reservoir is essentially isothermal and isoenthalpic. When the vapor pressure curve is reached, the pressure and temperature decline along this curve.

Point D in Figure 9-1 corresponds to a reservoir that originally existed at conditions of pressure and temperature above the critical values. As pressure declines due to production, a reservoir of this type would eventually become similar to reservoir A. These reservoirs usually do not exhibit pressure-temperature conditions that could result in crossing the vapor-pressure curve. Point E represents a reservoir whose initial condition of pressure is higher than the critical pressure. Due to production from the reservoir, pressure will decline, and the reservoir eventually will become similar to reservoirs C and B.

Table 9-1
Assessment of Geothermal Systems

Situation system	Technology	Environmental impact	Economics	Resource availability
Vapor-dominated	Established	Small	Attractive	Limited
Liquid-dominated	Partially established	Potentially large	Known but uncertain	Limited but significant in some areas
Dry (hot-rock) formations	Only partially developed	Unknown	Partially known	Potentially large
Geopressured (hot) accumulations	Only partially developed	Unknown	Partially known	Limited but significant in some areas

From the previous discussion and according to Whiting and Ramey (1969), it can be inferred that for performance prediction purposes, it is necessary to know mass production and enthalpy of the produced fluids. It has also been pointed out that the reservoir rock is an important potential source for energy recovery.

It is of interest to know in a general sense what are the state of technology development, environmental impact, economics, and availability of the different geothermal systems. Table 9-1 shows such type of information for present conditions (Davis and Golan, 1974). A striking fact drawn from this table is that our knowledge is more complete for those resources that are limited in extent. If geothermal energy is to be an important worldwide source of energy, the technology for energy recovery from dry (hot-rock) formations has to be fully developed. This will need an extensive research program to evaluate all aspects related to these systems, e.g., environmental impact, economics, and the necessary technology development. Such efforts are presently under way (Smith et al., 1973; Murphy, 1975).

Pressure Transient Analysis for Geothermal Wells

Transient pressure testing consists of recording the pressure variation versus time in the well or neighboring wells after the flow rate of the well(s) is changed, and, subsequently, estimating the reservoir and well properties. This technique estimates these data at the *in situ* conditions of pressure, temperature, and fluid saturation prevailing in the reservoir. From the analysis of pressure transient tests, among other data, the following information can be obtained:

1. Permeability thickness product (kh) in the drainage volume of the well and permeability (k)
2. Condition of the well, represented by the skin factor (s)
3. Average pressure within the drainage volume (p)
4. Porosity of the drainage volume (ϕ)
5. Pore volume of the reservoir (v_p) and its shape
6. Reservoir and fluid discontinuities (faults, etc.).

This information would be extremely useful to help analyze, improve, and forecast reservoir performance. Ramey (1975) presented an example of practical use of this data. He stated that quantitative information on these listed items would answer questions that are usually raised, such as: Is the low productivity of a well caused by plugging of the well, low formation permeability, or a low driving force and/or formation capacity for moving fluid into the well?

Thus, transient pressure testing is an extremely useful tool for reservoir diagnosis. Transient pressure testing is sometimes referred to as a practical application of reservoir engineering (Dake, 1978). In specific situations (Earlougher, 1977), it is indispensable to have correct well or reservoir analysis; for example, in defining near-wellbore and inter-well conditions as opposed to composite properties that would be obtained from steady-state tests.

To carry out a transient pressure analysis of field data, it is important to choose or obtain an adequate mathematical expression that can be used for interpretation and design of the test. This expression can be derived by properly combining the physical laws that describe the specific fluid flow problem in the reservoir, and considering the production condition of the well.

Transient pressure tests can be classified in two main types: *single well tests*, and *multiple well tests*. Single well tests, as the name indicates, are those that involve only one well, producer or injector. These tests require measuring the well's pressure response after a rate change has taken place. On the other hand, multiple well tests directly involve more than one well. In tests like this, a rate change in a well called "active" creates a pressure response in a neighboring "observation" well. Both pressure responses for single and multiple wells can be analyzed for reservoir properties. The most important well tests are:

1. *Single well tests*— drawdown tests, buildup tests, injectivity tests, falloff tests, and multiple-rate tests.
2. *Multiple well tests*— interference tests, and pulse tests.

Figure 9-2 shows a schematic representation of the variation of the mass flow rate w and pressure versus time during three different types of transient pressure tests. Figure 9-2a represents a drawdown test; Figure 9-2b, a buildup test; and Figure 9-2c, a two-rate test. The most simple pressure drawdown test consists of a series of bottomhole pressure measurements made over a period of time with the flow rate constant. Before the test, pressure throughout the reservoir should be uniform, i.e., static. If the constant rate production and the static pressure condition are not met, alternative methods are available (Matthews and Russell, 1967; Earlougher, 1977). Depending on the purpose for which they are performed, there are two types of drawdown tests: (a) short-term drawdown tests to estimate the formation permeability and wellbore condition; and (b) long-term or reservoir limit tests to estimate the reservoir volume.

A *pressure buildup test* consists of a series of shut-in bottomhole pressure measurements, p_{ws} , made immediately before and at times Δt after shut-in. Again, the most simple buildup test involves a well that produced with a constant flow rate, w, for a period of time before shut-in (Figure 9-2b). Buildup tests are conducted in wells for similar purposes as drawdown tests; namely to estimate the flow capacity of the formation, kh, in the drainage volume; to estimate the well condition, s; and to determine the average (static) pressure in the reservoir.

A *two-rate test* is a particular case of a multiple-rate test. It consists of a series of flowing bottomhole pressure measurements made at times Δt after a rate change of the well (Figure 9-2c). The two-rate tests provide information about formation permeability, wellbore condition, and average drainage volume pressure at the start of the test.

Whereas drawdown and buildup tests are conducted in production wells, their counterpart in injection wells are the injectivity and the falloff tests. An *injectivity test* consists of a series of bottomhole pressure measurements, p_{wf}, made over a period of time with the injection flow rate being constant. The type of information that can be obtained from the analysis of these tests is the same as from drawdown tests. A *falloff test* consists of a series of bottomhole pressure measurements made immediately before and at times Δt after stopping injection. Once more, the most simple falloff test is that where injection rate, w, is constant, until the well is shut-in at time t. The

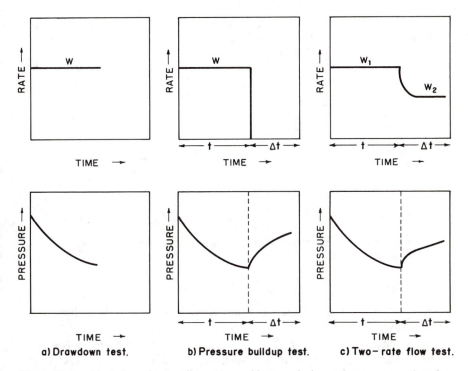

Figure 9-2. Variation of mass flow rate and bottom-hole pressure versus time for various pressure transient tests.

possible information to be obtained from these tests is the same as from pressure buildup tests.

Multiple-well transient tests consist of a series of pressure measurements of the pressure response in one or more observation wells, after a rate change on an active neighboring well. The most simple multiple-well transient test involves only one active and one observation well. Analysis could also be made of pressure response when there are more than one active and observation wells, but it will be more complicated. Figure 9-3 schematically illustrates the use of two wells in an interference or pulse test. The observation well is shut-in for pressure measuring purposes. As discussed by Matthews and Russell (1967), the name "pressure interference" comes from the fact that the pressure drop, caused by the active (producers or injectors) wells, at the shut-in observation well "interferes with" the pressure at the observation well.

In an interference test, the duration of the rate change is long, in contrast to the shorter duration of rate change in a pulse test, but the analysis

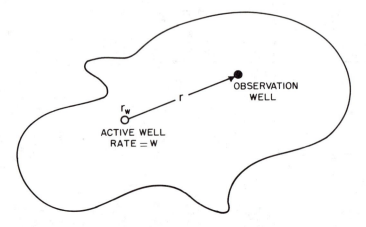

Figure 9-3. Schematic of an active and observation well in an interference test.

technique for reservoir properties is more elaborate. Multiple-well tests can give information about reservoir properties that cannot be obtained from ordinary single well tests. For instance, it is possible to estimate reservoir connectivity. A test like this may answer such questions as, Is the area of the reservoir closed to a well being drained by other wells and, if so, how fast? This test would also help to determine the preferential reservoir flow patterns. This could be accomplished by selectively opening wells that surround the shut-in well.

Because flow in a geothermal reservoir can be steam, hot water, or a mixture of steam and hot water, it is important to understand analysis techniques for these three different cases vis-á-vis some of the most used pressure transient tests.

Transient Pressure Analysis for Steam Wells

Due to the fact that single-phase (steam or hot water) flow in geothermal reservoirs is essentially isothermal (Whiting and Ramey, 1969), transient pressure analysis techniques are usually based on a strict analogy with the single-phase isothermal flow techniques developed by petroleum engineers and hydrogeologists. The laminar flow (a) of a slightly compressible fluid in a radial, horizontal, isotropic reservoir, (b) under the condition of small pressure gradients in the reservoir and applicability of Darcy's law, and (c) assuming that the rock and fluid properties are independent of pressure is expressed by the diffusivity equation (Muskat, 1938; Matthews and Russell, 1967; Earlougher, 1977). For convenience, the solution to this reservoir fluid flow problem is usually expressed in dimensionless form. The following groups have been defined by Ramey (1975) and Ramey and Gringarten (1975), where a and B are unit constants:

Dimensionless pressure for flow of steam—

$$p_D \ (r_D \ t_D \ ) \ = \ \frac{Mkh \ (p_i^2 \ - \ p^2)}{a \ w\mu \ Z \ T} \tag{9-1}$$

Dimensionless time—

$$t_D \ = \ \frac{B \ kt}{\phi \ \mu \ c_t \ r_w^2} \tag{9-2}$$

Dimensionless radial distance—

$$r_D = \frac{r}{r_w} \tag{9-3}$$

If a consistent set of units is used, then these constants will have a value equal to one. Table 9-2 shows some of the most used unit systems, and the corresponding values for a and B.

The pressure in the reservoir at any point in space and time can be estimated from Equation 9-1 if the particular p_D for the system under consideration is known. The dimensionless pressure p_D is a function of the type of system, infinite or finite, and of the boundary conditions, rate or pressure specified. For a well producing at a constant flow rate w, assuming that the conditions for the flow of a fluid of constant and small compressibility apply, and for infinite acting conditions, the p_D can be expressed by the line source solution (Earlougher, 1977):

$$p_D \ (r_D \ t_D \ ) = \ -\frac{1}{2} \ E_i \ \left(\ -\frac{r_D^2}{4t_D} \ \right) \tag{9-4}$$

For transient pressure analysis purposes, the pressure of interest is the wellbore pressure. For this condition, $r = r_w$, and combining Equations 9-1, 9-2, 9-3, and 9-4, and using the logarithmic approximation to the exponential integral, an expression for the wellbore pressure can be obtained as follows:

$$p_{wf}^2 = p_i - 1.1513a \ \frac{w\mu ZT}{Mkh} \left[\log\frac{kt}{\phi\mu \ c_t \ r_w^2} + \log\frac{4B}{\gamma} + \ 0.86859s \right] \tag{9-5}$$

Table 9-2
Absolute and Hybrid System of Units Used in Geothermal Reservoir Engineering

Variable	SI system*	Hybrid system
k	metre^2	md
h	m	m
p	Newton/metre^2 = pascal	kg$_f$/cm^2
w	kg/sec	ton/hr
v_{sc}	metre^3/kg	cm^3/gm
B	metre^3 $_{rc}$/metre^3 **	metre^3 $_{rc}$/metre^3 $_{sc}$
μ	kg/metre. sec	cp
t	sec	hours
ϕ	fraction	fraction
c_t	(Newton/metre^2)$^{-1}$	(kg/cm^2)$^{-1}$
r	metre	metre
a	26.1201	77.459 × 10^3
B	1	0.000348
δ	1/2 π	456.7869
ϵ	1/2 π	1/2 π
η	2.6465	27

* SI is the abbreviation for International System of Units.
** *rc* stands for reservoir conditions and *sc* for standard conditions.

where γ = 1.7810724, and log = base 10 logarithm.

This equation can be used for interpretation of a steam drawdown test. It can be observed that it describes a straight-line relationship between p_{wf}^2 and log t. Theoretically, a plot of flowing bottom hole pressure data versus the logarithm of time (usually called a "semilog plot") should be a straight line with slope m, given by the following expression:

$$m = \frac{1.1513 \, a \, w \, \mu \, Z \, T}{Mkh} \tag{9-6}$$

From this equation, the capacity of the formation can be obtained:

$$k \, h = \frac{1.1513 \, a \, w \, \mu \, Z \, T}{mM} \tag{9-7}$$

By rearranging Equation 9-5, an expression for the skin factor, s, can be obtained:

$$s = 1.1513 \left[\frac{p_i^2 - p_{1hr}^2}{m} - \log \frac{k}{\phi \mu \, c_t \, r_w^2} - \log \frac{4B}{\gamma} \right] \tag{9-8}$$

For pressure buildup testing, the bottomhole shut-in pressure of the well may be expressed by means of the principle of superposition for a well producing at a rate w until time t (Figure 9-2b), and at zero rate thereafter. Thus, at any time after shut-in, the wellbore pressure may be expressed as follows:

$$p_{ws}^2 = p_i^2 - a \frac{w\mu\, Z\, T}{Mkh} \left[p_D (t + \Delta t)_D - p_D (\Delta t_D) \right] \qquad (9\text{-}9)$$

Substituting Equation 9-5 into Equation 9-9:

$$p_{ws}^2 = p_i^2 - m \log \left(\frac{t + \Delta t}{\Delta t} \right) \qquad (9\text{-}10)$$

This equation can be used for interpretation of a buildup test. It describes a straight-line relationship between p_{ws}^2 and $\log (t + \Delta t) / \Delta t$. Theoretically, a plot of flowing bottomhole pressure data versus the $\log (t + \Delta t)/\Delta t$ should be a straight line with slope m given by Equation 9-6. Figure 9-4 shows a schematic Horner plot of pressure buildup data. The straight-line portion of the Horner plot can be extrapolated to $(t + \Delta t) / \Delta t = 1$, the equivalent to infinite shut-in time, to obtain an estimate of p_i. This is only valid for infinite acting systems during the flow period. After the exterior boundary affects production, the extrapolated pressure is p^*. This value can be used to estimate the average drainage volume pressure (Matthews et al., 1954). As pointed out by Matthews and Russell (1967), an equation describing the pressure behavior for an infinite acting reservoir may be immediately rewritten for the finite reservoir case by substituting p^* by p_i.

The skin factor, s, may also be estimated from pressure buildup analysis. An expression for this factor can be obtained by combining the expressions for flowing and shut-in pressures as given by Equations 9-5 and 9-10:

$$s = 1.1513 \left[\frac{p_{1hr}^2 - p_{wf}^2 (\Delta t = 0)}{m} - \log \frac{k}{\phi\mu\, c_t\, r_w^2} - \log \frac{4B}{\gamma} \right] \qquad (9\text{-}11)$$

In this equation, $p_{wf}^2 (\Delta t = 0)$ is the measured flowing bottomhole pressure immediately before shut-in. The p_{1hr}^2 pressure must be obtained from the straight line portion of the pressure buildup test 1 hour after shut-in, or from its extrapolation if the straight-line portion of the test has not been reached at this time. This also holds true for other types of tests, as for the drawdown test previously discussed.

The early deviation of a transient pressure test from a straight line may be caused, among other factors, by wellbore storage (van Everdingen and Hurst, 1949; Ramey, 1965; Agarwal, et al., 1970). This can be caused by expansion or compression of fluids in the wellbore and by liquid level

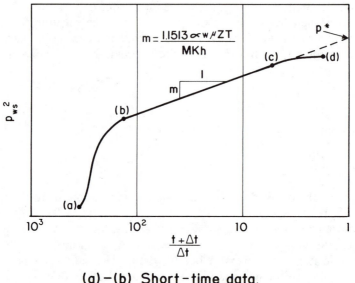

(a)-(b) Short-time data.

(b)-(c) Semilog straight line portion.

(c)-(d) Boundary effects.

Figure 9-4. Schematic representation of a Horner plot for pressure buildup data.

movements in the wellbore. This effect has to be properly considered for accurate test design and interpretation.

Example 9-1: Pressure Buildup Test Analysis—Horner Method

This example is a pressure buildup test analysis for the Geysers Steam Well C, presented by Ramey (1975). The reservoir data regarding this test were:

w	= 102.3 ton/hr	T	= 515°K
t	= 552 hr	M	= 18 g/g mole
r_w	= 0.122 m	m	= 210 $(kg/cm^2)^2$/log cycle
μ	= 0.0225 cp	$p^2_{1\ hr}$	= 1,047 $(kg/cm^2)^2$
c	= 0.037 $(kg/cm^2)^{-1}$	p^2_{wf}	= 262 $(kg/cm^2)^2$
Z	= 0.84	p^{*2}	= 1,333 $(kg/cm^2)^2$

Total depth = 1,120 m; open hole is below 155 m.

Figure 9.5 shows a semilogarithmic plot of test data. As previously mentioned, wellbore storage affects transient pressure behavior and, therefore,

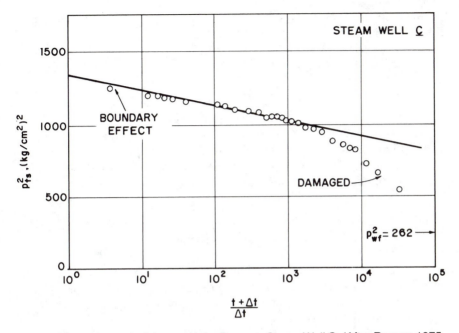

Figure 9-5. Horner buildup graph for Geysers Steam Well C. (After Ramey, 1975, Fig. 7, p. 1753)

should be considered in all transient pressure analyses. This will be considered later in this chapter. A straight line appears to start at a ratio approximately equal to 1000. This corresponds to a shut-in time of 0.55 hr. From the slope of the straight-line portion of the test pressure plot and Equation 9-7, the product kh may be estimated:

$$kh = \frac{1.1513\ (77.459 \times 10^3)\ (102.3)\ (0.0225)\ (0.84)\ (515)}{(210)\ (18)}$$

$$= 23{,}491 \text{ md-m}$$

The skin factor can be estimated by using Equation 9-11. Inasmuch as no data about the average drainage volume pressure at the start of the test, p_i, and porosity ϕ are given, it cannot be computed for this particular example.

Transient Pressure Analysis for Hot-Water Wells

As previously mentioned, the reservoir fluid flow of hot water can be described based on a strict analogy with the single-phase isothermal flow techniques. Once again, it is assumed that flow in the reservoir is approximately described by the diffusivity equation. For hot-water reservoir flow problems, the dimensionless pressure group has been defined by Ramey (1975) as follows:

$$p_D (r_D \ t_D) = \frac{kh (p_i - p)}{\delta \ v_{sc} \ wB \ \mu} \tag{9-12}$$

where δ is a conversion unit constant (Table 9-2).

For a well producing at a constant flow rate w, and considering that all the assumptions mentioned for the reservoir steam flow problem hold, Equation 9-12 can be combined with Equation 9-4, for conditions at the wellbore, $r = r_w$, to obtain an expression for the wellbore pressure:

$$p_{wf} = p_i - 1.1513\delta \ \frac{v_{sc} \ wB\mu}{kh} \left[\log \frac{kt}{\phi \ \mu \ c_t \ r_w^2} + \log \frac{4B}{\gamma} + 0.86859s \right] \tag{9-13}$$

This equation can be used for interpretation of a hot-water drawdown test. Theoretically, according to Equation 9-13, a plot of flowing bottomhole pressure data versus the logarithm of time should be a straight line, with slope m given by the following expression:

$$m = \frac{1.1513\delta \ v_{sc} w \ B \ \mu}{kh} \tag{9-14}$$

From this expression, the flow capacity of the formation can be obtained:

$$kh = \frac{1.1513\delta \ v_{sc} \ w \ B \ \mu}{m} \tag{9-15}$$

An expression for the skin factor, s, can be obtained by the rearrangement of Equation 9-13:

$$s = 1.1513 \left[\frac{p_i - p_{1hr}}{m} - \log \frac{k}{\phi\mu \ c_t \ r_w^2} - \log \frac{4B}{\gamma} \right] \tag{9-16}$$

The basis for pressure buildup analysis is the principle of superposition. Following a similar procedure as for the steam reservoir flow problem, an expression for the shut-in bottomhole pressure, p_{ws} can be written as follows:

$$p_{ws} = p_i - m \log \frac{(t + \Delta t)}{\Delta t} \tag{9-17}$$

This equation can be used for interpretation of a hot-water buildup test. Theoretically, according to Equation 9-17, a plot of flowing bottomhole shut-in pressure data versus the log $(t + \Delta t)/\Delta t$ should be a straight line with slope given by Equation 9-14. A schematic Horner plot of pressure buildup data would look like that shown in Figure 9-4 for a steam well, with p_{ws}^2 being replaced by p_{ws}. With regard to the extrapolation of the straight-line portion of the buildup curve, for infinite shut-in time, $(t + \Delta t)/\Delta t = 1)$, the pressure obtained is p^*. This can be used to estimate the average drainage volume pressure.

The skin factor, s, may also be estimated from pressure buildup analysis. An expression for this factor can be obtained by combining the expressions for flowing and shut-in pressures, given by Equations 9-13 and 9-17:

$$s = 1.1513 \left[\frac{p_{1hr} - p_{wf} (\Delta t = 0)}{m} - \log \frac{k}{\phi \mu \, c_t \, r_w^2} - \log \frac{4B}{\gamma} \right] \quad (9\text{-}18)$$

Everything that was discussed regarding p_{wf} $(\Delta t = 0)$ and p_{1hr}, for the steam flow problem, applies for this hot-water flow problem. With respect to the wellbore storage effect, it can affect a transient pressure test in much the same way as for a steam well (Gringarten, 1978).

The transient pressure analysis theory just presented for hot-water wells can be far more complex, if two-phase flow develops, either in the wellbore or in the formation. The petroleum engineering and hydrogeological methods of analysis have been shown to apply correctly to geothermal wells when no flashing occurs (Witherspoon, 1978). When flashing develops in the formation, however, they apparently cannot be used (Gulati, 1975; Garg, 1978). It was only recently (Rivera and Ramey, 1977; Gringarten, 1978) that information on flashing occurring in the wellbore became available. This is chiefly due to mechanical problems caused by the hostile, hot geothermal environment, and high flow rates of boiling fluids, making difficult the data gathering of bottomhole pressure, because the recording instruments could be easily damaged under these conditions.

Another type of test that has been successfully used in hot-water wells is *multiple-rate testing*. It is often inconvenient to shut-in a well for a pressure buildup survey because it involves loss of production and sometimes it is difficult, for a variety of reasons, to start production after the survey. Also, the previously described drawdown testing techniques of analysis require a constant flow rate. There are many cases where it is impractical or impossible to maintain a constant flow rate long enough, and this type of simplified pressure drawdown test cannot be carried out. Multiple-rate testing, therefore, appears frequently as an alternative for the determination of reservoir parameters. Multiple-rate tests may range (Earlougher, 1977) from one with uncontrolled, variable rate, to one with a series of constant rates, to one at constant bottomhole pressure conditions with a continuously changing flow rate. This type of test has been used with great success for hot-water wells (Rivera and Ramey, 1977; Gringarten, 1978; Saltuklaroglu and Rivera, 1978). One particular case of a multiple rate is when there are only two different flow rates. This type of test, which is called two-rate (Russell, 1963) simplifies testing and analysis. Multiple-rate tests provide information about formation flow capacity and well conditions, represented by the skin factor. The average drainage volume pressure prevailing at the beginning of the test can also be estimated.

To develop a general equation, the pressure test is divided into intervals during each of which production rate can be considered constant. The flow rate-time schedule is as follows:

$$
\begin{aligned}
w &= w_1, & 0 \le t \le t_1 \\
w &= w_2, & t_1 \le t \le t_2 \\
w &= w_3, & t_2 \le t \le t_3 \\
&\;\;\vdots \\
w &= w_N, & t_N - 1 \le t \le t_N
\end{aligned}
$$

This discretization of a possible continuously changing flow rate may be improved as the time intervals become smaller. The pressure drop during the time period N can be expressed by means of the principle of superposition and Equation 9-13:

$$
p_{wf} = -1.15138 \frac{v_{sc} B \mu}{kh} \sum_{j=1}^{N} (w_j - w_{j-1}) \log (t - t_{j-1}) + p_i
$$

$$
-1.15138 \frac{v_{sc} w_N B \mu}{kh} \left[\log \frac{k}{\phi \mu c_t r_w^2} + \log \frac{4B}{\gamma} + 0.86859 \, s \right] \quad (9\text{-}19)
$$

This expression can be rearranged in the following form:

$$
\frac{p_i - p_{wf}}{w_N} = 1.15138 \frac{v_{sc} B \mu}{kh} \sum_{j=1}^{N} \left[\frac{(w_j - w_{j-1}) \log (t - t_{j-1})}{w_N} \right]
$$

$$
+ 1.15138 \frac{v_{sc} B \mu}{kh} \left[\log \frac{k}{\phi \mu c_t r_w^2} + \log \frac{4B}{\gamma} + 0.86859 \, s \right] \quad (9\text{-}20)
$$

Equation 9-20 is the interpretation equation used for a general hot-water multiple-rate test. It describes a straight-line relationship of slope:

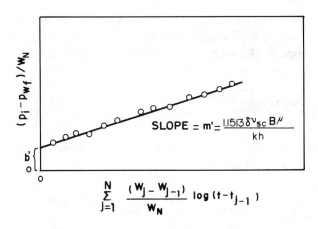

Figure 9-6. Schematic representation of a multiple-rate test.

$$m' = 1.15138 \frac{v_{sc} \, B \, \mu}{kh} \tag{9-21}$$

and intercept:

$$b' = m' \left[\log \frac{k}{\phi \mu \, c_t \, r_w^2} + \log \frac{4B}{\gamma} + 0.86859 \, s \right] \tag{9-22}$$

Multiple-rate pressure transient data follow a straight line when plotted in Cartesian coordinates as shown in Figure 9-6:

$$\frac{p_i - p_{wf}}{w_N} \quad v_s \quad \sum_{j=1}^{N} \frac{(w_j - w_{j-1})}{w_N} \quad \log (t - t_{j-1}) \tag{9-23}$$

Earlougher (1977) clearly stated that in order to plot it correctly, it is important that the rate corresponding to each plotted pressure point is w_N, the last rate that can affect that pressure. As time goes on, the number of flow rates may increase and the last rate may change; but each pressure is identified with the rate occurring when that pressure was measured. It should be clear that there may be more than one pressure reading associated with one specific rate. It is important to keep in mind that this multiple-rate theory assumes transient flow conditions throughout the whole test. This is due to the fact that the interpretation equation has been derived based on the line source solution, only valid for infinite acting systems (transient flow conditions). Thus, the separate flow periods should be of short duration so that transient flow will prevail at each rate through the whole test.

From the slope of the multiple-rate pressure curve, given by Equation 9-21, the formation permeability may be estimated as follows:

$$k = \frac{1.1513 \delta v_{sc} \, B\mu}{m' \, h} \tag{9-24}$$

Also, from the intercept of the pressure curve, given by Equation 9-22, the skin factor, s, can be estimated:

$$s = 1.1513 \left[\frac{b'}{m'} - \log \frac{k}{\phi\mu \, c_t \, r_w^2} - \log \frac{4B}{\gamma} \right] \tag{9-25}$$

Example 9-2: Multiple-Rate Test Analysis

This test, which was made for a hot-water well, ASAL 1, located in the French Territory of Afars and Issas (now Republic of Djibouti), was presented by Gringarten (1978). Flashing of the hot water occurred at some depth in the wellbore. The mass rate of flow variation is shown in Figure 9-7. The pressure data measured during the second 6-in. buildup test (flow period No. 4), was graphed (Figure 9-8) in a slightly different manner than that just described. Gringarten plotted p_{wf}, instead of the group $(p_i - p_{wf})/w_N$, and the time units used were minutes instead of hours. The pressure coordinate was changed because of uncertainty of the average drainage volume pressure for this well, p_i. Otherwise, the method of analysis is the same. Figure 9-8 shows a straight-line portion of slope $m' = 3.6 \times 10^{-3}$. Other data needed to analyze the test is $\mu = 0.2$ cp and $v = 1.25 \times 10^{-3}$ m³/kg, which is equivalent to $v_{sc}B$ in Equation 9-21. Substituting these values into Equation 9-21 and sc considering the change of units (Gringarten's eq. 5, p. 3) yields the following equation:

$$kh = \frac{0.228 \, v \, \mu}{m'} = \frac{0.228 \, (1.25 \times 10^{-3}) \, (0.2)}{3.6 \times 10^{-3}}$$

$$= \quad 15.9 \text{ darcy-meter}$$

The calculated average drainage volume pressure from the test is 76.6 bars (1 bar = 14.503 psi), which agrees reasonably with a measured value before the test of 77.4 bars.

It can be observed from Figure 9-8 that deviations from the straight line occur at short times (portion a-b, Figure 9-4) and at long times (portion c-d, Figure 9-4). It will be discussed that the short-time deviation (for buildup

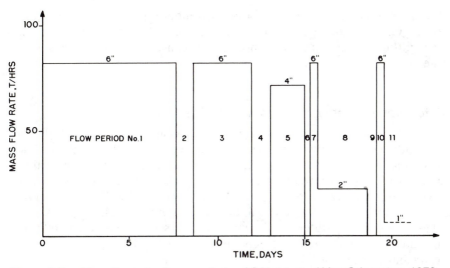

Figure 9-7. Mass flow rate changes during ASAL 1 tests. (After Gringarten, 1978, Fig. 2, p. 6, by permission of the S.P.E. of A.I.M.E.)

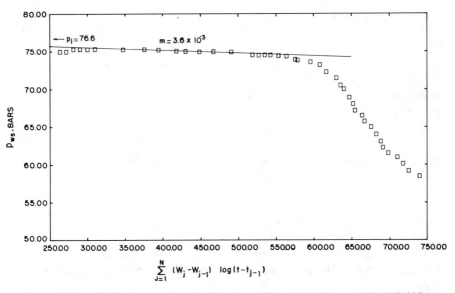

Figure 9-8. ASAL 1 pressure buildup test (6″ outlet, flow period #4) (After Gringarten, 1978, Fig. 7, p. 9, by permission of the S.P.E. of A.I.M.E.)

times less than 20 minutes) is caused by wellbore storage effects. For long times (buildup times greater than 6 hours), the data points deviate from the straight-line portion of the pressure curve because of steam condensation in the wellbore.

Interference Testing for Geothermal Reservoirs

Interference testing is preferred in some instances over single well tests depending on the type of information required (Rivera et al., 1978). It has been previously mentioned that important data like reservoir connectivity and porosity can be estimated from analysis of these tests. Reservoir connectivity is an important factor because the number of wells in a reservoir usually increases, causing mutual interference of wells. The application of these tests to hot-water geothermal reservoirs has been successfully reported in the literature (Witherspoon et al., 1978; Rivera et al., 1978; and Narasimhan et al., 1978).

Interference tests are usually analyzed by a type curve matching technique. The type curve used is the line source solution, for an infinite acting system, plotted on a log-log paper in terms of p vs $t_D/r_D{}^2$ (Ramey et al., 1973; Earlougher, 1977). Figure 9-9 presents this type curve for a well located in an infinite acting system. The type curve matching technique consists of plotting the observation well pressure drop, $\Delta p_i - p$, on the ordinate versus flowing time, t, on the abscissa of a log-log paper of the same size as in Figure 9-9. Normally, a tracing paper is placed over the type curve, and the major grid lines are traced for reference. The grid lines of Figure 9-9 (not shown) are used to plot actual data on the tracing paper. Next, the data plot is moved vertically and horizontally over Figure 9-9, keeping the grids of the type-curve and those of the data plot parallel to each other until the best match is obtained with the type-curve. A convenient match point is pierced and the values of $(\Delta p)_M$ and $(t)_M$ are read from the data plot. The corresponding points lying directly under this point on Figure 9-9 are $(p_D)_M$ and $(t_D)_M$. The formation permeability is estimated from the pressure match-point data and the definition of p_D is given by Equation 9-12:

$$k = \frac{\delta v_{sc} \, w \, B \, \mu}{h} \frac{(p_D)_M}{(\Delta p)_M} \tag{9-26}$$

In a similar way, the porosity compressibility product is estimated from the time match-point data and the definition of t_D and r_D by Equations 9-2 and 9-3:

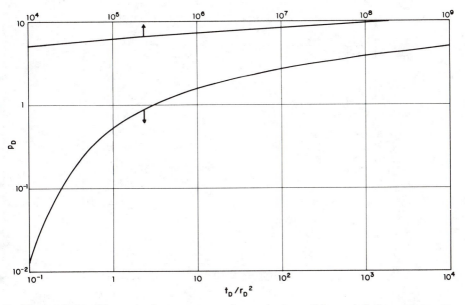

Figure 9-9. Dimensionless pressure for a single well in an infinite system, no wellbore storage, no skin. Exponential-integral solution. (After Earlougher, 1977, Fig. C.2, p. 194, by permission of S.P.E. of A.I.M.E.)

$$\phi c_t = \frac{B}{r^2}\frac{k}{\mu}\frac{(t)_M}{(t_D/r_D^2)_M} \qquad (9\text{-}27)$$

This type of curve analysis method is simple, rapid, and accurate, provided the conditions for applicability of the line source p_D solution are met, i.e., when $r_D = r/r_w > 20$ and $t_D/r_D^2 > 0.5$ (Earlougher, 1977).

*Example 9-3: Type Curve Matching Analysis of a Pressure
Interference Test*

This is a pressure interference test carried out in the liquid-dominated Cerro Prieto Geothermal Field, located in Mexico. These results are taken from a paper by Rivera et al. (1978), and involved one observation well (M-101) and four active wells (M-50, M-51, M-90, and M-91). The early pressure drop response at the observation well M-101 was analyzed, while the only active well was M-91 and the other three wells were shut-in. Figure

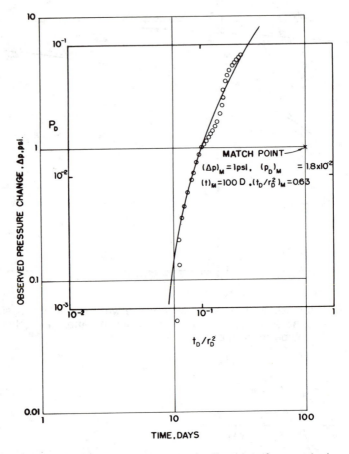

Figure 9-10. Type curve match of an interference test, active wells M-91, M-90, M-51, M-50. Observation well M-101. Cerro Prieto Geothermal Field (After Riviera et al., 1978, Fig. 5)

9-10 shows the type-curve match presented by these authors. Other related data for the test were:

$$w = 185.3 \text{ ton/hr}$$
$$v_{sc} = 1.043 \text{ cm}^3/\text{g}$$
$$B = 1.185 \text{ metre}_{rc}^3/\text{metre}_{sc}^3$$
$$\mu = 0.1017 \text{ cp}$$
$$r = 1550 \text{ m}$$

From the results of the match point data and from Equations 9-26 and 9-27, the following estimates for the reservoir parameters are obtained:

$$kh = 456.7869 \ (1.043) \ (185.3) \ (1.185) \ (0.1017) \frac{(1.8 \times 10^{-2})}{(1) \ (1/14.22)}$$

$$= 2{,}723 \ \text{md-m}$$

and

$$\phi c_t h = \left(\frac{0.000348}{(1{,}550)^2} \right) \left(\frac{2{,}723}{0.1017} \right) \left(\frac{(100) \ (24)}{0.63} \right) = 0.0148 \ \text{m/(kg/cm}^2).$$

Pressure Transient Analysis for Two-Phase Flow

Up to now, most of the discussion has been based on single-phase isothermal reservoir flow. Sometimes, however, flow conditions in the reservoir are such that two-phase flow can occur, causing the flow to be non-isothermal. Garg (1978a, 1978b) has recently presented a theory for analysis of two-phase pressure transient data in geothermal wells. He considered two different situations: (a) a reservoir that is originally two-phase everywhere, and (b) a reservoir that originally contained hot water, but after production the pressure drop resulted in a flashing front propagation away from the wellbore. He derived a diffusivity-type equation, which resembles quite closely the diffusivity equation. This type of analogy has been found for other reservoir fluid flow problems (Martin, 1959). Based on this similarity of the equations that describe the flow problem, Garg derived an expression for the flowing bottomhole pressure for the two above mentioned reservoir situations:

(a) Two-phase flow exists everywhere in the reservoir at initial conditions:

$$p_{wf} = p_i - 1.15138 \frac{w}{(k/v)_t h} \left[\log \frac{(k/v)_t t}{\phi \ \rho \ c_t r_w^2} + \log \frac{4B}{\gamma} \right] \qquad (9\text{-}28)$$

(b) Hot water is present everywhere in the reservoir at initial conditions, with a flashing-front propagating after two-phase pressure conditions are reached:

$$p_{wf} = p_s - \frac{\delta w}{2[(k/v)_t]_1 \ h} E_i(-\lambda^2) - 1.15138 \frac{w}{[\ (k/v)_t]_1 h}$$

$$\left[\log \frac{(k/v)_t t}{\phi \ \rho (c_t)_1 \ r_w^2} + \log \frac{4B}{\gamma} \right]$$

$(9\text{-}29)$

where: $(k/v)_t$ = total kinematic mobility = $(k/v)_w + (k/v)_s$
$\quad\quad (k/v)_w$ = kinematic mobility for the water = $k \ k_{rw} \ \rho_w/\mu_w$
$\quad\quad (k/v)_s$ = kinematic mobility for the steam = $k \ k_{rs} \ \rho_s/\mu_s$

λ = roots of eq. 13d of Garg's (1978b) paper
1 = subscript denoting the two-phase region

Equations 9-28 and 9-29 can be used for interpretation of a two-phase flow drawdown test. Theoretically, according to these equations, a plot of flowing bottomhole pressure data versus the logarithm of time should be a straight line with slope m given by the following expression:

$$m = \frac{1.1513\delta w}{(k/v)_t h} \tag{9-30}$$

From this expression, the total kinematic mobility can be estimated:

$$(k/v)_t = \frac{1.1513\delta w}{mh} \tag{9-31}$$

Recently, a sound theoretical basis for the estimation of the kinematic mobility from pressure interference testing has been suggested (Pruess et al., 1978). This is an extension of Garg's theory of analysis for two-phase flow.

Example 9-4: Two-Phase Flow Pressure Transient Test with a Propagating Flashing Front

This is a simulated test presented by Garg (1978b). Single-phase (hot water) conditions prevail for some time after the start of production, until eventually, the pressure level declines sufficiently to reach the two-phase flow conditions, and a flashing front propagates away from the well. Figure 9-11 shows a semilog plot of the results of this test. At early times, up to about 12 seconds, hot-water flow conditions prevail in the vicinity of the well. From this time on, two-phase flow conditions start to be operative and for a period of time related to numerical approximations, the curve is relatively flat. After the numerical problem is overcome, for late times, the pressure data follow approximately a straight line. The computed value of $(k/v)_t'$ obtained from the slope of the straight-line portion of the pressure curve of Figure 9-11, and Equation 9-31, agrees well with the actual range of values for this parameter.

Modern Well Test Analysis

According to Ramey (1970), modern well test analysis specifies that the well be tested for a period of time long enough to reach and define a proper "straight line," when pressure transient data are plotted on semilogarithmic

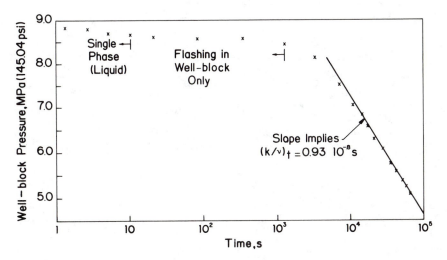

Figure 9-11. Simulated drawdown history *(g)*. Reservoir is initially single-phase everywhere ($p = 9.000$ MPa ~1,305.3 psi; $T = 573.15$K $= 572°$F). Absolute permeability k for this case is 0.01 μm^2 (~0.01 darcy) and the actual range of (k/v), values for points lying on the straight line is $(0.93\text{-}1.15)10^{-8}s$. (After Garg, 1978b, Fig. 7, P. 12, by permission of the S.P.E. of A.I.M.E.)

graph paper. It was not until the end of the '60s that pressure data recorded before the straight-line portion was reached started to be analyzed. It has since been realized that this early time data could be extremely helpful for test interpretation purposes. The early time data, also called "short-time data," are influenced by the effects of several factors, like wellbore storage, flow through perforations, partial penetration, and well stimulation such as fracturing or acidizing.

Short-time data can usually be analyzed by a log-log type curve matching technique similar to that previously described for interference testing. The type curve is a log-log graph of the solution of a specific reservoir fluid flow problem, plotted in terms of p_D vs t_D. The type curve should be chosen in such a manner as to closely represent the field situation. Perhaps, the most commonly used type curve to date has been that presented by Agarwal et al. (1970) and Wattenbarger and Ramey (1970) for the pressure transient behavior of a well, including wellbore storage and skin damage (Figure 9-12). This type curve can be used to approximately determine the beginning of the correct semilog straight line. The curves of Figure 9-12 show an early unit slope straight-line portion, representing wellbore storage dominated flow conditions, followed by a transition to the relatively flat curves for zero wellbore storage. The point where the curves including wellbore storage

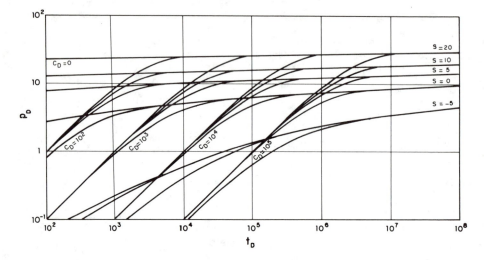

Figure 9-12. Dimensionless pressure for a single well in an infinite system, wellbore storage and skin included. (After Agarwal, Al-Hussainy, and Ramey. 1970. Graph courtesy H. J. Ramey, Jr.)

reach the zero wellbore storage curves represents the start of the correct semilog straight line. This time can be expressed (Ramey et al., 1973):

$$t_D = C_D (60 + 3.5 \, s) \tag{9-32}$$

where:

$$C_D = \frac{\epsilon \, C \, B \, v_{sc}}{\phi \, h \, c_t \, r_w{}^2} \qquad \text{for hot water} \tag{9-33}$$

and

$$C_D = \frac{\eta \, C' \, Z \, T}{M \, \phi \, h \, c_t \, r_w{}^2} \qquad \text{for steam.} \tag{9-34}$$

For a good semilog analysis of the pressure transient data, the duration of the test should be at least ten times the time given by Equation 9-32. This would provide a minimum of a log cycle of pressure field data, allowing a proper tracing of the straight-line portion of the pressure curve.

There are many type curves presently available for pressure analysis of short-time data. The difference between them is the well condition that they consider, e.g., a well with a horizontal fracture (Gringarten et al., 1974), a well with a vertical fracture (Gringarten et al., 1975; Cinco et al., 1978), etc.

Type curve matching techniques can be used, in combination with other data, to identify the near wellbore conditions and, as mentioned, to determine the correct start of the semilog straight line. With a properly estimated start of the semilog straight line, the results obtained from the semilog analysis are more accurate than those coming from log-log type-curve matching. As pointed out by Gringarten et al. (1975) a combination of type-curve matching techniques with conventional semilog analytical methods permits a highly confident analysis of field data.

Example 9-5: Modern Well Test Analysis of a Pressure Transient Test in a Steam Well

Example 9-1 showed the analysis of a pressure buildup test presented by Ramey (1975). The purpose of this example is to illustrate how type-curve matching techniques can be used to identify the correct start of the semilog straight line. Figure 9-13 shows a type-curve match of the buildup data to the type curve of Figure 9-12. Field data fairly matches the curve for $c_D = 10^2$ and $s = 5$. More important than this, however, is the fact that the data reaches the type-curve for $c_D = 0$ at Δt equal to 0.3 hour. This corresponds to a $(t + \Delta t)/\Delta t$ value of 1,841. Examination of Figure 9-5 shows that, according to type-curve analysis, the start of the straight line has been correctly identified. As previously pointed out, any other information in addition to the start of the straight line, should only be taken in a qualitative sense.

Simulation of Geothermal Reservoirs

The word "simulate" has been defined by Webster as "to assume the appearance of without the reality." To the reservoir engineer, the term reservoir simulation means the process of deducing the physical behavior of a real reservoir from the performance of a model. There are two basic types of models: (a) physical (for example, a laboratory sandpack), and (b) mathematical. A mathematical model of a physical system consists of a set of partial differential equations subject to certain simplifying assumptions, together with an appropriate set of boundary conditions, which describe the physical processes active in the reservoir (Peaceman, 1978; Coats, 1969; Crichlow, 1977).

Reservoir simulation applies the concepts and techniques of mathematical modeling to the analysis of the behavior of geothermal reservoir systems. Most often, the term reservoir simulation is used with regards to the hydrodynamics of flow within the reservoir, but in a more general sense it refers to the total geothermal system, which includes mainly the reservoir itself, the tubing, and the surface facilities. A simulator can be defined as a group of

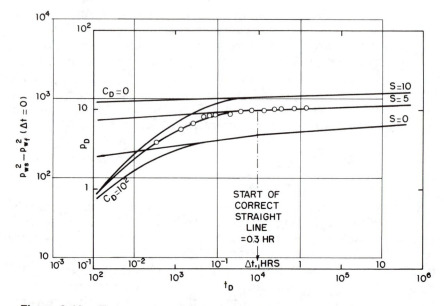

Figure 9-13. Type curve match of buildup data for steam well C to the type curve of Figure 9.12.

computer programs that implement the mathematical model in a digital computer.

The main purpose of simulation is to estimate the behavior of a geothermal reservoir under a variety of exploitation schemes. This is extremely advantageous because the reservoir can be produced only once, and the model can be produced or "run" as many times as needed at cost and over a short period of time. From observation of model performance, under a variety of producing conditions, the optimum exploitation conditions for the reservoir can be selected. Some of the information that can be obtained from reservoir simulation studies is:

1. The capability of a reservoir of producing significant quantities of energy over meaningful periods of time.
2. The number of wells and spacing required for optimum development of the reservoir.
3. The effect of the rate of production of the wells on total energy recovery.
4. Variation of fluid temperature with time.
5. Feasibility of implementation of an enhanced recovery process to recover additional heat.

Reservoir models range in complexity from simple ones for fairly homogeneous systems, where average values for reservoir properties, such as permeability and porosity, are adequate to describe their behavior, to those used for highly heterogeneous systems, where fragmentation into blocks (cells) is necessary. In order to choose a model that would represent a particular reservoir, a good understanding of the reservoir and a detailed examination of the data available (Odeh, 1969) is required. A model that fits a particular reservoir may not be appropriate for another reservoir, despite the apparent existence of similarities between them. As pointed out by Odeh (1969), a reservoir model is useful only when it reasonably matches the field situation. A general rule (Coats, 1969) that should be followed in reservoir simulation is to "select the *least* complicated model and largest reservoir description, which will allow the desired estimation of reservoir performance." In other words, the model to be used should be the simplest that duplicates reservoir behavior.

The discussion presented in this section includes a brief review of some of the simple reservoir models (zero dimension), and of the more complex (multidimensional) models, suited for numerical simulation.

Whiting and Ramey's Model (1969)

The model developed by Whiting and Ramey is zero dimensional because rock and fluid properties and pressure values are not a function of location in the reservoir. These parameters are calculated as average values for the whole reservoir. This type of models has also been called lumped-parameter models.

This model can be used for performance forecasting of reservoir behavior (Ramey et al., 1973). Where some field development already exists and production is in progress, a mathematical model for fluid flow in the reservoir can be postulated. The reservoir size and its productivity can be determined by matching measured production data (mass produced, enthalpy produced, reservoir pressure and temperature, etc.) with the corresponding parameters of the mathematical model. Once all model parameters and their relationships have been identified, the model can be used for performance production forecasting under different assumed exploitation schemes.

The model of the reservoir system developed by Whiting and Ramey is shown in Figure 9-14. The system contains rock, water, and steam. The reservoir system has a bulk volume V and a porosity of ϕ. The cumulative fluid production at any given time is W_p, whereas the cumulative heat production associated with this fluid production is Q_p. The model also considers heat loss, Q_L, and mass fluid loss, W_L, due to convection in, for

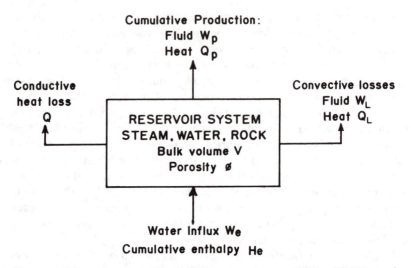

Figure 9-14. Schematic diagram of reservoir model (After Whiting and Ramey, 1969, Fig. 4, p. 895, by permission of the S.P.E. of A.I.M.E.)

example, hot springs, fumaroles, etc. The authors have discussed that heat loss, Q, at the reservoir boundary is negligible and should not significantly affect reservoir behavior. Water recharge, W_e, and its associated energy (cumulative enthalpy, H_e) are also considered.

The model of Whiting and Ramey considers the following basic assumptions (Ramey et al., 1973): (a) thermodynamic equilibrium (temperature of rock, water, and steam are equal); (b) pressure and saturation are uniform throughout the reservoir; and (c) uniform fluid production, which implies that fluid production comes from all parts of the reservoir.

A combined mass, energy, and volumetric balance gives the following expression:

$$W_p (H_p - E_c) + W_L (H_L - E_c) + Q$$

$$= W \left\{ E_i - E_c + \left[(1-\phi)/\phi \right] \left[x_i v_{si} + (1 - x_i) v_{wi} \right] \rho_r C v_r (T_i - T_c) \right\}$$

$$+ (H_e - E_c) (B/v_{we}) \Sigma Q_D (t_D) \Delta p_n \tag{9-35}$$

where: H_p = enthalpy of produced fluids
$\quad\quad\quad\quad$ W = initial mass of hot water and steam in reservoir bulk volume, V
$\quad\quad\quad\quad$ E = internal energy
$\quad\quad\quad\quad$ Q = conductive heat loss

ϕ = porosity of rock matrix
x = steam quality
v = specific volume
ρ_r = density of rock and contained fluids
C_{v_r} = specific heat at constant volume of reservoir rock and contained fluids
T = temperature
B = water recharge constant
$Q_D\,(t_D)$ = dimensionless cumulative recharge, corresponding to dimensionless time, t_D
Δp_n = pressure drop at any time n

The subscripts p, L, c, i, r, s, w, and e indicate produced, loss, current, initial, reservoir, steam, liquid water, and recharge values, respectively. The model given by Equation 9-35 can be simplified, according to the situation of a specific reservoir. For instance, for a reservoir containing only hot water, this equation reduces to:

$$(W_p + W_L)\,v_w = W\,(v_w - v_{wi}) + B\Sigma Q_D\,(t_D)\,\Delta p_n. \tag{9-36}$$

It is common to find that reservoirs are associated with adjacent aquifers. Due to production, reservoir pressure declines, causing the aquifer to react by yielding up water. If water recharge is influencing reservoir behavior, it has to be properly considered for accurate reservoir performance predictions. One of the most useful methods for calculating water recharge is that of van Everdingen and Hurst (1949). Miller et al. (1978) presented an excellent review of water recharge into geothermal reservoirs (see also Whiting and Ramey, 1969, p. 897).

Whiting and Ramey applied the model for liquid hot water (Equation 9-36) to the Wairakei geothermal reservoir in New Zealand. A least-mean-squares fit to the production history from 1956 to 1961 was used to obtain estimates for the initial water in place and the initial pressure. Once the optimum model parameters were determined, a prediction of reservoir performance through 1965 was made, indicating excellent agreement between measured and calculated values (Figure 9-15).

Brigham and Morrow's Model (1977)

Brigham and Morrow have presented a zero-dimensional model for vapor-dominated systems. They considered three cases regarding the distribution of hot water and steam. These authors assumed that the system is closed and that energy is derived from the rock mass itself.

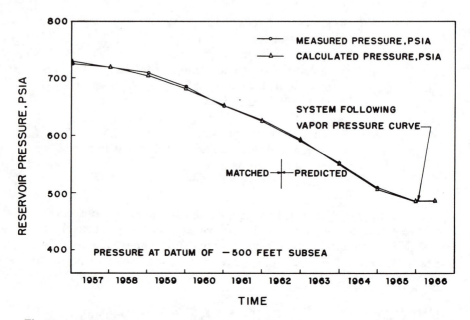

Figure 9-15. Prediction of geothermal reservoir performance. (After Whiting and Ramey, 1969, Fig. 5, p. 898, by permission of the S.P.E. of A.I.M.E.)

The first system is completely filled with steam with no water present. Flow in this system is essentially isothermal, because the heat capacity of the rock is large compared with that of steam. Then, in order to study the system's behavior only a mass balance needs to be taken. Under these reservoir flow conditions, a steam reservoir can be treated as an ordinary petroleum gas reservoir (Craft and Hawkins, 1959), where average reservoir pressure divided by the gas deviation factor, p/Z, is plotted versus cumulative production, resulting in a straight line. The intercept on the abscissa is equal to the original mass of fluid in place, m_{gi}. The equation can be presented as follows:

$$p_f/Z_f = (p_i/Z_i)\frac{(m_{gi} - \Delta m_g)}{m_{gi}} = (p_i/Z_i)\frac{(m_{gf})}{m_{gi}} \qquad (9\text{-}37)$$

where: p = reservoir pressure
m_g = mass of steam
Δm_g = mass of steam produced during a depletion step
i = subscript for the initial conditions of a depletion step
f = subscript for the end conditions of a depletion step

The reserve forecasting procedure, using Equation 9-37, will be used to compare predictions for the other two cases.

The second model considers that the steam zone is separated from an underlying liquid zone by a horizontal interface at which boiling takes place. As steam is produced, water boiling will take place, resulting in a liquid level drop; thus its name, the *falling liquid level model*. Flow in the steam zone is assumed to be isothermal, whereas on the water zone non-isothermal flow conditions would prevail. A mass and energy balance is made for the water zone.

The third model also considers that the steam zone has an underlying liquid zone, but assumes that liquid boiling takes place throughout the whole liquid zone, and that liquid level does not drop. This model is called the *constant liquid level model*. Consequently, in this system steam saturation will continuously build up within the liquid zone. The energy equation for this system resembles that of Whiting and Ramey (1969), with the exception that only steam flows out of the system. The authors have solved the falling liquid level and the constant liquid level models for three hypothetical reservoirs, having porosities of 0.05, 0.10, and 0.20. It was assumed that the volume of the steam zone was the same as the volume of the liquid zone. Calculations were also made, but not reported, for systems having a ratio of steam to liquid zone of 9, which showed results very close to those for a unity ratio.

Figure 9-16 shows the results of these authors for a steam falling liquid level system (their fig. 1). From the results of this figure and from other results shown by Brigham and Morrow, the following conclusions have been made. For low-porosity reservoirs, an extrapolation of the p/Z versus cumulative production plot will be too optimistic, whereas for high-porosity reservoirs it will be pessimistic. A porosity of about 0.10 will give approximately correct predictions. The constant liquid level model predicts higher recovery for a given reservoir pressure than the falling liquid level model. The presence of even a small water zone in the lower portion of the system is an important fraction of total system's mass, and can significantly affect the p/Z prediction procedure. Finally, they concluded that the steam zone of the reservoir remains isothermal whether or not there is a water boiling zone below the steam. This causes characterization problems, because pressure, temperature, and enthalpy measurements are not sufficient to determine the original state of the reservoir fluid system. An application of the p/Z versus cumulative production plot to actual field data has been presented by Ramey (1970a).

Other Zero-Dimensional Models

Another zero-dimensional model for the simulation of geothermal reservoirs has been presented by Martin (1975). His model is valid for single- and two-phase flows; it was derived following the logic behind solution gas drive

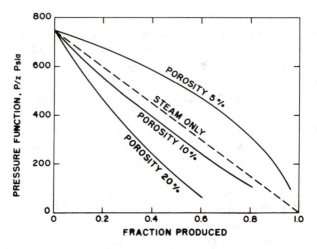

Figure 9-16. Pressure depletion vs. recovery, falling liquid level. (After Brigham and Morrow, 1977, Fig. 1, by permission of the S.P.E. of A.I.M.E.)

petroleum reservoirs. To the author's knowledge, this model has not received field application.

One more zero-dimensional model is that presented by Atkinson et al. (1978). Their model considered a steam and a water zone and the presence of carbon dioxide and other noncondensable gases in the steam zone. It also took into consideration the possibility that water could recharge the reservoir. This model was successfully evaluated, matching the production history of the Bagnore steam field in Italy.

Numerical Geothermal Reservoir Simulation

When important heterogeneities exist in the reservoir, rock and/or fluid properties (permeability, saturation, pressure, viscosity, etc.) vary in space, then simple zero-dimensional models can no longer be used for prediction of reservoir performance. A model that considers the variation of rock and fluid properties is called a distributed-parameter model. A reservoir where properties vary in space, is usually divided in blocks (cells), assigning to each one of them average values for rock and fluid properties. For a multicell model like this, the basic building block is the previously-described zero-dimensional model.

A mathematical model for describing fluid flow in a geothermal reservoir includes equations for the conservation of mass, momentum, and energy for water and steam, and for specifying the state of the system (Faust and Mercer, 1975; Thomas and Pierson, 1978). A review of the literature shows that several such models have been presented (Faust and Mercer, 1975; Thomas and Pierson, 1978; Coats, 1977; Brownell et al., 1975; Lasseter et al., 1975). These models differ from each other according to the assumptions implied. A model for the flow of hot water and steam, written for a general coordinate system, consists of the following equations:

$$\Delta \cdot \left[\frac{kk_{rw}\ \rho_w}{\mu_w}\ (\Delta p\ -\ \rho_w\ g) + \frac{kk_{rs}\ \rho_s}{\mu_s}\ (\Delta p\ -\ \rho_s\ g) \right]\ -q$$

$$=\ \frac{\partial}{\partial t}\ (\phi \rho_w\ S_w\ +\ \phi \rho_s\ S_s) \tag{9-38}$$

Energy balance for the water-steam-rock system can be presented as follows:

$$\Delta \cdot \left[\frac{k\ k_{rw}\ \rho_w\ H_w}{\mu_w}\ (\Delta p\ -\ \rho_w g)\ +\ \frac{k\ k_{rs}\ \rho_s\ H_s}{\mu_s}\ (\Delta p\ -\ \rho_s g) \right]$$

$$+\ \Delta\ (k\Delta T)\ -\ q_{HL}\ -\ q_H \tag{9-39}$$

$$=\ \frac{\partial}{\partial t}\left[(\phi\ \rho_w\ H_w\ S_w\ +\ \phi \rho_s\ H_s\ S_s\ )\ +\ (1-\phi)\ \rho_r\ H_r \right]$$

where:
k = absolute permeability
$k_{rs}\ (k_{rw})$ = relative permeability to steam (water)
$\rho_s\ (\rho_w)$ = steam (water) density
$\mu_s\ (\mu_w)$ = steam (water) viscosity
p = pressure
g = acceleration of gravity
q = mass (source) production rate
ϕ = porosity
$S_s\ (S_w)$ = steam (water) saturation
$H_s\ (H_w)$ = enthalpy of saturated steam (water)
K = thermal conductivity
T = temperature
q_{HL} = heat loss rate
q_H = enthalpy production rate
ρ_r = average rock-grain density
H_r = rock enthalpy

The pressure enthalpy approach followed in the previous mathematical model is one of the different possible avenues to be taken. Another approach followed by other authors uses density and internal energy as the unknown variables (Brownell et al., 1975; Lassater et al., 1975). The use of fluid pressure and enthalpy seems a logic approach, because they define the thermodynamic state of the system and are usually obtained in field operations.

In the derivation of Equations 9-38 and 9-39, the following assumptions have been made (Mercer et al., 1974; Faust and Mercer, 1975):

1. The reservoir is treated as a porous medium.
2. Capillary pressure effects are negligible. This means that the pressure in the water and steam phases are equal.
3. Thermal equilibrium exists among all phases: hot water, steam, and rock.
4. Validity of Darcy law for two-phase flow.
5. Thermal conductivity is a property of the medium.
6. The geothermal fluid is pure water.

Ignoring one or more of these assumptions will no doubt increase the accuracy of predictions in specific situations, but will also change the complexity and economics of the study. At the present time, there are geothermal mathematical models more complete than that given by Equations 9-38 and 9-39. An example is Coat's model (1977), where he considers capillary effects and the thermal dependence of relative permeability. Other recent models include the effect of inert gases (Zyvolski and Sullivan, 1978) and precipitation of dissolved salts on porosity and permeability (Todd et al., 1978).

In order to solve the two nonlinear partial differential equations given by Equations 9-38 and 9-39, it is necessary to assume some additional functional and algebraic relations between the variables involved. For a complete discussion of this subject, the paper by Coats (1977) should be referred to. As an example of the relations needed, the steam and water saturation must add to unity:

$$S_s + S_w = 1 \qquad (9\text{-}40)$$

To solve the mathematical model given by Equations 9-38 and 9-39 and its additional relations, the proper reservoir boundary conditions should be considered. Analytical solutions can be obtained by the classical methods of mathematical physics only for simple reservoir situations, such as homogeneous systems with regular boundaries (i.e., a single well in the center of a radial reservoir). Thus, the nonlinear partial differential Equations 9-38 and 9-39 describing the flow of fluids in geothermal reservoirs must be solved by approximation. If it is desired that approximate values of the solution to the mathematical model be obtained, then some numerical process is formulated that will produce these values after a finite number of calculations. With the advent of high-speed computers, it became increasingly desirable to produce algorithms that are applicable to a wide class of problems and simultaneously lead to increasingly more accurate approximations with an additional expenditure of computer time alone. Numerical methods have been extensively used because they meet these two criteria. These methods have proved to be highly successful for obtaining solutions to very complex reservoir situations. A numerical model (simulator), as previously defined, constitutes a group of computer programs that use numerical methods to obtain an approximate solution to the mathematical model. Further discussion of the numerical solution to the mathematical model is beyond the scope of this chapter. They include finite difference techniques, finite element methods, or a combination of the two.

Conclusions

The main purpose of this chapter was to provide a brief review of presently available techniques for prediction of geothermal reservoir behavior. It seems that predictions of reasonable accuracy can be made with actually available methods for the majority of geothermal systems saturated with fluids (convective systems). There is a lack of knowledge for other types of systems, however, such as the hot-dry formations. It is expected that geothermal reservoir engineering will continue to show advancement due to the strong research program currently under way.

Discussion of this chapter has included the practical aspects of reservoir engineering, such as well test analysis, and a brief presentation of mathematical reservoir simulation. The reservoir data obtained from pressure transient analysis best represent actual reservoir conditions, and are extremely useful as input for reservoir simulation studies. Reservoir simulation studies are initially carried out with simple models, which are continuously updated and refined as new production data becomes available. As a simple matter, the model used should be the simplest that duplicates reservoir behavior. A properly conducted reservoir engineering study would lead to the determination of the optimum exploitation conditions for the particular reservoir.

Nomenclature*

B = water-recharge constant, Reservoir Simulation Section

B = formation volume factor, metre$^3_{rc}$ /metre$^3_{sc}$

b' = intercept on semilog plot of transient-test pressure data normalized by rate, kg/cm^2/ (ton/hr), Equation 9-22

c = compressibility, (kg/cm^2)$^{-1}$

C = wellbore storage constant for hot water, ton/(kg/cm^2), Equation 9-32

C' = wellbore storage constant for steam, (ton/kg/cm^2), Equation 9-33

C_D = dimensionless wellbore storage constant, Equations 9-32 and 9-33

C_{vr} = specific heat at constant volume of reservoir rock and contained fluids

E = internal energy

g = acceleration of gravity

h = formation thickness, m

H = enthalpy

k = absolute permeability

k_r = relative permeability, fraction

m = slope of linear portion of semilog plot pressure transient data, kg/cm^2/cycle

m' = slope of the data plot for a multiple-rate test, kg/cm^2/ (cycle ton/hr), Equation 9-21

M = molecular weight, g/mole

p = pressure, kg/cm^2

q = mass (source) production rate

q_H = enthalpy production rate

q_{HL} = heat loss rate

Q = conductive heat loss

r = radius, m

s = van Everdingen-Hurst skin factor

t = time, hours

Δt = running testing time, hours

T = temperature, K

w = production rate, tons/hr

W = initial mass (hot water and steam in the reservoir), Equation 9-34

x = steam quality, fraction

z = real gas deviation factor

a = conversion factor, Equation 9-1

B = conversion factor, Equation 9-2

γ = constant equal to 1.7810724, Equation 9-5

δ = conversion factor, Equation 9-12

ϵ = conversion factor, Equation 9-32

η = conversion factor, Equation 9-33

μ = viscosity, cp

ν = specific volume, cm^3/g

ν_w = specific volume of saturated liquid water

ν_s = specific volume of saturated steam

ν_{wi} = initial specific volume of saturated liquid water

ν_{si} = initial specific volume of saturated steam

ρ = density

K = thermal conductivity

ϕ = porosity, fraction

*Unless otherwise stated, the units are those of the SI system. See Conversion Table 1-12.

Subscripts

c = current
D = dimensionless
e = recharge
f = flowing, force
i = initial
j = index
L = loss
M = match point in type-curve matching
N = last-rate interval in a multiple-rate flow test
p = produced
r = reservoir, rock
rc = reservoir conditions

s = shut-in, steam
t = surface
sc = standard conditions
t = total
w = wellbore, water
$1hr$ = data from straight-line portion of semilog plot at 1 hour of test time, extrapolated if necessary

Special functions:

Exponential integral
$$= E_i\,(-x)$$
$$= -\int_x^\infty \frac{e^{-u}}{u}\,du$$

References

Agarwal, R. G., Al-Hussainy, R., and Ramey, H. J., Jr., 1970. An Investigation of Wellbore Storage and Skin Effect in Unsteady Liquid Flow: I. Analytical Treatment. *Soc. Pet. Eng. J.* (Sept.), pp. 279-290.

Atkinson, P. G., Celati, R., Corsi, R., and Kucuk, F., 1978. Behavior of Two Component Vapor-Dominated Geothermal Reservoirs. Ann. California Regional Meeting, Soc. Pet. Eng., San Francisco, Ca., paper SPE No. 7132.

Brownell, D. H., Garg, S. K., and Pritchett, J. W., 1975. Computer Simulation of Geothermal Reservoirs. Ann. California Regional Meeting, Soc. Pet. Eng., Ventura, Ca., paper No. 5381.

Burnham, J. B., and Stewart, D. H., 1973. Recovery of Geothermal Energy from Hot, Dry Rock with Nuclear Explosives. In P. Kruger and K. Otte (Eds.) *Geothermal Energy Resources, Production and Stimulation.* Stanford University Press, Stanford, Ca., pp. 223-230.

Burst, J. F., 1969. Diagenesis of Gulf Coast Clayey Sediments and Its Possible Relation to Petroleum Migration. *Bull. Am. Assoc. Pet. Geologists*, 53 (1): 73-93.

Crichlow, H. B., 1977. *Modern Reservoir Engineering — A Simulation Approach.* Prentice-Hall, Englewood Cliffs, N. J.

Cinco-Ley, H., Samaniego V. F., and Domínguez, A. N., 1978. Transient Pressure Behavior for a Well with a Finite-Conductivity Vertical Fracture. *Soc. Pet. Eng. J.* (Aug.), pp. 253-264.

Coats, K. H., 1969. Use and Misuse of Reservoir Simulation Models. *J. Pet. Tech.* (Nov.), pp. 1391-1398.

Coats, K. H., 1977. Geothermal Reservoir Modelling. Ann. Fall Technical Conference, 52nd., Soc. Pet. Eng., Denver, Colo., paper No. 6892.

Craft, B. C., and Hawkins, M. F., 1959. *Applied Petroleum Reservoir Engineering.* Prentice-Hall, Englewood Cliff, N. J.

Dake, L. P., 1978. *Fundamentals of Reservoir Engineering. Developments in Petroleum Science*, 8, Elsevier, Amsterdam: 433 pp.

Davis, W. K., and Golan, S., 1974. The New Energy Sources Ind. Research. (Nov. 15), pp. 8-15.

Earlougher, R. C., Jr., 1977. *Advances in Well Test Analysis*. Monograph Series, Society of Petroleum Engineers of AIME, Dallas, Tex.: 264 pp.

Faust, C. R., and Mercer, J. W., 1975. Mathematical Modelling of Geothermal Systems. Second U. N. Symposium on the Development and Use of Geothermal Resources, San Francisco, Ca., Proc. Lawrence Berkeley Lab., University of California, Berkeley, Ca., pp. 1635-1641.

Garg, S. K., 1978a. Pressure Transient Analysis for Two-Phase Geothermal Reservoirs. *Trans. Geothermal Resources Council*, 2: pp. 203-206.

Garg, S. K., 1978b. Pressure Transient Analysis for Two-Phase (Liquid Water/Steam) Geothermal Reservoirs. Ann. Fall Technical Conference, 53rd., Soc. Pet. Eng., Houston, Tex., paper No. 7479.

Gringarten, A. C., 1978. Well Testing in Two-Phase Geothermal Wells. Ann. Fall Technical Conference, 53rd, Soc. Pet. Eng., Houston, Tex., paper No. 7480.

Gringarten, A. C., and Ramey, H. J., Jr., 1974a. Unsteady-State Pressure Distributions Created by a Well with a Single Horizontal Fracture, Partial Penetration or Restricted Entry. *Soc. Pet. Eng. J.* (Aug.), pp. 413-426.

Gringarten, A. C., Ramey, H. J., Jr., and Raghavan, R., 1974b. Unsteady-State Pressure Distributions Created by a Well with a Single Infinite Conductivity Vertical Fracture. *Soc. Pet. Eng. J.* (Aug.), pp. 347-360.

Gringarten, A. C., Ramey, H. J., Jr., and Raghavan, R., 1975. Applied Pressure Analysis for Fractured Wells. *J. Pet. Tech.* (July), pp. 887-892.

Gulati, M. S., 1975. *Pressure and Temperature Buildup in Geothermal Wells*. Stanford Geothermal Reservoir Engineering Workshop, Stanford, Ca., Proc.: Stanford University, SGP-TR-12, pp. 69-73.

Knapp, R. M., Isokrari, O. F., Garg, S. K., and Pritchett, J. W., 1977. An Analysis of Production from Geopressured Geothermal Aquifers. Ann. Fall Technical Conference, 52nd, Soc. Pet. Eng., Denver, Colo., paper No. 6825.

Lasseter, T. J., Witherspoon, P. A., and Lippmann, M. J., 1975. Multiphase Multidimensional Simulation of Geothermal Reservoirs. Second U. N. Symposium on the Development and Use of Geothermal Resources, San Francisco, Ca., Proc. Lawrence Berkeley Lab., University of California, Berkeley, Ca., pp. 1715-1723.

Martin, J. C., 1959. Simplified Equations of Flow in Gas Drive Reservoirs and the Theoretical Foundation of Multi-phase Pressure Buildup Analysis. *Trans. Am. Inst. Min. Metall. Eng.*, 216: pp. 309-311.

Martin, J. C., 1975. Analysis of Internal Steam Drive in Geothermal Reservoirs. *J. Pet. Tech.* (Dec.), pp. 1493-1499.

Matthews, C. S., Brons, F., and Hazebroeck, P., 1954. A Method for Determination of Average Pressure in a Bounded Reservoir. *Trans. Am. Inst. Min. Metall. Eng.*, 201: pp. 182-191.

Matthews, C. S., and Russell, D. G., 1967. *Pressure Buildup and Flow Tests in Wells*. Monograph Series, Society of Petroleum Engineers of AIME, Dallas, Tex.: 167 pp.

Mercer, J. W., Faust, C., and Pinder, G. F., 1974. Geothermal Reservoir Simulation. *Proc. of N.S.F.*, Conference on Research for the Development of Geothermal Energy Resources, Pasadena, Ca., pp. 256-267.

Miller, F. G., Cinco, H., Ramey, H. J., Jr., and Kucuk, F., 1978. Reservoir Engineering Aspects of Fluid Recharge and Heat Transfer in Geothermal Reservoirs. *Trans. Geothermal Resources Council*, 2: pp. 449-452.

Murphy, H. D., 1975. *Hydraulic-Fracture Geothermal Reservoir Engineering*. Stanford Geothermal Reservoir Eng. Workshop, SGP-TR-12, pp. 174-177.

Muskat, M., 1938. *The Flow of Homogeneous Fluids Through Porous Media.* McGraw-Hill Book Co., New York, N. Y.

Narasimhan, T. N., Schroader, R. C., Goranson, C. B., and Benson, S. M., 1978. Results of Reservoir Engineering Tests, 1977, East Mesa, California. Ann. Fall Technical Conference, 53rd, Soc. Pet. Eng., Houston, Tex., paper No. 7482.

Odeh, A. S., 1969. Reservoir Simulation. . . What Is It?. *J. Pet. Tech.* (Nov.), pp. 1383-1388.

Peaceman, D. W., 1977. *Fundamentals of Numerical Reservoir Simulation.* Elsevier, Amsterdam.

Pruess, K., Schroader, R. C., and Zerzan, J., 1978. *Studies of Flow Problems with the Simulator Shaft 78.* Stanford Geothermal Reservoir Engineering Workshop, Stanford, Ca., Proc.: Stanford University (in print).

Ramey, H. J., Jr., 1965. Non-Darcy Flow and Wellbore Storage Effects in Pressure Buildup and Drawdown of Gas Wells. *J. Pet. Tech.* (Feb.), pp. 223-233.

Ramey, H. J., Jr., 1970a. A Reservoir Engineering Study of the Geysers Geothermal Field. Testimony for the Trial of Reich and Reich vs Commissioner of the Internal Revenue, Tax Court of the U.S., 52 T. C. No. 74.

Ramey, H. J., Jr., 1970b. Short-Time Well Test Data Interpretation in the Presence of Skin Effect and Wellbore Storage. *J. Pet. Tech.* (Jan.), pp. 97-104.

Ramey, H. J., Jr., 1975. Pressure Transient Analysis for Geothermal Wells. Second U. N. Symposium on the Development and Use of Geothermal Resources, San Francisco, Ca., Proc. Lawrence Berkeley Lab., University of California, Berkeley, Ca., pp. 1749-1757.

Ramey, H. J., Jr., 1976. Practical Use of Modern Well Test Analysis. Ann. Fall Technical Conference, 51st, Soc. Pet. Eng., New Orleans, La., paper No. 5878.

Ramey, H. J., Jr., 1977. Petroleum Engineering Well Test Analysis-State of the Art. *Proceeding: Invitational Well Testing Symposium,* Oct. 19-21, Lawrence Berkeley Lab., Berkeley, Ca., pp. 5-9.

Ramey, H. J., Jr., Kruger, P., and Raghavan, R., 1973a. Explosive Stimulation of Geothermal Reservoirs. In P. Kruger and K. Otte (eds.), *Geothermal Energy Resources, Production and Stimulation.* Stanford University Press, Stanford, Ca., pp. 231-249.

Ramey, J. H., Jr., Kumar, A., and Gulati, M. S., 1973b. *Gas Well Test Analysis Under Water-Drive Conditions,* Amer. Gas Assoc., Arlington, Va.

Ramey, H. J., Jr., and Gringarten, A. C., 1975. Effect of High-Volume Vertical Fractures on Geothermal Steam Well Behavior. Second U. N. Symposium on the Development and Use of Geothermal Resources, San Francisco, Ca., Proc. Lawrence Berkeley Lab., University of California, Berkeley, Ca., pp. 1759-1762.

Richardson, J. G., and Stone, H. L., 1973. A Quarter Century of Progress in the Application of Reservoir Engineering. *J. Pet. Tech.* (Dec.), pp. 1371-1379.

Rivera, R. J., and Ramey, H. J., Jr., 1977. Application of Two-Rate Flow Tests to the Determination of Geothermal Reservoir Parameters. Ann. Fall Technical Conference, 52nd, Soc. Pet. Eng., Denver, Colo. paper No. 6887.

Rivera, R. J., Samaniego, V. F., and Schroader, R. C., 1978. Pressure Transient Testing at Cerro Prieto Geothermal Field. First Symposium on the Cerro Prieto Geothermal Field, San Diego, Ca., Proc.: Lawrence Berkeley Lab., University of California, Berkeley, Ca. (in print).

Russell, D. G., 1963. Determination of Formation Characteristics from Two-Rate Flow Tests. *J. Pet. Tech.* (Dec.), pp. 1347-1355.

Saltuklaroglu, M., and Rivera, R. J., 1978. *Injection Testing in Geothermal Wells.* Stanford Geothermal Reservoir Engineering Workshop, Stanford, Ca., Proc.: Stanford University (in print).

Smith, M., Potter, R., Brown, D., and Aamodt, R. L., 1973. Induction and Growth of Fractures in Hot-Rock. In P. Kruger and K. Otte (eds.), Geothermal Energy Resources, Production and Stimulation. Stanford University Press. Stanford, Ca., pp. 251-268.

Smith, M., Aamodt, R. L., Potter, R. M., and Brown, D. W., 1975. Man-Made Geothermal Reservoirs. Second U. N. Symposium on the Development and Use of Geothermal Resources, San Francisco, Ca., Proc. Lawrence Berkeley Lab., University of California, Berkeley, Ca., pp. 1781-1787.

Takahashi, P. K., Chen, B. H., Mashima, K. I., and Seki, A. S., 1975. State-of-the-Art of Geothermal Reservoir Engineering. *J. of the Power Div.,* A.S.C.E., 101 (July), pp. 111-126.

Thomas, K. K., and Pierson, R. G., 1978. Three Dimensional Geothermal Reservoir Simulation, *Soc. Pet. Eng. J.* (April), pp. 151-161.

Todd, L., Mercer, J. W., and Faust, C. R., 1978. *Simulation of Geothermal Reservoirs Including Changes in Porosity and Permeability due to Silica-Water Reactions.* Stanford Geothermal Reservoir Engineering Workshop, Stanford, Ca., Proc.: Stanford University (in print).

van Everdingen, A. F., and Hurst, W., 1949. The Application of the Laplace Transformation to Flow Problems in Reservoirs. *Trans. Am. Inst. Min. Metall. Eng.,* 186, pp. 305-324.

Wattenbarger, R. A., and Ramey, H. J., Jr., 1970. An Investigation of Wellbore Storage and Skin Effect in Unsteady Liquid Flow: II. Finite Difference Treatment. *Soc. Pet. Eng. J.* (Sept. 1970), pp. 291-297.

White, D. E., 1973. Characteristics of Geothermal Resources. In P. Kruger and K. Otte (eds.), *Geothermal Energy Resources, Production, and Stimulation.* Stanford University Press, Stanford, Ca., pp. 69-94.

White, D. E., Muffler, L. J. P., and Truesdell, A. H., 1971. Vapor Dominated Hydrothermal Systems Compared with Hot-Water Systems. Econ. Geol., V. 66, pp. 75-97.

Whitherspoon, P. A., Narasimhan, T. N., and McEdwards, D. G., 1978. Results of Interference Tests from Two Geothermal Reservoirs. *J. Pet. Tech.* (Jan.), pp. 10-16.

Whiting, R. L., and Ramey, H. J., Jr., 1969. Applications of Material and Energy Balances to Geothermal Steam Production. *J. Pet. Tech.* (July), pp. 893-900.

Zyvolski, G. A., and O'Sullivan, M. J., 1978. *Simulation of the Broadlands Geothermal Field, New Zealand.* Stanford Geothermal Reservoir Engineering Workshop, Stanford, Ca., Proc.: Stanford University (in print).

Jefferson Tester, Massachusetts Institute of Technology

10

Energy Conversion and Economic Issues for Geothermal Energy

Introduction and Scope

The issues associated with the economic utilization of geothermal energy are extremely complex because geothermal systems are site-specific: the characteristics of a particular geothermal site must be matched to appropriate energy conversion alternatives. This site-specific nature of geothermal systems is in contrast to more conventional technologies, such as electricity produced by nuclear, oil- or gas-fired systems. Major uncertainties in evaluating the economics of a geothermal system include estimating drilling costs, reservoir characteristics, and long-term performance of power conversion equipment. Thus, vast differences inevitably appear between economic assessments produced by overly optimistic promoters of a given technology and the assessments of fiscally conservative critics who may misjudge technical uncertainties and so overestimate the financial risks involved.

Because of these complicating factors and the highly capital intensive nature of geothermal development in general, individual organizations interested in commercialization will undoubtedly make their own very comprehensive and independent assessment of the technology as it applies to specific sites. Consequently, this chapter presents a general treatment of the subject in discussing energy conversion concepts and economic issues facing the growth of a commercial geothermal industry in the United States. Instead of considering specific cases in detail, the coverage will emphasize the importance of a wide variety of parameters as they influence the economics. Extensive reference will be made to the existing literature on this subject; the papers cited cover industrial as well as governmental research and development viewpoints and, in many cases, represent work under the

471

full or partial sponsorship of the United States Department of Energy and the Electric Power Research Institute (EPRI).

The key issues related to commercial feasibility can be divided into three major categories. The first concerns the resource, including such variables as the rock type and its properties, depths to the reservoir, fluid temperatures, and chemical composition. The second concerns the broad category of reservoir engineering and other technological issues, which would include production capacity, lifetime of the reservoir, the physical structure of the reservoir system, and the design and performance of the surface plant. The last category includes a wide range of financial and management issues including exploration, drilling and surface plant costs, selling price of the produced energy, the assessment of risk, anticipated rate of return, operating and maintenance costs, the type of ownership and a number of regulatory issues. For example, the anticipated rate of return and risk assessment by a major oil producer who may develop the geothermal field will be quite different from a private utility, which may purchase the geothermal steam or hot water and produce the power.

Investment cost factors for geothermal developments are discussed in detail. They are largely controlled by well drilling and completion costs associated with producing the fluid, and major equipment costs associated with processing the geothermal heat. Major equipment typically includes heat exchangers, condensers, turbine-generators, and pumps. Because well drilling and completion costs usually comprise 50-60% of the total capital costs, and frequently may be as much as 75%, they are the single most important cost item to be dealt with. An equally important factor is the real or perceived risk associated with development. The assessment of risk by the investor will establish values for the rate of return on equity capital and debt interest rates. The higher the risk, the higher the anticipated rate of return must be and the more difficult it is to secure bank loans.

With respect to the existing geothermal market, a general survey of electric, nonelectric, and cogeneration applications is considered. In the nonelectric category, both process and space heating applications are covered; and under cogeneration, a specific example involving the production and consumption of both process heat and electricity is included.

Many of the new alternate sources of energy currently being considered, such as geothermal, solar, and ocean thermal, share the common disadvantage of having an inherently low temperature. One technique of increasing their "quality" as well as their transportability is to convert their available energy into electric power. Unfortunately, the efficiency of this conversion process is limited by the temperature of the resource and by prevailing ambient conditions for heat rejection. Efficiencies for converting geothermal resources at temperatures below 200°C to electricity are substantially less than those of fossil-fuel-fired or nuclear powered plants.

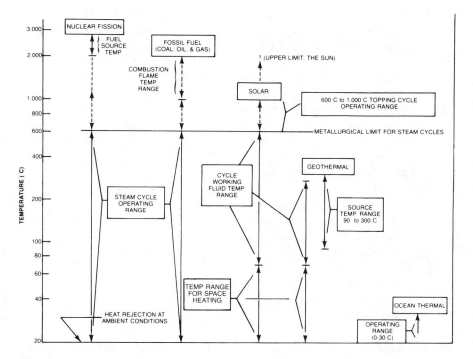

Figure 10-1. Operating temperature ranges for various energy options (note logarithmic scale for temperature).

Alternatively, this concept can be illustrated by considering the maximum operating temperature difference between the heat source and the heat sink, that is, the ambient environment where heat rejection ultimately occurs. Figure 10-1 shows operating temperature ranges for a number of new and established resource types. Note the logarithmic temperature scale in comparing the resources. The vast differences that obviously exist suggest totally different engineering strategies for resource recovery and utilization. Two major categories of source temperatures are evident: source temperatures resulting from combustion of fossil fuels or nuclear fission are markedly higher than those for geothermal, solar, and ocean thermal resources. The situation is actually less pronounced than shown since fossil fuel combustion flame temperatures ranging from 1,000 to 2,000°C are not normally utilized in commercial power systems at those temperatures because of materials limitations. For example, in steam-electric cycles operating temperatures are restricted to 600°C. This limitation would also be applied to conventional nuclear reactors, such as boiling-water or pressurized-water types; however, high temperature gas-cooled reactors might permit use of a Brayton gas turbine topping cycle using the reactor coolant, helium, as the prime mover. Other combined cycles involving potassium-steam, mercury-steam, or cesium-steam in topping and bottoming arrangements could extend the operating limit for fossil-fired and nuclear systems to 1,000°C.

The solar case is different; solar source (the sun) temperatures are essentially infinite on the scale of the figure. However, ignoring photovoltaics, maximum cycle operating temperatures are constrained similarly to fossil and nuclear systems. The high end of the temperature range corresponds to solar power tower-type concepts with concentrators used; at the low end are space and water heating systems.

Geothermal, low-temperature solar, and ocean thermal systems present enormous engineering challenges because of their low source temperatures and therefore small operating temperature differences. All conversion steps in the power utilization design must operate between the maximum source temperature and the ambient state:

300° to 90°C and ~ 20°C for geothermal
200° to 70°C and ~ 20°C for low-temperature solar
20° to 30°C and ~ 0 to 20°C for ocean thermal

The combination of a low thermodynamic efficiency and a restricted temperature range for cycle operation present abnormal constraints requiring innovative designs.

New technologies are developing for improving the practical thermodynamic efficiency of power production toward its ideal limits as well as for providing an economical process. For example, for a number of these low-temperature alternate energy sources, several hydrocarbons and their halogenated derivatives have been proposed as prime mover fluids, rather than steam. The properties of these organic compounds are considerably different from water in the low-temperature region, perhaps making them more suitable as working fluids. At the other extreme of end use, serious consideration is being given to nonelectric possibilities for geothermal energy. Recent estimates have shown that approximately 30 out of the 80 quads (10^{18} J or 10^{15} Btu) of the total amount of energy consumed annually in the United States are utilized at temperatures below 250°C. In addition, of the 30 quads, a substantial fraction is utilized in the temperature range of 50-80°C for both residential and industrial space heating applications (Reistad, 1957). Because of the high capital costs associated with a geothermal process heat or electric generation system, high load or capacity factors are desirable.

In either electric, nonelectric, or cogeneration applications smaller scale systems are more practical economic choices. For example, the optimal individual generating unit size for a given well distribution system in even an extensive geothermal field is probably below 100 megawatts electric [MW(e)] capacity. Similar criteria would apply to nonelectric and cogeneration systems. In any case, the geothermal option is usually considered for base load capacity.

The first section of this chapter, "Energy Conversion and Utilization," addresses a variety of energy conversion alternatives. Specific applications and cycle configurations for surface systems, as well as thermodynamic and engineering design criteria, are presented.

In the second section, "Economic Analysis Methods," several financial and regulatory aspects of geothermal development are outlined. These factors range from capital costs and financial arrangements, to operating and maintenance costs, to the management and organizational structure as well as a discussion of a wide variety of tax and regulatory issues. In addition, a discussion of economic methods of analysis and investment criteria is included. A review will be made of several decision criteria and economic modeling approaches including the capital recovery factor, fixed charge rate, discounted cash flow, and levelized annual cost approaches. In the third section, "Exploration, Drilling and Reservoir Development Costs," specific resource-related cost factors are presented parametrically with the major concern being drilling and completion costs for the reservoir system. Then, costs associated with the surface conversion equipment are discussed in the section titled "Surface Conversion Equipment Costs." Empirical correlations for heat exchanger and condenser costs as well as turbine and pump systems are included.

In the next two sections, specific economic factors influencing power production, nonelectric heat production, and cogeneration systems are presented. In addition, several specific case studies are reviewed.

In the last section, the commercial feasibility issues facing geothermal development are summarized. Economic estimates for existing and projected geothermal systems will be compared to conventional energy supplies and costs.

Energy Conversion and Utilization

Applications and Configurations of Surface Systems

Many end use options exist for geothermal heat utilization. Of these, electric power generation and space and process heating have been applied to practical commerical operations throughout the world (Kruger and Otte, 1973; Berman, 1976; 2nd U.S. Symposium on the Development and Use of Geothermal Resources Proceedings, 1975; Armstead, 1978; DiPippo, 1980; and EPRI, 1978). This section is designed to introduce the engineering components and thermodynamic issues associated with geothermal power and heating systems. In order to evaluate the capital costs required to construct a geothermal plant, as well as the geothermal fluid requirements (kg/s per MW(e)), it is necessary to characterize the performance of the plant and size of its components. For a more detailed treatment of engineering component design criteria, interested readers should consult Kestin et al., 1980; Armstead, 1978; Wahl, 1977; and Milora and Tester, 1976.

Conversion Cycle Configurations

Various electric conversions and direct-use cycles are illustrated in Figures 10-2 — 10-9. A Rankine cycle utilizing dry or saturated steam expansion in a low-pressure condensing turbine with direct contact condensation is presently used to generate power for vapor-dominated resources such as those at The Geysers, California and Lardarello, Italy (see Figure 10-2). Assuming particulate matter can be removed before entering the turbine, conventional turbine designs and materials of construction can be used. The multi-stage flashing cycle also depicted in Figure 10-2 is used for liquid-dominated resources, such as those at Wairakei, New Zealand, and Cerro Prieto, Mexico. In this case, saturated steam is created by flashing the geothermal fluid at the surface to a lower pressure, followed by expansion in a condensing steam turbine with the unflashed liquid fraction either reinjected or discarded. The performance of the steam cycles shown in Figure 10-2 is significantly lowered by the presence of noncondensable gases in the geothermal fluid. As Wahl (1977) pointed out, an extraction pump or a vacuum ejector is required to remove noncondensable gases, such as CO_2, to maintain the turbine exhaust at subatmospheric conditions. If noncondensable gas concentrations are too high, indirect binary or two-phase flow conversion cycles such as those depicted in Figures 10-3, 10-4, and 10-5 might be the only feasible alternatives.

Binary-fluid cycles are closed-loop Rankine cycles that involve a primary heat exchange step where heat from the geothermal fluid is transferred to a secondary working fluid, which then expands through a turbogenerator and is condensed with heat finally rejected to the environment in a cooling tower or similar device. The closed cycle is completed by pumping the secondary fluid up to the maximum cycle operating pressure. Compounds considered for these applications include lower-molecular-weight hydrocarbons such as isobutane and propane, and their halogenated derivatives (fluorocarbons), such as CH_2F_2, $CHCl_2F$, and C_2ClF_5, and ammonia (NH_3).

The performance of such binary cycles is insensitive to noncondensables contained in the primary geothermal fluid. Furthermore, because these non-aqueous secondary working fluids have low-temperature vapor densities considerably larger than steam, much smaller and less expensive turbines for

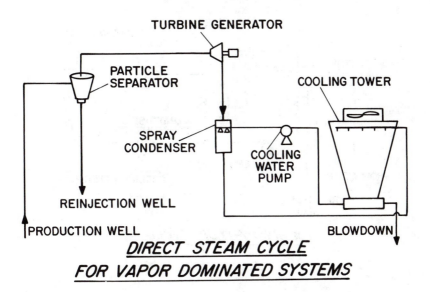

DIRECT STEAM CYCLE
FOR VAPOR DOMINATED SYSTEMS

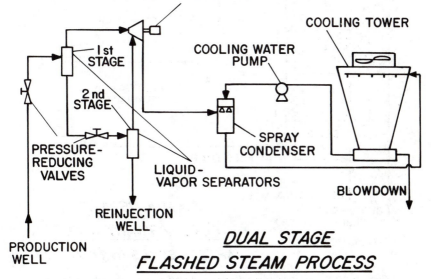

DUAL STAGE
FLASHED STEAM PROCESS

Figure 10-2.   Commercially-operated geothermal power conversion cycles.

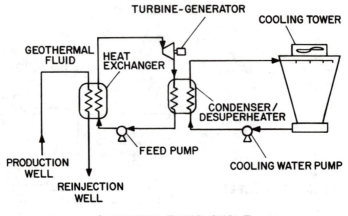

A. BINARY-FLUID CYCLE

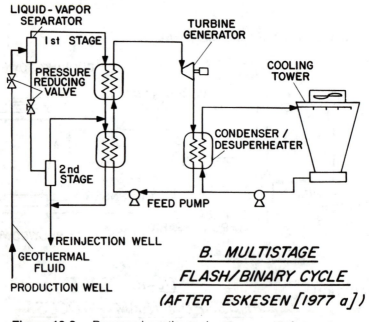

B. MULTISTAGE FLASH/BINARY CYCLE
(AFTER ESKESEN [1977 a])

Figure 10-3. Proposed geothermal power conversion systems.

(figure continued at top of next page)

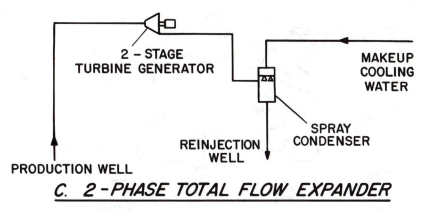

C. 2-PHASE TOTAL FLOW EXPANDER

Figure 10-3. (continued) Proposed geothermal power conversion systems.

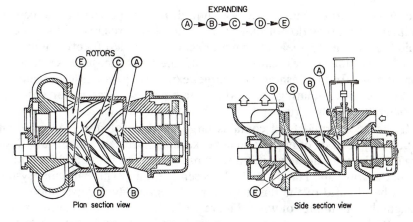

Figure 10-4. Helical screw expander for two-phase expansions. The working fluid enters at A. The effective passage expands in volume as the fluid traverses the passage in the sequence B-C-D-E. (By permission of Hydrothermal Power Co., Ltd., Mission Viejo, Calif.)

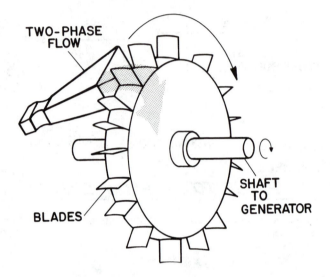

Figure 10-5. Axial-flow impulse turbine schematic for two-phase expansion. (By permission of A.L. Austin, Lawrence Livermore Laboratory, Livermore, Calif.).

the same power output would be required. Flashing cycles are, of course, simpler in that they do not require a primary heat exchanger or secondary fluid feed pump. This leads to a major economic trade-off between flashing and binary fluid cycles centered around the capital costs of certain key components. In the flashing case, large, expensive low-pressure steam turbines play a dominating role, while for binary systems major costs are associated with the primary heat exchanger and condenser.

If the geothermal fluid contains large amounts of dissolved material, which may corrode or deposit on heat exchange surfaces, a multistage flash-organic binary cycle may offer a reasonable solution (Holt and Ghormley, 1976; Schapiro and Hajela, 1977; Eskesen 1977a,b). This particular system is really a hybrid of the flashed steam and binary-fluid cycles and can be designed in a number of ways. Figure 10-3B shows one alternative, a two-stage flash unit to produce steam for vaporizing an organic fluid which then expands through a turbine to produce power in a closed Rankine cycle. Other designs such as those discussed by Holt and Ghormley (1976) may involve using a dual cycle system where the flashed steam fraction drives a turbogenerator and the remaining liquid fraction heats the secondary working fluid in a binary cycle.

Cycles such as those proposed by Eskesen (1977a,b) and Schapiro and Hajela (1977) may successfully deal with the deposition — scaling problem but are less efficient than multistage, direct steam flashing systems, which may be practicable for these cases if dissolved gas contents are low. Sperry Research is also investigating a unique binary system that involves a downhole heat exchanger and organic turbine-driven pump in a "gravity head cycle" for resources in 120° to 200°C range (McBee and Mathews, 1979).

Austin and associates at the Lawrence Livermore Laboratory (Austin 1978, 1980) and Armstead (1978) have evaluated total flow, two-phase expansion systems for geothermal applications. Total flow expanders as depicted in Figure 10-3C may be particularly useful for high-temperature, high-salinity brines such as those found in the Niland-Salton Sea area of Imperial Valley, California. Because a large portion of the initial fluid in the flashed system is not converted into steam for expansion in a turbine, a substantial fraction of the geothermal fluid is unusable for power production. The total flow process avoids this waste by expanding the entire geothermal fluid through a suitable device. As yet unachieved engine efficiencies of 70% or greater will be required to improve on the present performance of multistage, flashed steam and organic binary cycles. Nonetheless, the inherent simplicity of total flow machines suggests their future capital costs will be lower than the alternatives.

As Austin (1978, 1980) and Armstead (1978) point out, several engine concepts are possible. These include impulse-reaction, positive displacement, and impulse machines. The helical screw expander as shown in Figure 10-4 is a positive displacement machine under development by the Hydrothermal Power Company and the Jet Propulsion Laboratory (McKay, 1977). Positive displacement expanders will be large in comparison to comparable condensing steam turbines, so their size may limit central power station use. But for smaller-scale [$\sim\leqslant 50$ MW(e)] non-condensing applications, positive displacement expanders may be quite practicable. At the Livermore Laboratory an axial-flow impulse turbine concept was investigated. These machines require the development of efficient, two-phase expansion nozzles and of efficient turbine blading for transfer of the fluid momentum to the rotating shaft (Figure 10-5). Because practical experience has been mostly with steam turbine expansion on the vapor-rich side of the two-phase envelope, conventional concepts regarding efficiency losses due to moisture (liquid droplets) in the vapor phase might be totally misleading when applied to the liquid-rich region (Austin, 1980). Although many questions remain to be answered before total flow systems become a commercial reality, the eventual payoff in capital cost reduction should warrant further research and development.

With the possible exception of small binary-fluid power plants operating in Japan and the USSR (Kruger and Otte, 1973; Moskvicheva, 1971; and

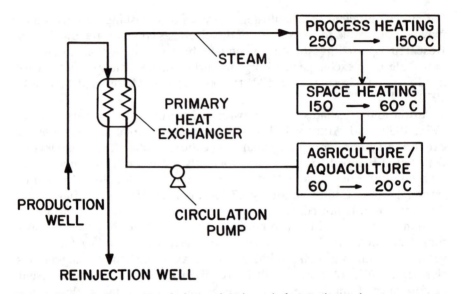

Figure 10-6. Indirect cascaded, nonelectric cycle for geothermal energy source.

Moskvicheva and Popov, 1970), direct steam injection and steam flashing cycles are the only ones in actual commercial service today for geothermal resources. Interest in organic binary and multistage flash-organic binary cycles has really blossomed in the past five years, not only for geothermal applications but also for many other low-temperature heat source systems, including solar, ocean thermal, waste heat recovery, and bottoming cycles. Holt and Ghormley (1976) and Eskesen (1977a) imply that organic binary systems represent state-of-the-art technology requiring modest developments, including the design of larger capacity hydrocarbon units, control and valving systems, turbine seals for hydrocarbon vapors, and improved resistance to corrosion, erosion, and scaling of certain components.

Even with low conversion efficiencies, electric power production from geothermal resources continues to receive considerable attention for existing and proposed sites in the world. In part, this is because the inherent base load character of a geothermal operation is coupled to a frequently remote location from an existing load center. This is certainly true at The Geysers field in California, where the San Francisco load center for the over 900 MW(e) generated is approximately 129 km from the field. Although electrical transmission costs are generally lower than for direct fluid transportation, commercial nonelectric applications for space heating can involve hot fluid transport for several kilometers, as demonstrated in Iceland, USSR,

Japan, and Boise, Idaho (Kruger and Otte, 1973; Kunze, 1977; Lund, 1976; and EPRI, 1978). Hausz (1976), Meyer and Hausz (1977), and Karkheck (1978) examined the question of hot fluid transmission and conclude that large-scale transport in some cases may be competitive at distances greater than 100 km.

Nonelectric systems can be described very simply as shown in Figure 10-6, where a primary heat exchange loop is depicted with a cascaded utilization system. If the fluid is of good quality, it may be transported directly, avoiding the primary heat exchange step. Recycling of the geothermal fluid may not be required in this case. Extensive use is presently being made of geothermal fluids ranging from 40°C to 200°C for process heat for the mining and pulp and paper industries, district space heating, therapeutic baths, absorption air conditioning, and agricultural and aquacultural purposes (see Table 10-1; EPRI, 1978; and Howard, 1975). In all cases, the end use requirements and geothermal fluid characteristics must be considered in designing the system.

Geothermal cogeneration systems are also likely to be attractive options for some process industries and private utilities. Figure 10-7 shows a conceptual cogeneration system for a large Kraft and modified sulfite process pulp and paper manufacturing operation. In this case, approximately 50-75 MW(e) may be required for operating pumps, paper machines, and other mechanical drives. In addition, an extensive cascaded process steam system with up to 1×10^6 kg/hr of steam ranging in pressure from 1 to 10 bars is required. Depending on the actual paper mill involved, the mix of electric versus nonelectric loads can vary. Another interesting cogeneration concept that can be applied to seasonal variations in heating, cooling, and electricity load demands is described in Figure 10-8. Dual and topping — bottoming cycle configurations are shown for summer and winter operation. To improve the effective overall capacity factor in the summer, more electricity is needed for air conditioning and refrigeration, and when this demand drops in the winter, the bottom portion of the cycles are replaced with a heat exchanger supplying energy to a district heating system. A number of European cities already use this arrangement with a fossil-fired system.

The thermodynamic conversion of both fossil and geothermal energy to electricity is improved by the hybrid cycle shown in Figure 10-9 (see Anno, et al., 1977; Kingston, et al., 1978; and Khalifa, 1980b). In this concept, geothermally supplied heat is used for feedwater preheating followed by a fossil-fired unit supplying additional heat before the steam enters the turbo-generator. This arrangement can significantly improve efficiency by reducing coal and geothermal fluid consumption (per MW(e) basis); but has the disadvantage of being a more complex and a more capital intensive enterprise and may require proximity to both geothermal and coal resources for commercial feasibility.

Table 10-1

Existing and Potential Major Non-electric Uses for Geothermal Energy

Application	Temperature range, °C	Steam requirement, kg steam/ kg product	Total annual U.S. consumption (1967) 10^{15} J	Geothermal Status
Space heating	80-50	0.4	30526	Existing installations throughout world, e.g., Klamath Falls, OR; Boise, ID; USSR; Japan; Iceland; Hungary; France; Czechoslovakia; Argentina
Water heating	120-80	0.5	2389	
Refrigeration and air conditioning				
Ammonia absorption	≥180	—	} 1540	
Lithium bromide absorption	> 70	—		
Chemical industry				
Chlorine-caustic soda production using diaphragm cells	≥120	8 (NaOH)	69	Potential use
Soda ash production with Solvay process	≥120	2	57	Potential use
Alumina production from bauxite with Bayer process	150	7	52	Potential use
Heavy water production by hydrogen sulfide-water isotope exchange	170	10000	?	Potential use
Desalination by multiple-effect distillation of seawater	120	0.08	—	Pilot plant in Chile
By-product recovery of minerals from geothermal brines via evaporation	≥120	10-2	—	Boric acid recovery at Larderello, Italy
Sulfur mining with Frasch process	≥120		60	Potential use
Concrete processing and drying	110	2-1		Existing facility in Reykjavik, Iceland

Table 10-1 continued

Application	Temperature range, °C	Steam requirement, kg steam / kg product	Total annual U.S. consumption (1967) 10^{15} J	Geothermal Status
Diatomaceous earth drying	170	3-2	—	Existing facility in Namafjall, Iceland
Petroleum refining	250-175 (20%) 175-150 (40%) 150-125 (40%)	—	} 960	Potential use
Food processing				
Freeze drying via ammonia absorption refrigeration	180°C	4-2		Potential use
Canning	140	2-1	} 900	Existing small scale efforts
Drying	140-120	2-1		
Sugar refining	130	6 (beet) 1.7 (cane)		
Pulp and paper processing				
Kraft or sulfite pulp digestion	200-175 (70%)	3-4	} 1005	Existing cogeneration facility in New Zealand
Dissolving pulp	175-150 (30%)	4		
Agricultural				
Crop drying	60°C			Operational on a small scale throughout the world, e.g., Hungary, Iceland, Italy, U.S.A.
Greenhouse heating	60°C			
Aquacultural				
Fish farming	20°C			Existing system in Japan

Sources: Lindal (1973), Howard (1975), Reistad (1975), and UNESCO (1970).

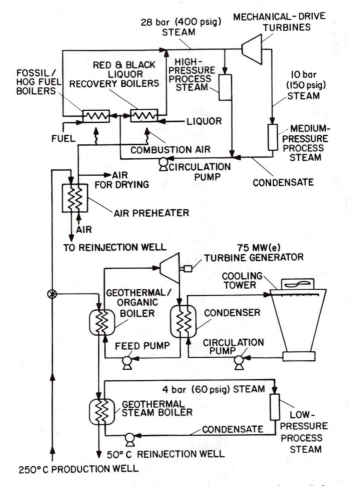

Figure 10-7. Proposed geothermal cogeneration system schematic for a pulp and paper mill.

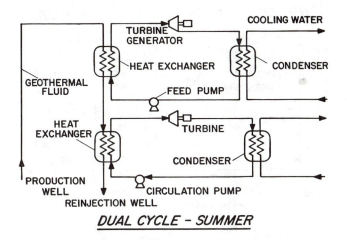

DUAL CYCLE - SUMMER

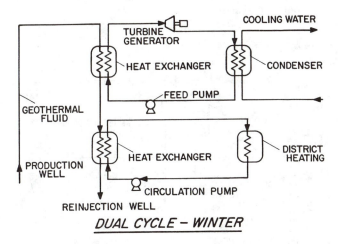

DUAL CYCLE - WINTER

Figure 10-8. Proposed geothermal cogeneration systems.

(figure continued on next page)

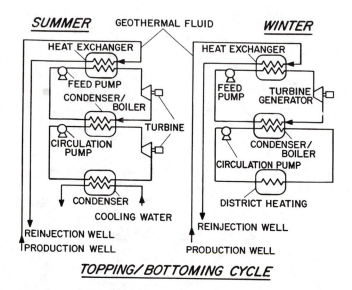

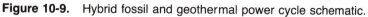

Table 10-8 (continued). Proposed geothermal co-generation systems.

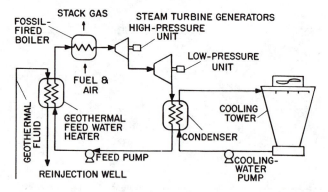

Figure 10-9. Hybrid fossil and geothermal power cycle schematic.

Market for Geothermal Energy

A frequent concern is the potential market for geothermal-produced electricity or heat. Even with the inherent remoteness of many hydrothermal systems, electricity can be generated at the site and tied into an existing grid system at an acceptable cost. Geothermal power plants have smaller capacities (~50-100 MW(e)) and, consequently, can be used as base load units in much larger total generating capacity systems that presently employ fossil and nuclear-fired plants. Of the 450,000 MW(e) generating capacity in the U.S., only 930 MW(e) is supplied by geothermal energy. There is unquestionably a market for both new and replacement capacity that can compete economically.

The situation with the nonelectric market is quite different. First, geothermal space and process heat systems may require some retrofitting by the consumer. For example, a fuel oil or natural gas furnace usually is designed with in-house storage or is attached to an existing distribution network. Thus, the use of geothermal hot water or steam will require a completely new fluid distribution system within the community.

Reistad (1975) discusses energy usage as a function of utilization temperature and concludes that about 40% of annual fossil energy consumption in the U.S.A. is severely degraded thermodynamically. Typically, 1,000 to 1,500°C fossil combustion temperatures are used to produce space and process heat at temperatures below 250°C. The distribution of energy consumed versus temperature is illustrated by the fractional and cumulative distribution functions developed by Reistad (1975), as shown in Figures 10-10 and 10-11. The fractional function is simply represented by the derivative of the cumulative function. A wide variety of applications are contained in the 40% or 30 quads of energy used below 250°C. Table 10-1 summarizes some of the major ones. Detailed discussions of how geothermal energy is now used or may be used in the future can be found in the literature. (See References at end of chapter.) In addition to Reistad's important work (1975), papers by Howard (1975), the Electric Power Research Institute (1978), Keller and Kunze (1976), Schultz et al. (1977), Armstead (1978), and Reistad et al. (1977) are of interest as well as several of the conference proceedings cited at the end of this chapter.

Studies of other direct heating applications are relevant to geothermal utilization because of the low temperatures associated with such systems. These have included (1) fossil-and nuclear-fired cogeneration systems (Inter. Atomic Energy Agency, 1977 and Resource Planning Associates, 1977); (2) the use of thermal discharges from power plants (Miller et al., 1971; Beall, 1970; and Beall and Samuels, 1971); (3) discharges with underground storage (Hausz, 1976; and Meyer et al., 1977); and (4) solar (Linn, 1979, and Intertechnology Corporation, 1977).

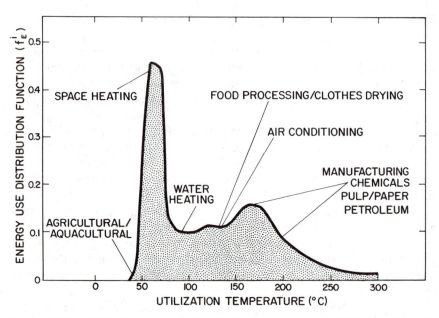

Figure 10-10. Fractional energy use distribution as a function of utilization temperature. (Adapted by permission from Reistad, 1975, and Intertechnology Corp., 1977).

Thermodynamic Criteria

Efficiency and Available Work

The second law of thermodynamics imposes stringent limitations on the production of electricity from a low-temperature geothermal heat source. Inasmuch as the temperature of a condensed geofluid lowers as one extracts useful work, a somewhat specialized ideal process must be constructed for calculating the maximum amount of work that can be produced. In one scheme, an infinite number of infinitesimally small reversible Carnot heat engines would be required. The maximum amount of work would then result from taking the geothermal fluid at wellhead conditions (temperature T_{gf}, and pressure, P_{gf}) and allowing heat to be removed through these Carnot engines to produce work and reject an amount of heat to the environment at temperature T_o. This process would be continued until the geofluid reached the so-called dead state or ambient condition (T_o, P_o). The

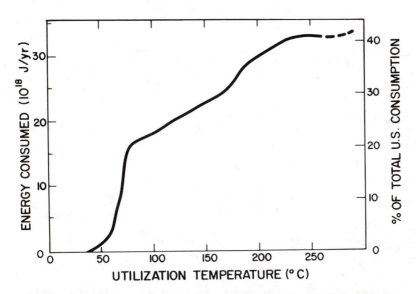

Figure 10-11.   Estimated total energy consumed in the U.S.A. in 1979 (80 × 10¹⁸J) as a function of utilization temperature. (Adapted by permission from Reistad, 1975, and Intertechnology Corp., 1977).

total maximum work can then be expressed by a quantity called the availability B as follows:

$$W_{\max} = \Delta B = (\Delta H - T_o\Delta S) \left] \begin{array}{c} (T_{gf}, P_{gf}) \\ (T_o, P_o) \end{array} \right.$$ (10-1)

where ΔH = enthalpy difference between state (T_{gf}, P_{gf}) and state (T_o, P_o);

  ΔS = entropy difference between state (T_{gf}, P_{gf}) and state (T_o, P_o)

   The availability is plotted in Figure 10-12 for saturated steam and water over a range of T_o from 25 to 45°C. This maximum work quantity can then be compared to the actual amount of work produced by any real power conversion process (W_{net}).

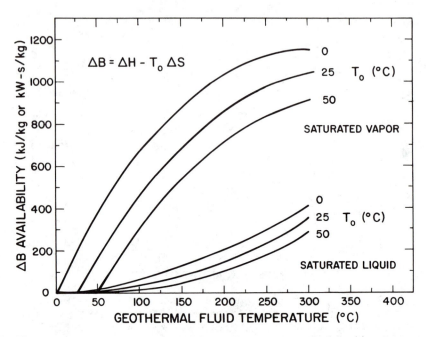

Figure 10-12. Maximum available work from a saturated liquid or vapor water source. T_0 is the ambient heat rejection temperature.

Comparisons of this type are usually achieved by defining a cycle efficiency (η_{cycle}), which represents the net useful work, W_{net}, obtained from the system divided by the amount of heat transferred from the geothermal fluid, Q_h. As the cycle efficiency decreases, the amount of heat rejected to the environment increases. For an ultimate sink of 25°C, with a geothermal heat source of 50°C, cycle efficiencies would be less than 5%. As the temperature increases to 100°C, the efficiency would be ~8%; at 150°C, ~12.5%; and at 200°C, ~17.5%. A typical range of cycle efficiencies is given in Figure 10-13 as a function of geothermal fluid and condensing temperatures.

An alternative approach to using cycle efficiency would be to compare directly the real work to the maximum possible work ΔB by defining a utilization efficiency η_u as follows:

$$\eta_u = \frac{W_{net}}{\Delta B} \tag{10-2}$$

which can be approximated by simplifying the expression for ΔB

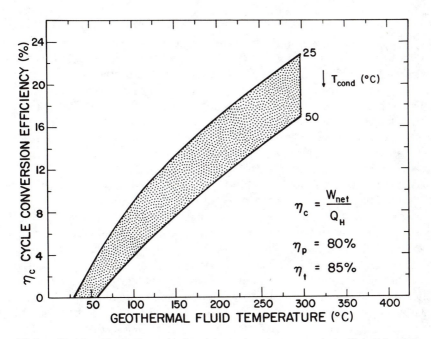

Figure 10-13.   Typical range of anticipated cycle conversion efficiencies for geothermal power plants.

$$\eta_u = \frac{P}{\left(T_{gf} - T_o - T_o \ln \left[\dfrac{T_{gf}}{T_o} \right] \right) \dot{m}_{gf} C_p} \tag{10-3}$$

where **P**   = net power extracted

$\dot{m}_{gf}$ = geothermal fluid flow rate

C_p = heat capacity.

Then, η_u is a direct measure of the effectiveness of resource utilization, because for a fixed T_{gf}, higher values of η_u correspond to lower well flow rates for a given power output. In contrast, the cycle efficiency, η_{cycle}, is a measure of how efficiently the transferred geothermal heat is converted into work.

If in this heat transfer process fluid is cooled to some intermediate temperature T_{gf}^{out}, then η_{cycle} becomes:

$$\eta_{cycle} = \frac{P}{\dot{m}_{gf} C_p \left(T_{gf}^{in} - T_{gf}^{out} \right)} \tag{10-4}$$

When a small amount of heat is being extracted from the geothermal fluid ($T_{gf}^{in} \cong T_{gf}^{out}$), η_{cycle} will be proportionally greater than η_u, because the resource is being utilized poorly. This is true because the availability of the discarded or reinjected fluid at temperatures above ambient is not considered as a thermodynamic loss in calculating η_{cycle}. The utilization efficiency concept is particularly useful in comparing binary-fluid and flashing cycle performance for the same resource conditions. Ideally, it is desirable to keep η_u as high as possible; however, there are limitations in the efficiency of work-producing machinery (turbines and pumps) as well as in the heat transfer systems associated with generating power.

Because the maximum availability of the geofluid (ΔB) increases as the resource temperature (T_{gf}) increases and as the sink temperature (T_o) decreases, high values of T_{gf} and low values of T_o should be maintained if possible. Because of the relatively low geothermal gradients that prevail in the eastern United States, there are economic limitations to maintaining a high T_{gf}, as discussed in subsequent sections. However, some areas have more favorable conditions for heat rejection, particularly where large quantities of water exist. This situation should be contrasted with prevailing conditions in the arid western United States, where the lack of water may require the use of dry cooling towers or direct air-cooled condensers. There are further limitations to these minimum and maximum temperatures. For example, any heat transfer step will require reasonable temperature differences to keep the equipment sizes at practical levels. Temperature differences are very important in the tradeoff between fluid utilization and the economics associated with producing power. (See "Engineering Design Criteria" and "Power Plant Performance — Related Effects" later in this chapter for further discussion.)

Thermodynamic Properties

Accurate data on the thermodynamic properties of proposed fluids are required to calculate electric power cycle performance. Heat capacity at constant pressure in the ideal gas state, C_p^*, vapor pressure, P^{sat}, pressure-volume-temperature (PVT) behavior, enthalpy and entropy changes, and liquid density at saturation, ρ_L^{sat}, can be expressed with semi-empirical equations. For example, a modified form of the Martin-Hou equation of state with 21 parameters has been used for calculating PVT properties (Martin, 1967; Milora and Tester, 1976; and Milora, 1973). This equation is accurate for densities above the critical point, a requirement for supercritical Rankine cycle calculations, and has been successfully applied to predicting the PVT properties of many fluorocarbons. Milora (1973) and Milora and Combs (1977) have also generated Martin-Hou parameters for isobutane and ammonia.

One disadvantage of the Martin-Hou equation is that it does not provide an explicit value of the compressed liquid and vapor densities at saturation. Thus, companion empirical equations for saturated vapor pressure and saturated liquid density are required in addition to the ideal gas-state heat capacity. By using a modified Benedict-Webb-Rubin (BWR) equation of state as developed by Starling (1973) some of these complications can be avoided. The Starling equation has 10 adjustable parameters and has been used extensively for correlating the PVT properties of many light hydrocarbons. Similar to the Martin-Hou equation, which is implicitly formulated in reduced specific volumes, the Starling equation is implicit in density, and iterative calculations must be used to obtain density roots for a given temperature and pressure. The smallest and largest roots correspond to the vapor and liquid densities at those conditions. As noted by Eskesen (1977a), an additional iterative procedure employing enthalpy and entropy departure formulation with the criteria of liquid-vapor equilibrium can be used to generate P^{sat}, ρ_L^{sat}, ρ_v^{sat}, and the enthalpies and entropies of vaporization. This requires only the Starling equation and an empirical representation of C_p^*.

Starling et al. (1974, 1977, 1978) and Holt and Ghormley (1976) have considered the use of organic mixtures in the design of binary-fluid cycles. Their treatment of the thermodynamic properties of the mixtures uses a modified BWR approach with a corresponding states formulation (Reid et al., 1977, and Meissner and Seferian, 1951).

In addition to organic fluids, the properties of water and steam containing significant quantities of noncondensable gases, such as CO_2, and/or dissolved materials, such as NaCl, SiO_2, KCl, and HCO_3^- — CO_3^{2-}, need to be specified. This involves temperature-enthalpy variations for heat exchange, as well as temperature-pressure-enthalpy-entropy relations for turbine expansions. Eskesen (1977a, 1980a,b) presents an elaborate treatment of the effects of noncondensables and dissolved salts on the thermodynamic properties of geothermal brines. In general, the presence of large amounts of dissolved, noncondensable gases in geothermal brines may prevent the use of condensing steam turbine cycles because of the excessive extractive pump work required to remove gas from the condenser. Dissolved salts not only present a potential scaling and corrosion problem but also reduce the specific enthalpy of the brine (J/kg) and subsequently reduce the amount of steam produced in a flash step. For the properties of essentially pure water and steam, tabular, numerical, and/or graphical representations such as those presented by Keenan and Keys (1968) can be used.

Derived properties such as entropy and enthalpy are obtained by suitable differentiation and integration of the semi-empirical equations for C_p^*, P^{sat}, $P = f(T,v)$, ρ_L^{sat}. An equation format is preferred to tabular by many investigators when iterative calculations are routinely performed (Eskesen,

1977; Milora and Tester, 1976; and Starling et al. 1974, 1977). Some investigators have used graphical or table look-up approaches (Shepherd, 1977).

Working Fluid Choice

Hydrocarbons, fluorocarbons, and other organic working fluids have been examined for potential use in low-temperature power conversion cycles as a replacement for water. For example, Milora and Tester (1976) selected refrigerants [R-22 ($CHClF_2$), R-600a (isobutane, i-C_4H_{10}), R-32 (CH_2F_2), R-717 (ammonia, NH_3) RC-318 (C_4F_8), R-114 ($C_2Cl_2F_4$) and R-115 (C_2ClF_5)] in their studies to provide a range of properties including critical temperature and pressure and molecular weight. All of these compounds have relatively high vapor densities at heat rejection temperatures as low as 20°C and would result in very compact turbines in comparison to steam turbines of similar capacity. Shepherd (1977) compares the performance of R-21 ($CHCl_2$), R-600a (isobutane) and Fluorinol (tri-fluoro-ethanol). Eskesen (1977a,b) considers a number of hydrocarbons including propane (R-290), a propylene (R-1270), isobutane, n-butane (R-600), iso-pentane, and n-pentane as well as R-115, R-22, R-32, and R-717.

Other factors besides desirable thermodynamic properties frequently determine practical working fluid choices. These include fluid thermal and chemical stability, flammability, toxicity, materials compatibility (corrosion), and cost (Milora and Tester, 1976). Unquestionably, the one major disadvantage of hydrocarbons, such as propane, pentane, and isobutane is their flammability, which necessitates costly explosion-proof equipment and ventilating systems. Below 300°C, the rate of dissociation of ammonia to hydrogen and nitrogen is relatively slow and should not limit its use in power cycle applications. The use of fluorocarbons is much more problematic. Eskesen (1977) cautioned those who may consider using fluorocarbons at temperatures above 125°C, because of potentially unstable behavior, especially in the presence of oil, steel, or copper, which may catalyze chemical decomposition. Extensive evaluation of a number of fluorocarbons for geothermal, solar, waste heat, recovery, and automotive applications by the Allied Chemical (Murphy, 1979) and Monsanto (Miller, Null, and Thompson, 1973) companies suggests, however, that acceptable operation may be possible at temperatures well above 200°C.

Thermal stability studies of fluorocarbons have been largely directed at their applications in hermetically-sealed, domestic and industrial refrigeration systems, which are expected to operate with little or no maintenance for 20-year periods (Callighan, 1977; Parmelee, 1968). Thus, these applications have imposed very stringent requirements and it is improper to evaluate fluorocarbon working fluids for power cycle applications on the same basis as for refrigeration systems. Hydrocarbon molecules nearly saturated with fluorine tend to be the most stable. For example, RC-318 (C_4F_8) and R-115 (C_2ClF_5) have been tested for extended periods at 200°C in the presence of iron, steel, and copper with no observable decomposition (Murphy, 1979). Very stable behavior is also observed when chlorine atoms are not present, for example in the case of R-32 (CH_2F_2). Even R-113 (CCl_3F_3), however, has been shown to be stable at 200°C for 10,000 hrs when exposed to carbon steel (Nichols, 1979). Another factor to consider is the important effect oil seems to have on thermal stability. Oil exposure occurs in most refrigeration systems. In geothermal power cycles fluorocarbon pressures are always above one atmosphere, so any leakage that occurs around seals, etc. will be out-leakage. Furthermore, as has been common practice in fossil-fired steam cycles, boiler feed makeup is used to replace lost or contaminated steam. Although costs associated with fluorocarbons are higher, some level of makeup should be economically feasible.

Engineering Design Criteria

Any real process for generating electricity or heat has inefficiencies or nonreversible steps that result in an amount of work less than W^{max} given in Equation 10-1. For example, neither the turbine nor the feed pump operates at 100% efficiency, and frictional losses in the various cycle components reduce the efficiency further. Any heat exchange step destroys availability. This step results in finite irreversibilities due to the finite temperature difference between each fluid necessary to transfer heat. Ideally, one could obtain more work from this process by operating a series of Carnot heat engines across any finite temperature difference present in the conversion process. Optimal operation with minimum irreversibility occurs when the heat capacities, C_p, of both streams are constant. This situation produces a balanced exchanger in which the temperature difference, ΔT, between streams can be kept uniform. As the working fluid operating pressure is increased above its critical pressure, distinct preheating, boiling, and superheating sections are avoided, the C_p becomes more uniform, and ideal operation is approached. Thus, the critical properties of alternate working fluids relative to the geothermal source temperature are important in the initial screening process. Efficient use of the resource may be necessary for commercial feasibility if reservoir development costs are high, i.e., if drilling costs are high and/or reservoir productivities are low. Efficient utilization will result when:

1. Most of the heat is extracted from the geothermal fluid before disposal or reinjection.
2. Temperature differentials across heat transfer surfaces are maintained at minimum practical levels.

3. Turbines and feed pumps are carefully designed for optimum efficiency.
4. Heat is rejected from the thermodynamic cycle at a temperature near the minimum ambient temperature T_o.

Whether or not these conditions can be met depends largely upon the choice of thermodynamic working medium, the geothermal fluid temperature, and the temperature of the coolant (water or air) to which the power plant rejects waste heat. For example, if waste heat is to be rejected from the thermodynamic cycle at a constant temperature by a condensing vapor (as in condition 4 above), then the compound's critical temperature must be greater than the temperature of the power plant coolant. As discussed later in "Power Cycle Performance," fulfillment of conditions 1 and 2 suggests the use of supercritical Rankine cycles. Thus, the critical temperature of the working fluid should also be below the maximum geothermal fluid temperature (Milora and Tester, 1976).

Heat Exchange and Prime Mover Fluid Production

As illustrated in Figures 10-2 through 10-9, a variety of heat transfer steps may be required in a geothermal electric or heating system to produce a suitable prime mover fluid. For example, for a binary-fluid cycle, a primary heat exchanger operating as a combined pre-heater, boiler, and superheater at subcritical pressures or as a vaporizer at supercritical pressures will be required to remove heat from the geothermal fluid and generate the organic vapor for injection into the turbine (see Figure 10-3). In a flashing cycle, prime mover steam is produced by reducing the pressure of the fluid in one or more steps. In both cases, the design and component sizing of the heat exchanger or flashing unit will require a specification of an appropriate duty factor for each unit (J/h, or kg steam generated/h) and temperature and pressure differences.

Because power conversion efficiencies are low, the amount of heat transferred may be 5 to 15 times greater than the power produced. Large heat exchangers will be required and their impact on cost is significant. For example, a 100-MW(e) geothermal plant with a 12% cycle efficiency requires about 60,000 m^2 (650,000 ft^2) of heat exchanger surface area. Thus, multiple, parallel units each having surface areas of 20,000 to 60,000 ft^2 would be utilized.

For a primary heat exchanger in a binary cycle, the overall heat exchange surface area, A, can be expressed as a function of the total heat load, Q, an appropriate mean $\overline{\Delta T}$, and an effective overall heat transfer coefficient, U:

$$A = \frac{Q}{U\Delta T} \qquad\qquad (10\text{-}5)$$

A practical ΔT should be selected to keep the surface area and, consequently, the cost at a reasonable level. In some cases, the exchanger would reach a "pinched" condition whereby ΔT would approach zero at one or more locations. Then, a minimum ΔT would have to be specified for any potential pinch points. These concepts are applied to a number of specific cases in the later discussions of power cycle thermodynamics and economics.

To estimate the surface area of any exchanger, realistic values for heat transfer coefficients of both fluids, tube wall resistance, and fouling factors may be required. Because many different geothermal brine compositions and working fluids operating in sub- and supercritical modes are under consideration, no uniquely optimum design can be proposed. This includes the type of heat exchanger, flow configuration, tube size, and baffling arrangements.

Inasmuch as the chemical compositions and temperatures of geothermal fluids vary widely from site to site, it is extremely difficult to estimate potential scaling or fouling problems for primary heat exchangers unless field tests are performed. For example, the 300,000 ppm total dissolved solids (TDS) and 370°C temperatures of Salton Sea geothermal brines in the Imperial Valley are considerably more difficult to deal with than the brines at Cerro Prieto, Mexico, with <20,000 ppm TDS and at 300°C temperatures (Kruger and Otte, 1973). Calcium carbonate ($CaCO_3$) and silica (SiO_2) scales have been observed in a number of natural hydrothermal systems.

Hot dry rock systems introduce another factor in that the circulating water is not indigenous to the reservoir. Dissolution products build up in the circulating fluid and might also present scaling problems, particularly at higher reservoir temperatures.

The nucleation and growth of scale deposits are complex and controlled by many factors including supersaturation, CO_2 partial pressures, contents and types of various ions (e.g., Mg^{+2}, Al^{+3}, $Fe^{+2 \text{ or } +3}$), hydrodynamic conditions (as specified by the Reynolds number, N_{Re} or by the presence of rapid acceleration or deceleration of the fluid), and heat transfer rates.

As discussed by Dart and Whitbeck (1980), only certain types of heat exchanger designs are appropriate for geothermal applications. These include shell and tube exchangers, fluidized bed exchangers, and direct contact exchangers. Direct contact and fluidized bed systems may be the only viable candidates when scale depostion rates are high. But even with moderate scaling problems, careful design and operation of shell and tube units are required. For example, geothermal fluid may be routed through the tube side. U-tube bundles would be avoided to ease cleaning and reduce material costs.

The conceptual design and sizing of shell and tube units have been considered by a number of investigators (Holt and Ghormley, 1976; Dart and Whitbeck, 1980; Milora and Tester, 1976; Pope et al., 1978; Giedt, 1975; Eskesen, 1977a; and Hankin et al., 1976). In almost all cases, the approaches have been standard following the classical design principles described by Perry and Chilton (1973), McAdams (1974), Fraas and Ozisik (1965), Kern (1950), and Rosenhow and Hartnett (1973). In a rather unique approach, Pope et al. (1978) suggest that the fouling factor can be "optimized." Differences between heat transfer coefficients under sub- and supercritical flow conditions may be important and inasmuch as data and analyses have been limited to just a few studies (for example, Bringer and Smith, 1957; Sastry, 1974; and Koppel and Smith, 1961), system designs should be based on actual data for a fluid of interest.

In a typical shell and tube exchanger design, individual shell and tube side coefficients would be calculated from existing empirical correlations with specified fluid velocities to limit pressure drops and correction factors applied to the shell side coefficient for baffle configuration, baffle leakage, and tube bundle bypassing effects. Fouling factors should be used and in severe cases of fouling would actually consitute the controlling resistance. In cases treated by Milora and Tester (1976), Eskesen (1977a), and Dart and Whitbeck (1980), fouling on the geothermal fluid side was assumed to be equivalent to a range of coefficients from 570 to 2,480 W/m²K (100 to 500 Btu/hr ft² °F) between cleaning periods. Because the organic working fluids used will be relatively free of contamination, equivalent fouling coefficients of 10^4 to 10^5 W/m²K (2,000-20,000 Btu/hr ft² °F) would be appropriate. For typical shell and tube designs, pressure drops would be limited to 1 to 10% of the turbine inlet pressure with tube diameters ranging from 1.90 cm (0.75 in.) to 2.54 cm (1 in.). Overall coefficients with supercritical flow on the shell side and pressurized liquid geothermal fluid in the tubes would range from 740 to 900 W/m²K (130-160 Btu/hr ft² °F). In subcritical designs, separate preheat, boiling, and superheat sections would be used with coefficients applicable to each section calculated. For most of the geothermal heat exchange systems, however, overall coefficients will generally range from 400 to 1,400 W/m²K (70 to 250 Btu/hr ft² °F). Consequently, because of the relatively large heat duties and low values of effective ΔT, surface area requirements are large.

When fouling problems are extremely severe, direct contact and fluidized bed systems may provide designs for maintaining high heat transfer coefficients between the organic working fluid and the geothermal brine (Wahl, 1977; Boehm et al., and Allen et al., 1976; and Urbanek, 1978). In direct contact systems, more efficient heat transfer occurs when a vaporizing organic working fluid is in contact with the geothermal brine. The absence of heat transfer surfaces avoids scaling altogether so that performance should not decline with time. Even though organic fluids with very low solubilities in water can be selected, finite losses will occur and chemical stripping of the brine to recover the organic fluid may be required before it is reinjected or discarded. Elimination of noncondensable gases will be necessary as will careful consideration of additional pumping requirements if geothermal fluid pressures do not match organic fluid design pressures. Fluidized bed exchangers utilize particle motion past surfaces to improve heat transfer rates by reducing film thicknesses and by impeding the nucleation and growth of scale. Research studies have attempted to establish optimum particle sizes and compositions (Allen et al., 1976). Commercial size units for either direct contact or fluidized bed heat exchange with geothermal fluids have not yet been built, but considerable testing is underway. A very complete discussion of both concepts is contained in the monograph edited by Kestin et al. (1980).

When steam is the prime mover fluid and is produced by flashing and used directly in a condensing turbine as shown in Figure 10-2, obviously no primary heat exchange step is required (see Armstead, 1978 for more details). In its place, pressure reduction is used to produce a saturated vapor fraction. In a typical installation, the first stage of flashing is done at the wellhead with a two-phase mixture produced, which is piped to a separator system. Frequently, the latter is a cyclone unit, where the vapor (steam) and liquid brine are separated. The steam fraction will be transmitted to the turbine-generator unit located in a central area, and the liquid fraction can be reinjected or discarded or flashed to a lower pressure to produce an additional amount of steam for turbine injection. This process can be continued as many times as economically feasible. A two-stage flash system may produce 20 to 30% more available work than a single stage unit at the same total geothermal fluid flow rate. But as Eskesen (1980a) points out, the improved performance of multistage systems must be balanced against the additional complexity and cost of utilizing the secondary steam. For example, in a two-stage system, dual admission steam turbine units may be the economical choice. Operating difficulties encountered by such factors as high intrusion losses at the secondary admission points, however, may require a practical choice of separate turbine units to accept first- and second-stage steam flows.

Turbines

In selecting nonaqueous working fluids for power cycle applications, turbine sizes must be small to reduce costs, because of the economic tradeoff between the additional heat exchange surface area required for binary-fluid cycles and the much larger and more costly turbines required in steam flashing systems.

It is important to operate turbines at high efficiency. A similarity analysis of performance shows that turbine efficiency is essentially controlled by two dimensionless numbers involving four parameters: *Blade pitch diameter; rotational speed; stage enthalpy drop;* and *volumetric gas flow rate* (Baljé, 1962; Eskesen, 1977a; Shepherd, 1977). This assumes that Mach and Reynolds numbers effects can be neglected. For operation at maximum turbine efficiency the relationship among these parameters is specified; therefore, turbine sizes and operating conditions and, consequently, costs can be estimated. For fluid screening purposes, a generalized figure of merit, ξ, which scales directly with turbine size was developed by Milora and Tester (1976). The ξ is expressed as an explicit function of the fluid's molecular weight m, critical pressure P_c, and reduced saturated vapor energy density $(h_{fg})_r/(v_g)_r^{sat}$. The h_{fg}/RT_c is the reduced latent heat $(h_{fg})_r$ and $(v_g)^{sat}_r = v_g/v_c$ is the reduced gas specific volume evaluated at the condensing temperature, T_{cond}. Table 10-2 presents values of ξ for water and a number of hydrocarbon and fluorocarbon working fluids studied by Shepherd (1977), Eskesen (1977a,b), and Milora and Tester (1976).

The organic fluids described in Table 10-2 obviously offer a significant reduction in turbine size. For example, a 100-MW(e) capacity plant designed for a 150°C liquid-dominated resource would require numerous large turbine exhaust ends with steam flashing, whereas a single exhaust end would be possible in an ammonia binary cycle (Troulakis, 1968). It should again be emphasized that organic turbines have not been manufactured with capacities larger than about 1-MW(e). In contrast, low pressure steam turbines for geothermal operation with 50-MW(e) capacities or greater have been commercially produced for a number of years. In fact, only one organic binary cycle using geothermal water at 80°C has been in commercial operation (Moskvicheva, 1971; and Vymorkov, 1965). Several small systems (~100 kVA) have been built by Barber-Nichols for geothermal demonstrations, including an isobutane binary, a direct-contact isobutane binary, and a R-114 binary (Nichols, 1979; and Olander et al., 1979). A number of other small systems have been constructed for solar Rankine cooling and waste heat recovery systems using hydrocarbon and fluorocarbon working fluids. Nevertheless, Holt and Ghormley (1976) and Eskesen (1977a, 1980a) suggest that axial and radial flow turbines designed for capacities up to 50 MW(e) with nonaqueous fluid service will require only modest engineering developments, particularly in comparison to other new energy alternatives. For example, Eskesen (1980a,b) discusses the need for careful design to keep inlet and exhaust pressure losses at a minimum as well as some development of seals. At this point, one should be optimistic about the technical and economic feasibility of organic Rankine systems.

Table 10-2
Turbine Size Figure of Merit ξ at Various Condensing Temperatures (T$_{cond}$)

		ξ	
Compound	Formula	T$_{cond}$=26.7°C(80°F)	T$_{cond}$=37.8°C(100°F)
R-717	NH$_3$ (ammonia)	0.177	0.133
R-32	CH$_2$F$_2$	0.223	0.173
R-1270	C$_3$H$_6$ (propylene)	0.258	0.204
R-290	C$_3$H$_8$ (propane)	0.327	0.249
R-22	CHClF$_2$	0.411	0.308
R-115	C$_2$ClF$_5$	0.649	0.487
R-600a	C$_4$H$_{10}$ (isobutane)	0.881	0.734
R-600	C$_4$H$_{10}$ (n-butane)	1.114	0.711
RC-318	C$_4$F$_8$	1.628	1.170
Isopentane	C$_5$H$_{12}$	3.210	2.490
n-pentane	C$_5$H$_{12}$	3.770	2.639
R-21	CHCl$_2$F	1.962	1.413
Fluorinol	C$_2$Cl$_3$H$_2$OH		
	(trichloroethanol)	5.090	3.902
R-114	C$_2$Cl$_2$F$_4$	2.246	1.604
Water	H$_2$O	30.71	17.47

After Milora and Tester, 1976

$$\xi \propto \text{Turbine exhaust end area} \quad = \quad \frac{\sqrt{m}\ (v_g^{sat})_r}{P_c\ \ (h_{fg})_r}\ T_{cond}(g/g\ mole)^{1/2}bar^{-1}$$

Detailed discussions of steam and organic turbine design criteria are contained in recent studies by Eskesen (1980a,b), Shepherd (1977), and Milora and Tester (1976). Shepherd (1977) and Milora and Tester (1976) consider generalized design criteria including a detailed similarity analysis of performance adapted from Baljé (1962). In particular, the interrelationship between turbine physical characteristics, such as stage volumetric flow capacity (V_i), rotational speed and stage pitch diameter (D_p), and blade height (h), and the thermodynamic properties of the working fluid, expressed as stage enthalpy drop (ΔH_i), are discussed in the context of optimizing efficiency.

The Eskesen study is perhaps the most comprehensive work on the subject available today where performance characteristics, flow configurations, stage design and sizing, selection of materials of construction, and valve and control design are presented in detail for both axial flow steam and organic turbines. In comparing the performance of low-pressure steam turbines to organic units of similar capacity, the large differences in fluid densities at exhaust conditions, moisture content during expansion, and stage enthalpy drops or isentropic expansion work lead to significant differences in design. For example, in low-pressure condensing steam turbines it is necessary to increase exhaust end flow capacity, at the expense of efficiency, by allowing the ratio of blade height to pitch diameter (h/D_p) to exceed its optimum performance value of 0.1. In large capacity units, blade lengths approximately one-third as long as the last stage pitch diameter are commonly found (Eskesen, 1977a, Table 19). In addition, wet expansions in geothermal flashing systems can lead to severe turbine bucket erosion by liquid droplet impingement (Salisbury, 1950, 1974). Furthermore, wet expansions also cause a decrease in efficiency. Consequently, for geothermal steam turbines, moisture separators at each stage are commonly used. These problems can be avoided completely in organic binary systems by proper selection of working fluids and operating conditions to ensure that the turbine expansion is outside of the liquid-vapor region.

Pumps

Pumping requirements for geothermal systems span four major areas: (1) downhole pumps for fluid production; (2) fluid reinjection pumps; (3) power cycle feed pumps; and (4) geothermal fluid transport pumps. A complete survey and assessment of existing geothermal pumping equipment (Nichols and Malgieri, 1978) concluded that conventional multistage centrifugal designs can be used for commercial-sized applications (2)-(4) in most cases. Reinjection pumps may present some problems where highly corrosive and erosive geothermal brines are used. Unquestionably, the most challenging problems are concerned with downhole pumping, however, because of the severe environmental and design constraints. Downhole applications include high-temperature, high-salinity brine resources with restricted wellbore sizes, typically with 32 to 100 kg/s (500 to 1,500 gpm) flow capacity required at depths of 2,000 ft or more. Downhole pumping may be required to increase production flow to economically acceptable levels, as well as to prevent flashing within the wellbore, which may in some cases lead to carbonate scaling and CO_2 release. In hot dry rock reservoirs, downhole pumps may be used to reduce formation water losses by lowering effective downhole pressures.

Four general types of downhole pumping systems have been proposed. These include (1) shaft-driven systems; (2) submersible electric motor-driven multistage, vertical centrifugal designs, which are commercially available and have been tested by a number of companies; (3) steam- or organic-fluid-driven, downhole turbine pumps; and (4) hydraulically-driven systems.

As cited by Nichols and Malgieri (1978), manufacturers of line-shaft driven units include Peerless, Johnston, Bryan Jackson, Floway, and Layne and Bowler. LMC and TRW (Reda Division) produce electric motor-driven units. Both types have practical limitations. The complexities associated with deeply-located shaft pumps, driven mechanically on the surface, are obvious. Line-shaft bearings for high temperature service are also another weak structural factor. Electrically-driven submersibles present serious problems when operating temperatures are above 180°C, including protection of the motor from heat, moisture, and the brine itself.

In two other systems, steam- or organic-fluid-driven, downhole turbine pumps (Sperry Research, 1977; Nichols et al., 1977; and McBee and Matthews, 1979) and hydraulically-driven systems (Sunstrand and Worthington, in: Nichols and Malgieri, 1978), the problems of shaft or electric motor-driven units are replaced with a fairly complex concentric wellbore piping system (Sperry Research, 1977). The downhole turbine system, however, may offer some distinct advantages for high temperature operation at high efficiencies in that a small fraction of the available thermal energy of the brine is used to vaporize water or an organic fluid, which then expands through a downhole turbine to provide power for geothermal brine pump (Matthews, 1976). A three concentric pipe string arrangement is required to accommodate the geothermal brine flow; organic fluid or water injection; organic vapor or steam turbine exhaust; and a water line for bearing lubrication. The turbine prime mover is powered by a separate loop using steam or organic fluid with a surface condenser and feed pump to avoid contamination with the brine.

Although the steam turbine pump concept was successfully demonstrated in short field tests conducted in December, 1976 at temperatures above 200°C using a test well at Heber, California, problems were encountered with well productivity, downhole boiler performance, and some sand erosion on the leading edge of the pump impellers. Sperry Research (1977) suggests that a 10,000-h life-time should be possible with satisfactory operation in the case of resources having temperatures above 180°C (360°F) and 240°C (460°F), and commercial development is underway. For resources having temperatures of 180°C or less, steam turbine pumps are inefficient and require large exhaust areas; consequently, organic working fluids offer an alternative (McBee and Matthews, 1979).

One aspect of power cycle feed pumps should be emphasized. In organic binary systems, pumping requirements can be an order of magnitude or more greater than those for conventional boiler feed pumps in steam cycle applications. High pump efficiencies, therefore, should be maintained whenever possible. Typical efficiencies encountered range from 70-80% with 85% as a probable upper design limit. Organic feed pumping requirements for a specified capacity plant can be estimated using a generalized correlation developed by Milora and Tester (1976), which relates the power required to the compressibility factor at the critical point; pump and cycle efficiencies; reduced condensing temperature; and reduced pressure drop across the pump.

Condensers and Heat Rejection

Because conversion efficiencies are low, a large fraction of heat transferred will be ultimately rejected to the environment. For example, a geothermal plant operating with a cycle efficiency of 10% will reject about 3 to 6 times as much heat as a similar capacity fossil- or nuclear-fired unit. Consequently, costs associated with condenser-desuperheaters and cooling towers in geothermal systems are substantial, and strongly influence conditions for economic feasibility. In addition, seasonal and diurnal variations of ambient dry bulb temperatures can significantly affect cycle performance (Khalifa, 1980). If dry cooling is used, cycle condensing temperatures will be higher than for wet cooling and will change seasonally causing changes in the power output of the plant. Because of the smaller size of geothermal units [<100 MW(e)], it might be desirable to operate with a floating power output. Khalifa has shown that with a seasonal variation of ±15°C, the power loss associated with an increase in ambient temperature is fully compensated by the power gain associated with an equivalent decrease in ambient temperature. For this fluctuation of ±15°C and a 200°C geothermal resource, Khalifa estimated that the power output would vary by ±32%.

Because costs associated with producing the geothermal fluid are at least equal to or greater than the plant costs themselves, a premium is placed on optimizing cycle performance by using lower ambient temperatures when and where environmental conditions permit. For example, for a 200°C liquid resource, a decrease in condensing temperature from 49 to 27°C increases the potentially available work by as much as 40%. Methods of vapor condensation and ultimate heat rejection to the environment present difficulties. First, the inherent differences that exist between the large, low-temperature vapor densities of selected nonaqueous working fluids described earlier and the smaller densities of condensing steam create different design criteria. It may be economically undesirable to build and operate steam turbines at condensing temperatures much below 37.8°C (100°F) because of the large turbine exhaust end areas required. Consequently, condensation and heat rejection systems can be designed around these conditions. If low ambient temperatures exist at the site, however, organic binary systems may be able to operate more efficiently by condensing at temperatures below 37.8°C. This particular situation does exist in several parts of the world, for example, the Valles Caldera KGRA of New Mexico, U.S.A., where the average ambient temperature is ~3°C.

Another concern is the availability of water for cooling. In many cases, fossil and nuclear plants are located near abundant water supplies (estuaries, lakes, oceans). Of course, geothermal plants have to be located at the resource regardless of the water situation. Even so, if evaporative cooling is used, water consumption rates are high. For example, a 50-MW(e) geothermal plant operating at 10% efficiency would consume cooling water at a rate of approximately 250 kg/s (3,800 gpm). In many existing plants, cooling water is provided by the geothermal fluid itself. For example, evaporative cooling at The Geysers consumes approximately 80% of the geothermal fluid, so only 20% is reinjected (Kruger and Otte, 1973). For liquid-dominated and geopressured reservoirs, total reinjection may be required to avoid subsidence or for other environmental or regulatory reasons. Furthermore, the brine may have to be treated before reinjection. Normally, hot dry rock systems will also operate with total reinjection (Tester and Smith, 1977). For these resources, auxiliary makeup water will be required for evaporative cooling and formation losses. If cooling water is not readily available, more costly wet/dry or completely dry cooling designs will be used.

Robertson (1978, 1980) in his definitive treatise on waste heat rejection from geothermal power stations, cited a number of design options that are possible. These include direct contact spray condensers; multipass water-cooled surface condensers; direct air-cooled condensers; spray ponds; and wet, wet/dry, and dry cooling towers of forced, induced, or natural draft design. Some of these are schematically represented in Figures 10-14 through 10-19. Robertson discusses these designs in great detail, and interested readers should refer to his comprehensive review of this area of importance to low temperature power conversion.

The most significant engineering issue concerning heat reinjection is the selection of an optimum design and operating conditions for the condenser/desuperheater and heat dissipation equipment. This choice will depend on power cycle type; working fluid and geothermal fluid properties; fluid reinjection requirements; dry and wet bulb ambient temperatures; water availability and cost; and environmental and regulatory constraints.

For direct steam injection and flashing cycles it has been common practice to use low-level or barometric-leg, direct-contact condensers with mechanical draft evaporative cooling towers as, for example, at The Geysers, and Cerro Prieto (Armstead, 1978; and Kruger and Otte, 1973). Direct contact systems are simple in design, inexpensive, and easy to maintain (see Figure 10-14). As with any direct contact system, approach temperatures are excellent and performance is not affected by fouling as would be the case with the vertical, water-cooled condenser shown in Figure 10-15.

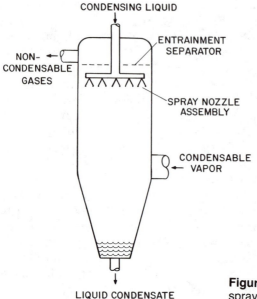

CONDENSING LIQUID

ENTRAINMENT
SEPARATOR

NON-
CONDENSABLE
GASES

SPRAY NOZZLE
ASSEMBLY

CONDENSABLE
VAPOR

LIQUID CONDENSATE

Figure 10-14. Direct contact
spray condenser schematic.

Once-through cooling systems, spray ponds, and similar systems with
water-cooled condensers will probably be only remote possibilities for most
geothermal applications because of their environmentally unacceptable
conditions: high water usage, discharged heat in waterways, impingement of
fish on inlet screens for once-through systems, and excessive land use
requirements for spray cooling systems. Even evaporative wet cooling
towers, present environmental problems, e.g., water droplet drift, liquid
disposal from blowdown, icing, and fog formation. The choice between
mechanical (forced) and natural draft evaporative towers is largely based on
prevailing meteorological conditions at the site and cost (see Figures 10-16
and 10-17). For example, as Robertson (1978) points out, low average
wet-bulb-temperatures and high relative humidities, large capacity systems,
and problems with ground-level fogging might favor the use of natural draft
systems. Although their capital costs are lower, the auxiliary power requirements and greater complexity of forced draft systems will cause operating
and maintenance costs to be much larger. The large imposing, hyperbolic
shape of typical natural draft designs can create visual pollution problems in
some areas.

For many power cycle operations evaporative cooling has been used
commonly and will no doubt continue to be employed in future designs.
When total reinjection is required and water is scarce or expensive, how-

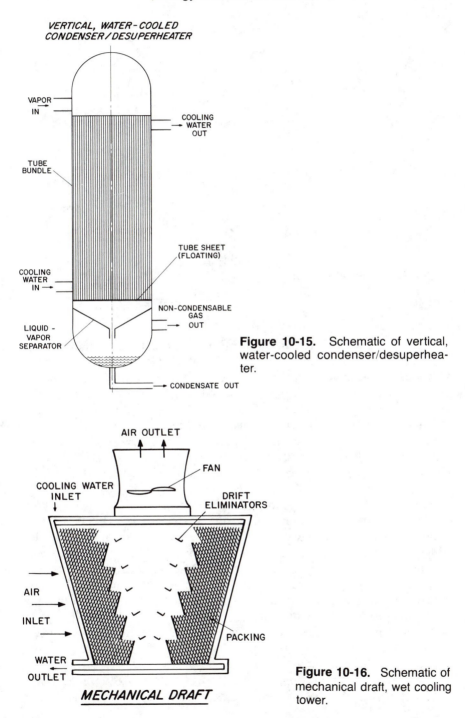

Figure 10-15. Schematic of vertical, water-cooled condenser/desuperheater.

Figure 10-16. Schematic of mechanical draft, wet cooling tower.

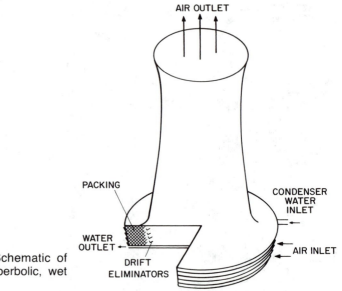

Figure 10-17. Schematic of natural draft, hyperbolic, wet cooling tower.

ever, dry and wet/dry cooling are the only alternatives. Although high capital costs have prevented widespread use of completely dry systems in the U.S., many European systems have been operating successfully (Miliaras, 1974). At least one major 330-MW(e) facility is operational in the U.S. in Gillette, Wyoming (Robertson, 1980), however, and more might be anticipated in future designs in arid regions. Both direct and indirect dry cooling systems are possible. As shown in Figure 10-18, extended-surface, forced-draft, air-cooled exchangers, which serve as working fluid condensers/desuperheaters as well as heat dissipation units, would be typical of direct dry cooling systems employed today.

Indirect units involve an intermediate condensing step, which may be itself of direct contact or shell and tube design, followed by heat dissipation in an air-cooled, extended surface unit. The major disadvantage, besides higher cost, of dry systems is that performance of the power cycle now depends on dry-bulb temperature fluctuations. If floating power capacity is an acceptable operating mode for small base-load units, then it is conceivable that variations in power output will not be a serious detriment to using totally dry systems. A small 60-MW(e) prototype system of this type, using an R-114 binary cycle, has been constructed for use at the Fenton Hill Hot Dry Rock demonstration site, which has a $3 \pm 20°C$ ambient dry-bulb temperature (Olander et al., 1979).

An alternative to totally dry systems is a wet/dry cooling tower, as shown in Figure 10-19, which combines advantageous features of each type. For

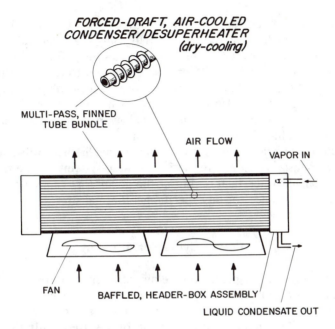

Figure 10-18. Schematic of extended surface, forced-draft, air-cooled condenser/desuperheater (dry-cooling).

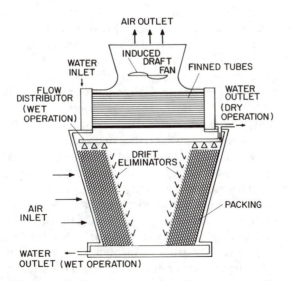

Figure 10-19. Mechanical draft wet/dry cooling tower schematic. (Adapted by permission from Robertson, 1978).

example, low condensing temperatures can be maintained during peak periods with high dry-bulb temperatures by using supplemental evaporative cooling. In addition, water can be conserved during periods of low dry-bulb temperature. Fogging and drift are also better controlled. Wet/dry towers, however, cost more and are somewhat more difficult to operate and maintain. Robertson (1980) also discussed other modifications of wet/dry systems. These include deluge cooling of dry towers during peak load periods and phased cooling for diurnal fluctuations.

The outline of heat reinjection alternatives that has been presented in this section emphasizes that careful design of these systems will be required to minimize busbar (electrical output at the generator) generating costs and optimize performance for any specific geothermal site.

Power Cycle Performance

Proper perspective is placed on the overall problem of power conversion of geothermal resources if one keeps in mind the relatively small operating range of temperature that exists. Most of the resources being considered have wellhead fluid temperatures of 250°C or less, frequently 150°C or less. Given that heat rejection ultimately occurs at ambient conditions of 25 to 35°C, this leaves only about 225° to 115°C of temperature differential for operating a cycle to produce power (see Figure 10-1). Because of these generally low temperatures and small differentials, even the maximum amount of power produced by an ideal, reversible Carnot cycle is small. Consequently, for economic reasons, a premium is placed on designing systems to operate efficiently. As discussed earlier in the section on "Thermodynamic Criteria — Efficiency and Available Work," the utilization efficiency η_u provides a good measure of performance for evaluating cycle designs.

In principle, η_u can assume any value between zero ($\mathbf{P}=0$) and unity ($\mathbf{P}=\dot{m}_{gf}W^{\max}$). In practice, however, its value is determined from economic considerations by balancing the cost of obtaining the heat (drilling and piping costs) against the cost of processing it (heat exchangers, condensers, turbines, pumps) to generate electricity in the power station. If η_u is small, then the resource is being utilized poorly and a large investment in wells is required (cost per unit power $\to\infty$ as $\eta_u\to0$). On the other hand, in the case of utilization of the full potential of the resource, then total well costs will be lower, but the required investment in highly efficient power conversion equipment will be large (cost per unit $\to\infty$ as $\eta_u\to1$). The economic optimum occurs when η_u assumes some intermediate value; for example, at The Geysers a value of 0.55 is typical at $T_0=26.7$°C (80°F). In the examples that follow, the emphasis is placed on maximizing η_u within reasonable limits because the relative cost of obtaining geothermal heat may be high, particularly for low-temperature, liquid-dominated resources. The η_u will be expressed in terms of performance of major plant components, including heat exchangers, condensers, turbines, and pumps.

In a typical analysis of cycle performance, a set of equations is developed to describe the work and heat flow rates to and from the major plant components (Eskesen, 1977a; Shepherd, 1977; Milora and Tester, 1976; Holt and Ghormley, 1976; and Pope et al., 1977). The net power, P, of the plant is specified as is the ambient heat rejection temperature. Heat exchanger performance is also regulated by limiting minimum approach temperatures to some specified value. Air or liquid-cooled condenser conditions are also controlled in a similar manner to specify the condenser outlet working fluid temperature at some value above T_0. After selecting a working fluid and a geothermal fluid inlet temperature, the major independent design variables are maximum cycle operating pressure, P, at the turbine inlet, condensing temperature, and heat exchanger approach temperatures.

Space limitations prohibit an extensive discussion of the many excellent studies of cycle performance that have appeared in recent years. Interested readers should examine the literature cited in Table 10-3 in detail. In many of these studies, detailed cycle calculations have been performed to examine the effects of the cycle operating conditions cited earlier as well as other design and resource parameters, such as turbine and pump efficiencies and geothermal fluid composition, particularly levels of salinity and noncondensable gas concentration. In each case, a utilization efficiency can be determined to evaluate and compare cycle performance under different operating conditions.

Other approaches have defined analogous parameters to compare performance. For example, Eskesen (1977) used specific output (work produced/mass of geothermal brine, J/kg or Btu/lb). An example of this approach is shown in Figure 10-20, where a cycle efficiency-specific output diagram is shown for a propane binary cycle operating over a range of brine temperatures from 190°C (375°F) to 275°C (525°F) and turbine inlet pressures from 52 bars (750 psia) to 241 bars (3,500 psia). Since the critical pressure of propane is 42.6 bars (617 psia), cycles considered are supercritical. Single and multiple parametric effects have been explored using computer computational techniques (Milora and Tester, 1976; Eskesen, 1977a,b; Starling et al., 1977, 1978; Holt and Ghormley, 1976; Pope et al., 1977).

A specific example is included here for a 150°C liquid-dominated resource at a condenser outlet temperature, T_{cond}, of 26.7°C and with R-115 as the working fluid. This resource condition may be typical of many in the U.S. where cooling water is available with low ambient, wet-bulb temperatures (~3°C). For this case, heat exchanger and condenser performance is speci-

Table 10-3
Studies of Low-Temperature Electric Power Conversion Cycles

Investigators	Geothermal reservoirs examined	Power cycle designs	Working fluids
1. Aikawa and Soda (1975)	Liquid-dominated: Hatchobaru, Japan	Two-stage flash, dual-admission turbine	Water
2. Anderson (1973)	Liquid-dominated: general	Binary cycle	Isobutane
3. Anno, Dore, Grijalva, Lang, and Thomas (1977)*	Liquid-dominated: Roosevelt Hot Springs, Utah; Coso, East Mesa and Long Valley, California	Hybrid fossil-geothermal cycle	Water
4. Aronson (1961)	General	Topping — bottoming cycle	Water — ammonia, etc.
5. Austin et al. (1973, 1980)	Liquid-dominated, high salinity: general	Total flow expander	Water
6. Bechtel Corporation (1975, 1976)* and Hankin et al., (1975, 1977)*	Liquid-dominated: Heber and Niland, California	Multistage steam flashing and binary cycles	Water, isobutane
7. Bloomster et al. (1975a, b, 1976)* Walter et al., (1975, 1976)	Liquid-dominated: general	Multistage steam flashing and binary cycles	Water, hydrocarbons and fluorocarbons
8. Chou et al., (1973, 1974)	Liquid-dominated: 150-250°C	Binary regenerative cycle	Isobutane
9. Cortez, Holt, and Hutchinson (1973)*	Liquid-dominated: Imperial Valley, California	Binary cycle	Isobutane
10. Dan et al. (1975)*	Vapor-dominated: The Geysers, California (180°C, 8 bar)	Direct steam injection cycle	Water

Table 10-3 continued

Investigators	Geothermal reservoirs examined	Power cycle designs	Working fluids
11. Elliott (1976)	Liquid-dominated: general	Multistage flash, flash/binary, binary and total flow cycles	Water, isobutane
12. Eskesen (1977a,b,* 1980a,b)	Liquid and vapor dominated: general (150-300°C; with and without CO_2: high and low salinity)	Multistage flashing and steam injection, binary, dual-flash binary cycles	Water, hydrocarbons, fluorocarbons; including pentane, isobutane, butane, propane, propylene, R-32, R-114, R-115, R-22 and ammonia
13. Gault et al. (1976)*	Geopressured with methane; Gulf Coast, U.S.A. (163°C, 138 bar)	Hydraulic turbine, flashing and binary cycles	Water, isobutane
14. Green, Pines, Pope, et al., (1975, 1976, 1977)*	Liquid-dominated: general (100 to 300°C, 0.3% non-condensables)	Multistage steam flashing and binary cycles — multiparameter optimization	Water, hydrocarbons and fluorocarbons; including isobutane, and propane, R-22, R-113, and ammonia
15. Guiza (1975)	Liquid-dominated: Cerro Prieto, Mexico (344°C, 5.3 bar)	Single-stage flashing cycle	Water
16. Hansen (1964)	Liquid-dominated: general (200°C)	Multistage flashing cycle	Water
17. Holt and Ghormley (1976)*	Liquid-dominated: Heber, California (180°C); Raft River, Idaho (150°C); Valles Caldera, New Mexico (260°C)	Multistage flashing and binary hybrid (flashed steam and organic) cycles	Water, light hydrocarbons (isobutane, propane, and isopentan and mixtures)

Table 10-3 continued

Investigators	Geothermal reservoirs examined	Power cycle designs	Working fluids
18. House, Johnson, and Towse (1975)*	Geopressured with methane; Gulf Coast, U.S.A.	Hydraulic and total flow cycles	Water
19. Jonsson, Taylor, and Charmichael (1969)*	Liquid-dominated: Iceland conditions	Binary fluid cycle	Fluorocarbons (R-21 in particular)
20. Khalifa (1980b)*	Liquid-dominated: general	Hybrid fossil — geothermal systems	Water
21. Kingston et al. (1978)*	Liquid and vapor dominated: Western US	Hybrid fossil — geothermal systems	Water
22. Kunze et al. (1975, 1976)*	Liquid-dominated: Raft River Idaho (150°C)	Dual boiler, binary cycle	Isobutane
23. Lavi (1973, 1975)	Ocean thermal	Binary cycle	Ammonia
24. McKay (1977)	Liquid-dominated, high salinity	Helical screw expander	Water
25. Milora and Tester (1976)*	Liquid-dominated and hot dry rock: general (100 to 300°C)	Multistage flashing, single fluid, dual, and topping and bottoming binary cycles; Brayton cycles	Water, hydrocarbons, and fluorocarbons; including R-22, R-114, R-115, RC-318, R-32, isobutane, and ammonia
26. Moskvicheva (1971), Popov (1970), and Vymorkov (1965)	Liquid-dominated: Paratunka River, USSR (80°C)	Binary cycle — 750 kW(e)	Fluorocarbon R-12
27. Olander et al. (1979)	Hot dry rock — Fenton Hill, NM	Binary cycle with floating power and dry cooling	R-114

Table 10-3 continued

Investigators	Geothermal reservoirs examined	Power cycle designs	Working fluids
28. Ramachandran et al. (1977)*	Liquid and vapor-dominated: California	Direct steam injection, flashing multistage flash — binary, and binary cycles	Water, hydrocarbons
29. Sapre and Schoeppel (1975)*	General	Flashing and binary cycles	Water, isobutane
30. Shapiro and Hajela (1977)*	Liquid-dominated: general high and low salinity brines (160-200°C)	Multistage flashing, binary, and multistage flash — binary cycles	Water, isobutane, R-113
31. Shepherd (1977)	Liquid-dominated: general (150-323°C)	Binary cycles	Isobutane, R-21, Fluorinol (trifluoroethanol)
32. Starling et al. (1974, 1977, 1978)*	Liquid-dominated: general (150-205°C)	Binary cycles with mixtures	Hydrocarbon mixtures
33. Swearingen (1977)	Liquid-dominated: high salinity; general	Binary cycles with specialized packed bed heat transfer	Hydrocarbon mixtures
34. Tester, Morris, Cummings, and Bivins (1979)*	Hot dry rock resources: general (100-300°C)	Binary cycles	Hydrocarbons and fluorocarbons
35. Walter et al. (1975, 1976)*	Liquid-dominated	Binary cycles	Hydrocarbons, fluorocarbons
36. Wilson et al. (1975, 1977)*	Geopressured US Gulf Coast	Flashing with hydraulic turbines, methane recovery process	Water, methane

See Kestin et al. (1980) and Armstead (1978) for excellent reviews of current technology and hardware.
*Economic factors considered.

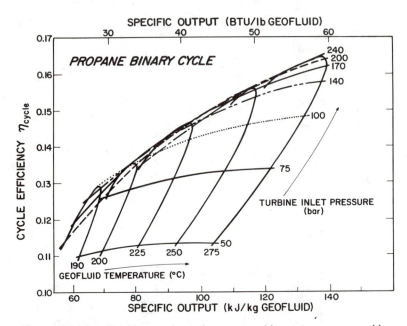

Figure 10-20. Rankine cycle performance with propane as a working fluid. Effect of geothermal fluid temperature and turbine inlet pressure shown with $\eta_p = \eta_t = 0.85, \Delta T_{pinch} = 14°C$ and $T_{cond} = 37.8°C$ (100°F). (Adapted by permission from Eskesen, 1977a).

fied by limiting minimum approach temperatures to 10°C and fixing dry turbine stage efficiencies at 85% and feed pump efficiencies at 80%. In addition, the working fluid outlet temperature is specified at 15°C below the geothermal fluid inlet. Consequently, the mass flow rate ratio of geothermal to working fluid and the geothermal fluid reinjection temperature, as determined by the primary heat exchanger pinch condition, are fixed, because turbine inlet and condensing temperatures are fixed. Alternatives are also possible. For example, Eskesen (1977a) fixes the geothermal fluid flow rate, iterates to establish the pinch point, and then determines the turbine inlet temperature. He repeats this procedure until a maximum in specific output is reached. A preliminary analysis indicates that these approach temperature values are within the range commonly experienced for near optimum economic performance (see "Power Plant Performance — Related Effects" later in this chapter).

Temperature/enthalpy diagrams for four different operating pressures are presented in Figure 10-21 for a R-115 binary cycle to illustrate the dramatic effect that pressure has on cycle performance. In going from a subcritical cycle at 27.5 bars (Case A, $P_r = 0.87$) to a supercritical cycle at 80.1 bars

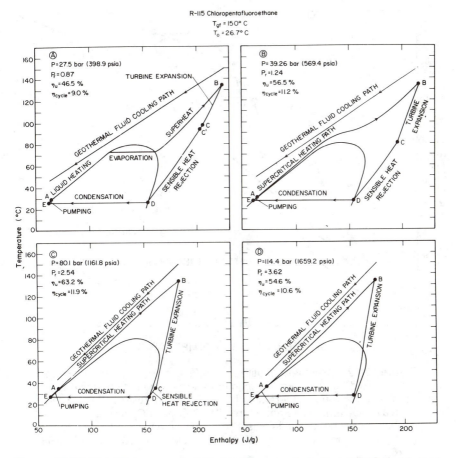

R-115 Chloropentafluoroethane
$T_{gf} = 150°$ C
$T_o = 26.7°$ C

Figure 10-21. R-115 binary cycle optimization case study for a 150°C liquid-dominated resource with an average heat rejection ambient temperature of 3°C and a condensing temperature of 26.7°C. Temperature-enthalpy diagrams for geothermal, organic working, and cooling fluids shown at different reduced cycle pressures with $\eta_p = 0.80$ and $\eta_t = 0.85$ (Adapted by permission from Milora and Tester, 1976).

(Case C, $P_r = 2.54$), the utilization efficiency η_u increases from 46.5 to 63.2%. This improvement is due primarily to a more uniform heat capacity at the higher pressure and a reduction in the amount of sensible heat rejection (desuperheat). At supercritical pressures there is no phase change and the working fluid heating path can be maintained almost parallel to the geothermal fluid cooling path. As the pressure is increased to 114.4 bars (Case D), cycle performance declines to an η_u of 54.6%. This is caused by the less than ideal efficiencies of the turbine and pump components (85 and

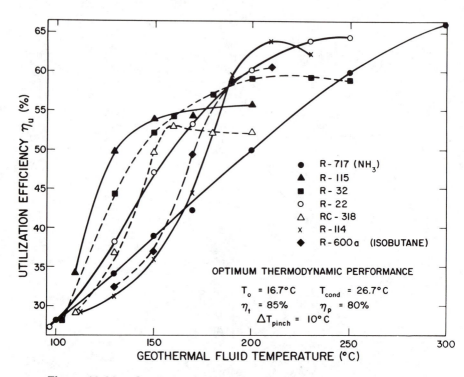

Figure 10-22.   Geothermal utilization efficiency η_u as a function of geothermal fluid temperature for optimum thermodynamic operating conditions. A condensing temperature of 26.7°C was used with a 10°C approach to an average ambient temperature of 16.7°C. A 10°C minimum approach on the primary heating side was also used with an 85% turbine and 80% feed pump efficiency. (Adapted by permission from Milora and Tester, 1976).

80%, respectively) and the larger component work requirements associated with high-pressure operation.

As seen in Figures 10-20 and 10-21, for any given working fluid, there is an optimum set of operating conditions yielding a maximum η_u for particular geothermal fluid and heat rejection temperatures. In screening potential working fluids, some knowledge of the magnitude of η_u and how it changes would be useful. Computer optimizations for seven working fluids, which were conducted for geothermal fluid temperatures ranging from 100 to 300°C, are shown in Figure 10-22 for a heat sink temperature of 16.7°C. At each point, cycle pressures were varied until an optimum was determined at that temperature. One observes a characteristic maximum η_u at a particular resource temperature, which is different for each fluid but generally falls within the range of 55 to 65%. Figure 10-23 combines data from a number of

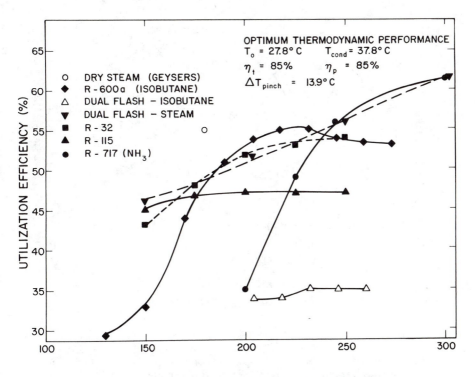

GEOTHERMAL FLUID TEMPERATURE (°C)

Figure 10-23.   Utilization efficiency η_u as a function of geothermal fluid tempera-
ture. $\Delta T_{HE} = 15°C$; $\Delta T_c = 10°C$; $T^c_{pinch} = 5°C$; $\Delta T^{HE}_{out} = 15°C$; $\Delta P/P$ primary heat
exchanger $= 0.10$; $\Delta P/P$ condenser/desuperheater $= 0.05$. (Adapted by permis-
sion from Milora and Tester, 1976, and Eskesen, 1977, with optimum thermodynamic performance indicated.)

sources to compare maximum utilization efficiencies for different conversion methods, including single- and dual-stage, flashing, binary, direct steam
injection, and dual flash/binary cycles.

Hansen (1971) and Eskesen (1980a) also review the performance of steam
flashing cycles in detail. The output of a multistage cycle primarily depends
on initial fluid temperature and composition; flashing conditions of each
stage; $i(P^{sat}_i, T_i)$; turbine efficiencies; and condensing temperature. Eskesen
has shown that a CO_2 composition of 3% by weight in original fluid gives rise
to an 8% loss in power output for a dual-stage flash system operating on a
230°C liquid resource.

Generally, the work produced by a given flash stage is proportional to the
amount of vapor created by the isenthalpic throttling step (see Figure 10-2),
and the enthalpy difference that the vapor experiences when it is isentropi-

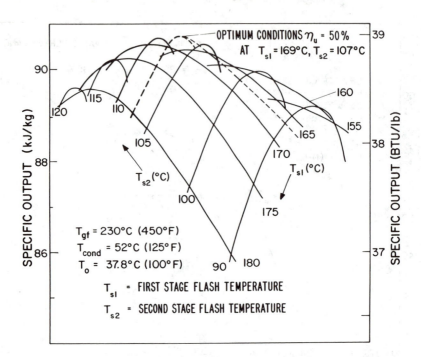

Figure 10-24.   Effect of flashing conditions on the performance of a dual flash-dual admission steam turbine cycle operating on a 230°C, CO_2 free brine containing 2.5 wt % NaCl. (Adapted by permission from Eskesen, 1977a, 1979a.)

cally expanded to the condensing temperature. If the temperature of the flashed fluid is only slightly below the wellhead temperature, then the fraction of vapor produced will be small, whereas the isentropic enthalpy difference will be large. The opposite is true if the fluid is flashed to just above the condensing temperature. Optimal performance in terms of power output will result at some intermediate flashing temperature. The performance of a two-stage flashing system is shown in Figure 10-24 for a 230°C CO_2-free brine with a 52°C condensing temperature and an average turbine efficiency of 75% under wet expansion conditions. Maximum output with η_u = 0.58 occurs when the first flashing stage is at 167°C (T_{s1}) and the second at 107°C (T_{s2}). An analytical approach developed by Milora and Tester (1976) results in 161°C for T_{s1} and 103°C for T_{s2}.

Increasing the number of stages does improve efficiency, but at a decreasing rate as more stages are added. For example, going from one to two stages increases output by 20-30%; going from two to three stages increases output by 15% (Eskesen, 1980b).

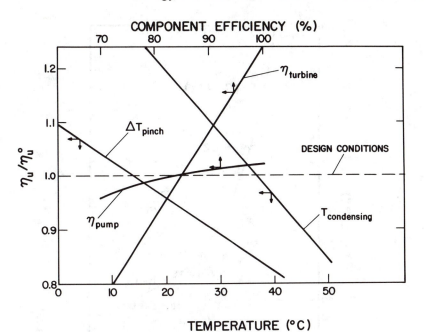

Figure 10-25. Effect of cycle operating conditions on performances including pump and turbine efficiencies (η), primary heat exchanger pinch point approach temperature (ΔT_{pinch}), and condensing temperature (T_c) for a dual flash—n-butane binary cycle. (Adapted by permission from Eskesen, 1977).

The large effects that cycle component efficiencies and pinch and condensing temperatures can have on performance are shown in Figure 10-25 for a dual flash — n-butane binary cycle. Because the calculations leading to Figures 10-20 to 10-25 are complex and time consuming, a simpler technique for preliminary binary fluid evaluation would be useful for performance comparisons with flashing systems. An evaluation is obtained by plotting in Figure 10-26 the temperature T^* at maximum η_u minus the fluid's critical temperature (T_c) as a function of the reduced, ideal gas state heat capacity ($C_p^*/R = (\gamma/(\gamma-1)$, where $\gamma = C_p/C_v$ and R is the gas constant). The data for several fluids studied fit the empirical equation (Milora and Tester, 1976):

$$T^* - T_c = 790/(C_p^*/R) \tag{10-6}$$

The quantity ($T^* - T_c$) is effectively the degree of superheat above the critical temperature for optimum performance and can be correlated to changes in properties associated with changes in molecular weight. Lower-molecular-weight compounds, such as R-22 (M = 86) and R-717 (ammonia;

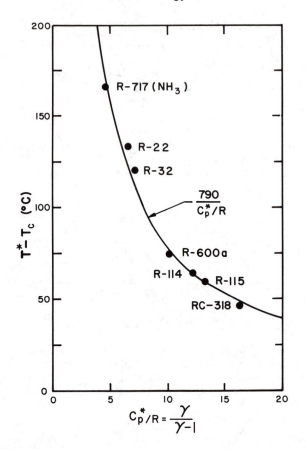

Figure 10-26. Generalized correlation for the degree of superheat above the critical temperature for optimum geothermal fluid utilization as a function of the ideal gas state reduced heat capacity, (Reprinted by permission from Milora and Tester, 1976).

$M = 17$) have fewer molecular degrees of freedom resulting in lower values of C_p^*/R and higher values of superheat above the critical point. As molecular weight increases, more degrees of freedom exist, C_p^*/R increases, and $(T^* - T_c)$ decreases. An empirical procedure of this type should be valuable for preliminary working fluid assessments, because γ and T_c are commonly known properties of many compounds.

Economic Analysis Methods

This section introduces the general issues related to the commercialization of geothermal energy. These include the estimation and financing of capital and operating and maintenance costs associated with the power plant and the reservoir, as well as the management and structure of the organization itself. In addition, revenue factors, such as the type of product and its selling price relative to the competitive market and regulatory and tax issues, are also considered. Different methods of assessing commercial feasibility are discussed and compared with respect to their treatment of cash flow, revenues, risk, and rate of return on invested capital. Methods of estimating effective busbar or energy selling prices for economic viability are also compared in this section.

Supply (Cost) Factors

The utilization of geothermal resources is very capital intensive. This is enhanced by the requirement that a large fraction of exploration, field drilling and development costs are incurred long before the power plant is built and revenues appear. The logical separation between reservoir-related or fluid production costs, including exploration, drilling, and completion, and surface plant-related costs have led to an unusual management and organizational structure for geothermal developments. In current U.S. practice, private firms (typically oil companies) produce the geothermal steam or hot water and then sell it directly to a public or private utility, which uses it to generate electricity. In these cases, a higher risk premium is placed on the field development, which typically results in significantly higher anticipated rates of return and higher equity-to-debt ratios for the producer as compared to the power generator. Wholly-owned ventures result in a completely different situation. Because of the flexibility of controlling both the fluid production "fuel" cost and power generation cost financing, lower effective rates of return and interest rates would be anticipated, particularly if this were a regulated investor-owned or public utility. The type of ownership obviously has a major impact on the magnitude of busbar costs.

Field development capital costs are difficult to estimate accurately because they are based on many complex and variable factors. For example, individual production or reinjection well costs can vary widely even in a given field depending on drilling difficulties. Fluid production and reinjection rates also differ among wells in a specific field. Techniques used for estimating total development costs for a field are discussed in more detail in the subsequent section on resource-related costs. Surface plant costs, although subject to some contingencies, are easier to estimate than field development costs.

Operating and maintenance $(O + M)$ costs for geothermal developments usually constitute only a small fraction of the total. This is partly because in the case of separate ownership of the reservoir and power plant, the effective "fuel" costs incurred by a private utility by purchasing steam or hot water from the producer are not treated as $O + M$ costs. Commonly, a long-term (20-40 yr) fluid supply contract has been negotiated between the utility and the producer, which provides a sufficient rate of return on capital invested in developing the field. In cases where handling the geothermal fluid is more difficult because of the presence of large quantities of corrosive, erosive, or fouling components, such as chlorides, silica, carbonates, or large concentrations of noncondensable gases, such as CO_2 and H_2S, O + M costs may be proportionally higher than for those cases with relatively benign fluids. Typical O + M costs would range from 0.1 to 0.4 ¢/kWh (1980 dollars) depending on these factors. Another factor in estimating O + M costs is system performance during the lifetime of the reservoir/plant complex. In the later stages, maintaining design power outputs and high capacity factors may be difficult for a number of reasons. These include thermal drawdown of the reservoir and deteriorated heat exchanger and turbine performance, which may become severe enough to require increased levels of O + M to provide sufficient revenues during late periods in the lifetime of the plant or reservoir.

The question of water availability and its effect on the design and operation of heat rejection systems is an important one. For example, in semi-arid regions of the U.S., total wet cooling may be practically as well as economically impossible. Modified wet/dry or dry systems may be required with the possibility of floating power output explored to optimize performance on a daily and annual basis.

Geothermal plants, because of their inherent low thermal efficiency when compared to fossil fuel and nuclear units, are constrained by markedly different economy of scale effects. Power developers can easily show a significant reduction in capital costs per installed kW for nuclear and fossil-fired units with increasing capacity from 100 to 1000 MW(e). Geothermal power plants are restricted to much smaller sizes for optimal economics. A primary consideration is fluid transportation to the generating station. Because of low conversion efficiencies, large flow rates of water or steam are required per MW(e) produced. For any plant size over a few MW(e) multiple production wells are required. Fluid transmission costs will depend on well spacing, flow rate per well, and plant size which will determine total flow requirements. These costs need to be balanced against the capital costs per kW(e) of the plant itself to reach an optimum size. For many locations around the world this appears to be somewhere between 30 and 100 MW(e) (DiPippo, 1980a,b; and Bloomster, 1975b). The terminal value of a geothermal reservoir/plant system is frequently difficult to estimate *a priori*. For instance, the economic lifetime of the plant may be 30 years, whereas the reservoir itself may have the potential of sustained operation well beyond that period. How one treats the terminal value of the reservoir is thus a very important question. Prediction of long-term reservoir performance is very difficult, so *a priori* knowledge of potential, significant terminal values is not usually available. Consequently, field developers and producers almost always neglect terminal value benefits in their initial economic analysis of a specific field.

Demand (Revenue) Factor

The anticipated selling prices for the product are of equal importance to capital costs, equity rates of return, and interest rates. In the case of electricity, busbar generating costs for alternative fossil-fuel-fired and nuclear plants are crucial in establishing whether geothermal energy systems can compete. In most applications, geothermal systems are being considered as alternatives for additional rather than replacement base load capacity. Their costs must compete with new unit capital costs and fuel costs for nuclear or fossil-fired systems. Escalation of capital costs for nuclear and coal-fired plants exceeded inflation rates significantly in the past decade, indicating the increased level of pollution abatement and/or safeguard systems which have been required by more stringent state and federal regulations. This, coupled with the rather uncertain picture of oil and natural gas availability and prices as well as the question of environmental acceptability of nuclear and coal-fired systems, creates an optimistic future for geothermal development where the resource is of high quality. These issues are discussed later ("Comparison with Conventional Energy Supplies") and comparative cost estimates for geothermal, fossil, and nuclear energy are presented.

Governmental Constraints and Involvement

Taxes

There are at least three important tax components: revenue, ad valorem or property, and income taxes (Wagner, 1978; and Eisenstat, 1978). *Revenue* or severance taxes are normally computed as a fraction of the gross revenue or value of the geothermal fluids produced. Each state has its own formulae for computing the tax rate, but it typically varies from 3 to 8% of utility gross receipts.

Property taxes are somewhat more complicated. For example, California increases its property tax rate as a given geothermal reservoir is proven to reflect its increased value. Normally, property taxes are assessed as some mill rate times one-third of the non-depreciated property value. The methods used for valuation are nonuniform and frequently speculative. This can lead to negative effects on geothermal development because property taxes are levied long before the actual assessed income value has been realized (Wagner, 1978).

The third form of tax is *income tax,* computed as a percentage of taxable annual income. Federal, state, and local taxes may be of this type. It has been common practice to assume that the combined rate is roughly 51% of the net taxable income unless special situations exist.

Tax Incentives and Credits

A number of tax incentives and investment tax credits have recently been enacted by federal statutes. These were carefully reviewed by several authors at the Geothermal Resources Council meeting on the "Commercialization of Geothermal Resources" (1978b). The National Energy Act passed by Congress on October 15, 1978 initiated major changes in existing tax treatments for geothermal development. Effective December 31, 1977, deductions are allowed for intangible drilling costs for geothermal wells using the same criteria applied to oil and gas drilling.

Intangible drilling costs represent nondepreciating costs and may be as much as 70% or more of the total drilling costs. Expensing of intangible costs in the year that they are incurred increases cash flow by creating a tax deferral (Ramachandran et al., 1977).

A 22% depletion allowance declining to 15% in 1985 is also included for hydrothermal and hot dry rock resources. In addition, a 10% depletion allowance is permitted for geopressured methane recovery during the period October 1, 1978, through December 31, 1984. Geopressured methane prices are also deregulated. The depletion allowance effectively lets the field developer or producer take a certain percentage of his revenue from the sale of geothermal water or steam to reduce his net taxable income. This method of tax reduction was in effect for the oil and gas industry until 1975. The depletion allowance is considered a preference item and deductions associated with preference items are limited by certain minimum tax provisions (Wagner, 1979).

Ramachandran et al. (1977) consider the effects of tax incentives on geothermal development in California in detail. An additional 10% investment tax credit for a business installing geothermal equipment increased the total credit to 20%. Aside from these financial incentives, two important regulatory issues were covered (Stephens, 1978):

1. Expansion of the Federal Energy Regulatory Commission's (FERC) authority to order interconnection and wheeling (sale and power exchanges) and capacity expansions and the request of small power producers and cogenerators.
2. Exemption of small utilities (<30 MW(e) capacity) from the Federal Power Act and the Federal Public Utility Holding Company Act restrictions.

Royalties

Royalty payments to governmental or private land owners is common practice for existing hydrothermal developments in the U.S. On the federal land leases, royalty payments are typically 10% of gross revenues. Private rates are variable but may be as high as 20%. Rex (1978) pointed out that because of the high incidence of new geothermal discoveries on federally owned lands, particularly in the western U.S. potential royalty income to the U.S. government is high. For a 1,000-MW(e) development with electricity priced at 4.5 ¢/kWh at the busbar (1980 $), royalty payments would amount to $2.7 billion over a projected 30-year project life. At this point it is difficult to predict how royalties will be treated for hot dry rock resources because of its decentralized character. Tester et al. (1979) suggest that there may be no royalties for hot dry rock systems located in broadly distributed heat flow regions; but on the boundaries of KGRA hydrothermal deposits some payment may be required.

Geothermal Loan Guarantee Program

One very positive effect that the government has had on geothermal development has been in the direct aid provided by underwriting loans for new geothermal projects. The present program covers 75% of the fixed project capital cost with private sector support providing the remaining 25% (Woods, 1978). Even with this substantial government support, the 25% "risk" may be perceived to be too high in many cases. Stephens (1978) pointed out that "a 90% guaranty for publicly-owned utility projects may be appropriate." One shortcoming of the program is that the government loan guarantee covers only fixed costs; thus, the risk of resource depletion or extended power plant outage would create severe problems for the private investor (Beim, 1978; Woods, 1978).

Environmental Regulations

The present situation in the U.S. regarding emission regulations for geothermal power plants indicates an increasing concern over noncondensable gases and H_2S abatement in particular. This is certainly the case at The Geysers and for future developments in the Valles Caldera of New Mexico. Water use, including net consumption and reinjection programs, can be critically important in arid regions or where the potential for groundwater contamination exists. In any case, the costs associated with required abatement and monitoring systems can represent a significant fraction of the total capital investment (EPRI, 1978, 1979; Ramachandran et al., 1977).

Cost Models

Perhaps one of the most challenging problems faced by geothermal energy technologists is to develop straightforward, unambiguous, and reliable methods to estimate busbar generating costs for electricity or selling prices for geothermally-produced energy used as heat. In addition to the difficulties of estimating capital costs associated with drilling and power plant components, the qualitative treatment of such things as risk, rate of return on equity, interest rates, revenues, and taxes is clouded with many alternate approaches, which can eventually lead to a wide range of estimated costs for commercially-feasible operation.

A study by El-Sawy et al. (1979) lucidly illustrated this point by comparing electricity generation cost estimates prepared by sixteen different organizations including private utilities, oil and gas companies, U.S. Department of Energy R&D contractors, and engineering design companies. In each case, resource properties were used to derive engineering cost estimates, which were then combined with a number of financial parameters in an economic model to produce an estimate of levelized busbar cost. Using the same sets of resource, power plant and financial (taxes, discount and interest rates, and capital costs) assumptions, ranges of 3.8 to 7.0 ¢/kWh and from 2.5 to 5.2 ¢/kWh for two different resources resulted from estimates from the sixteen organizations. In effect, the approaches described in the El-Sawy et al. (1979) study encompass the current state of the art of economic modeling for geothermal systems.

Because it will be impossible to cover every detail of costing methods and models in this section, interested readers should consult the following additional papers which are specifically directed toward geothermal and solar energy economics: Bloomster, 1980; Banwell, 1975; Bloomster et al., 1975a; Doane et al., 1976; Ramachandran et al., 1977; Cummings and Morris, 1975; Greider, 1975; Golabi and Scherer, 1977; Geothermal Resources Council, 1978; Juul-Dam and Dunlap, 1975; Lang and Heidt, 1979. A number of general references dealing with economic theory are also of interest: Sternmole 1974; Abdel-Aal and Schmelzlee, 1976; Baumal, 1965; and Newendrop, 1975. Engineering cost approaches were presented by Peters and Timmerhaus, 1980; Happel and Jordan, 1975; and Rudd and Watson, 1968.

The basic idea of any model is to combine capital and operating costs, reservoir and power plant operating parameters, and regulatory and financial data and constraints to produce an estimate of minimum economic feasibility conditions. This will determine the price a public or private utility or steam/hot water producer will have to charge its customers to cover all costs, including principal and interest payments on bank loans used to secure capital, return on equity to shareholders, ad valorem and income taxes, and operating and maintenance costs. For utilities, purchase costs of geothermal steam or hot water ("fuel" costs) must be recovered. The selling price for electricity or energy that meets these conditions can then be referred to as a "breakeven" cost. In many economic circles, this breakeven cost is regarded as a levelized busbar cost.

Levelized costs can be determined by a number of financial methods. These include yearly discounted cash flow analysis, net present value methods, and fixed-charge rate approaches. By using discounted cash flow (DCF), money flows occurring in different time periods can be equated. For example, to compare costs (C) at their present value (PV), the following discounting scheme can be used:

$$PV \ = \ \frac{C}{(1+r)^t} \qquad (10\text{-}7)$$

where t = number of time periods between the present and when the cost was incurred

r = effective discount rate in each period.

Fixed-charge rate (FCR) methods usually determine an annual cost as a fraction of the present value of the total capital investment. This then is superimposed on other annual costs, such as operating and maintenance and fuel costs, to arrive at a generating cost estimate. In order to describe the basic concepts, three separate topics are treated below: methods used to estimate capital costs; fixed-charge rate methods; and intertemporal discounted cash flow analysis employed for hot dry rock systems (Cummings and Morris, 1979).

Capital Costs

The total capital cost of a geothermal power plant has two major components: the *power conversion equipment* and the *wells and distribution system* that supply the geothermal fluid. In some cases, where the fluid producer and power generator functions are under separate ownerhsip, it would be more appropriate to consider capital costs associated with reservoir development first to establish a "fuel" cost (per Btu or J basis). Then the power operator, a public or private utility, would treat these "fuel" costs as operating costs and superimpose them into an appropriate annualized capital cost for the surface power plant to arrive at a busbar price. In any case, accurate estimates of total drilling and installed equipment costs are required.

One common technique for doing this uses a factored estimate approach to determine the installed cost from the purchased cost of the major components. Peters and Timmerhaus (1980), Rudd and Watson (1968), and Happel and Jordan (1975) describe the methodology as applied to preliminary estimates of chemical process and manufacturing plant costs. For geothermal systems, the total capital investment, ϕ, can be expressed as a function of the total purchased equipment cost, ϕ_E, including heat exchangers, turbines, generators, pumps, condensers, and cooling towers; drilling and completion costs, ϕ_W; site exploration and land acquisition costs, ϕ_S; and fluid distribution and transmission costs ϕ_F. Using a factored estimate approach, ϕ can be expressed as:

$$\phi = (1 + \Sigma f_i) f_E^l \phi_E + f_W^l \phi_W + f_F^l \phi_F + f_S^l \phi_S \tag{10-8}$$

where f_i is the direct cost fraction of ϕ_E necessary to cover costs of equipment installation, buildings, and structures, in-plant piping, instrumentation, etc.; and f_E^l, f_W^l, f_F^l, and f_S^l are indirect cost factors that include engineering design and legal fees, overhead and escalation, environmental and regulatory fees, and contingency. Table 10-4 provides appropriate ranges for these direct and indirect cost factors. As can be seen, installed costs are larger than purchased by factors between 2.5 and 4.0. More details on actual component cost estimates are presented later in the section on "Exploration, Drilling, and Reservoir Development Costs" and in the section on "Surface Conversion Equipment Costs."

Fixed Charge Rate

The annualized fixed charge rate (FCR) approach produces a levelized busbar cost estimate by simplifying a discounted cash flow analysis. The total capital investment is simply multiplied by the FCR to give an equivalent annual cost. Although straightforward, the FCR incorporates a number of complex financial parameters including, rates of return on equity capital, debt interest rates, income and ad valorem taxes, insurance, and depreciation. Consequently, there are a number of methods to estimate the FCR (Doane et al., 1976; El-Sawy et al., 1979; and Bloomster, 1980). In any case, the FCR is a function primarily of the capital recovery factor (CRF) and tax credits. The CRF represents a levelized annual payment, expressed as a fraction of the principal, necessary to fully repay a loan (principal and interest) over a specified period. If i is the annual interest rate and n is the number of years for repayment, then the CRF is equal to:

$$CRF = \frac{i}{1 - (1+i)^{-n}} \tag{10-9}$$

In this instance, the FCR becomes (Bloomster; 1980; Doane et al., 1979).

Table 10-4
Direct and Indirect Cost Factors for Geothermal Installations

Surface Power Plant — Direct Cost Factors

| | f_i fraction of equipment cost ϕ_E | | | |
	Vapor dominated	Liquid-dominated	Geopressured (incl. methane)	Hot dry rock
Major equipment installation	0.10	0.10	0.10	0.10
Instrumentation	0.12	0.14	0.14	0.14
In-plant piping	0.08	0.05	0.10	0.05
Insulation	0.03	0.03	0.02	0.03
Foundations	0.06	0.06	0.06	0.06
Structures and buildings	0.08	0.08	0.08	0.08
Fireproofing and exploration-proof equipment	0.01	0.01-0.05*	0.05	0.01-0.05*
Electrical (switch yard)	0.06	0.06	0.06	0.06
Environmental controls	0.12	0.12	0.10	0.03
Total $(1+\Sigma f_i)$	1.66	1.65-1.69	1.71	1.56-1.60

Indirect Cost Factors

	Equipment f_E^i	Well f_W^i	Exploration f_S^i	Fluid distribution f_F^i
Engineering and legal fees	0.17	0.05	0.01	0.08
Overhead and escalation	0.30	0.15	0.15	0.20
Contingency	0.13	0.13	0.13	0.13
Environmental and regulatory fees	0.10	0.05	0.02	0.10
Total $(f_i^i = 1 + \Sigma f_i^i)$	1.70	1.38	1.31	1.51

After Milora and Tester, 1976
*High values result when flammable working fluids are employed.

$$FCR = CRF + f(CRF, \text{tax credits}) \qquad (10\text{-}10)$$

The *FCR* approach can be modified to account for different rates of return for debt and equity capital and other factors including operating and maintenance costs and insurance. In geothermal projects in the U.S., the *FCR* can vary from ~0.12 to 0.18 for the utilities portion of the investment to 0.15 to 0.30 for the producer or field developer's part to account for a presumed higher risk and subsequent higher rate of return or shorter payback period.

Intertemporal Discounted Cash Flow Analysis

Inasmuch as geothermal reservoirs are depletable and redrilling or additional drilling may be required during the lifetime of the surface power plant, proper management of the reservoir-plant system is important (see "Resource Related Effects"). Cummings and Morris (1979) developed an intertemporal discounted cash flow model for optimizing after-tax net benefits for hot dry rock systems. This model illustrates the concepts of optimal management of a depletable resource by taking into consideration the various power plant and reservoir design and operating choices that can be selected to maximize the return on investment. These include:

1. Drilling and redrilling strategies, mainly when and how deep the hot dry rock wells should be drilled for a given temperature gradient (ΔT).
2. Reservoir flow rate. In order to alter the thermal drawdown characteristics of fixed-size reservoir, the mass flow rate (kg/s) could be changed in each decision period.
3. Plant design temperature. The performance of the power plant (MW(e) per kg/s of geofluid flow) could be optimally matched to the resource characteristics; whereas the relative power plant, to drilling capital costs.

The model employed in their reseach requires as input basic information about the hot dry rock system under consideration. In particular, the effective reservoir heat transfer surface area and average geothermal temperature gradient must be specified. In addition, values for the price of electricity, load factors, plant size, design reservoir flow rate, and plant design temperature must be determined. The optimal time path of management decisions for this system is then determined (optimal defines that management strategy, drilling activity, or flow rate changes that maximize the present value of after-tax return, ϕ, to the operators). This objective is expressed mathematically by Tester et al. (1979) as follows:

$$\text{Maximize } \phi = \sum_{n=1}^{N} \left[\sum_{t=1}^{\tau} (R_{n,t}) (1 + r)^{-(t+\tau(n-1))} \right]$$

$$- \left[C_S + C_F + C_E + \sum_{n=1}^{N} C_{d,n} (1 + r)^{-\tau(n-1)} \right]$$

$$(10\text{-}11)$$

$$- 0.51 \left[\sum_{n=1}^{N} C_{d,n}(1 + r)^{-\tau(n-1)} - \frac{C_{d,n}}{(1 + r)^{\tau N}} \right]$$

$$+ 0.51 \, (C_S + C_F + C_E) \left[\frac{(1 + i)^{\tau N}}{(1 + r)^{\tau N}} - \frac{1}{(1 + r)^{\tau N}} \right]$$

where

N	=	the number of decision periods over the assumed life of the HDR system
τ	=	the number of years in a decision period
r	=	the opportunity cost of capital (real rate of return)
i	=	the real debt rate of interest
$R_{n,t}$	=	revenues net of revenue tax and operating cost for the tth year of the nth decision period
C_E, C_S, C_F	=	surface plant (E) site acquisition (S) and fluid distribution (F) cost discounted to present value at year 0 (start of plant operation)
$C_{d,n}$	=	drilling costs in period n.

Equation 10-11 is structured in such a manner that the system is examined at each decision period. After exploring the ramifications of each management option for every state of the system, the option that maximizes the objective function is then selected. By doing this for all decision periods over the life of the system, this equation enables determination of the optimal time path of such decisions and, hence, the most economic way of operating the system.

In this formulation, the revenues and various costs are each discounted to the beginning of electricity production or plant operation ("the present") by an appropriate discount rate. Then the difference between revenues and total costs is calculated. The present discounted value of revenues is the double-summed term just to the right of the equality in Equation 10-11. The remaining terms represent the present discounted value of costs and is composed of separate terms for surface plant costs, drilling costs, income taxes on equity, and income tax deduction *credits* on interest payments to debt holders.

The reader should appreciate at the outset that all costs and revenues in the model are transformed into "real" constant year- 1980 dollars. These are dollars that, no matter when they are received or spent, have the same real purchasing power as the present dollar does. The use of this concept allows one to automatically determine the impacts of general price inflation or deflation from the analysis. This approach, however, does assume that there is no change in the real relative prices of the purchased inputs of HDR technology relative to its marketed output (electricity).

For a given set of resource, reservoir performance, and power plant characteristics, Equation 10-11 can be used to calculate the present value of net (after-tax) profits for a specified price for electricity or heat. By altering these characteristics in a systematic way, sensitivity and parametric effects can easily be explored to establish ranges of conditions that must be met for commercial feasibility.

The question of what busbar price is required to reach economic feasibility is certainly an important one. Because of the mathematical structure of the model, however, it must be determined indirectly. In the intertemporal model described, the busbar price of electricity, p, is specified as a parameter in the revenue term, as are geothermal gradient, ∇T, and design flow rate before the dynamic programming algorithm begins computation. When the net profits or benefits from Equation 10-11 equal zero, this break-even point is reached for a given set of conditions including price, i.e., $p = p^*$. The price p^* is, therefore, referred to as the break-even price. At this point, the revenues produced from the sale of electricity are enough to just cover all costs, including interest payments of debt, return on equity and income, and ad valorem taxes, where both the revenues and costs are evaluated in terms of present discounted value.

The actual application of this methodology to geothermal systems is discussed later in a section entitled "Economics of Electric Power Production," which follows two separate sections covering capital costs of exploration, drilling, and reservoir development and power plant equipment costs.

Exploration, Drilling and Reservoir Development Costs

Several particular site-specific characteristics are important in establishing the economic quality of the resource. For hydrothermal and geopressured systems, geothermal fluid temperature, pressure, composition of fluids, reservoir depth, production rates, and rock type are the major ones. Power plant design and conversion efficiency are critically controlled by fluid temperature and composition as described earlier, whereas drilling costs depend on depth and rock type. In hot dry rock systems, fluid chemistry is important as well as reservoir temperature and depth, which can be related to the average geothermal temperature gradient at the site. Higher gradients will result in shallower, less expensive individual well costs. Thus, the gradient is perhaps the most distinguishing feature that determines hot dry rock resource quality at a particular site. The distribution of gradients across the U.S. is presented in the section on resource potential in Chapter 2.

Exploration and Land Acquisition Costs

Locating and evaluating the resource can be very difficult and expensive particularly in regions that have not been developed or proven or where surface manifestations of geothermal activity are absent (see Chapters 3 and 4). Consequently, site exploration and evaluation account for a certain fraction of the costs. These will probably involve the use of a number of reconnaissance measurements or surveys (geologic, hydrologic, and geophysical) aimed at locating hot water or rock under the surface. In addition, other costs include land acquisition by purchasing or leasing geothermal rights, shallow exploratory drilling, and deep drilling to evaluate the potential of the reservoir. Estimates of these costs are given in Table 10-5. Readers are cautioned that since exploration and land acquisition costs can vary considerably from location to location, the values in Table 10-5 should be used only as a guide. The impact of these costs, however, are significant because they are incurred 10 or 20 years before any power is produced at a site.

Drilling and Completion Costs

Drilling costs for deep exploratory, production, and reinjection wells are difficult to estimate because they are subject to many unforeseen contingencies. In general, they comprise the largest single cost component contributing to the capital intensive nature of geothermal development. In addition, their effect is magnified similar to exploration costs because they are in-

Table 10-5
Typical Geothermal Exploration and Site Acquisition
Cost Estimates
[200 MW(e) Capacity]

| | Cost in $1000 (1980) | | |
| | Hydrothermal | | Hot dry rock |
	Developed low-risk area	Undeveloped high-risk area	
1. Power plant site purchase	150	150	150
2. Leased well sites	750-7500*	2400**	150†
3. Geological, hydrologic, and geophysical surveys for site reconnaissance	0	2220	158
4. Geophysical surveys for site selection	300	2220	158
5. Shallow exploration drilling and temperature holes	350	1400	320
6. Deep reservoir drilling and Reservoir and well testing	1050	2100	2860
Totals	2600-9350	10490	3796

Based on Grieder (1973, 1975), Tester, Morris, Cummings, and Bivins (1979), and Altseimer (1976). An inflation rate of ~10% per year was used during the period 1975-1980.

*1 Area required — 7500 acres for 200 MW(e) at $100-$1000/acre.

**32 Areas required as prospects — 240,000 acres for 200 MW(e) at $10/acre.

†10 Acres required as prospects — 15,000 acres for 200 MW(e) at $10/acre.

curred before any revenues from power generation appear. In effect, they can be viewed as replacing the fuel cost of a typical nuclear or fossil-fired power plant, which is treated as an operating cost. In some cases, drilling is very difficult because of the presence of hard, fractured rock formations, lost circulation of drilling fluids, hole collapse, high temperatures, and the presence of corrosive and erosive geothermal fluids that increase drill bit wear (Altseimer, 1976). Furthermore, a need frequently exists for directional drilling at high temperatures and sophisticated well completion programs, involving special cement formulations, casing materials, etc., (see Chapter 5). Most of the directional drilling to date in geothermal systems has involved whipstocking techniques or downhole motor-driven drills, which present some design and material problems for operation at high temperature (Maurer et al., 1977).

Well and casing diameters must be large enough to avoid excessive pressure drops in the flowing fluid. Proper design will require some prior knowledge of anticipated flows and gas-to-liquid ratios.

Flow rates typically range from 27 to 50 kg/s (60 to 110 lb/s) for artesian, liquid-dominated aquifers and from 5 to 23 kg/s (11 to 50 lb/s) for vapor-dominated aquifers with wells of approximately 20 to 30 cm (8 to 12 in.) in diameter.

Higher flows with larger pressure drops in liquid-dominated systems frequently result in premature flashing within the wellbore. This reduces the fluid temperature and can lead to solids precipitation in the casing and further limit well capacity. Downhole pumps not only could enhance flow but also recover the lost hydrostatic head, and, thus, prevent flashing that occurs in most high-temperature wells even at low flow rates.

A hot dry rock system differs considerably from the other natural geothermal systems in that the fluid flow rate from the downhole reservoir can be controlled to create a situation where thermal conduction through the walls of the fracture limits the heat transfer. Thus, by using a sufficiently large surface area system, much higher flow rates may be achieved, that is 225 kg/s (500 lb/s), either in a self-pumping or externally-pumped mode, although rates of 40 to 100 kg/s would be more reasonable. The hot dry rock system also permits deeper drilling to obtain higher temperatures in areas having uniform temperature gradients.

Costs for both production and reinjection wells (if required) should be considered and, depending on the nature of the reservoirs, these might be considerably different. For example, in a dry-steam or direct-steam flashing system, part of the steam might be utilized for evaporative cooling and not returned to the aquifer. On the other hand, if a two-hole circulating system is developed for hot dry rock reservoirs, the cost for production and reinjection wells will be essentially the same. Another important factor is the uncertainty of obtaining a successful production well even when drilling in a proven geothermal field. Unsuccessful wells cost less than completed production wells, because they do not need extensive surface plumbing or casing and, in some cases, they can serve as reinjection wells.

Given the above constraints, individual well cost is controlled mainly by rock-type, well diameter, and depth. Figure 10-27 shows an exponential dependence of completed well costs as a function of depth for boreholes of 25-35 cm (10-14 in) diameter in geothermal formations (Milora and Tester, 1976). The depth dependence was selected to parallel the behavior that has been observed for oil and gas wells with the geothermal wells costing between 2 to 4 times more than an oil or gas well drilled to the same depth. The well costs cited include site preparation, drill rig mobilization/demobilization, bits, casing, cementing, cores, geophysical logging, drilling fluids,

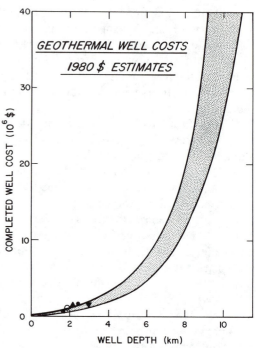

Figure 10-27. Predicted geothermal completed well costs. An average annual escalation rate of 10% between 1975 and 1980 was used to normalize all costs to 1980$.

directional surveys, reservoir stimulation, and the rental charges for the rig itself. It was necessary to utilize the extensive oil and gas data as a means of extrapolating the rather limited set of geothermal well cost data available. For example, costs for wells drilled in The Geysers and Imperial Valley region of California exist for wells ranging in depths typically from 2 to 3 km (Altseimer, 1974, 1976; Greider, 1973; Chappel et al., 1979; Newson et al., 1977; Kestin et al., 1980; and Kruger and Otte, 1973). In addition, some preliminary costs have been cited for hot dry rock wells drilled in granite in New Mexico (Milora and Tester, 1976). Actual geothermal well cost data normalized to 1980 dollars are presented with the predictions in Figures 10-27 and 10-28. The range of predicted costs for a given depth shown represents about one standard deviation about the mean. Exceptions to this trend are common as numerous problems with lost circulation, hole collapse, fishing operations, drillstring twistoffs, excessive bit wear, etc. can lead to very high individual well costs. (See Chapter 5.) It is hoped, however, that in developing a large field, where many wells are being drilled, cost averages would be satisfactory for economic assessment purposes.

Readers interested in expanding and updating the geothermal well cost data base should obtain dry hole data from oil and gas drilling contractors

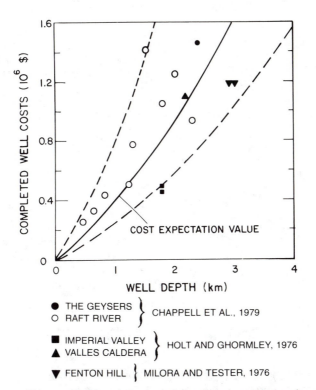

Figure 10-28. Actual and predicted geothermal completed well costs for depths below 4 km. Solid line indicates average estimate at a given depth, while the dotted lines represent a range of approximately one standard deviation about the mean.

and contact geothermal developers directly for new cost information. In the U.S. these would include Union Oil, Standard Oil of California, Phillips Petroleum, Republic Geothermal, and AMAX. The Joint Association Survey results as they appear (see references) and organizations such as the Geothermal Resources Council, Davis, California or the Geothermal Division of the U.S. Department of Energy also issue reports dealing with well costs.

The oil and gas well cost data base used for geothermal well cost extrapolation literally represents thousands of wells drilled throughout the U.S. (continent and offshore). Annual summaries of these costs are published by the Joint Association Survey of the American Petroleum Institute, Independent Petroleum Association of America, and the Mid-Continent Oil and Gas Association (1973). Figure 10-29 was developed using this data as well as studies conducted by Mathematica (Bee Dagum and Heiss, 1968), Altseimer (1974, 1976), and the Continental Drilling group (Shoemaker,

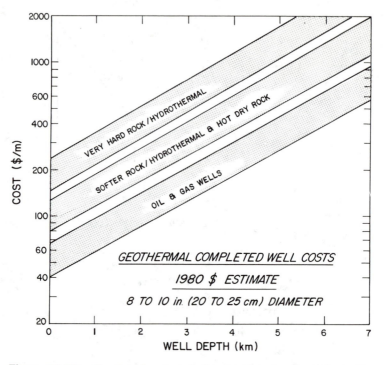

Figure 10-29. Predicted well costs for completed wells. (Adapted by permission from Milora and Tester, 1976).

1975) with existing geothermal data. The model assumes that the functional relationship of cost with depth will be the same for geothermal wells as it is for oil and gas wells. The semilogarithmic correlation shown in Figure 10-29 illustrates the extrapolation method. Methods for updating these cost estimates to current dollars are given in Appendix F, "Cost Inflation Estimation Techniques," at the end of this chapter.

For geothermal wells greater than 4 km in depth, well costs might be expected to converge toward oil and gas well costs. This is because difficulties encountered in drilling deep wells in general are so great that differences between costs of geothermal and oil and gas wells are minimized (Altseimer, 1976).

A recent comprehensive study conducted by Republic Geothermal, Inc. (Nicholson et al., 1979) suggests several aspects of drilling deep hot dry rock (HDR) wells that may result in economic advantages over oil and gas wells to the same depth. Well costs predicted from this model vary linearly with depth showing a slightly higher cost than the exponential model of Figure 10-27 for depths less than 3 km (10,000 ft) and considerably lower costs at

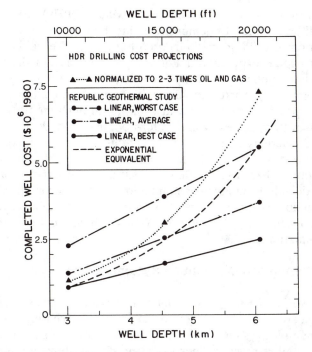

Figure 10-30. Comparison of the Republic Geothermal Inc. drilling cost model (Nicholson et al. 1979) and the exponential, oil and gas well based, geothermal well cost model (Figure 10-27) for hot dry rock wells in crystalline basement formations.

depths much greater than 3 km (see Figure 10-30). The actual reasons for this behavior center around anticipated bit penetration rate and lifetime and the absence of drilling mud and hole stability problems critical to very deep oil and gas wells in sedimentary formations. The RGI study anticipated "normal" bit performance in deep environments (2-5 m/h, 70-100 m lifetime with rotary drilling) for HDR wells, whereas oil and gas well drilling in deep systems typically show a strong decline in penetration rate. Because HDR wells in crystalline rock will be drilled with water rather than mud, several million dollars are saved by avoiding elaborate mud programs. Furthermore, the four deep wells (3-5 km) drilled by the Los Alamos National Laboratory have been very stable requiring a much more modest casing program than that for sedimentary wells of similar depth.

Improvements of bit performance could have a very large impact on reducing geothermal development costs (Newson et al., 1977). For directional as well as conventional drilling, the use of high-temperature downhole turbodrills may result in reduced costs (Maurer, 1979). Extensive field testing will be required to demonstrate this, however.

In addition to drilling, methods used to produce the fluid may require significant capital expenditure. For example, in the hot dry rock case, artificial stimulation may be costly, involving complex fracturing techniques to produce multiple-fracture systems with proper flow control. This will be required to provide uniform flow distribution through the fractured reservoir. In hydrothermal reservoirs, water-flooding methods may be used to enhance productivity and will require careful placement and operation of production and injection wells as described.

In addition, when artesian flow is insufficient, reservoirs may require surface or downhole pumping or stimulation to ensure economically acceptable production or reinjection rates. This will increase capital and operating costs. See Chapter 5, "Geothermal Well Drilling and Completion," for more information.

Fluid Gathering and Distribution Systems Costs

Costs associated with geothermal fluid gathering, transportation, and distribution are largely dependent on the cost of purchasing and installing pipe and pumping systems if required. For developments in the field, fluid production and reinjection lines would be placed above ground with the production lines insulated. Underground systems would typically be used for district heating applications, which could require expensive retrofitting (Kunze, 1977) operations for conversion of existing heating systems. Piping costs depend primarily on the following factors:

1. Distance, as determined by well spacing and power plant location for field developments. Also, distance between the field and the end user for non-electric applications.
2. Diameter of pipes, as determined by allowable pressure drop and mass flow rate.
3. Degree of insulation, to reduce temperature losses in production lines.
4. Position, above and below ground.
5. Fluid composition, heat exchange from highly saline brines to clean water may be required to avoid special materials.

Surface Conversion Equipment Costs

The type of end-use will obviously control the capital cost of the surface power plant. Frequently, for electric systems, because of the high costs associated with drilling and completing the reservoir (in some cases 60 to 80% of the total investment), a premium is placed on designing and operating the power plant near its thermodynamic limiting efficiency. This will involve proper selection of the power cycle, working fluid, and operating conditions to match the initial geofluid temperature and composition, anticipated thermal drawdown, and prevailing ambient heat rejection conditions.

Capital costs for three major equipment types are of prime importance in determining the installed cost of a geothermal electric power plant:

1. Heat exchange equipment (primary heat exchangers, condensers, and cooling towers).
2. Pumps: downhole, power cycle feed, and surface geofluid circulation and reinjection pumps.
3. Turbines and generators.

Empirical techniques for estimating costs of these components are presented in this section. These methods can then be used to calculate the total plant capital investment given a specified power cycle design as described previously using the factored-estimate approach. Methods for estimating inflation factors to recalculate costs in current year dollars are presented in Appendix F at the end of this chapter.

Heat Exchange Equipment

The cost of heat exchangers and condensers required in binary-fluid and flashing cycles can be determined by knowing the heat exchange surface area, pressure rating, and materials required. The engineering procedures needed to specify heat exchanger design and calculate heat transfer areas were described previously. Because fluidized-bed and direct contact exchangers are still under development for geothermal applications, cost data are inadequate at this time. Consequently, the remaining discussion concentrates on shell and tube and finned air-cooled exchangers and cooling towers.

Normally, a power law scaling function would be used for estimating costs, but for a 100-MW(e) plant the surface area requirements for both the primary heat exchanger and the condenser are so large ($>1,000,000$ ft^2) that there is essentially no economy of scale above a certain size and multiple units having surface areas of 20,000 to 35,000 ft^2 are required. Cost estimates were obtained by contacting exchanger manufacturers, and recommendations for both carbon steel shell and tube and air-cooled units are presented in Figure 10-31. Even for a fixed design, cost estimates based on a number of manufacturers' quotes can vary by 50% or more. Because heat exchangers represent such a large fraction of the total plant cost, careful consideration should be given to obtaining accurate cost figures directly from the manufacturer. Figure 10-31 should be used as a guide to show relative rather than

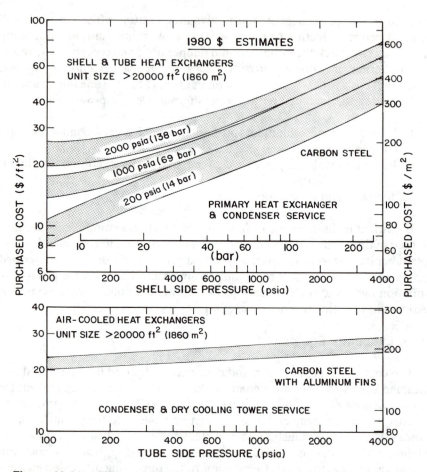

Figure 10-31. Estimated heat exchanger costs as a function of shell and tube side pressure.

absolute costs. As seen in the figure, increasing shell and tube side pressure ratings can have a very large effect on costs. This is particularly important for supercritical operation with organic binary Rankine cycles. Fraas and Ozisik (1965), Peter and Timmerhaus (1980), and Robertson (1980) recommend relative cost scaling factors for materials of construction other than carbon steel. Approximate values are cited in Table 10-6. Although corrosion problems can be reduced by using special alloys, the increase in price can be significant.

Direct air-cooled units may be used in conjunction with condensers employing dry cooling (see Robertson, 1980 and Olander et al., 1979). When conditions permit, wet and wet/dry cooling towers may also be used for heat

Table 10-6
Relative Cost Factors for Different Materials of Construction for Swell and Tube Heat Exchange Service with Carbon Steel Shells

Tube side material	Fe	Cr	Ni	Mo	Cu	Ti	Zn	Relative cost
		Approximate Composition (%)						
Carbon steel	>98	—	—	—	—	—	—	1.0
Copper or brass	—	—	—	—	100-70	—	0-30	1.5
Austenitic stainless steel 304	>65	19	10	—	—	—	—	2.0
Austenitic stainless steel 316	>62	17	12.5	2.5	—	—	—	2.3
Nickel	—	—	100	—	—	—	—	2.3
Copper-nickel (70-30)	—	0	30	0	70	0	—	2.6
Titanium*	—	—	—	—	—	100	—	3.4
Carpenter-20*	>40	20	34	2.5	3.5	—	—	?
Ferritic stainless (E-BRITE 26-1 steel)*	>72	26	<0.5	1.0	<0.2	—	—	?
Hastelloy C*	—	16	>67	15	—	<0.5	—	?

Sources: Fraas and Ozisik (1965), Peters and Timmerhaus (1968), Robertson (1980), Airco
        Vacuum Metals, and Carpenter Technology Corporation.
*May be required for extremely corrosive geothermal environments.

rejection. In these cases, the power cycle would probably use a shell and tube design for the condenser using cooling water to condense the steam or binary fluid.

Wet and wet/dry cooling tower costs are frequently based on their capacity expressed in gallons per minute or kilograms per second of cooling water (Peters and Timmerhaus, 1980; Happel and Jordan, 1975; Perry and Chilton, 1973). Figure 10-32 presents approximate ranges for these costs obtained from the above sources as well as directly from manufacturer's estimates. The upper section of the range for wet and wet/dry units corresponds to less favorable atmospheric and operating conditions — 30°C (85°F) or greater wet bulb, 17°C (30°F) terminal difference, and 2.8°C (5°F) approach; whereas the lower section corresponds to more favorable conditions — 15°C (60°F) or less wet bulb, 8°C (15°F) terminal difference, and a 5.6°C (10°F) approach.

In some cases of direct steam flashing or when large amounts of desuperheating are required, spray condensers might be used to reduce costs because of the simpler design and improved heat transfer by direct contact. Low temperature water sources are frequently used for spray condensers

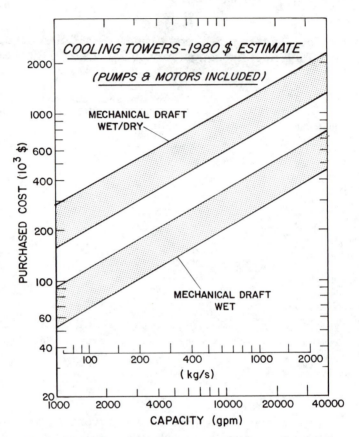

Figure 10-32. Estimated wet and wet/dry cooling tower costs as a function of cooling water flow capacity.

with evaporative or dry cooling towers employed in the absence of such sources. Partial removal of organic working fluid superheat can also be obtained by direct spray condensing with a stream of subcooled organic liquid.

Pumps

Downhole, surface injection, and power cycle feed pump costs will depend primarily on their rated power capacity (horsepower or MW), pressure rating, materials of construction, and design. For example, downhole electric motor-driven pumps are more expensive *per se* than centrifugal circulating or reinjection pumps of similar capacity located on the surface, because more stringent requirements are placed on seals and bearings and because of

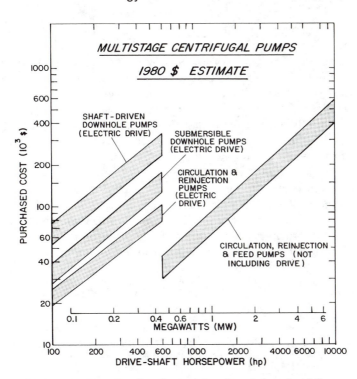

Figure 10-33. Predicted costs for downhole, circulation, and reinjection pumps.

the geometric constraints of the wellbore. Approximate costs as a function of drive power/capacity are given in Figure 10-33. The higher cost figure at a given power refers to high-pressure operation at 138 to 207 bars (2,000 to 3,000 psia) whereas the lower cost refers to lower pressures of 34 to 69 bars (500 to 1,000 psia). Because most binary-fluid cycles require feed pumping in excess of 1-MW, turbine drives rather than electric motor drives would be the logical economic choice. These turbines would utilize the same fluid as was used in the main turbogenerator unit. Equation 10-12 (introduced in the next section) could be used for cost estimating purposes with size specifications determined by the procedures outlined in the section on turbine and pump design criteria.

Turbines and Generators

Low-pressure steam turbines have been in service for a number of years using fluids from both liquid- and vapor-dominated fields. Actual costs until recently, however, have been difficult to evaluate because of two major factors: the cost of new or prototype units are either too low to allow market penetration, or production units in a commercially-mature line may be priced to compete rather than to reflect actual manufacturing costs.

Eskesen (1977a,b; 1980a) has recently sorted out the major cost factors for geothermal steam turbine service, and provided some conceptual cost estimates for hydrocarbon turbines applied to particular resources.

Because the fluids involved (other than water) are not presently commercially used in large turbines, cost estimates were based on a model developed by the Barber-Nichols company of Denver, Colorado (Nichols, 1975), which was first described by Milora and Tester (1976). Comparison between actual manufacturers' costs and those predicted by the Barber-Nichols model are given in Figure 10-34 for steam and nonaqueous fluid service. Turbine costs are scaled as a function of exhaust end size, D_p (pitch diameter); blade tip speed, $D_p N$ (N = rotary speed in revolutions per second); number of stages, n_s; number of exhaust ends, n_e; and maximum operating pressure. Below 540°C (1,000°F), temperature does not affect turbine cost significantly.

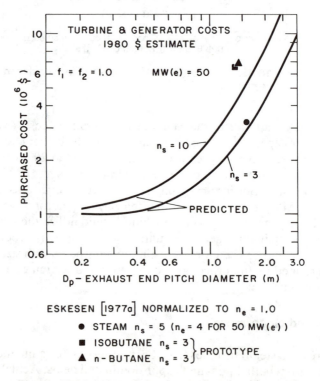

ESKESEN [1977a] NORMALIZED TO n_e = 1.0

- ● STEAM n_s = 5 (n_e = 4 FOR 50 MW(e))
- ■ ISOBUTANE n_s = 3 ⎤
- ▲ n - BUTANE n_s = 3 ⎦ PROTOTYPE

Figure 10-34. Actual versus predicted turbine costs.

Turbine and generator costs include the purchased equipment cost, and can be expressed in equation form as follows:

$$\text{Turbine } \phi_T \text{ (in 1980 \$)} = n_e \left[1 - 0.04(n_e - 1)\right] \\ f_2 \left[128{,}814 n_s f_1 D_p^{2.1} + 175{,}990 D_p^3 + 88{,}570 D_p^2\right] \tag{10-12}$$

$$\text{Generator } \phi_G \text{ (in 1980 \$)} = 329{,}400 \left[\frac{\text{MW(e)}}{10}\right]^{0.7} \tag{10-13}$$

Generator costs do have some economy of scale as expressed in Equation 10-13, where MW(e) is the rated electric output of the unit in megawatts. Pitch diameters should be expressed in meters when using Equation 10-12. The factor f_1 corrects for blade tip speed effects, and f_2 for pressure effects (see Figure 10-35). The first term in the square bracket $(128{,}814 n_s f_1 D_p^{2.1})$ accounts for stage costs and includes the rotor and stator charges, as well as a fraction of the casing and shaft costs associated with each stage length. The second term $(175{,}990 D_p^3)$ accounts for the remainder of the casing associated with the inlet and exhaust plenums and control valves, which scale as the cube of D_p. The third term $(88{,}750 D_p^2)$ covers the precision components, which include labyrinth seals, radial and thrust bearings, and the remainder of the shaft. The multiplier term $[1 - 0.04(n_e - 1)]$ gives the economy of scale associated with multiple exhaust ends. Equations 10-12 and 10-13 apply for power ratings of 1-MW to 100-MW with blade height to pitch diameter (h^*/D_p) ratios of 0.03 to 0.11, pitch diameters ranging from 0.5 to 3 m, and for $n_e \leqslant 4$ for tandem configurations on the same shaft. In the case of low-pressure steam turbines, h^*/D_p exceeds 0.11 in the last stages and, consequently, the costs predicted by Equation 10-12 should be increased by a factor of 1.25.

For most cases of practical interest employing alternate working fluids for geothermal applications, the blade tip speed will be less than 300 m/s and f_1 will be unity. The rapid increase in f_1 for speeds greater than 300 m/s is primarily caused by increased cost of the higher strength alloys required. The increase in f_2 with increasing pressure is mainly due to increased casing wall thickness. This becomes important under supercritical conditions for many of the working fluids considered in this study.

Economics of Electric Power Production

Resource-Related Effects

Lithology (rock type), depth to the reservoir, composition of fluids, pressure, temperature, and ambient conditions are of major importance in establishing the economic quality of the resource at a particular site. These

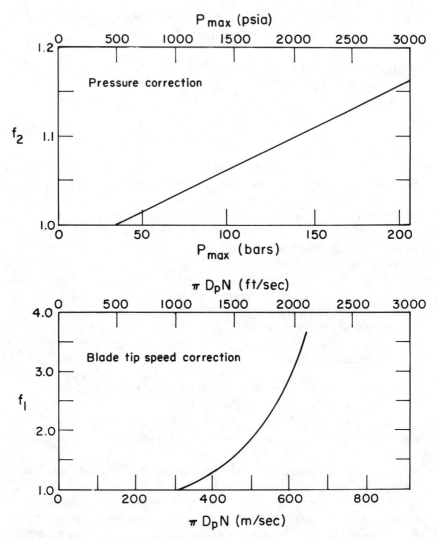

Figure 10-35. Turbine cost factors for pressure and blade tip speed. (Reprinted by permission from MIT Press).

parameters control drilling costs and determine power conversion and end-use efficiencies. Quantifying this assessment of quality for natural hydrothermal and geopressured systems is difficult. However, the differences between the very high-grade reservoirs throughout the world that are under commercial development versus lower-grade systems still awaiting development are easy to appreciate. For example, this situation exists in California in comparing the quality of The Geysers resource to that of the Salton Sea area.

For hot dry rock systems, commercial attractiveness in terms of resource-related parameters centers around the magnitude of the average geothermal gradient. High gradient areas ($>40°C/km$) will obviously have lower reservoir development costs than lower gradient regions, for the same reservoir conditions (temperature, size, etc.). Because of the strong dependence of drilling costs on depth, the difference between a $20°C/km$ and a $60°C/km$ gradient can result in as much as a factor of 20 in individual well costs to the same depth. As Cummings and Morris (1979) point out, lower gradients lead to lower optimal reservoir design temperatures. If constant design temperatures were selected, then busbar generating costs for break-even conditions would be proportionally higher than indicated in Figure 10-36. The parametric effects of gradient on optimal reservoir design temperatures can be examined by starting with the base case conditions cited in Table 10-7, (see Figure 10-37).

In many cases, a choice will have to be made among several power conversion options when electricity is the end-product. Table 10-8 lists favorable and unfavorable resource conditions for flashing, binary-fluid, and total flow systems. For efficient utilization, lower reservoir and ambient temperatures will favor binary-fluid cycles because of the higher densities of organic vapors relative to water and the smaller steam fractions that result in flashing. One should keep in mind, however, that of these options only flashing has been commercially demonstrated for plant capacities greater than 50-MW(e).

Reservoir Performance-Related Effects

With initial fluid temperatures and composition defined, well productivity and lifetime will determine the number of wells required for a given installed plant capacity. The most desirable approach would be to maintain a constant output temperature while maximizing the mass flow rate of fluid through the reservoir. As shown previously, the required geothermal fluid flow rate per MW(e) can be estimated once the efficiency of the electric power generation or heat utilization scheme is defined. For example, Figure 10-37 shows typical flow rates that might be required for binary-fluid plants. Because

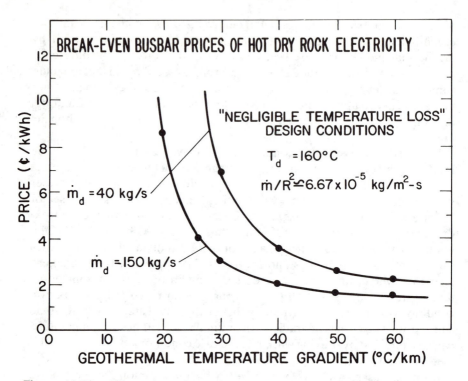

Figure 10-36. Effect of geothermal gradient on break-even busbar generating costs for hot dry rock systems with the reference design conditions listed in Table 10-7.

reservoir depletion for natural hydrothermal, geopressured, and hot dry rock systems is anticipated, some consideration should be given to the proper design and management of the combined power plant/reservoir system.

Only recently has the issue of reservoir depletion been considered in detail (Tester et al., 1979; Cummings and Morris, 1979; El-Sawy et al., 1979; Hankin et al., 1977). For illustrative purposes, the effect of reservoir drawdown on hot dry rock (HDR) systems is considered. Thermal drawdown rates for fractured HDR reservoirs will depend on: accessible fracture surface area, A; mass flow rate, $\dot{m}_W$; distribution of fluid across the fractured rock surface; and thermal properties of the rock (density, heat capacity and conductivity). A simplified approach to estimating reservoir performance would assume that a certain fraction, η, of the recoverable power, corresponding to uniform flow across the face of an ideal plane fracture, could be extracted. By solving the transient problem of one-dimensional heat conduction from the rock into the fracture face, the recoverable power, $P(t)$ in

Table 10-7
Hot Dry Rock Conceptual Reference Design Case Conditions
Exponential Drilling Costs Used (Figures 10-40 and 10-41)

Rate of return on equity..............................	12.0%
Rate of interest on debt	9.0%
Number of fractures	6
Fracture radius......................................	300 m
Geothermal gradient..................................	40°C/km
Plant design temperature	160°C
Design well-flow rate per pair of wells	75 kg/s
Capacity factor......................................	0.85
Plant lifetime	30 Years
Plant capacity.......................................	50 MW(e)
Operation and Maintenance	1.3 mills/kWh
Contingency	13%
Working capital	10% of surface plant cost
Taxes	51% of taxable income

From Cummings and Morris, 1979

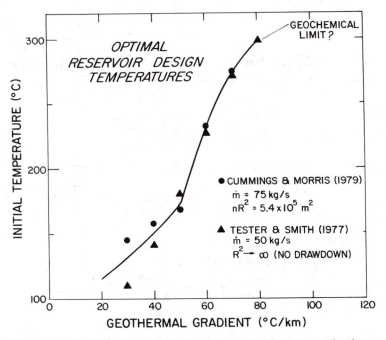

Figure 10-37. Effect of geothermal temperature gradient on optimal reservoir design temperature for hot dry rock systems with the reference design conditions listed in Table 10-7.

Table 10-8
Effect of Resource Conditions on Power Cycle Choice

Cycle type	Favorable	Unfavorable
Flashing	High wellhead temperature (>200°C)	Large concentrations of non-condensables
	High ambient temperatures (>35°C)	High relative well costs
Binary-fluid	Low wellhead temperatures (<150°C)	High cycle pressures may be required
	Low ambient temperatures (<30°C)	Working fluid instability and/or flammability
	Large concentrations of noncondensables; high relative well cost	Scaling and corrosion may be a problem
Total flow	Scaling problems	High relative well costs (poorer conversion efficiency)
	Corrosive brines	
	High ambient temperatures (>30°C)	Large concentrations of non-condensables

J/s, for uniform flow can be expressed as (see McFarland and Murphy, 1976; McFarland, 1974; and Wunder and Murphy, 1978):

$$\mathbf{P}(t) = \eta \dot{m}_W C_W \ (T_i - T_{\min}) \ \text{erf} \left(\sqrt{\frac{(\lambda \rho C)_r}{t}} \ \frac{\pi R^2}{m_W C_W} \right) \tag{10-14}$$

where: $A = \pi R^2$ = area of one face of the fracture, m^2
$\quad\quad C_w$ = heat capacity of water = 4,200 J/kgK
$\quad\quad C_r$ = heat capacity of granite = 1,000 J/kgK
$\quad\quad \dot{m}_w$ = water flow rate through the fracture, kg/s
$\quad\quad R$ = fracture radius, m
$\quad\quad t$ = time, s
$\quad\quad T_i$ = mean initial rock temperature, °C
$\quad\quad T_{\min}$ = fluid reinjection temperature, °C
$\quad\quad \lambda_r$ = thermal conductivity of granite, 3.0 W/m K
$\quad\quad \rho_r$ = rock density, ~2,500 kg/m³

McFarland and Murphy (1976) compared $\mathbf{P}(t)$ to estimate values that account for non-uniform flow across the accessible fracture area. Fluid buoyancy and convection effects within an ideal fracture as well as transient

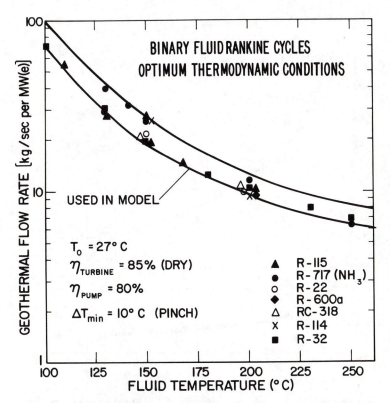

Figure 10-38. Geothermal fluid flow rate requirements per megawatt as a function of fluid temperature.

conduction of heat through the surrounding rock are treated in a numerical solution of the four coupled two-dimensional, nonlinear partial differential equations describing continuity, fluid momentum, and rock and fluid energy balances. Depending on the location and separation of fluid injection and recovery points within the fracture and the internal fracture permeability (gap width versus radius), the recovered fraction of power η may vary from 0.4 to 0.9 depending on the degree of buoyant circulation.

Equation 10-14 shows that the relative power $[\mathbf{P}(t)/\mathbf{P}(t=0)]$ depends directly on the error function of

$$K\left[R^2/(\dot{m}_W \sqrt{t}\,) \right] \qquad (10\text{-}15)$$

for constant rock and fluid properties, where $K = \pi\sqrt{(\lambda\, \rho C)_r/C_W^2}$

Consequently, reasonably accurate predictions of reservoir lifetime can be made for specified ideal fracture sizes and flow rates. Figure 10-39 presents

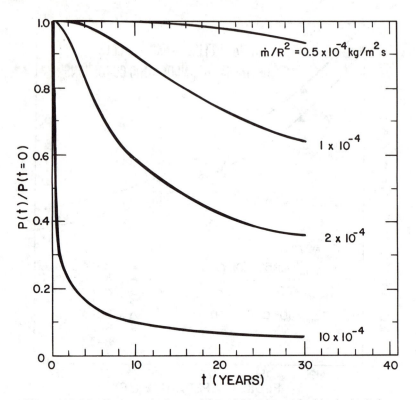

Figure 10-39. Parametric thermal drawdown curves for ideal, single fracture, hot dry rock reservoirs. Fractional thermal power history as a function of the mass loading parameter (m_w/R^2).

parametric results for the thermal power ratio versus time, t, using different values of $\dot{m}_w/R^2$ to generate a family of curves for a granitic single-fracture reservoir. For cases where large stable fractures cannot be produced, smaller, multiple, parallel fractures may be used to generate the required surface area to maintain an acceptable reservoir lifetime (Gringarten et al., 1975; Tester and Smith, 1977; and Raleigh, 1974). Because of the low thermal conductivity of granite, the penetration depth of the thermal wave is small and fractures spaced horizontally by 50 m or more avoid thermal interference over a 20- to 30-year period.

Cummings and Morris (1979) utilized this multiple-fracture concept to explore the effects of reservoir capacity and lifetime. Using the reference design conditions of Table 10-7 as a starting point, they examined the effects of changing the mass flow rate and the fracture surface area (which is proportional to a number of fractures) on the break-even busbar price using

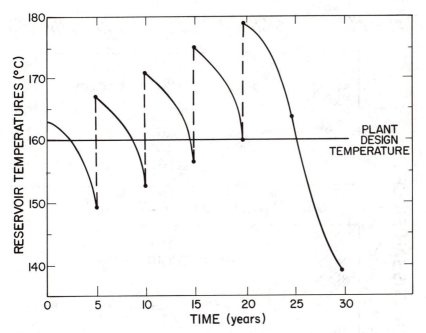

Figure 10-40. Reservoir temperatures as a function of thermal drawdown and redrilling phases for optimal management; base case conditions presented in Table 10-7. (Adapted by permission from Cummings and Morris, 1979).

the intertemporal discounted cash flow optimization model described previously. For the base case, significant drawdown occurs during each 5-year decision period with $\dot{m}_w/R^2 = 1.4 \times 10^{-4}$ kg/m²s, as shown in Figure 10-40. For optimal management, redrilling to a new region of hot (undepleted) rock is warranted for the first five 5-year periods. By applying this strategy, a minimum break-even condition results when the selling price of electricity is 4.3¢/kwh (p^*). The effect of flow rate, with $\dot{m}_w/R^2$ maintained constant to provide the same thermal drawdown rate, and the effect of drawdown rate, obtained with $\dot{m}_w = 75$ kg/s and the number of fractures varied to vary the effective radius ($R_e^2 = nR^2$) or surface area, are shown in Figure 10-41. In addition, in Figure 10-41 the effect of gradient is also depicted for the base case condition for comparison purposes.

Power Plant Performance — Related Effects

As discussed previously, several parameters related to power cycle operation affect costs. For example, for binary cycles, the optimal choice of

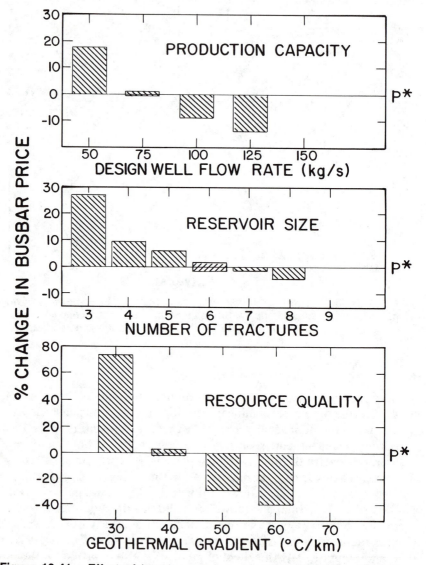

Figure 10-41. Effect of hot dry rock resource and reservoir performance parameters on break-even busbar price; variations from base case conditions presented in Table 10-7. (Adapted by permission from Cummings and Morris, 1979).

approach temperatures and cycle pressure of the primary heat exchanger and condenser depend on the cost of producing the geothermal fluid relative to the plant costs. Using an idealized approach for a liquid-dominated resource, cost-optimized temperature differences are shown for a binary cycle in Figure 10-42. For this case, ΔT_1 is the average temperature difference between the geothermal fluid and the organic working fluid in the primary heat exchanger and ΔT_2 is the average difference between the coolant and condenser/desuperheater temperature (see Figure 10-3A). The parameters C_1 and C_2 represent the costs per unit surface area ($/m^2$) of the primary heat exchanger and condenser/desuperheater, whereas U_1 and U_2 are the effective overall heat transfer coefficients (W/m^{2}K). The C_{gf} is the cost per unit of well flow rate ($/kg/s$) and reflects the cost of drilling, production and reinjection wells. For given values of C_1, C_2, U_1, U_2, and C_{gf}, optimized values of ΔT_1 and ΔT_2 are obtained from Figure 10-41. As well costs decrease relative to either condenser or primary heat exchanger costs, a limiting maximum value of ~65°C is reached for either ΔT. At the other extreme of very large relative well costs, low values of ΔT_1 and ΔT_2 result in optimal economics as one approaches the thermodynamic limiting conversion efficiency.

In order to show how cycle operating pressures may affect costs, Milora and Tester (1976) examined two case studies for a liquid-dominated (150°C) and a hot dry rock (250°C) resource. A discussion of the 150°C hydrothermal resource is presented to illustrate the important effects. In this case, a 100-MW(e) capacity R-32 (CH$_2$F$_2$) cycle was examined with constant, isothermal 45 kg/s artesian well flows for a 20-year plant lifetime assumed. A low-temperature (16.7°C) cooling water supply was also assumed. A minimum approach temperature of 10°C was assumed for the primary heat exchanger and condenser/desuperheater along with a dry-stage turbine efficiency of 85% and a feed pump efficiency of 80%. The working fluid turbine inlet temperature was 135°C, with a condensing temperature of 26.7°C. Reduced cycle pressures from 0.87 to 3.64 were studied using the computerized power cycle analysis code developed by Milora and Tester (1976). The exponential well cost model was used (Figures 10-27 and 10-29) as were the empirical equipment cost functions described previously. Total installed costs and their breakdown are shown in Figure 10-43.

Optimum economic conditions were achieved at a supercritical reduced cycle pressure P_r of 1.24 using R-32 as the working fluid. This is very close to the thermodynamic optimum ($\eta_u = 52.2\%$), which occurred at $P_r = 1.78$ (see Figure 10-22). It is reasonable to assume that operating pressures less than those corresponding to a maximum η_u would lower the total cost because of lower heat exchanger costs associated with lower pressure operation (see Figure 10-31). This is contingent on η_u not decreasing significantly

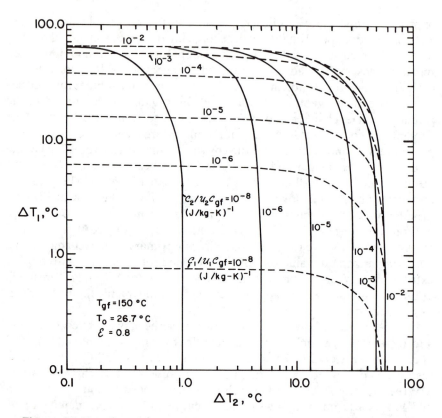

Figure 10-42. Cost-optimized heat exchanger (ΔT_1) and condenser (ΔT_2) temperature differences for 150°C geothermal wellhead temperature showing the relative effects of heat exchanger, condenser, and drilling costs. (Reprinted from Milora and Tester, 1976, by permission from MIT Press).

from the maximum value. At the economic optimum, η_u has only decreased to 50.6%. Figure 10-43 illustrates this effect by showing the component cost breakdown. Heat exchanger and condenser cost decreases from $640/kW to $505/kW as P_r decreases from 1.78 at the thermodynamic optimum to 1.24, whereas the well cost is essentially constant at $880/kW over this range. When P_r values are large and η_u decreases markedly, well and pump costs begin to dominate the economics. In contrast, turbogenerator costs remained relatively constant at $40 to $50/kW. The turbine component cost actually increased from $12/kW at a P_r of 1.78 to $20/kW at a P_r of 3.64, but the 100-MW(e) generator at $31/kW was still the major fraction of the turbogenerator unit cost.

Having established optimal cycle pressures and approach temperatures, the most important variables controlling surface plant costs are geothermal fluid temperature and composition and condensing temperature. For rela-

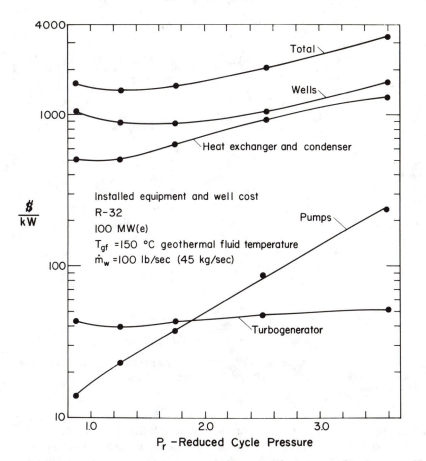

Figure 10-43.   Equipment and well cost break-down for a R-32 binary-fluid cycle with a 150°C liquid-dominated resource and heat rejection to a sink at 16.7°C. Costs (1976$) are a function of reduced cycle pressure. (Adapted from Milora and Tester, 1976, by permission from MIT Press).

tively brine-free fluids with low concentrations of non-condensable gases, binary plant costs ($/kW(e) installed) will decrease with increasing fluid temperature as shown in Figure 10-44. This decrease is primarily due to increased cycle efficiency as temperature increases, thereby decreasing heat exchange area requirements.

Financial-Related Effects

Levels of assumed risk, as expressed in the equity rate of return (r) and the debt of interest rate (i), and drilling costs are perhaps the most important

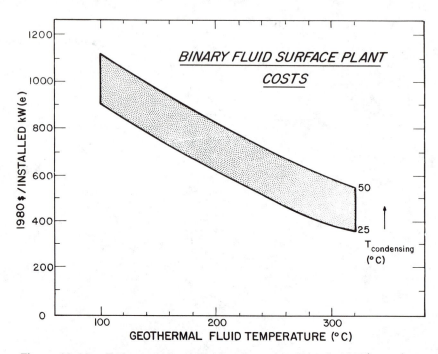

Figure 10-44.   Estimated binary-fluid surface plant costs in 1980$ as a function of geothermal fluid and condensing temperatures.

parameters controlling overall commercial feasibility. Their effect as well as plant capacity factor, lifetime, and size, are given in Figure 10-45 as percent changes around the base case conditions for the hot dry rock study described previously (see Table 10-7). Selling price of the produced energy is another very important factor. Its effect can be explored either by calculating a selling price corresponding to break-even conditions and comparing it with the current market price, or by using current market prices to determine whether or not estimated revenues are sufficient to balance costs.

Case Studies

A large number of studies have been conducted in recent years to assess the commercial feasibility of hydrothermal, geopressured, and hot dry rock resources. Many of these are cited in Table 10-3. Of particular interest for hydrothermal vapor- and liquid-dominated systems are the papers that appear in the 1st and 2nd U.N. Symposium on the Development and Use of Geothermal Resources, 1971, 1975; and the 11th and 12th Intersociety Energy Conversion Engineering Conferences, 1976, 1977. In addition, stu-

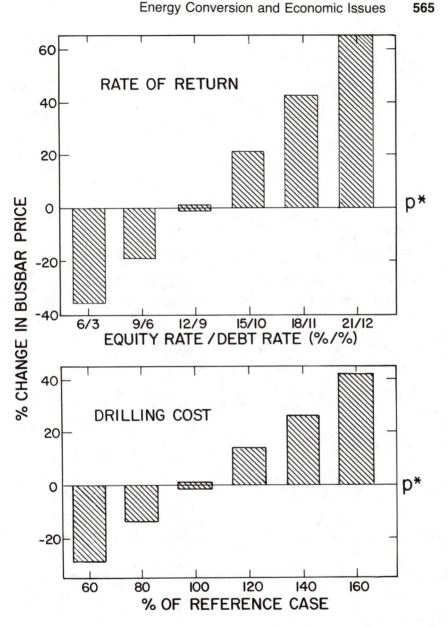

Figure 10-45. Effect of various financial and economic assumptions on break-even busbar prices for hot dry rock systems; variations from base case conditions presented in Table 10-7. (Adapted by permission from Cummings and Morris, 1979).

(figure continued on next page)

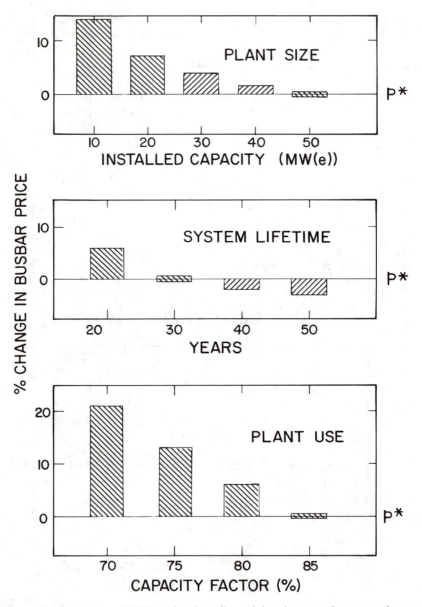

Figure 10-45. (continued) Effect of various financial and economic assumptions on break-even busbar prices for hot dry rock systems; variations from base case conditions presented in Table 10-7. (Adapted by permission from Cummings and Morris, 1979).

dies by Hankin et al. 1977; Eskesen, 1977a,b; Holt and Ghormley, 1976; and Ramachandran et al. 1977 examined several particular sites in the western U.S.A. Generic economic modeling of geothermal resources has been under development at the Lawrence Berkeley Laboratory (Green and Pines, 1975; and Pope et al. 1977 and Battelle Pacific Northwest Laboratories (Bloomster et al. 1975, 1976). Additional current information on the economic assessment of hydrothermal systems can be found in Kestin et al. 1980.

As shown in Table 10-3, very few comprehensive studies of geopressured and hot dry rock resources exist. Important contributions include those by Wilson (1977) on geopressured resources and by Cummings et al. 1979; Cummings and Morris, 1979; Lang and Heidt, 1979; and Tester et al. 1979 on hot dry rock reservoirs.

Many approaches have been utilized in the studies cited. These include factored estimate techniques to estimate capital costs, combined with appropriate annual fixed charge rates and load factors to calculate busbar generating costs. Capital recovery factor, discounted cash flow, and present value approaches are also commonly used to define break-even busbar costs.

As the MITRE study so vividly pointed out (El-Sawy et al., 1979), even with similar capital costs for drilling and power plant components and reservoir characteristics, widely different estimates of busbar costs can result. Consequently, some caution should be exercised before using the results of any case study as the sole basis for financial investment or cost comparison with other energy alternatives.

Economics of Nonelectric Heat Production

Because of the wide variety of end-uses in the nonelectric heat production area, economic factors cannot be uniquely defined and only some generalizations can be made. For example, issues relevant to geothermal district heating may not be completely applicable to industrial process heating. Costs will depend on the following major parameters: resource quality (fluid quality, temperatures, depths, and gradients); reservoir performance (lifetime and capacity); load or capacity factor; and fluid transmission and distribution costs. The first two of these factors parallel the criteria applied to electricity production with two exceptions. Inasmuch as end-use efficiencies are much higher, lower temperatures ($<150°C$) can be utilized effectively. This will lead to commercial development of lower-grade resources ($\Delta T < 40°C/km$). Secondly, fluid composition may have a larger impact on cost if an indirect heat exchange step and/or reinjection can be avoided in the direct use cycle.

Nonelectric systems have very high capital costs in comparison to more conventional fossil-fired installations. Consequently, the annual demand for heat, expressed as a load or capacity factor, is critical in determining the competitiveness of geothermal direct use. For example, space heating has a relatively low capacity factor in the most populated parts of the world. For example, in New York the annual load factor — the actual heat consumed per year divided by the maximum amount of heat that could be provided if the geothermal system operated continuously throughout the year — would range from 0.2 to 0.3. In contrast, the extensive district space heating development in Iceland is in part made more attractive by an annual load factor of 0.5 to 0.6 (EPRI, 1978, and Armstead, 1978).

If reservoir drawdown is significant, direct use systems that may require a minimum temperature or a specific steam enthalpy may demand augmentation. For example, partially depleted reservoir fluids may be mixed with fluids from hotter reservoirs to produce desired temperatures, or fossil-fired peaking boilers or heaters could be used.

One other major factor involves the magnitude of the heating demand and the distance between the geothermal reservoir and the load center. Acceptable distances will be primarily dependent on drilling and alternative energy costs. If drilling costs are low and costs of alternatives are high, very long transmission distances may result in economic systems. Distances of 10-30 km appear feasible for a number of situations.

Another issue concerns whether or not retrofitting is required. Although common in Europe, centralized district heating systems using fossil resources are rare in the U.S. with the exception of parts of New York City, many college and university campuses, and government complexes. In addition, many houses are not equipped for hot water heating. Consequently, retrofitting to install in-house heating systems, as well as a fluid distribution network will be costly requiring street excavation and building modifications.

There are a number of excellent references that consider the economics of nonelectric applications in more detail than presented here; for example, the *Proceedings of the NATO-CCMS Conference on the Economics of Direct Uses of Geothermal Energy,* held in June, 1977, in Washington, D.C. The computerized model GEOCITY developed at Battelle Northwest Laboratories (McDonald et al., 1977) for district heating is of particular interest because of its careful documentation and its applicability to a wide variety of climatic, location, and resource characteristics and financial and tax assumptions.

Case Studies

Iceland, Hungary, Japan, USSR, New Zealand, Paris, and the U.S. cities Boise (Idaho) and Klamath Falls (Oregon) have extensive geothermal district heating systems in operation today (Armstead, 1978; EPRI, 1978;

Bloomster et al., 1978; and Kruger and Otte, 1973). For instance, Iceland and New Zealand have a combined capacity of over 600-MW (thermal). In view of the current escalation of fossil fuel costs and relatively inexpensive reservoir development costs in certain areas, these systems can be very attractive financially. Future developments for geothermal district use systems are being planned by a number of countries partly in response to the current world fuel situation.

Readers interested in learning more about potential application should consult EPRI, 1978; Keller and Kunze, 1976; Howard, 1975; Reistad, 1975; Bodvarsson et al., 1974; Delisle et al., 1975; and Geothermal Resources Council, 1978. Other specific case studies are worth mentioning because of their comprehensive treatment. These include presentations by Arnorsson et al., 1975; Bodvarsson and Zoega, 1961; Lienau, 1978; Kunze, 1977; and Lund, 1977.

Summary of Commercial Feasibility Issues and Prospects

Geothermally-produced electric power and heat are already a commercial reality in many parts of the world. For example, the Lardarello dry steam fields in Italy began commercial power generation in the early 1900s. Similarly, development of the U.S. vapor-dominated field at The Geysers in California has proceeded from the early 1960s to a present installed capacity of over 930 MW(e). In addition, a number of liquid-dominated fields have been also commercially developed for electricity generation and for space or process heating. These include Klamath Falls, Oregon; Wairakei, New Zealand, Cerro Prieto, Mexico; and a number of areas in Iceland, the Philippines, and the USSR. The remaining vast potential of natural hydrothermal, geopressured, and hot dry rock reservoirs, however, have not been developed. A partial explanation for this has been that alternatives such as oil, coal, and natural gas were actually or presumed to be less expensive than geothermal. Because of recent changes in worldwide fossil fuel availability and supply, cost escalations have been rapid. Consequently, financial incentives for developing geothermal energy are on the increase.

The future situation is much more uncertain. Many proven high-grade hydrothermal systems exist throughout the world (see Chapters 2, 3, and 4). Their commercial development rate, however, has been slowed by a number of factors; some are based on financial problems of raising sufficient investment capital, whereas others are focused on institutional, regulatory, or environmental constraints. In the case of geopressured and hot dry rock, major technical feasibility questions concerning the productivity and lifetime of these reservoirs remain. A large research and development effort sponsored mainly by the U.S. Department of Energy with some private support is examining geopressured systems for hot water and methane recovery (Samuels, 1979). Hot dry rock is currently being examined in a large project as Fenton Hill, New Mexico supported jointly by the U.S. Department of Energy, Japan, and West Germany. Another large project located in Cornwall, England is supported jointly by the European Economic Community (EEC) and the U.S. Department of Energy. Other smaller projects are underway in Sweden, Switzerland, Germany, and Italy. The fate of geopressured and hot dry rock development in the near term will to a large degree depend on the outcome of these important projects.

Important Issues

In evaluating the potential commercial feasibility of any geothermal reservoir, care must be exercised to consider all pertinent issues. As described previously, some of these factors are site-specific and relate to the characterisitics of the resource and reservoir, others depend on the design and operation of the power plant, and still others depend on a wide variety of financial and management issues. Table 10-9 lists the major factors and issues in these areas. Any one of these can greatly influence the ultimate commercial viability of a specific geothermal prospect. For example, uncertainty, perceived or real, in the productivity of the reservoir including mass

Table 10-9
Major Economic Factors Affecting Geothermal Development

Resource-related
Depth to the reservoir (geothermal gradient)
Rock type and temperature
Fluid composition
Liquid-to-gas ratio
Heat rejection conditions
Water availability

Reservoir and plant engineering-related
Mass flow rate (well productivity)
Thermal — pressure drawdown (lifetime)
Conversion efficiency [(lb/hr per kW(e) or kg/s per MW(e)]

Financial-related
Drilling and plant capital costs
Equity discount rate
Debt interest rate
Selling price of products (¢/kWh or $/MMBtu)
Tax treatment and royalty payments

flow rates, temperatures, and/or chemical composition of the fluid will lead to a sluggish investment climate that demands high rates of return. Government intervention in the form of investment tax credits, loan guarantees, or actual cost sharing may help.

The inherent differences between generating power in a fossil- or nuclear-fueled versus a geothermal facility cannot be avoided. Confidence in a given geothermal prospect certainly increases as wells are drilled and produced. But it may take years or even decades, based on past history, to develop a field to its maximum commercial potential. Environmental regulations can also have a large impact on the rate of development. For example, H_2S emission standards and abatement system acceptance is a major factor in growth at The Geysers, California. Fear of water table depletion or of aquifer contamination by reinjection are other major environmental issues that can influence the commercial attractiveness of a geothermal venture.

Comparison with Conventional Energy Supplies

One major difference that distinguishes geothermal energy from other new alternatives such as solar, fusion, and wind is that commercially acceptable prices for electric power and heat have already been achieved. For example, electricity produced by Pacific Gas and Electric at The Geysers is less expensive than that from any other source in California, except existing hydroelectric. A similar situation exists for space heating applications in Klamath Falls, Oregon. Whether this trend continues for new fields in the U.S. and elsewhere will in fact depend on how fossil and nuclear costs escalate. A breakdown of capital equipment and fuel costs ranges is presented in Table 10-10 for geothermal, fossil-fuel, and nuclear electric power plants. Costs are listed in 1980 dollars and represent new installed capacity estimates. At the low end of the geothermal costs for vapor-dominated systems, there is no question that they can compete with current alternatives. Liquid-dominated systems such as those in the Imperial Valley, California, and the Valles Caldera, New Mexico areas certainly are competitive at the low-end estimate with some question at the high end. Although geopressured and hot dry rock systems are still under development to demonstrate technical feasibility, reasonable assumptions with regard to drilling and reservoir development costs result in a competitive range of estimated busbar prices.

In the nonelectric and cogeneration area, the eventual commercial geothermal potential may be very large. Direct use avoids the problems with low electric conversion efficiency and essentially upgrades the market value of the resource. For instance, geothermal heat available at 150°C could be utilized directly with ~80% efficiency, whereas conversion to electricity

Table 10-10
Base Load Busbar Cost Estimates for New Generating Capacity in the U.S.A.
(1981 $)

Resource Type	Installed power plant costs, $/kW	Annual power plant costs*, ¢/kWh	O + M, ¢/kWh	Well or fuel cost, ¢/kWh	Total generating cost, ¢/kWh	References (see below)
Nuclear	1200	2.9	0.1	0.4	3.4	1,2,10
Oil	600-800**	1.5-1.9	0.1	6.0 ($36/bbl)	7.6-8.0	1,2
Coal	600-1000**	1.5-2.4	0.2	1.2 ($30/ton)	2.9-3.8	1,2
Hydrothermal (vapor)†	300	0.8	0.1	1.3	2.1	1,3
Hydrothermal (liquid)††						
Flashing	500-800	1.2-1.9	0.3	1.7-2.7	3.2-4.9	1-3
Binary fluid	550-950	1.3-2.3	0.3	1.5-2.5	3.1-5.1	
Geopressured§ (incl. Methane at 40 SCF/bbl)	875-750	2.1-1.8	0.4	1.8-3.8‖	4.3-6.0	4-7
Hot Dry Rock	550-950	1.3-2.3	0.3	1.8-4.2	3.4-6.8	8,9

The above data are adapted from estimates provided by the following references:

1. Ramachandran et al. (1977)
2. Milora and Tester (1976)
3. Holt and Ghormley (1976)
4. Greider (1973)
5. Wilson et al. (1975) and Wilson (1977)
6. Samuels (1979)
7. Bloomster and Knutsen (1976)
8. Cummings and Morris (1979)
9. Tester, Morris, Cummings and Bivins (1979)
10. Rossin and Rieck (1978)

*Based on a 17% annual fixed charge rate; 80% load factor (7000 hr/yr at capacity).

**Higher capital costs include more advanced pollution abatement systems.

†Current projections for The Geysers (see ref. 1 above).

††150-200°C resources with well flow rates from 100-300 lb/sec; reinjection required; ~0% non-condensable gases.

§40,000 bbl/day production, 155°C, 2000-psi geopressured resource.

‖Includes credit for methane recovered.

would be only ~10% efficient. Consequently, the cost of producing the fluid $/kW (thermal) utilized would be greatly reduced over the electric case. A major concern in direct use, of course, is the distance between the field and the user. Fluid transmission costs can quickly dominate the economic picture if sites are located more than a few kilometers away. Hot dry rock systems may have some distinct advantages over natural hydrothermal and geopressured resources in this regard, because they are not restricted to specific locations.

Worldwide Developments: Present and Future

Koenig (1973), Rex and Howell (1973), Rex (1978), DiPippo (1980a,b), Roberts and Kruger (1979), and the Electric Power Research Institute (EPRI, 1978) carefully reviewed the status of worldwide geothermal developments. In Table 10-11, current and future installed capacities by country for electric and nonelectric applications are presented. Future capacities are based on questionnaires and should be viewed with some caution. Even with the constraints and impediments cited earlier, however, the implication of significant growth for the geothermal industry in the coming years is clear. This is perhaps best illustrated by the recent activities in the Philippines.

Of the over 400,000 MW of electrical generating capacity in the U.S., less than 0.3% currently has a geothermal source. Although, this percentage will not increase significantly in the coming decades, as shown by the predictions cited in Table 10-11, the displacement of oil consumption by geothermal energy will be important. For example 30,000 MW(e) of oil-fired electric capacity consumes approximately 1 million barrels of oil per day. According to numerous estimates (Roberts and Kruger, 1979; DiPippo, 1980 a,b; and EPRI, 1978) the U.S. will have greater than 3,000 MW(e) of geothermally produced electricity by the year 2000, so that it could have an important role in liquid fuel conservation by replacing existing gas-or oil-fired capacity. At least nine specific sites are under private or private/government development. These include: (1) The Geysers — Calistoga area in California, (2) Imperial Valley, California, (3) Valles Caldera, New Mexico, (4) Roosevelt Hot Springs, Utah, (5) Hilo, Hawaii, (6) Raft River, Idaho, (7) Klamath Falls, Oregon, (8) Gulf Coast region of Texas and Louisiana, and (9) Fenton Hill, New Mexico. The first seven sites are liquid-dominated systems with the exception of The Geysers. A number of companies are involved with the development of The Geysers — Calistoga field. The present 930 MW(e) electric capacity is expected to grow to 1,700 MW(e) by 1985 (Rex, 1978). At the present time, it is the only commercial operation in the U.S. Rapid development of numerous other sites, however, is anticipated during the period 1980-2000 where demonstration size plants [10-50 MW(e)] should be

Table 10-11
Geothermal Electric Generating and Nonelectric Capacities

Country	Installed (1981)		Future Projections by 2000	
	Electric MW(e)	Nonelectric* MW(t)	Electric MW(e)	Nonelectric* MW(t)
United States	930	17	~2,000	~3,400
Philippines	500	4.9	765	?
Italy	420	24	800	38
New Zealand	203	196	400	380
Japan	165	2,900	~2,000	3,100
Mexico	150	—	150	680
El Salvador	95	—	180	—
Iceland	32	410	150	570
USSR	5	4,860	?	?
China (Taiwan)	1.9	0.6	?	5
Turkey	0.5	0.2	400	?
Nicaragua	—	—	150	?
Costa Rica	—	—	100	—
Guatemala	—	—	100	—
Honduras	—	—	100	—
Panama	—	—	60	—
Argentina	—	—	20	—
Portugal	—	—	30	—
Spain	—	—	25	10
Kenya	—	—	30	—
Indonesia	—	—	30-100	—
Thailand	—	—	~10	—
Canada	—	—	—	10
England	—	—	—	~2
France	—	24	—	490
Hungary	—	1,050	—	?
Czechoslovakia	—	93	—	?
Yugoslavia	—	4.9	—	58
Total	2,502 MW(e)	9,585 MW(t)		

From DiPippo (1980a,b), Roberts and Kruger (1979) and EPRI (1978)
*Capacities less than 1 MW(t) are not listed.

placed on-line in the East Mesa [58 MW(e)], Heber [50 MW(e)], North Brawley [100 MW(e)], and Salton Sea [10 MW(e)] areas of the Imperial Valley; the Baca field in the Valles Caldera [50 MW(e)]; and Roosevelt Hot Springs [50 MW(e)]. Geopressured reservoirs in the Gulf Coast area, U.S., are undergoing extensive assessment with field tests underway to determine reservoir productivity and lifetime, dissolved methane concentration, and subsidence effects (Wilson, 1977; Samuels, 1979).

The Fenton Hill, New Mexico site is the first site in the U.S. aimed at exploiting hot dry rock resources. Prototype heat extraction experiments have been conducted by the Los Alamos National Laboratory under U.S. D.O.E. sponsorship to form a fractured reservoir in low-permeability crystalline basement rock at depths of 3km (Tester and Albright, 1979). Although these experiments have demonstrated the technical feasibility of many facets of hot dry rock systems, large effective heat transfer areas with acceptable flow impedance and distribution through fractured zones have yet to be proven. Although the first reservoir system developed in 1977 had only 8,000 m^2 of effective surface and was too small to be of commercial value, workover operations including additional hydraulic fracturing were successful in enlarging the system to greater than 50,000 m^2 (Murphy, et al., 1980). Furthermore, this area consists of essentially vertical fractures separated by about 300 m between primary injection and recovery points. They can be viewed as building-block or module fractures for the next phase of the Los Alamos project aimed at producing a commercial-sized reservoir with multiple, parallel fractures from inclined wellbores separated vertically by 300 m. This 2-hole reservoir should have about 1 million m^2 of active surface area at an average depth of 3.5 km ($\sim$250°C) to result in a 50 MW(t) capacity with extended lifetime.

Other major field developments in the United Kingdom in Cornwall and in West Germany will play considerable roles in evaluating the feasibility of hot dry rock (Batchelor et al., 1978; Pearson, 1980; Batchelor, 1980; and Kappelmeyer et al., 1979).

If one accepts that the technical reservoir engineering problems associated with hydrothermal, geopressured, magma, and hot dry rock resources can be solved, perhaps the most critical economic factors affecting the growth of geothermally produced energy will include drilling and plant capital cost escalation relative to alternatives, the pricing of competitive fossil and nuclear fuels, tax treatment and royalties, and finally the assessment of risk and required rate of return to stimulate investment.

In any situation there are no simple methods of predicting commercial viability. Each site must be examined in detail with every issue considered. It is clear, however, that economically competitive geothermal development will continue worldwide; the only uncertainty concerns its rate of growth.

References

Abdel-Aal, H.K. and Schmelzlee, R., 1976. *Petroleum Economics and Engineering an Introduction,* Marcel Dekker, Inc., New York.

Aikawa, K. and Soda, M., 1975. Advanced Design in Hatchobaru Geothermal Power Station, *2nd United Nations Symposium on the Development and Use of Geothermal Resources,* San Francisco, CA. *3,* pp. 1881-1888, May.

Allen, C.A., Grimmett, E.S., and Wagner, K.L., 1977. Fluidized Bed Heat Exchangers for Geothermal Applications. *Proc. of 11th Intersociety Energy Conversion Engineering Conference,* State Line, NV, pp. 761-767, September.

Altseimer, J.H., 1976. Technical and Cost Analysis of Rock-Melting Systems for Producing Geothermal Wells, Los Alamos National Laboratory Report LA-6555-MS, Los Alamos, NM, November.

Altseimer, J.H., 1974. Geothermal Well Technology and Potential Applications of Subterrene Devices — A Status Review. Los Alamos National Laboratory report LA-5689-MS, Los Alamos, NM, August.

Anderson, J.H., 1973. The Vapor-Turbine Cycle for Geothermal Power Production. In: *Geothermal Energy,* Stanford Univ. Press, Stanford, CA.

Anno, G.H., Dore, M.A., Grijalva, R.L., and Thomas, F.J., 1977. Site-Specific Analysis of Hybrid Geothermal/Fossil Power Plants, Pacific-Sierra Research Corp. report 705, April.

Armstead, H.C.H., 1978. *Geothermal Energy,* J. Wiley & Sons, New York.

Arnold, E., 1970. *Steam Tables in SI Units,* London, U.K.

Arnovsson, S., Raguars, K., Bendiktsson, S., Gislason, G., Thorhallsson, S., Bjornsson, S., Gronvold, K., and Lindal, B., 1975. Exploitation of Saline High Temperature Water for Space Heating. *2nd UN Symposium on the Development and Use of Geothermal Resources,* San Francisco, CA, May.

Aronson, D., 1961. Binary Cycle for Power Generation. *Proc. of Amer. Power Conf.,* 23.

Austin, A.L., Higgins, G.H., and Howard, J.H., 1973. The Total Flow Concept for Recovery of Energy from Geothermal Hot Brine Deposits. Lawrence Livermore Laboratory report UCRL-51366, Livermore, CA, April.

Austin, A.L., 1980. Status of the Development of the Total Flow System for Electric Power Production. In: *A Sourcebook on the Production of Electricity from Geothermal Energy,* J. Kestin (ed.). U.S. Government Printing Office, Washington, D.C. (January). Also in report C00-4051-21/CATMEC/15 Brown University, Providence, RI, April 1978.

Baljé, O.E., 1962. A Study on Design Criteria and Matching of Turbomachines, Parts A & B. *J. Eng. for Power,* Trans. ASME:83, January.

Banwell, C.J., 1975. Geothermal Energy and Its Uses: Technical, Economic, Environmental, and Legal Aspects. *2nd United Nations Symposium on the Development and Use of Geothermal Resources,* San Francisco, CA, *3,* pp. 2257-2267, May.

Barber-Nichols Engineering Co., Arvada, CO, Contract #LP8-5395D, 60 kW Thermal Electric Generating Unit R-114, with the Los Alamos National Laboratory (1979).

Barkman, J., 1979. Personal communication, Republic Geothermal, Santa Fe Springs, CA, April.

Barr, R.C., 1975. Geothermal Energy and Electrical Power Generation, *2nd United Nations Symposium on the Development and Use of Geothermal Resources,* San Francisco, CA, *3,* pp. 1937-1941, May.

Barr, R.C., 1975. Geothermal Exploration Strategy and Budgeting. *2nd United Nations Symposium on the Development and Use of Geothermal Resources,* San Francisco, CA, *3,* pp. 2269-2271, May.

Batchelor, A.S., et al., 1978. Camborne School of Mines Geothermal Exploitation Research, 1st Annual Report, October 1977 — December 1978. Camborne School of Mines, Redruth, Cornwall, UK.

Batchelor, A.S., 1980. Hot Dry Rock Well Stimulation at a Depth of 2,000 m in Granite, Camborne School of Mines Geothermal Exploitation Research bimonthly circular, Camborne School of Mines, Redruth, Cornwall U.K., Nov.

Beall, S.E., 1970. Agricultural and Urban Uses of Low-Temperature Heat. In: *Beneficial Uses of Thermal Discharges,* S.P. Mathur and R. Stewart, (eds.) New York State Department of Environmental Conservation, Albany, NY, pp. 185-202.

Beall, S.E., 1971. Waste Heat Uses Cut Thermal Pollution. *Mech. Eng.* 93, 15.

Beall, S.E. and Samuels, G., 1971. The Use of Warm Water for Heating and Cooling Plant and Animal Enclosures. Oak Ridge National Laboratory report ORNL-TM-3381, Oak Ridge, TN.

Bee Dagum, E.M. and Heiss, K-P., 1968. An Econometric Study of Small and Intermediate Size Diameter Drilling Costs for the United States. PNE-3012 (Vol. 1 and 2) prepared for the U.S. Atomic Energy Commission by Mathematica, Princeton, NJ, June.

Bechtel Corporation, 1975. Electric Power Generation Using Geothermal Brine Resource for a Proof-of-Concept Facility ERDA Grant AER74-19931 A01, San Francisco, CA, May.

Bechtel Corporation, 1976. Conceptual Design 50 MWe (Net) Geothermal Power Plants at Heber and Niland, California. Final report #SAN-1124-1, San Francisco, CA, October.

Beim, D.O., 1978. Project Financing of Geothermal Development. In: *Commercialization of Geothermal Resources.* Geothermal Resources Council, Davis, CA, November.

Berman, E.R., 1975. *Geothermal Energy,* Noyes Data Corporation, Park Ridge, NJ.

Bloomster, C.H., 1980. Economic Considerations. In: *Sourcebook on the Production of Electricity from Geothermal Energy,* J. Kestin (ed.), U.S. Government Printing Office, Washington, D.C., January.

Bloomster, C.H., Fassbender, L.L., and McDonald, C.L., 1978. Geothermal Energy Potential for District and Process Heating Applications in the U.S. — An Economic Analysis. In: proceedings of the NATO-CCMS Conference on the Economics of Direct Uses of Geothermal Energy, U.S. Department of Energy report CONF-770681, Washington, D.C., July.

Bloomster, C.H., Cohn, P.D., DeSteese, J.G., Huber, H.D., LaMori, P.N., Shannon, D.W., Sheff, J.R., and Walter, R.A., 1975a. GEOCOST: A Computer Program for Geothermal Cost Analysis. Battelle Northwest Laboratory report BNWL-1888, UC-13 Richland, WA, February.

Bloomster, C.H., 1975b. An Economic Model for Geothermal Cost Analysis. *2nd United Nations Symposium on the Development and Use of Geothermal Resources,* San Francisco, CA, *3,* pp. 2273-2282, May.

Bloomster C.H. and Engel, R.L., 1976. The Potential Benefits of Geothermal Electricity Production from Hydrothermal Resources. Battelle Pacific Northwest Laboratories report BNWL-2001, Richland, WA, June.

Bloomster C.H. and Knutsen, C.A., 1976. An Analysis of Electricity Production Costs from the Geopressured Geothermal Resource. Battelle Pacific Northwest Laboratories report BNWL-2192, Richland, WA, February.

Bodvarsson, G., Boersma, L., Couch, R., David, L. and Reistad, G., 1974. Systems Study for the Use of Geothermal Energies in the Pacific Northwest. Oregon State University report RLO-2227-T19-1, Corvallis, OR.

Bodvarsson, G. and Zoega, J., 1961. Production and Distribution of Natural Heat for Domestic and Industrial Heating in Iceland. Paper G/37, *Proceedings of the U.N. Conference on New Sources of Energy,* Rome, Italy, *3,* p. 449.

Boehm, R., Jacobs, H., Bliss, R., and Kelly, D., 1976. Direct Contact Heat Exchangers for Geothermal Power Plants. *Proceedings of 11th Intersociety Energy Conversion Engineering Conference,* State Line, NV, pp. 754-760, September.

Bringer, R.P. and Smith, J.M., 1957. Heat Transfer in the Critical Region. *AIChE J.* v. 3, p 49.

Bupp, C., Devian, J.C., Donsimoni, M-P., and Treitel, R., 1975. The Economics of Nuclear Power. *Technology Review,* 77 (4), 14.

Callighan, J.A., 1971. Thermal Stability Data on Six Fluorocarbons, Heating/Piping/Air Conditioning. Reinhold Publishing Corp., pp. 119-126, September.

Cataldi, R., DiMario, P., and Leardini, T., 1973. Application of Geothermal Energy to the Supply of Electricity in Rural Areas, *Geothermics* 2 (1), pp. 3-16.

Cerini, D.J., 1977. A Two-Phase Rotary Separator Demonstration System for Geothermal Energy Conversion. *Proceedings of 12th Intersociety Energy Conversion Engineering Conference*, Washington, D.C., Vol 1, pp. 884-892.

Chappell, R.N., Prestwich, S.I., Miller, L.G., and Ross, H.P., 1979. Geothermal Well Drilling Estimates Based on Past Well Costs, Geothermal Resources Council *Transactions, 3,* 99.

Cheremisinoff, P.H., and Morresi, A.C., 1976. *Geothermal Energy Technology Assessment,* Technomic Publishing, Westport, CT.

Chou, J.C.S., 1973. Regenerative Vapor-Turbine Cycle for Geothermal Power Plant. *Geothermal Energy* 2, 21.

Chou, J.C.S., Ahluwalia, R.K., and Woo, E.Y.K., 1974. Regenerative Vapor Cycle with Isobutane as Working Fluid. *Geothermics* 3 (3), pp. 93-99.

Cortez, D.H., Holt, B., and Hutchinson, A.J.L., 1973. Advanced Binary Cycles for Geothermal Power Generation. *Energy Sources* 1 (1), 74.

Cummings, R.G., Morris, G.E., Tester, J.W., and Bivins, R.L., 1979. Mining Earth's Heat: Hot Dry Rock Geothermal Energy. *Technology Review 81*(4), pp. 57-78, February.

Cummings, R.G., and Morris, G.E., 1979. Assessing the Economics of Producing Electricity from Hot Dry Rock Geothermal Resources: Methodology and Analyses. Electric Power Research Institute and Los Alamos National Laboratory report, November.

Dan, F.J., Hersam, D.E., Kho, S.K., and Krumland, L.R., 1975. Development of a Typical Generating Unit at The Geysers Geothermal Project. *2nd United Nations Symposium on the Development and Use of Geothermal Resources, 3,* San Francisco, CA, pp. 1949-1958, May.

Dart, R.H., and Whitbeck, J.F., 1980. Brine Heat Exchangers. In: *Sourcebook on the Production of Electricity from Geothermal Resources,* J. Kestin (ed.), U.S. Government Printing Office, Washington, D.C., January.

Delisle, G., Kappelmeyer, O., and Haend, R., 1975. Prospects for Geothermal Energy for Space Heating in Low Enthalpy Areas. *2nd United Nations Symposium on the Development and Use of Geothermal Resources,* San Francisco, CA, pp. 2283-2289, May.

DiPippo, R., 1980a. Geothermal Power Plants Around the World. In: *A Sourcebook on the Production of Electricity from Geothermal Energy,* J. Kestin et al. (eds.), U.S. Government Printing Office, Washington, D.C., March.

DiPippo, R., 1980b. *Geothermal Energy as a Source of Electricity: A Worldwide Survey of the Design and Operation of Geothermal Power Plants,* U.S. Government Printing Office, Washington, D.C., January.

Doane, J.W., O'Toole, R.P., Chamberlain, R.G., Bos, P.B., and Maycock, P.D., 1976. The Cost of Energy from Utility-Owned Solar Electric Systems. Jet Propulsion Laboratory report 5040-29, ERDA/JPL-1012-76/3, Pasadena, CA, June.

Dosher, T.M., Osborne, R.H., Wilson, T., Rhee, S.W., et al., 1978. The Economics of Producing Methane from Geopressured Aquifers Along the Gulf Coast. Petroleum Engineering Dept. report, Univ. of Southern California, Los Angeles, CA, March.

Eisenstat, S.M., 1978. Federal Tax Treatment of Geothermal Exploration and Production. In: *Commercialization of Geothermal Resources.* Geothermal Resources Council, Davis, CA, November.

Electric Power Research Institute, 1978. Geothermal Energy Prospects for the Next 50 Years, EPRI report ER-611-SR, Palo Alto, CA, February.

Electric Power Research Institute, *Proceedings of 3rd Annual Meeting on Geothermal Energy,* Monterey, CA, (June 1979).

El-Sawy, A.N., Leigh, J.G., and Trehan, R.K., 1979. A Comparative Analysis of Energy Costing Methodologies. The Mitre Corporation report MTR-7689, McLean, VA, February.

Elliott, D.G., 1976. Comparison of Geothermal Power Conversion Cycles, proceedings of 11th Intersociety Energy Conversion Engineering Conference, State Line, NV, pp. 771-777, September.

Eskesen, J.H., 1977a. Study of Practical Cycles for Geothermal Power Plants. General Electric Co report COO-2619-1 UC-66, Contract No. EY-76-C-02-2619, April.

Eskesen, J.H., 1977b. Cost and Performance Comparison of Flash Binary and Steam Turbine Cycles for the Imperial Valley, California. *Proceedings of 12th Intersociety Energy Conversion Engineering Conference,* Washington, D.C., pp 842-849.

Eskesen, J.H., 1980a. Flashed Steam-Steam Turbine Cycles. In: *Sourcebook on the Production of Electricity from Geothermal Energy,* J. Kestin (ed.), U.S. Government Printing Office, Washington, D.C., January.

Eskesen, J.H., Kestin, J., et al., 1980b. Binary Cycles. In: *Sourcebook on the Production of Electricity from Geothermal Energy,* J. Kestin (ed.), U.S. Government Printing Office, Washington, D.C., January.

Fearnside, T.A. and Cheney, F.C., 1972. Fast Estimate of Power Plant Costs, ASME paper, Power and Management Divisions, New York.

Finn, D.F.X., 1975. Price of Steam at The Geysers. *2nd United Nations Symposium on the Development and Use of Geothermal Resources,* 3, San Francisco, CA, pp. 2295-2300, May.

Fraas, A.P. and Ozisik, M.N., 1965. *Heat Exchanger Design,* J. Wiley and Sons, New York.

Gault, J., Hall, J.W., Wilson, J.S., Michael, H., Shepherd, B.P., Underhill, G. and Rios-Castellon, L., 1976. Preliminary Analysis of Electric Generation Utilizing Geopressured Geothermal Fluids. *Proceedings 11th Intersociety Energy Conversion Engineering Conference,* State Line, NV, pp. 790-797, September.

Geothermal Resources Council, 1978a. *Proceedings of Direct Utilization of Geothermal Energy: A Symposium,* San Diego, CA, DOE Contract No. EY-76-S-03-1340, Jan. 31-Feb. 2.

Geothermal Resources Council, 1978b. Commercialization of Geothermal Resources, Davis, CA, November.

Geothermal Resources Council, 1977-1979. Davis, CA, *Transactions,* Vols 1, 2, 3.

Giedt, W.H., 1975. The Geothermal Binary Cycle: Heat Exchanger Area Requirements and Initial Costs. Lawrence Livermore Laboratory report UCRL-51912, Livermore, CA, September.

Golabi, K. and Scherer, C.R., 1977. Optimal Extraction of Geothermal Energy, University of California Los Angeles report UCLA-ENG-7715, June.

Goldsmith, K., 1975. Economic Aspects of Geothermal Development. *2nd United Nations Symposium on the Development and Use of Geothermal Resources,* San Francisco, CA, 3, pp. 2301-2303, May.

Green, M.A. and Pines, H.S., 1974. Calculation of Geothermal Power Plant Cycles Using Program GEOTHM. *2nd United Nations Geothermal Energy Symposium,* San Francisco, CA, 3, pp. 1965-1972, May.

Greider, R., 1973. Economic Considerations for Geothermal Exploration in the Western United States. Presented at the *Symposium of Colorado Department of Natural Resources,* Denver, CO, December.

Greider, R., 1975. Status of Economics and Financing of Geothermal Energy Power Production, *2nd United Nations Symposium on the Development and Use of Geothermal Resources,* San Francisco, CA, 3, pp. 2303-2314, May.

Gringarten, A.C. et al., 1975. Theory of Heat Extraction from Fractured Hot Dry Rock, *J. Geophys. Res., 80* (8), 1120.

Guiza, J., 1975. Power Generation at Cerro Prieto Geothermal Field. *Proceedings of 2nd United Nations Symposium on the Development and Use of Geothermal Resources,* San Francisco, CA, *3,* pp. 1976-1978, May.

Hankin, J.W., Beaulaurier, L.O., and Comprelli, F.O., 1975. Conceptual Design and Cost Estimate for a 10-MWe(net) Generating Unit and Experimental Facility Using Geothermal Brine Resources. *2nd United Nations Symposium on the Development and Use of Geothermal Resources,* San Francisco, CA, *3,* pp. 1985-1997, May.

Hankin, J.W., Hogue, R.A., Cassel, T.A.V., and Fick, T.R., 1977. Effect of Reservoir Temperature Decline on Geothermal Power Plant Design and Economics. *Proceedings of 12th Intersociety Energy Conversion Engineering Conference,* Washington, D.C., pp. 870-876.

Hansen, A., 1964. Thermal Cycles for Geothermal Sites and Turbine Installation at The Geysers Power Plant, California. In: *Geothermal Energy, Proc. U.N. Conf. on New Sources of Energy,* Rome, Italy, *3,* pp. 365-379, August 21-31.

Happel, J., and Jordan, D.G., 1975. *Chemical Process Economics, 2nd ed.,* Marcel Dekker, New York.

Harlow, F.H., and Pracht, W.E., 1972. A Theoretical Study of Geothermal Energy Extraction. *J. of Geo. Res. 77,* p. 7038.

Hausz, W., 1976. Annual Storage: A Catalyst for Conservation, presented at the International Total Energy Congress, Copenhagen, Denmark, October 4-8.

Hausz, W., 1975. Seasonal Storage in District Heating. *District Heating,* p 5, July, August, September.

HDR Staff, 1978. Hot Dry Rock Geothermal Energy Development Project, Annual report FY 1977, Los Alamos National Laboratory report LA-7109-PR, February.

Heller, L., 1965. New Power Station System for Unit Capacities in the 1000 MW Order, Acta. Techn. Hung. 50, pp. 93-123.

Holt, B., and Ghormley, E.L., 1976. Energy Conversion and Economics for Geothermal Power Generation at Heber, California, Valles Caldera, New Mexico, and Raft River, Idaho — Case Studies. Electric Power Research Institute report EPRI ER-301, Topical report #2, Palo Alto, CA, November.

House, P.A., Johnson, P.M., and Towse, D.F., 1975. Potential Power Generation and Gas Production from Gulf Cost Geopressured Reservoirs. *2nd United Nations Symposium on the Development and Use of Geothermal Resources,* San Francisco, CA, *3,* pp. 2001-2006, May.

Howard, J.H., (ed.), 1975. Present Status and Future Prospects for Nonelectrical Uses of Geothermal Resources. Lawrence Livermore Laboratory report UCRL-51926, Livermore, CA, October.

Intersociety Energy Conversion Engineering Conference, Proceedings of the 11th, 1976. State Line, NV, September.

Intersociety Energy Conversion Engineering Conference, Proceedings of the 12th, 1977. Washington, D.C., August.

International Atomic Energy Agency, 1977. *Urban District Heating Using Nuclear Heat, Proceedings of an IAEA Advisory Group,* Vienna, Austria.

Intertechnology Corporation, 1977. Analysis of the Economic Potential of Solar Thermal Energy to Provide Industrial Process Heat. U.S. ERDA Contract EY-76-C-02-2829.

James, R., and Meidav, T., 1977. Thermal Efficiency of Geothermal Power. *Geothermal Energy, 5(4),* 8, April.

Joint Association Survey, (1972) of the U.S. Oil and Gas Producing Industry, 1973. *Section I, Drilling Costs; Section II, Expenditures for Exploration, Development and Production,* November.

Joint Association Survey (1973) of the U.S. Oil and Gas Producing Industry., 1975. *Section I, Drilling Costs,* February.

Jonsson, V.K., Taylor, A.J., and Charmichael, A.D., 1969. Optimization of Geothermal Power Plant by Use of Freon Vapour Cycle. Timarit-VF1, p. 2.

Juul-Dam, T., and Dunlap, H.F., 1975. Economic Analysis of a Geothermal Exploration and Production Venture. *2nd United Nations Symposium on the Development and Use of Geothermal Resources,* San Francisco, CA, *3,* pp. 2315-2324, May.

Kapplemeyer, O., et al., 1979. Status Report of Hot Dry Rock Activities in the Federal Republic of Germany. *Proceedings of 2nd Information Conference on Hot Dry Rock,* Santa Fe, NM, September.

Karkheck, J., 1978. Some Considerations for Geothermal District Heating. *Proceedings of NATO-CCMS Conf. on the Economics of Geothermal Energy,* U.S. Department of Energy report CONF-770681, pp. 305-317, July.

Keenan, J.H., and Keyes, R.G., 1968. *Steam Tables,* J. Wiley and Sons, New York.

Keller, J.G., and Kunze, J.F., 1976. Space Heating Systems in the Northwest-Energy Usage and Cost Analysis. Idaho National Engineering Laboratory report ANCR-1276, Idaho Falls, ID, January.

Kern, D.Q., 1950. *Process Heat Transfer,* McGraw-Hill, New York.

Kern, D.Q., 1966. Heat Exchanger Design for Fouling Service. *Proceedings of the 3rd International Heat Transfer Conference,* Chicago, IL, *1,* 170, August.

Kestin, J., et al. (eds.), 1980. *A Sourcebook on the Production of Electricity from Geothermal Energy,* U.S. Government Printing Office, Washington, D.C., March.

Khalifa, H.E., 1980a. Effect of Fluctuations in Ambient Temperatures. In: *A Sourcebook on the Production of Electricity from Geothermal Energy,* J. Kestin et al. (eds.) U.S. Government Printing Office, Washington, D.C., pp. 656-663, March.

Khalifa, H.E., 1980b. Hybrid Fossil-Geothermal Power Plants. In: *A Sourcebook on the Production of Electricity from Geothermal Energy,* U.S. Government Printing Office, Washington, D.C., pp. 471-504, March.

Kingston, Reynolds, Tetsom and Allardice Ltd., 1978. City of Burbank Hybrid Geothermal/ Coal Power Plant. Report on Geothermal Resource Requirements and Utilisation, U.S. Dept. of Energy grant EG-77-G-03-1572, Auckland, New Zealand, July.

Klei, H.E., and Maslan, F., 1976. Capital and Electric Production Costs for Geothermal Power Plants. *Energy Sources 2*(4), pp. 331-345.

Koppel, L.B., and Smith, J.M., 1961. Turbulent Heat Transfer in the Critical Region. *Proceedings of Conference on International Development in Heat Transfer,* American Society of Mech. Engineers, part 1, pp. 585-590.

Kruger, P., and Otte, C., (eds.), 1973. *Geothermal Energy,* Stanford University Press, Stanford, CA.

Kunze, J.F., 1977. Geothermal Space Heating the Symbiosis with Fossil Fuel. *Proceedings of 12th Intersociety Energy Conversion Engineering Conference,* Washington, D.C., Vol 1, pp. 810-815.

Kunze, J.F., Miller, L.G., and Whitbeck, J.F., 1975. Moderate Temperature Utilization Project in the Raft River Valley. *2nd United Nations Symposium on the Development and Use of Geothermal Resources,* San Francisco, CA, 3, pp. 2021-2030, May.

Kunze, J.F., Whitbeck, J.F., Miller, L.G., and Griffith, J.L., 1976. Making Electricity from Moderate Temperature Fluids. *Geothermal Energy,* 4(10), 7, October.

Lang, B., and Heidt, F.D., 1979. Economic Analysis for MAGES. Dornier report DS-ERT-2/79, January.

Lavi, A., (ed.), 1973. *Proceedings of Solar Sea Power Plant Conference and Workshop.* Carnegie Mellon Univ., Pittsburgh, PA, June.

Lavi, A., 1975. Solar Sea Project. Report NSF/RANN/SE/GI-39114/PR/74/6, Carnegie Mellon University, Pittsburgh, PA, January.

Lienau, P.J., 1978. Space Conditioning with Geothermal Energy. In: *Commercialization of Geothermal Resources.* Geothermal Resources Council pp. 23-26, November.

Lindal, B., 1973. Industrial and Other Applications of Geothermal Energy. In: *Geothermal Energy: Review of Research and Development.* Paris, UNESCO, LC No. 72-97138, pp. 135-148.

Linn, J.K., 1979. Analysis of Collectors for Heat Applications. Sandia Laboratories report SAND 78-1977, Albuquerque, NM, February.

Lund, J.W., 1976. The Utilization and Economics of Low Temperature Geothermal Water for Space Heating. *Proceedings of 11th Intersociety Energy Conversion and Engineering Conference,* State Line, NV, Vol 1, pp. 822-827.

Lydersen, A.L., Greenkorn, R.A., and Hougen, O.A., 1955. Generalized Thermodynamic Properties of Pure Fluids. Engineering Exp. Station, Report No. 4, Univ. of Wisconsin, Madison, WI, October.

Martin, J.J., 1967. Equations of State. *Ind. Eng. Chem.,* 59, pp.34-52.

Maslan, F., Gordon, T.J., Deitch, L., 1975. Economics and Social Aspects of Geothermal Energy Resource Development. *2nd United Nations Symposium on the Development and Use of Geothermal Resources,* San Francisco, CA, *3,* pp.2325-2331, May.

Matthews, H.B., 1976. Geothermal Energy Control System and Method. United States Patent 3,938,334 (February 17, 1976), and other patents 3,967,448; 3,824,793; 3,898,020; 3,988,896; 3,910,050; 3,939,659; 3,905,196.

Maurer, W.C., Nixon, J.D., Matson, L.W., and Rowley, J.C., 1977. New Turbodrill for Geothermal Drilling. In: *Proceedings of 12th Intersociety Energy Conversion Engineering Conference,* Washington, D.C., Vol 1, pp.904-911.

Maurer, W.C., 1979. LASL Turbodrill Economics. Maurer Engineering report TR79-11, Houston, TX, April.

McAdams, W.H., 1974. *Heat Transmission,* 3rd Ed., McGraw-Hill, New York.

McBee, W.D., and Mathews, H.B., 1979. Low Temperature Energy Conversion Systems. U.S. Dept. of Energy Contract #ET-78-C-02-4633.

McDonald, C.L., Bloomster, C.H., and Schute, S.C., 1977. GEOCITY: A Computer Code for Calculating Costs of District Heating Using Geothermal Resources. Battelle Northwest Laboratory report NNWL-2208, Richland, WA, February.

McFarland, R.D., 1975. Geothermal Reservoir Models — Crack Plane Model. Los Alamos National Laboratory report LA-5947-MS, Los Alamos, NM, April.

McFarland, R.D., and Murphy, H.D., 1976. Extracting Energy from Hydraulically-Fractured Geothermal Reservoirs. *Proceedings of 11th Intersociety Energy Conversion Engineering Conference,* State Line, NV.

McKay, R.A., 1977. The Helical Screw Expander Evaluation Project. *Proceedings of 12th Intersociety Energy Conversion Engineering Conference,* Washington, D.C., Vol 1, pp. 899-903.

Megill, R.E., 1971. *An Introduction to Exploration Economics,* Petroleum Publ. Co., Tulsa, OK.

Meidav, T., 1974. Geothermal Opportunities Bear a Closer Look. *Oil and Gas J.,* 72, p.102.

Meidav, T., 1975. Time is of the Essence in Developing Geothermal Energy. *Oil and Gas J.,* 73, p. 168.

Meissner, H.P., and Seferian, R., 1951. P-V-T Relations of Gases, *Chem. Eng. Progr.,* 47, p.579.

Meyer, C.F., Hausz, W., Ayres, B.L., and Ingrom, H.M., 1977. Role of the Heat Storage Well in Future U.S. Energy Systems. General Electric report GE76TMP-27A, Santa Barbara, CA, December.

Miliaras, E.S., 1974. *Power Plants with Air-Cooled Condensing Systems,* MIT Press, Cambridge, MA.

Miller, A.J., Payne, H.R., Lackey, M.E., Samuels, G., Heath, M.T., Hagen, E.W., and Savolanen, A.W., 1971. Use of Steam-Electric Power Plants to Provide Thermal Energy to Urban Areas. Oak Ridge National Laboratory report ORNL-HUD-14, Oak Ridge, TN.

Miller, D.R., Null, H.L., and Thompson, Q.E., 1973. Optimum Working Fluids for Automotive Engines, Monsanto Research Corporation reports APTD 1563, 1364, 1365, Vol. 3, June.

Milora, S.L., 1973. Application of the Martin Equation of State to the Thermodynamic Properties of Ammonia, Oak Ridge National Laboratory report ORNL-TM-4413, Oak Ridge, TN, December.

Milora, S.L., and Tester, J.W., 1976. *Geothermal Energy as a Source of Electric Power*, MIT Press, Cambridge, MA.

Milora, S.L., and Combs, S.K., 1977. Thermodynamic Representation of Ammonia and Isobutane. Oak Ridge National Laboratory report ORNL-TM-5847, Oak Ridge, TN.

Milora, S.L., STATEQ: A Nonlinear Least-Squares Code for Obtaining Martin Thermodynamic Representations of Fluids in the Gaseous and Dense Gaseous Regions. Oak Ridge National Laboratory report ORNL-TM-5115, Oak Ridge, TN (to be published).

Moskvicheva, V.N., 1971. Geopower Plant on the Paratunka River. USSR Academy of Sciences.

Moskvicheva, V.N., and Popov, A.E., 1970. Geothermal Power Plant on the Paratunka River. *Proceedings of the UN Symposium on the Development and Utilization of Geothermal Resources*. Pisa, Italy, 2 p. 1567.

Murphy, H.D., and Aamodt, R.L., et al., 1980. Preliminary Evaluation of the Second Hot Dry Rock Geothermal Energy Reservoir: Results of Phase 1, Run Segment 4, Los Alamos National Laboratory report LA-8354-MS, Los Alamos, NM, May.

Murphy, K., 1979. Personal Communication. Allied Chemical, Wayne, NJ., April.

NATO-CCMS Conference on the Economics of Direct Uses of Geothermal Energy 1978. *Proceedings, U.S. Department of Energy Report CONF-770681*, NATO-CCMS Report No. 66, Washington, D.C., July.

Neill, D.T., 1976. Geothermal Powered Heat Pumps to Produce Process Heat. *Proceedings of 11th Intersociety Energy Conversion Engineering Conference*, State Line, NV, Vol 1, pp. 802-807.

Newendorp, P.D. *Decision Analysis for Petroleum Exploration*, The Petroleum Publ. Co., Tulsa, OK.

Newsom, M.M., Barnett, J.H., Baker, L.E., Narnado, S.G., and Polito, J., 1977. Geothermal Well Technology, Drilling and Completion Program Plan. Sandia Labs report SAND77-1630, Albuquerque, NM, November.

Nichols, K.E., 1975. Turbine Prime Mover Cost Model Empirical Equation, P.O. 1/Y-49428V, Barber-Nichols Engineering Co., Arvada, CO, March.

Nichols, K.E., Prigmore, D., Matthews, H., and Halat, J., 1977. Design and Field Test of a Steam Powered Downhole Geothermal Pump. *Proceedings of 12th Intersociety Energy Conversion Engineering Conference*, Washington, D.C., pp. 877-883.

Nichols, K.E., and Malgieri, A., 1978. Technology Assessment of Geothermal Pumping Equipment. Final report, U.S. Dept. of Energy Contract #EG-77-C-04-4162, Barber-Nichols Engineering, Arvada, CO, July.

Nichols, K.E., 1979. Personal communication. Barber-Nichols Engineering Co., Arvada, CO, April.

Nicholson, R., and Verity, T., et al., 1979. Industrial Assessment of the Drilling, Completion, and Workover Costs of Well and Fracture Subsystems of Hot Dry Rock Reservoirs. Republic Geothermal, Inc., Santa Fe Springs, CA.

Olander, R.G., Nichols, K.E., and Tester, J.W., 1979. Utilization of Hot Dry Rock Geothermal Energy: Power Plant Design Considerations. *Proceedings of 3rd National Congress, Pressure Vessel and Piping Division*, Amer. Soc. of Mech. Eng., San Francisco, CA, June.

Parmelee, H.M., 1968. Sealed-Tube Stability Tests on Refrigeration Materials. ASHRAE Semiannual Meeting, Chicago, IL, January.

Pearson, C.M., 1980. Permeability Enhancement by Explosive Initiation in the South-West Granites, with Particular Reference to Hot Dry Rock Energy Systems, Ph.D. thesis, Camborne School of Mines, Redruth, Cornwall, U.K., Sept.

Perry, R.H., and Chilton, C.H., 1973. *Chemical Engineers Handbook,* 5th ed., McGraw-Hill, NY.

Peters, M.S., and Timmerhaus, K.D., 1980. *Plant Design and Economics for Chemical Engineers,* 3rd ed., McGraw-Hill, NY.

Peterson, R.E., 1975. Economic Factors in Resource Exploration and Exploitation, *2nd United Nations Symposium on the Development and Use of Geothermal Resources,* San Francisco, CA, *3,* pp. 2333-2338, May.

Pines, H.S., and Green, M.A., 1976. The Use of Program GEOTHM to Design and Optimize Geothermal Power Cycles. *Proceedings of 11th Intersociety Energy Conversion Engineering Conference,* Lake Tahoe, NV, pp. 836-842, September.

Pope, W.L., Pines, H.S., Silvester, L.F., Green, M.A., and Williams, J.D., 1977. Multiparameter Optimization Studies of Geothermal Energy Cycles. *Proceedings of 12th Intersociety Energy Conversion Engineering Conference,* Washington, D.C., pp. 865-869.

Pope, W.L., Pines, H.S., Fulton, R.L., and Doyle, P.A., 1978. Heat Exchanger Design — Why Guess a Design Fouling Factor When It Can Be Optimized? ASME paper presented at Winter Annual Meeting, November.

Ramachandran, G., et al., 1977. Economic Analysis of Geothermal Energy Development in California. Vols. 1 and 2. Stanford Research Inst., project ECU 5013, Menlo Park, CA, May.

Raleigh, C.B., et al., 1974. Multiple Hydraulic Fracturing for the Recovery of Geothermal Energy (abstract), *EOS Trans.* AGU, *55* (4), 4026.

Reid, R.C., Prausnitz, J.M., and Sherwood, T.K., 1977. *The Properties of Gases and Liquids,* 3rd Ed., McGraw-Hill, NY.

Reistad, G.M., 1975. Potential for Nonelectrical Applications of Geothermal Energy and Their Place in the National Economy, *2nd United Nations Symposium on the Development and Use of Geothermal Resources,* San Francisco, CA, Vol 3, pp. 2155-2164.

Reistad, G.M., Yao, B., and Gunderson, M., 1977. A Thermodynamic Study of Heating with Geothermal Energy. *Proceedings of the ASME* Winter Annual Meeting, Atlanta, GA, paper No. 77-WA/Ener-1, November 27-December 2.

Resource Planning Associates, 1977. The Potential for Cogeneration Development in Six Major Industries by 1985. RPA Reference #RA 77-1018 report HCP/M60172-01/2 Cambridge, MA, December.

Rex, R.W. and Howell, D.J., 1973. Assessment of U.S. Geothermal Resources. In: *Geothermal Energy,* Kruger and Otte (eds.), Stanford University Press, Stanford, CA.

Rex, R.W., 1978. The U.S. Geothermal Industry in 1978. Presented at the Hot Dry Rock Geothermal Information Conference, Santa Fe, NM sponsored by LASL, April 19-20.

Roberts, V. and Kruger, P., 1979. Utility Industry Estimates of Geothermal Electricity. Geothermal Resources Council *Transactions,* 3, p. 581.

Robertson, R.C., 1980. Waste Heat Rejection from Geothermal Power Stations. Oak Ridge National Laboratory report ORNL/TM-6533, Oak Ridge, TN (1978) and in: *A Sourcebook on the Production of Electricity from Geothermal Energy,* J. Kestin et al. (eds.), U.S. Government Printing Office, Washington, D.C., March.

Rohsenow, W.M. and Hartnett, J.P., *Handbook of Heat Transfer,* McGraw-Hill, NY.

Rossin, A.D. and Rieck, T.A., 1978. Economics of Nuclear Power, *Science 201,* August.

Rudd, D.F. and Watson, C.C., 1968. *Strategy of Process Engineering,* J. Wiley and Sons, NY.

Salisbury, J.K., 1950. Steam Turbines and Their Cycles. Krieger Publishing Co., Huntington, NY, (reprinted 1974 with supplemental chapters).

Samuels, G., 1979. Geopressure Energy Resource Evaluation. Oak Ridge National Laboratory report ORNL/PPA-79/2, Oak Ridge, TN, May.

Sapre, A.R. and Schoeppel, R.J., 1975. Technological and Economic Assessment of Electric Power Generation from Geothermal Hot Water. *2nd United Nations Symposium on the Development and Use of Geothermal Resources,* San Francisco, CA, *3,* pp 2343-2350, May.

Sastry, V.S., 1974. An Analytical Investigation of Forced Convection Heat Transfer to Fluids Near the Thermodynamic Critical Point. ASME 74-WA/HT-29.

Schapiro, A.R. and Hajela, G.P., 1977. Geothermal Power Cycle Analysis. *Proceedings of 12th Intersociety Energy Conversion Engineering Conference,* Washington, D.C., pp 857-864.

Scherer, C.R. and Golabi, K., 1978. Geothermal Reservoir Management, University of California, Berkeley, report UCB/SERL #78-5, February.

Schultz, R.J., Swink, D.G., and Kunze, J.F., 1977. Direct Applications of Geothermal Brine in Process Application: *Proceedings of the 83rd AIChE National Meeting,* paper 16B, Houston, TX, March.

Scott, D.L., 1976. *Financing the Growth of Electric Utilities,* Praeger Publishers, New York.

Sesonske, A., 1973. Nuclear Power Plant Design Analysis. U.S. Atomic Energy Commission report TID-26241, Nat. Tech. Info. Service, Springfield, VA.

Seth, R.G. and Steiglemann, W., 1972. Binary-Cycle Power Plants Using Dry Cooling Systems, Part I, Technical and Economic Evaluation. The Franklin Institute Research Laboratories final report F-C3023. Philadelphia, PA, January.

Shepherd, D.G., 1977. A Comparison of Three Working Fluids for the Design of Geothermal Power Plants. *Proceedings of 12th Intersociety Energy Conversion Engineering Conference,* Washington, D.C., pp. 832-841.

Shoemaker, E.M., (ed.), 1975. Continental Drilling. Report of the Workshop on Continental Drilling, Abiquiu, New Mexico, by the Carnegie Institution, Washington, D.C., June.

Sperry Research, 1977. Feasibility Demonstration of the Sperry Down-Well Pumping System. Final report contract No. EY-76-C-02-2838. U.S. Dept. of Energy, SCRC-CR-77-48, Sudbury, MA, July.

Starling, K.E., 1973. *Fluid Thermodynamic Properties of Light Petroleum Systems,* Gulf Publishing Co., Houston, TX, pp. 194-205.

Starling, K.E., Fish, L.W., Iqbal, K.Z., and Yieh, D., 1974. Advantages of Using Mixtures as Working Fluids in Geothermal Binary Cycles. *Proceedings of Oklahoma Acad. Science, 63rd Annual Meeting,* Durant, OK.

Starling, K.E., Fish, L.W., Iqbal, K.Z., and West, H.H., 1977. The Use of Mixture Working Fluids in Geothermal Binary Power Cycles. *Proceedings of 12th Intersociety Energy Conversion Engineering Conference,* Washington, D.C., pp. 850-856.

Starling, K.E., Sleipcevich, C.M., Fish, L.W., Goin, K.M., Aboud-Foutoh, K.H., Kumar, K.H., Lee, T.J., Milani, J.S., and Zemp, K.L., 1978. Development of Working Fluid Thermodynamic Properties Information for Geothermal Cycles: Phase I. Univ. of Oklahoma report ORO-5249-2, Norman, OK.

Stephens, R.C., 1978. Geothermal Energy Legislation: The National Energy Act and New Proposals. In: *Commercialization of Geothermal Resources.* Geothermal Resources Council, Davis, CA, November.

Sternmole, F.J., 1974. *Economic Evaluation and Investment Decision Methods,* Investment Evaluations Corp., Golden, CO.

Swearingen, J.S., 1977. Power from Hot Geothermal Brines. *Chem. Eng. Progress,* 73(7), 83, July.

Swink, D.G., 1976. Potential for the Use of Intermediate Temperature Geothermal Heat in Food Processing. *Proceedings of 11th Intersociety Energy Conversion Engineering Conference,* State Line, NV, Vol 1, pp. 808-814.

Tester, J.W. and Smith, M.C., 1977. Energy Extraction Concepts for Hot Dry Rock Geothermal Systems. *Proceedings of the 12th Intersociety Energy Conversion Engineering Conference,* Washington, D.C., p. 816, August.

Tester, J.W., Morris, G.E., Cummings, R.G., and Bivins, R.L., 1979. Electricity from Hot Dry Rock Geothermal Energy: Technical and Economic Issues. Los Alamos National Laboratory report LA-7603-MS, Los Alamos, NM, January.

Tester, J.W. and Albright, J.N., (eds.), 1979. Hot Dry Rock Energy Extraction Field Test: 75 Days of Operation of a Prototype Reservoir at Fenton Hill. Los Alamos National Laboratory report LA-7771-MS, April.

Troulakis, S., 1968. Steam Ammonia Power Cycle Study. Report No. STA-19, DeLaval Turbine Inc., Trenton, NJ.

UNESCO, 1970. *Geothermics,* United Nations Symposium on the Development and Utilization of Geothermal Resources, *2,* Special Issue No. 2, Pisa, Italy.

United Nations, 1975. *Proceedings of 2nd United Nations Symposium on the Development and Use of Geothermal Resources,* San Francisco, CA, May 20-29.

Urbanek, M.W., 1978. Development of Direct Heat Exchangers for Geothermal Brines, DSS Engineers, Inc., final report Lawrence Berkeley Laboratory report LBL-8558, Berkeley, CA.

Vaux, H.J., Jr. Nakayama, B., 1975. The Economics of Geothermal Resources in the Imperial Valley: A Preliminary Analysis, California Water Resources Center, Davis, CA, No. 153, November.

Vymorkov, B.M., 1965. First Geothermal Power Stations. *Vestn. Akad. Nau SSSR, 10,* 32-39.

Wagner, S.C., 1979. Taxation of Geothermal Energy. Paper presented at *An Introduction to Exploration and Development of Geothermal Resources,* Geothermal Resources Council, Davis, CA, May 8-9.

Wagner, S.C., 1978. A Review of Western State Tax Law Applied to Geothermal Development. In: *Commercialization of Geothermal Resources,* Geothermal Resources Council, Davis, CA, November.

Wahl, E.F., 1977. *Geothermal Energy Utilization,* John Wiley and Sons, New York.

Walter, R.A., Bloomster, C.M., and Wise, S.E., 1975. Thermodynamic Modelling of Geothermal Power Plant, Battelle Northwest Laboratory, report BNWL-1911, UC-13 Richland, WA, November.

Walter, R.A. and Wilson, S.W., 1976. Economic Optimization of Binary Fluid Cycle Power Plants for Geothermal Systems. *Proceedings of 11th Intersociety Energy Conversion Engineering Conference,* State Line, NV, pp. 731-738, September.

Wilson, J.S., 1977. A Geothermal Energy Plant. *Chem. Eng. Progress.* 73(11), pp. 95-98.

Wilson, J.S., et al., 1975. Economic Analysis of the Use of Texas Geopressured Resources for the Production of Electrical Power. *Proceedings of First Geopressured Geothermal Energy Conference,* M. H. Dorfman and R. W. Deller (eds.) Univ. of Texas, Austin, TX, June 3-4.

Woods, J.H., 1978. Structuring of Geothermal Development Loans. In: *Commercialization of Geothermal Resources,* Geothermal Resources Council, Davis, CA, November.

Wunder, R. and Murphy, H.D., 1978. Thermal Drawdown and Recovery of Singly and Multiply Fractured Hot Dry Rock Reservoirs. Los Alamos National Laboratory Report LA-7219-MS, April.

Acknowledgements

The technical guidance and encouragement provided by my colleagues especially R. B. Duffield, the late A. G. Blair, Kenneth E. Nichols, J. H. Eskesen, G. J. Nunz, R. B. Brownlee, A. W. Laughlin, G. Heiken, R. L. Aamodt, L. M. Edwards, R. M. Potter, G. E. Morris, R. G. Cummings, R. E. Bivins and J. C. Rowley are greatly appreciated. Of special importance were the contributions by S.L. Milora to quantifying the state of the art in low temperature power conversion technology. In addition, R. J. Hanold was very helpful in supplying information concerning pumps for geothermal applications. I am also deeply grateful to my wife Sue; Doris Elsner, Barbara Ramsey, and Frani Sierra for their assistance in preparing the manuscript for publication. The support of the Division of Geothermal Energy of the U.S Department of Energy is also acknowledged.

Appendix F

Cost Inflation
Estimation Techniques

Various strategies can be adopted for updating and refining estimates. For geothermal systems the problem is twofold: one concern is estimating well drilling and completion and reservoir development costs while the other primarily deals with purchased and installed cost estimating for major pieces of hardware, including geothermal fluid piping, gas-liquid separators, heat exchangers, condensers, turbines, generators and pumps. The painstaking process of obtaining actual cost quotes from equipment manufacturers or drilling contractors would obviously be required for final designs and can be updated as needed. In the absence of this requirement, for example, for preliminary designs ($\pm 20\%$ accuracy) a simpler procedure is possible. As recommended by Peters and Timmerhaus (1980) a cost indexing technique can be used, where the cost in present day dollars is related to the original cost by a simple ratio of the cost indexes for the years involved. For example,

$$\text{Cost of item in [1982 \$]} = \text{cost of item in [1975 \$]} \left[\frac{\text{index in 1982}}{\text{index in 1975}} \right]$$

The challenge now is to find a proper cost index to use. Peters and Timmerhaus (1980) suggest several: the *Marshall and Swift All Industry and Process-Industry Equipment* indexes, the *Engineering News-Record* index, the *Nelson Refinery Construction* Index, and *the Chemical Engineering* plant cost index. *The Monthly Labor Review* also provides good estimates of labor cost inflation for updating installation costs.

For certain conventional components, the existing indexes, perhaps some weighted average of several, would be most appropriate. Equipment such as heat exchangers, cooling towers, condensers, pumps and electric generators would fall into this category. In fact, 1979 $ estimates provided by Peters and Timmerhaus (1980) should be very useful for these. More specialized equipment tailored to geothermal use will require more effort. For such things as low-pressure steam turbines and organic fluid turbines direct manufacturer costs or more detailed algorithms such as those presented in Chapter 10 for turbine costs should be used with best estimates of inflation factors.

In the absence of other data or more sophisticated approaches, estimates provided in Chapter 10 used the following method. The following combined cost index for 1965-1982 was developed averaging the *Marshall and Swift* (1967 basis process industry only), *Engineering News Record, Nelson refinery construction,* and *Chemical Engineering* plant cost indexes (1965 = 100):

Year	Index
1965	100
1966	104
1967	108
1968	114
1969	122
1970	132
1971	144
1972	154
1973	163
1974	183
1975	202
1976	216
1977	231
1978	247
1979	266
1980	285
1981	305
1982	332

Index

Index of Authors